Three Years' War
Annotated and Illustrated

Christiaan de Wet, June 1902. He is forty four years old at the end of the war.

Volume 5 of the series, *The Last Horse War*

Series compiled, annotated, illustrated and expanded by Payton Fireman, JD.

ISBN: 978-1-7320124-8-6

Front Cover Photo Credits: [4]

Back Cover Photo Credits: [5] [6] .[7]

Frontispiece Photo Credits:[8]

[1] Please see the Bibliography for license links and other information regarding the material in this volume.

Original Dutch Cover and Text

De strijd tusschen Boer en Brit.—The struggle between Boer and Brit.

C. R de Wet.

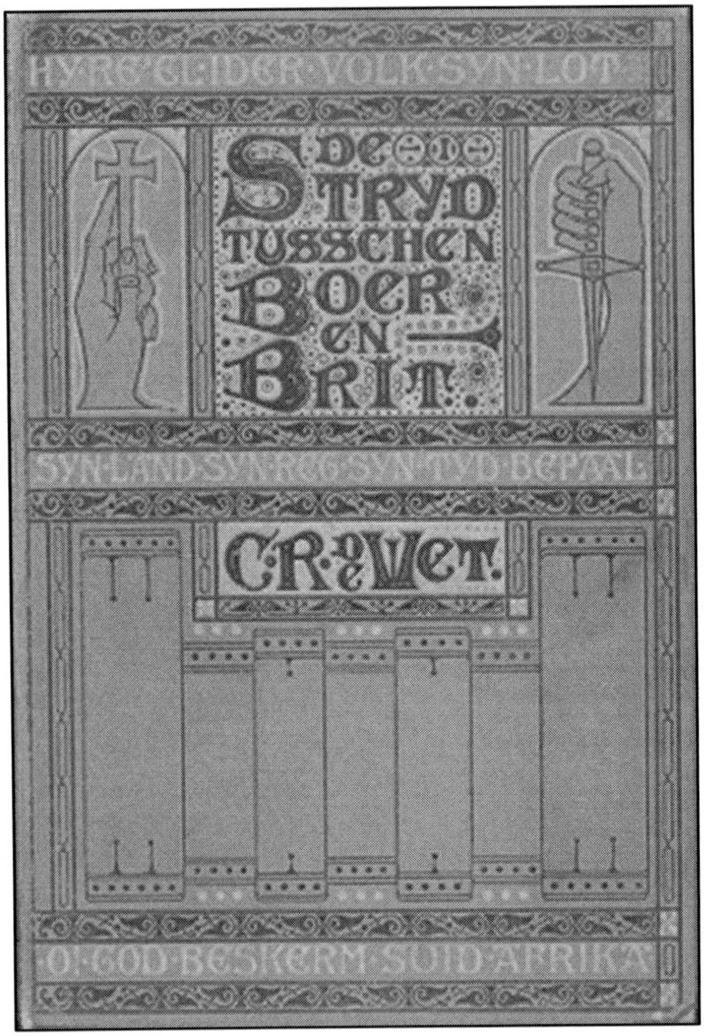

Original Dutch Language Cover.[9]

Hy re' El Ider Volk Syn Lot.—He is the leader of his people's fate.

Syn land Syn Reg Syn Tyd Bepaal.—Our land Our Rule Our Time Determine.

O! God Beskerm Suid Afirka—O! God Protect South Africa.

To

MY FELLOW SUBJECTS

OF

THE BRITISH EMPIRE

English Translation Cover

THREE YEARS' WAR

CHRISTIAAN RUDOLF DE WET

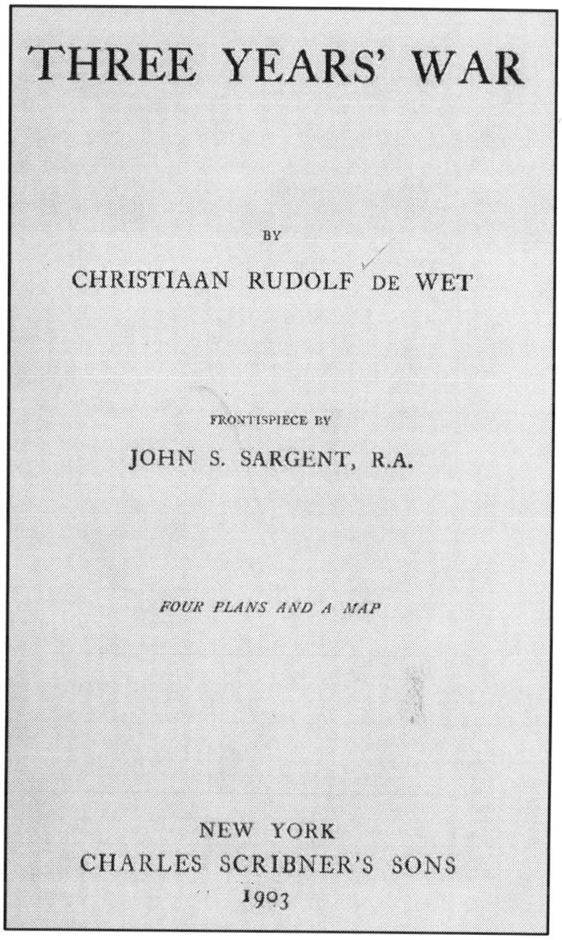

THREE YEARS' WAR

BY

CHRISTIAAN RUDOLF DE WET

FRONTISPIECE BY

JOHN S. SARGENT, R.A.

FOUR PLANS AND A MAP

NEW YORK
CHARLES SCRIBNER'S SONS
1903

Harvard College Library

VE RI TAS

From the Library of
Ernest Lewis Gay
Class of 1897

Given by his Nephew
GEORGE HENRY GAY
June 15, 1927

Donation Recognition.[10]

Portrait of the Author by John S. Sargent, R.A.

General de Wet's Autograph.

Contents

General Christiaan de Wet and staff 1900.[112]

:

[2] PF: Ihave to imagine that one or more of Christiaan's sons are in this picture. My guess is that two of them could be the two young men, one seated and one standing on the viewer's right side of the picture.
"Amongst the commandeered was I; and thus, as a private burgher, I entered on the campaign. With me were my three sons—Kootie, (Jacobus) Isaac, and Christiaan."

PF Preface

This work is already jam packed with pages so I'll keep this short. If I thought 'Commando' was hard to track as to where events were occurring, this work was even more difficult.

There are only a few first-person accounts by senior commanders of the Boer armies. So far as can be found, Louis Botha did not write any extensive memoir about his service and command during the war.[3] [4] Neither did Jacobus de la Rey, although his wife, Jacoba did.[5] Commandants Christiaan de Wet and Ben Viljoen are the two most senior military commanders who wrote extensive first person accounts in the immediate aftermath of the conflict. Both memoirists have an engaging style of writing with amazing experiences to relate. They also provide names, dates and other thousands of granular details that, as always, tie in to the other narratives in this series. By necessity we will traverse the same battles and employ many of the same images again in this volume.

I've wondered for a long time what it is I do. I annotate, illustrate, map; I'm a book builder, but there's a little more, since by now I know where the other players are and what the outcomes are. There is some insight provided and occasionally I'm within a stones' throw of a Greek Chorus. I compile journalistic briefs or title searches, like I would for any legal work I was involved in. This series of books has been a title search on an industrial scale.[6] What is the history and who owns the facts. In title work there is also the mapping and the platting. In these works title to information, location and action belongs to the narrator with the best evidence and first person testimony. This work has turned out to be more of a scholars' edition with all the mapping I've included.

With respect to maps, I've processed and reprocessed the Times' History maps to the limit of my ability and still they are only barely adequate. The best online version is the University of Toronto set of copies.[7] [8] These files, in color, have more discernible detail than the enhanced black and white versions in this work.[9] I would suggest having a pencil to circle locations on the maps the reader wants to be able to refer to easily.

Christiaan is not only a fine raconteur but he must have had a diary or access to his despatches in order to compose his work. It was the specificity of dates and places that prompted me to follow his journey, map wise, as best I could. I made the maps as large as I could to assist with legibility. Feel free to ignore or read past them.

There is an amazing amount of some inconsistent spelling that I am unable to resolve. The map makers and the British in general used different spellings in their maps than the Afrikaans map makers did. The same holds true for towns and locations in general as well as identifications of people and places in the captions to photographs.

With respect to the extensive use of the compiled Boer War Chronology. I have not substantially altered the jingoistic language of the compilers. In some cases it is somewhat amusing to read their take on events. They can be quite the spin doctors.[10] Also the chronology made different spelling choices too.

Christiaan's Appendices are transcriptions of the actual peace negotiations which are found nowhere else so far as I can determine. They are also reproduced in volume 9; The 'A Century of Wrong and Injustice' tract.

See Steyn's reply to Kitchener's August 1901 Proclamation in chapter 30 for the finest and most succinct statement of cause for the conflict.

I hope you enjoy this work. It has been my pleasure to assemble this material.

[3] There are several biographies of Louis Botha and they are listed in the Bibliography along with some research on the topic of memoirs.

[4] Research relating to whether Louis Botha ever published any extensive first person memoir of his Boer War experiences. See: https://www.angloboerwar.com/forum/11-research/33431-louis-botha-s-memoirs

[5] See her Dutch work "A Woman's Wanderings and Trials During the Anglo-Boer War" translated by Lucy Hotz, in the bibliography.

[6] It is an evolution of my work as an attorney and land title abstractor in West Virginia. My first jobs after law school were to run title searches in various county courthouses. Going daily to whatever county where I had a job and then hoisting the 20-pound deed books up onto the tables while consulting the Grantor and Grantee indexes in order to acquire a title chain. Going back fifty, a hundred, or two hundred years the original Patent in 1775.

[7] In general, look for libraries in northern climates when searching for the best copies of scans. The rate of paper oxidation, combustion, over time is slower.

[8] See the Bibliography, Infra for: The Times' History of the War in South Africa–1902 by Amery, L. S. (Leopold Stennett).

[9] They can be exported from the PDFs as PNGs and further modified to extract more on-screen definition. Most Times History maps were created by Spottiswoode & Co. Ltd.

[10] All adapted from Creswicke, South Africa and the Transvaal War Vols. 2–7 and With Roberts to the Transvaal by C. N. Robinson.

Maps.

Orange Free State

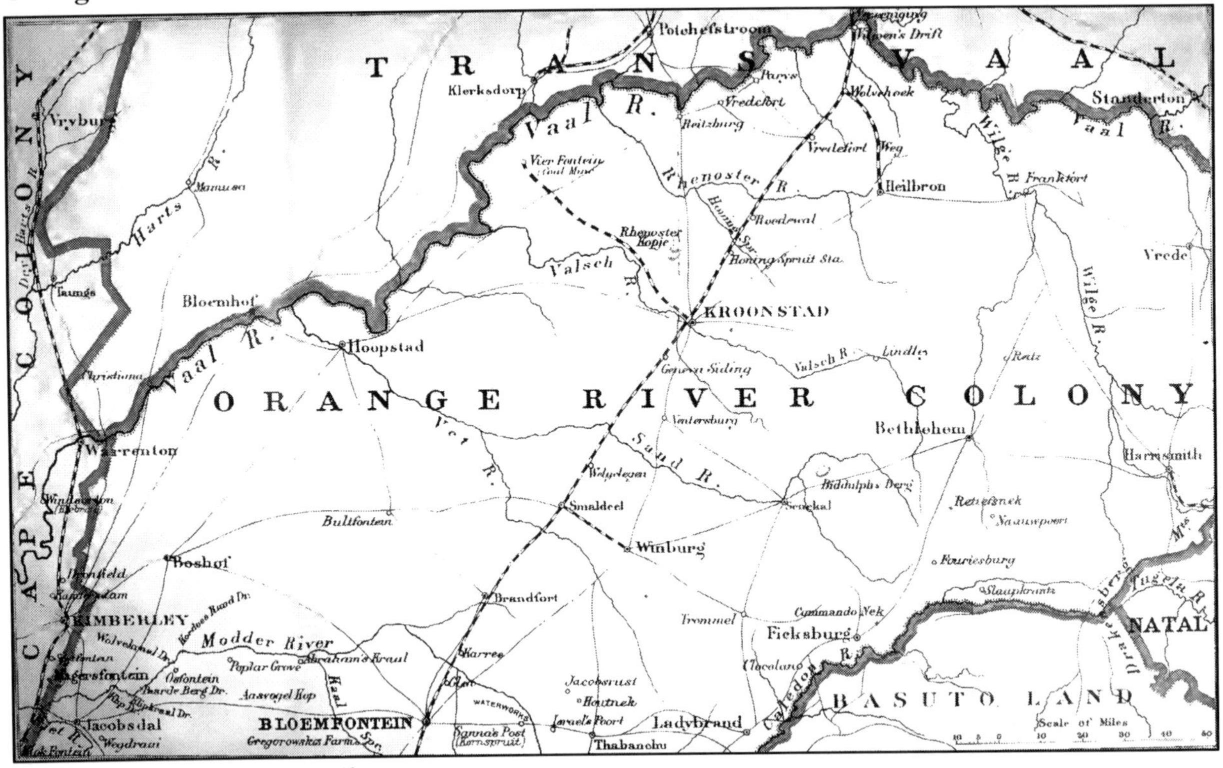

Orange River Colony North.[12]

Titled: Orange River Colony South.[13]

Ian Hamilton's Line of March

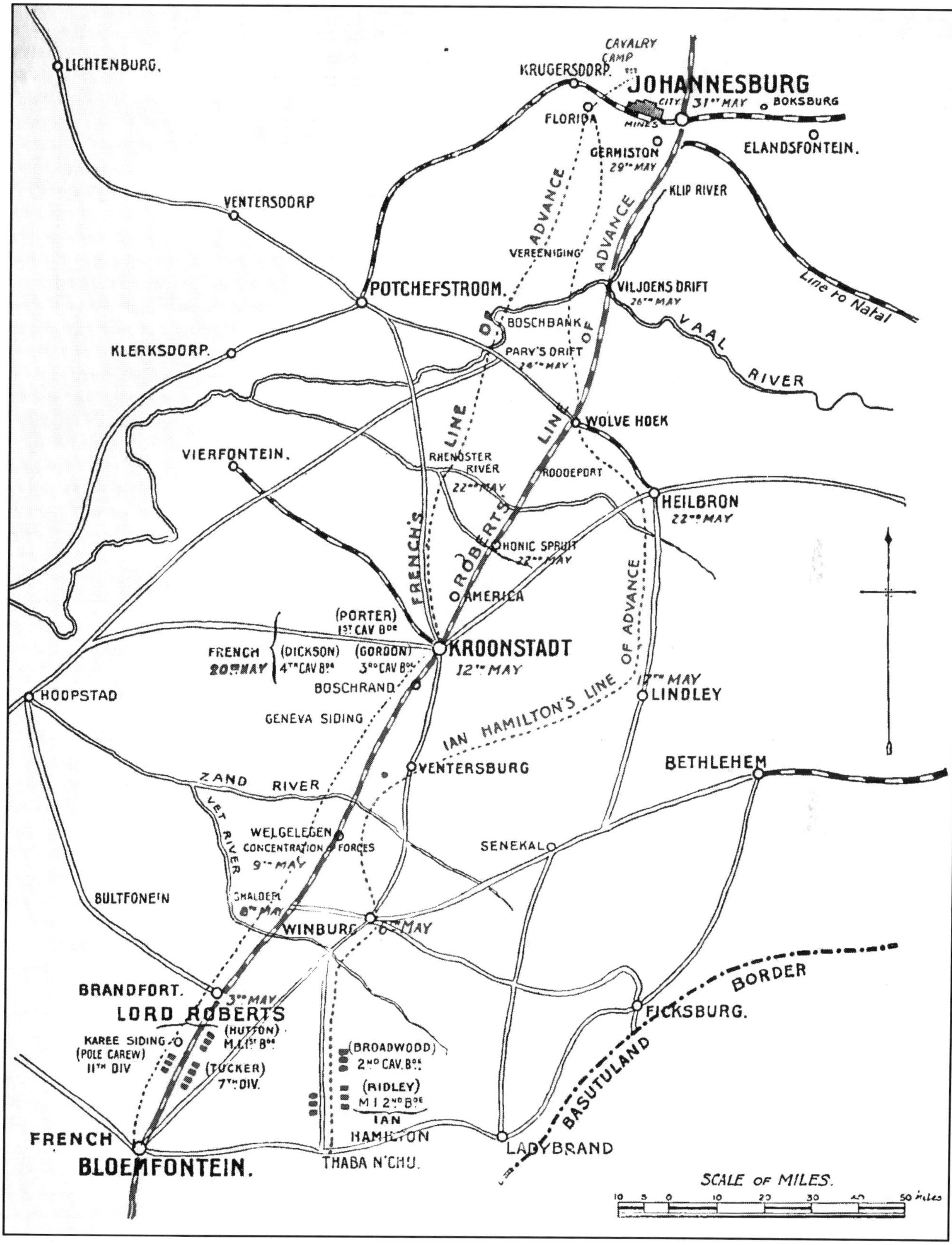

Hamilton's Line of March with dates.[14]

Hillegas Map

This map contains details of the primary areas where De Wet conducts his operations.

Hillegas Map(s).[15]

Davitt Map

This map identifies a large number of locations in the area of the Orang Free State.

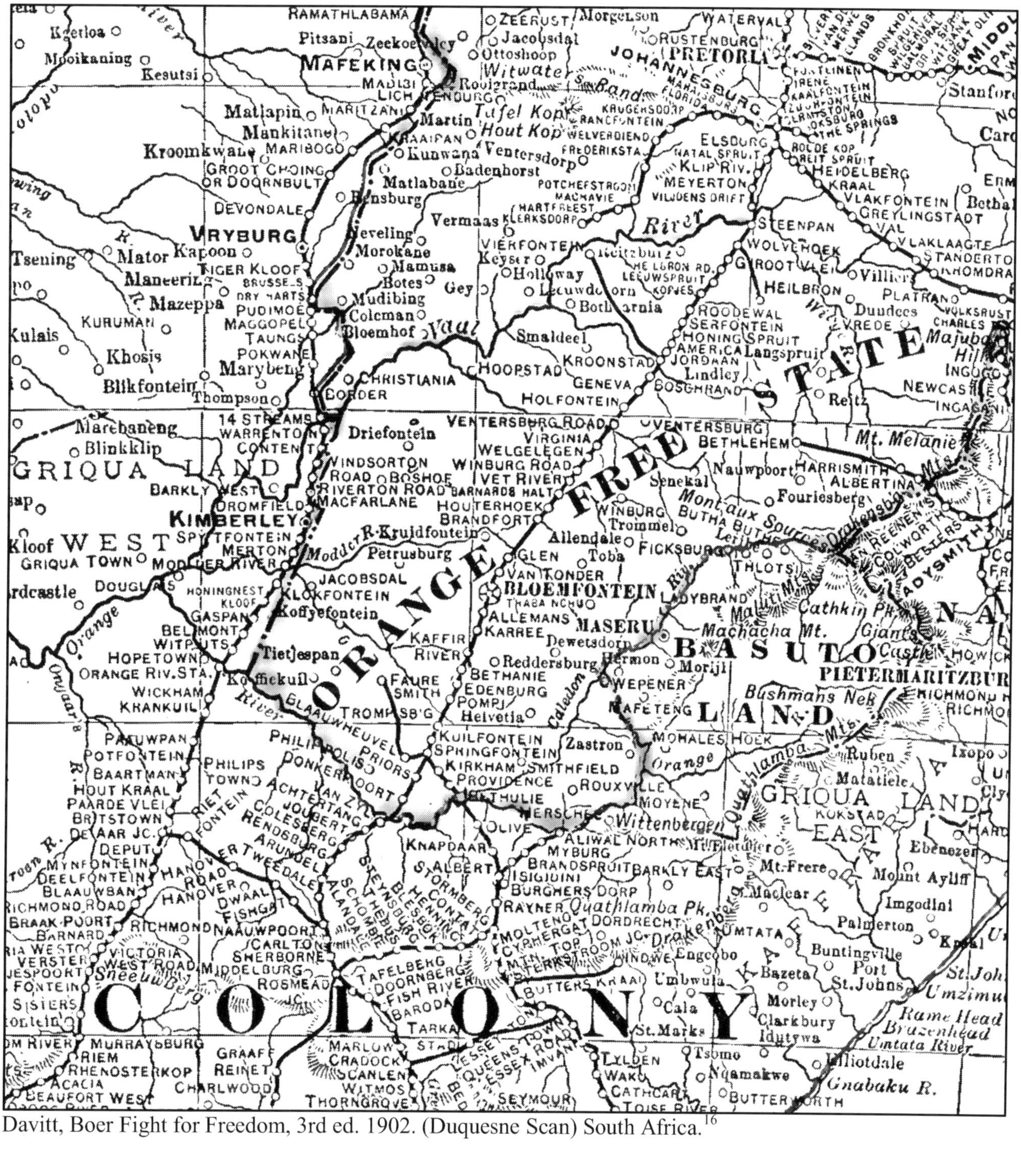

Davitt, Boer Fight for Freedom, 3rd ed. 1902. (Duquesne Scan) South Africa.[16]

Times' Orange Free State

The following maps are from Times' History volumes 4 and 5 as indicated by the endnote references for each. Once cited by footnote or endnote additional, portions reproduced to show close ups of landmarks are not re-cited in the work. With apologies for the overall quality. This is the best that could be accomplished with the images available as noted in the preface and its footnotes.

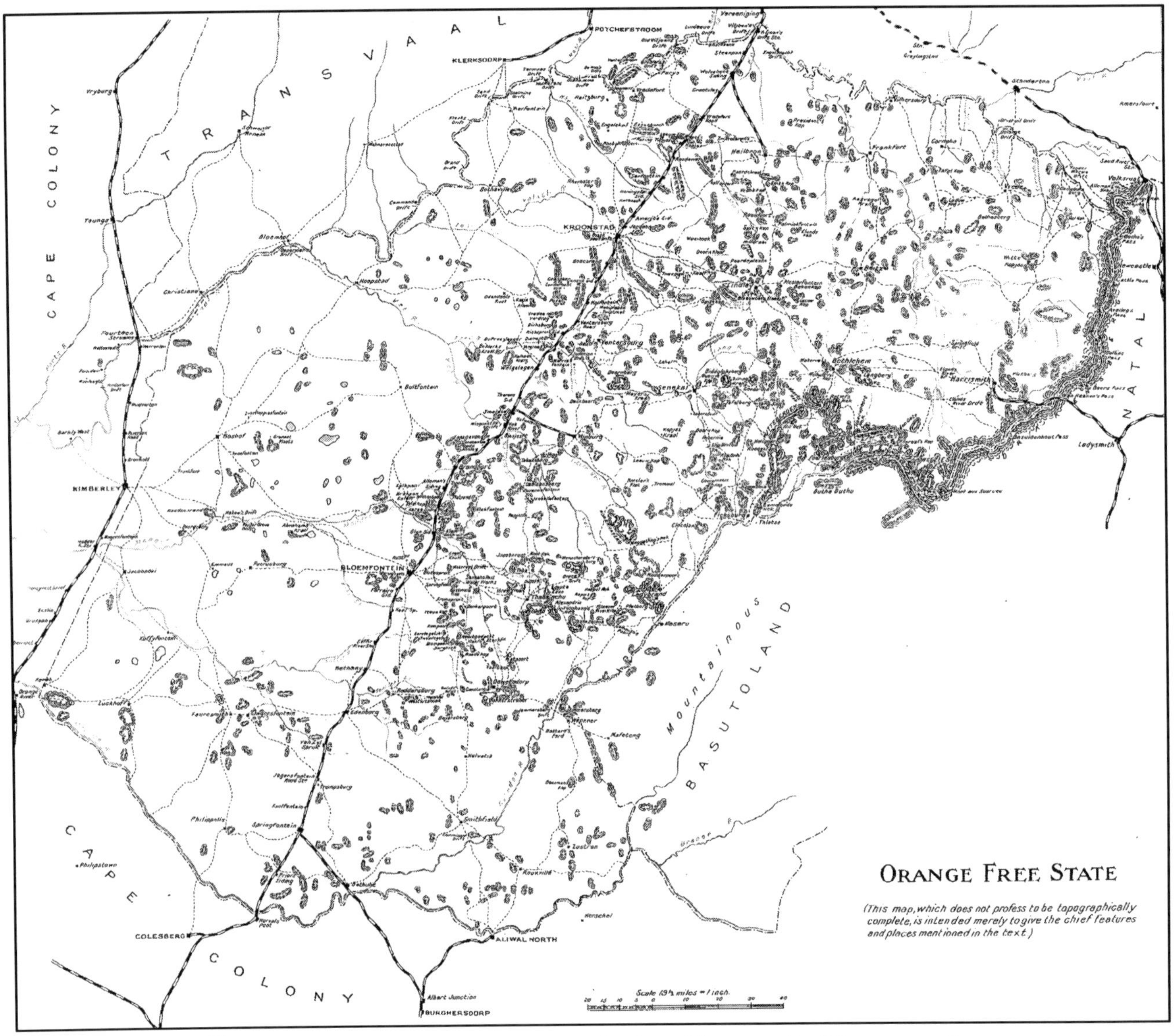

Orange Free State.[17] Portions of this map are used extensively to illustrate De Wet's narrative.

The First De Wet Hunt.

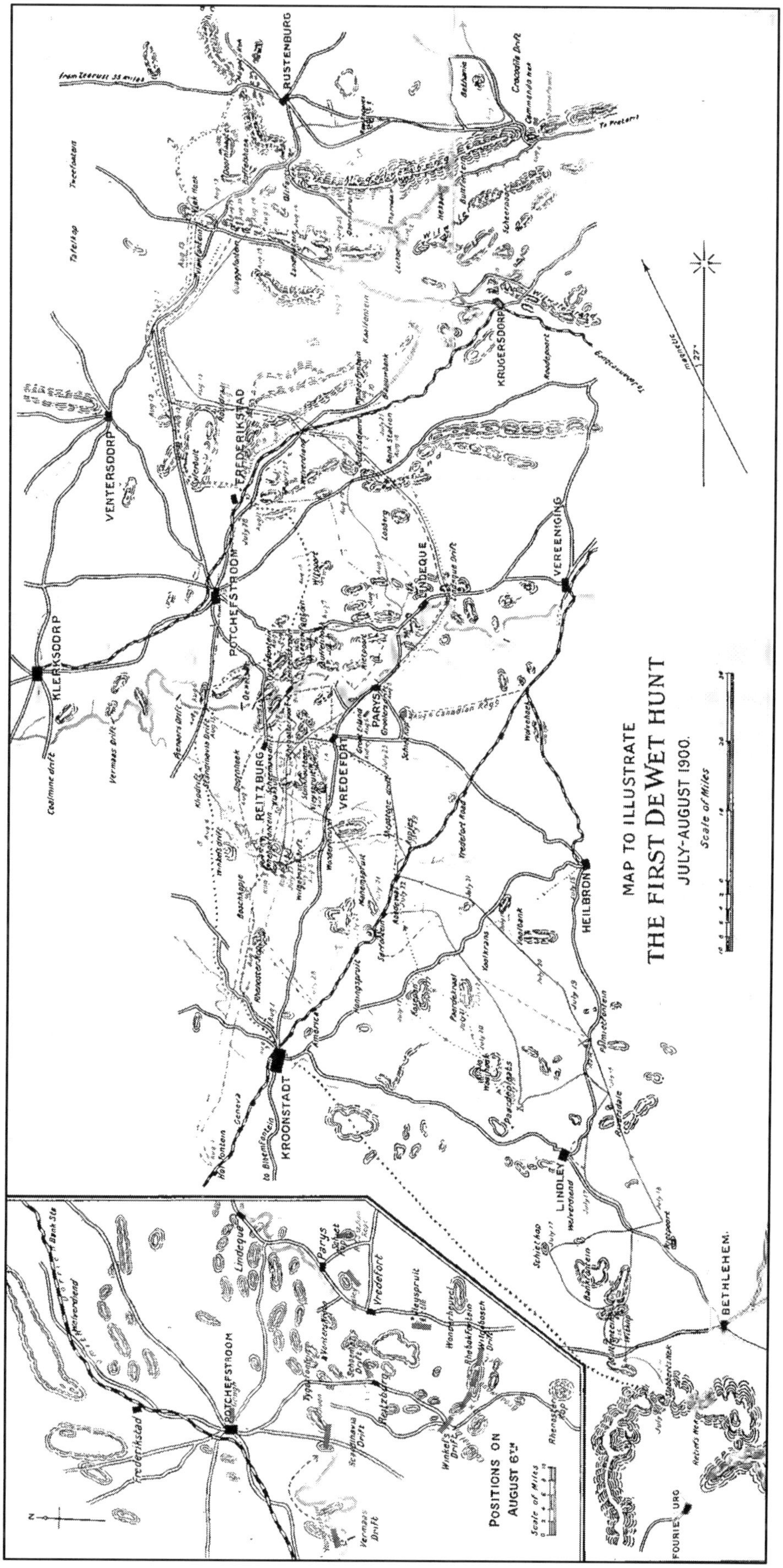

The First De Wet Hunt.[18]

Second De Wet Hunt

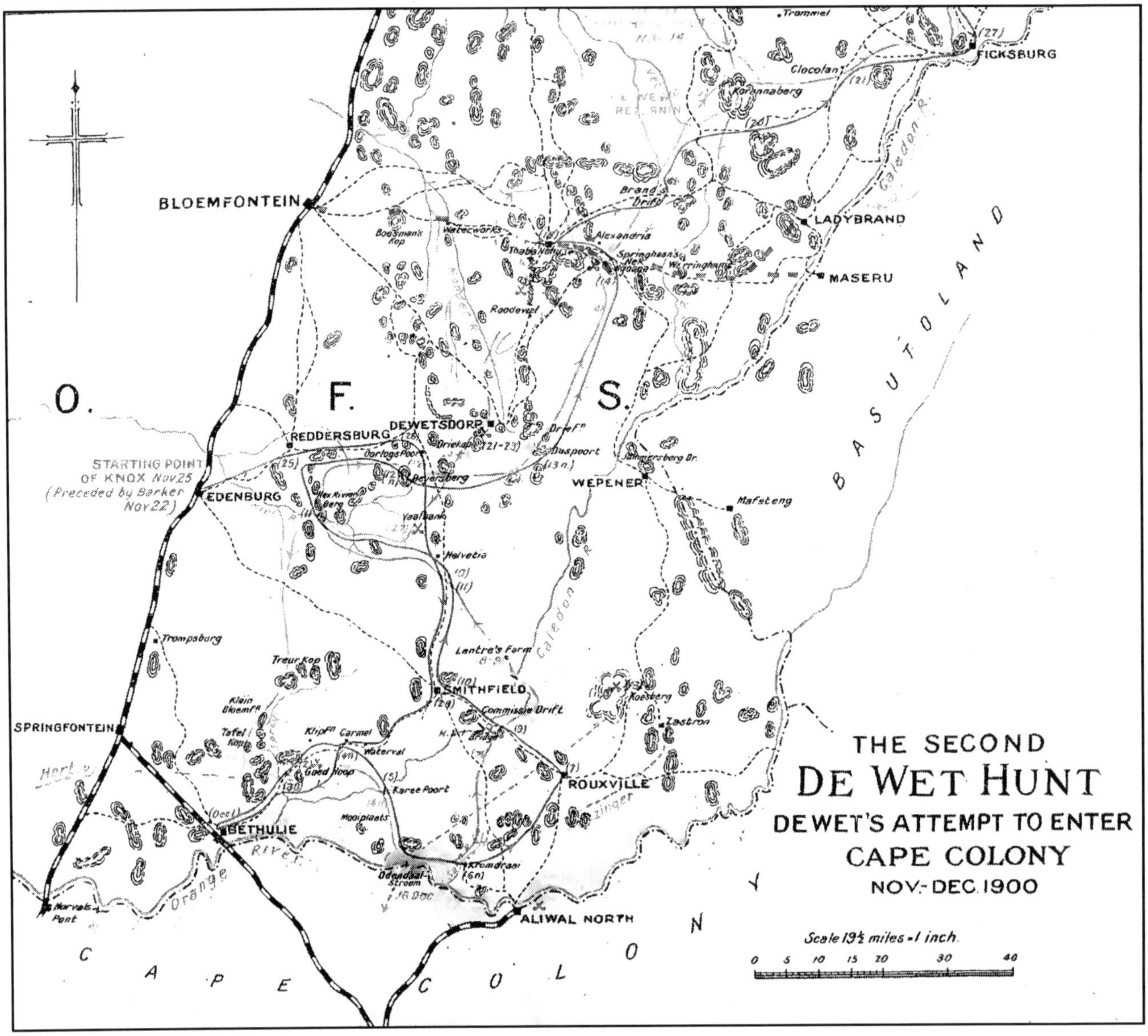

Second De Wet Hunt.[19]

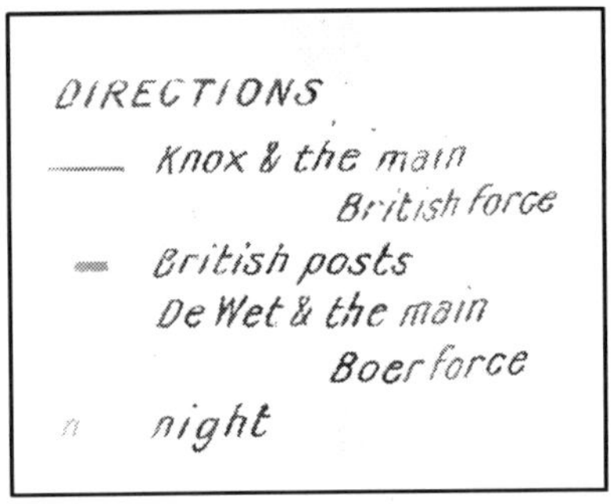

Third De Wet Hunt

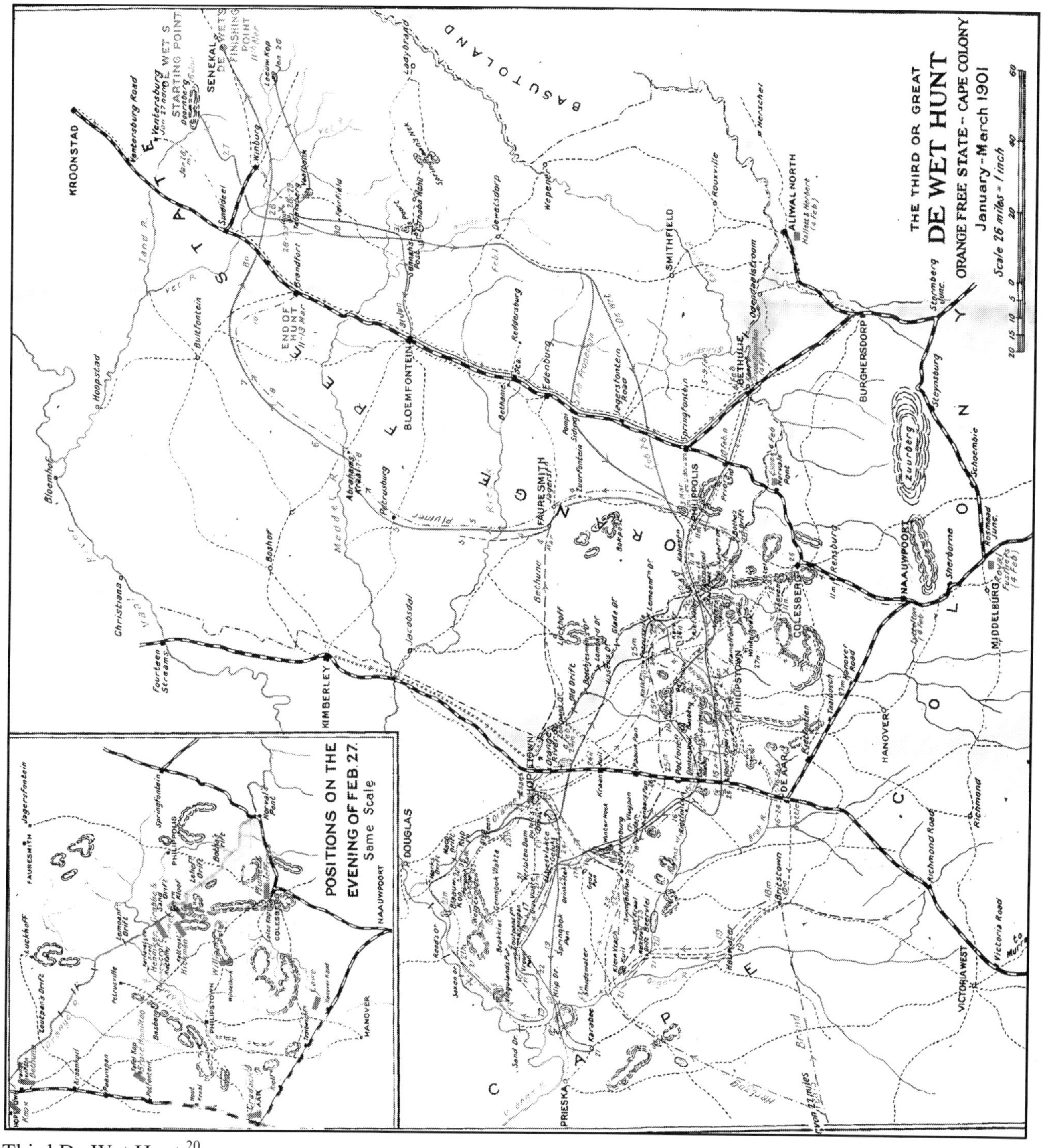

Third De Wet Hunt.[20]

Feb 27th

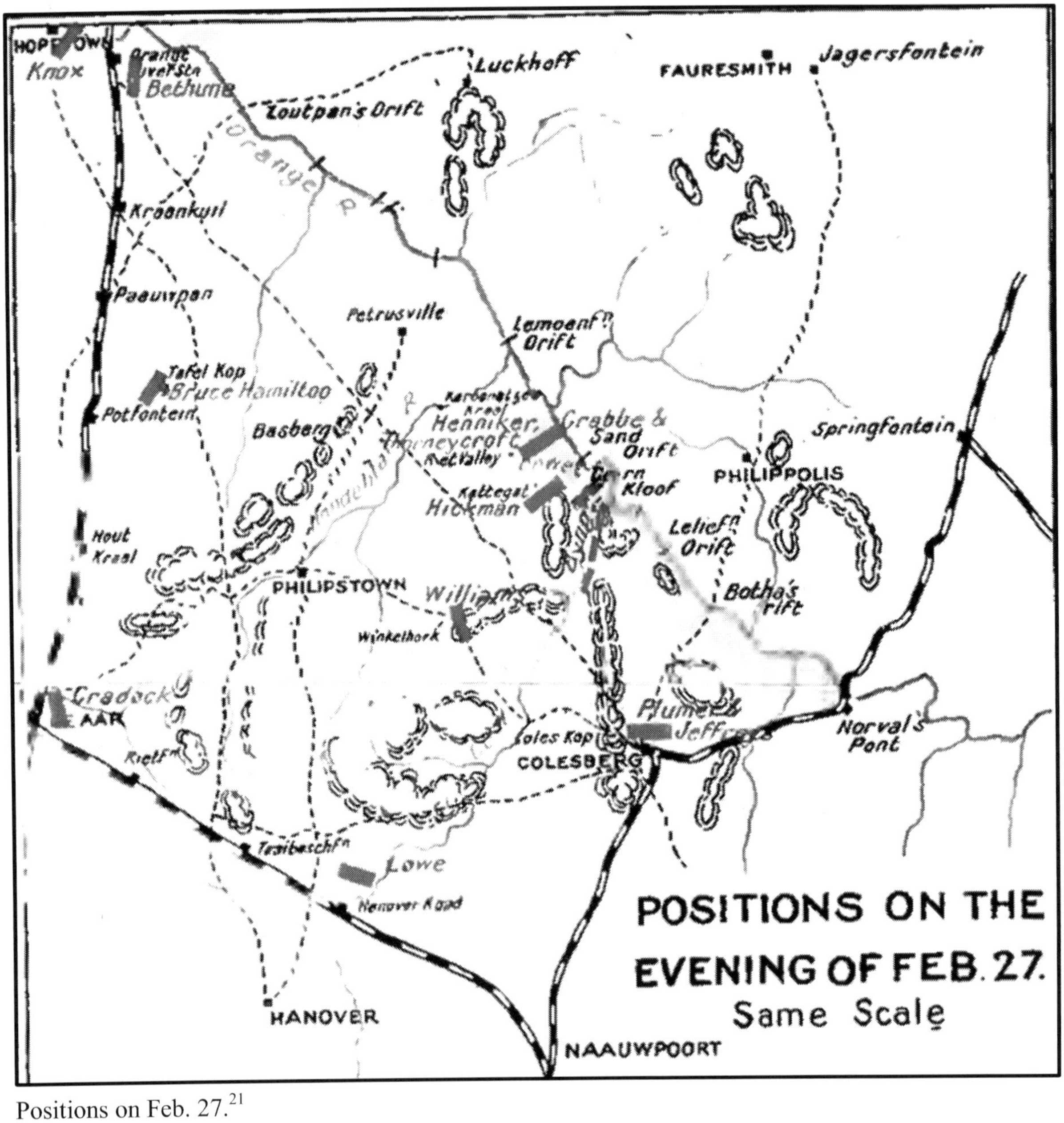

Positions on Feb. 27.[21]

Last Campaign Nov 1901

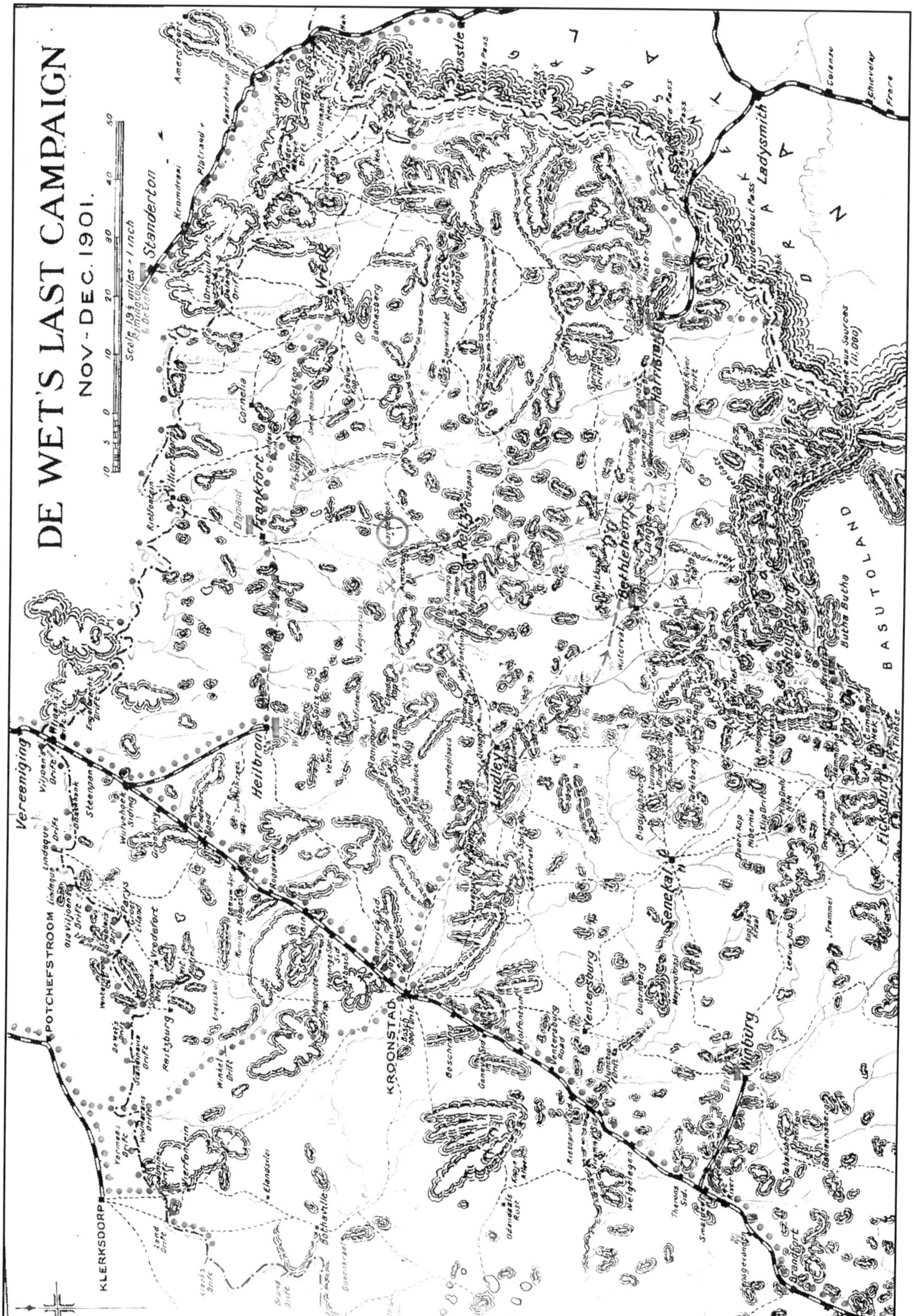

Last Campaign Nov 1901.[22]

Walker Map

This map identifies generalized topographic features and the names of rivers.

Map By: Geo. H. Walker & Co. Boston. Copyright 1899, Library of Congress Maps and Charts # 2877.[23]

Drives in OFS

Drives in OFS. See Times' History Toronto Copy for best online version [24, 25]

Christiaan's Preface

By way of introduction to my work I wish, dear reader, to say only this short word: "I am no book-writer."—But I felt that the story of this struggle, in which a small people fought for liberty and right, is rightly said, throughout the civilized world, to be unknown, and that it was my duty to record my personal experiences in this war, for the present and for the future generations, not only for the Afrikander people, but for the whole world.

Not only did I consider this my duty, but I was encouraged to write by the urgings of prominent men among my people, of men of various nationalities and even of several British officers.

Well, dear reader, I hope that you will not feel disappointed in reading these experiences, as it is not in me, as is perhaps sometimes the case with historical authors, to conjure up thrilling pictures—imaginary things—and put them together merely to make up a book or to make a name for themselves. That be far from me! In publishing my book (although it is written in simple style) I had one object only, viz., to give to the world a story which, although it does not contain the whole of the truth, as regards this wondrous war, yet contains nothing but the truth.

The original has been written by me in Dutch, and I can therefore not be answerable for its translation into other languages.

C. R. de Wet.

Free State Commandants at the beginning of the conflict.

Left: Dickson (State Aty.) Swanpoel, Steenkamp, Ferreira, Fick, Poigieter, Wessels Du Toit, Van der Mewe.
Middle: Albrecht, Du Plessis, Nel, Pres. Steyn, Olivier, J. Prinsloo, De Villiers.
Lower: M. Prinsloo, Van zyl, Du Plooy, Lubbe, Naudi.[26] Identifications from this copy.[27]

1. I Go on Commando as a Private Burgher

In the month of September, 1899, the burghers of the Orange Free State were notified, under the Commando Law, to hold themselves in readiness to go on active service at the shortest possible notice.

Before proceeding any further I should like to explain that portion of the Commando Law which dealt with commandeering. It stipulated that every burgher between the ages of sixteen and sixty must be prepared to fight for his country at any moment; and that, if required for active service, he must provide himself with a riding-horse, saddle and bridle, with a rifle and thirty cartridges—or, if he were unable to obtain a rifle, he must bring with him thirty bullets, thirty caps, and half a pound of powder—in addition he must be provisioned for eight days. That there should have been an alternative to the rifle was due to the fact that the law was made at a time when only a few burghers possessed breech-loading rifles—achterlaaiers,[11] as we call them.

With reference to the provisions the law did not specify their quality or quantity, but there was an unwritten but strictly observed rule amongst the burghers that they should consist of meat cut in strips, salted, peppered, and dried, or else of sausages and "Boer biscuits."[12] With regard to quantity, each burgher had to make his own estimate of the amount he would require for eight days.

It was not long after they were notified to hold themselves ready that the burghers were called up for active service. On the 2nd of October, 1899, the order came. On that day the Veldtcornets, or their lieutenants, visited every farm and commandeered the men.

Amongst the commandeered was I; and thus, as a private burgher, I entered on the campaign. With me were my three sons—Kootie, Isaac, and Christiaan.

The following day the men of the sub-district of Krom Ellenborg, in the district of Heilbron—to which I belonged—mustered at Elandslaagte Farm. The Veldtcornet of this sub-district was Mr. Marthinus Els, and the Commandant of the whole contingent Mr. Lucas Steenekamp. It soon became known that the War Commission had decided that our commando was to proceed as rapidly as possible to the Natal frontier, and that with us were to go the troops from Vrede and Harrismith, as well as some from Bethlehem, Winburg, and Kroonstad. Carrying out these orders, we all arrived at Harrismith six days later.

Commando life now began in real earnest.

The eight days during which the burghers had to feed themselves were soon over, and now it was the duty of the Government to provide for them.

It may be interesting to mention here that the British commissariat differed greatly from ours. Rations were served out daily to their troops. Each soldier received the same quantity and the same quality as his comrade. Our methods were very different, except as regards flour, coffee, sugar, and other articles of that nature. The British soldier, for instance, received his meat ready cooked in the form of bully-beef (blikkiescost we called it), whilst the burgher received his meat raw, and had to cook it as best he could.

Before I leave this subject I may be forgiven if I describe the method of distributing meat to the burghers. After it had been cut up, the Vleeschkorporaal[13] handed out the pieces—a sufficiently responsible task, as it proved, for, as the portions differed much in quality, it became of the first importance that the Vleeschkorporaal should be a man whose impartiality was above suspicion. To avoid any temptations to favoritism, this useful personage used to turn his back on the burghers, and as the men came up in turn he would pick up the piece of meat which lay nearest to hand and, without looking round, give it to the man who was waiting behind him to receive it.

This arrangement should have been satisfactory to all, but it sometimes happened that some burgher, whom fortune had not favoured, made no effort to conceal his discontent, and thus squabbles frequently occurred. Then the Vleeschkorporaal, fully convinced of his own uprightness, would let his tongue go, and the burgher who had complained was a man to be pitied. But such quarrels only occurred early in the campaign. By the time that the Vleeschkorporaal had been a few weeks at his work he had gained a considerable knowledge of human nature, and the injustice of his fellows no longer troubled him. Accordingly he allowed the complaints of the

[11] Literally Backloaders as in breech loading rifle as opposed to a muzzle loader. But this term is also applied and used in another context; to servants, the people who watch your back, help you. See Also: achterrijder, or achterryder A mounted groom or other attendant; a military servant, or batman. agterryer - definition of agterryer in A Dictionary of South African English - DSAE

[12] Small loaves manufactured of flour, with fermented raisins instead of yeast, and twice baked.

[13] Officer in charge of the meat—literally, Flesh-corporal.

men to go in at one ear and at once to come out at the other. The burghers, too, soon became convinced of the foolishness of their conduct, and learnt the lesson of content and forbearance.

As I have already stated, the burgher had to boil or roast his own meat. The roasting was done on a spit cut in the shape of a fork, the wood being obtained from a branch of the nearest tree. A more ambitious fork was manufactured from fencing wire, and had sometimes even as many as four prongs. A skillful man would so arrange the meat on his spit as to have alternate pieces of fat and of lean, and thus get what we used to call a bout span.[14]

The burghers utilized the flour supplied to them in making cakes; these they cooked in boiling fat, and called them stormjagers[15] or maagbommen.[16]

Later on, the British, finding that by looting our cattle they could get fresh meat for nothing, were no longer forced to be content with bully-beef. They then, like ourselves, killed oxen and sheep; but, unlike us, were very wasteful with it. Often, in the camping places they had vacated, we found the remains of half-eaten oxen, sheep, pigs, and poultry.

But I shall not go further into this matter. I leave it to other pens to describe how the British looted our property, wantonly killed our cattle, and devastated our farms. In the course of this narrative my intention is to mention only those cases which I saw with my own eyes. The reader, perusing them, may well pause in surprise and cry out, "Can such things be possible?" To such a question I have only one answer—"They actually occurred, and so my only course is to record them."

But enough of these digressions. Let me return to my proper subject—the story of my own experiences and doings in the great struggle which took place between Boer and Briton. As I have already said, I had been commandeered, and, together with the other burghers of the Heilbron commando, had just reached Harrismith, on the road to the south-eastern frontier.

During our stay there the other commandos, in obedience to Commando Law, joined us, and we proceeded to elect a Commander-in-Chief. The Commandants present were [Lucas] Steenekamp, of Heilbron; Anthonie Lombaard, of Vrede; C. J. De Villiers, of Harrismith; Hans Nandé, of Bethlehem; Marthinus Prinsloo, of Winburg; and C. Nel, of Kroonstad. The result of the voting was that Prinsloo was chosen for the supreme command.

Then the burghers of Winburg selected Mr. Theunissen as their Commandant. He fulfilled his duties admirably, until he was made a prisoner of war. This happened when he was leading a courageous attack at Paardeberg in order to relieve General Piet Cronje.

From Harrismith our commando advanced to within six miles of the Natal-Free State frontier, and camped not far from Bezuidenhoutspas, in the Drakensberg. This imposing range of mountains, which then formed the dividing line between Boer and British territory, slopes down gently into the Free State, but on the Natal side is very steep and precipitous.

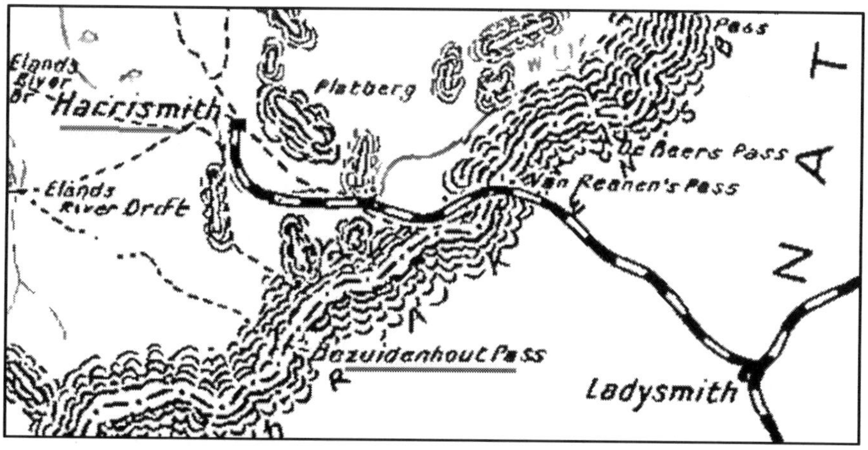

Harrismith and Bezuidenhout Pass. PF: I will subsequently omit 'Note' and simply refer to locations.

[14] Literally, a team of oxen which are not all of the same colour.

[15] Storm-hunters; so-called from being rapidly cooked.

[16] Stomach-bombs—a reflection on their wholesomeness. PF: In the USA, Doughnuts are known as 'Gut Grenades'

An example of the Great Escarpment. Giants Castle in the Drakensberg Range.[28]

The day after we had elected our Commander-in-Chief [M. Prinsloo] I was sent by Commandant Steenekamp, with a small detachment of burghers, to the Natal frontier. I saw nothing of the English there, for they had abandoned all their positions on the frontier shortly before the beginning of the war. When I returned in the evening I found that the burghers had chosen me, in my absence, as Vice-Commandant[17] under Commandant Steenekamp.

It was at five o'clock on the afternoon of that day—the 11th of October, 1899—that the time, which the ultimatum allowed to England, expired. The British had not complied with the terms which the South African Republic demanded—the time for negotiations had passed, and war had actually broken out.

On this very day martial law was proclaimed by the Governments of the two Republics, and orders were given to occupy the passes on the Drakensberg. Commander-in-Chief Prinsloo despatched Steenekamp that night to Bezuidenhoutspas. Eastwards from there the following commandos were to hold the passes:—Botha's pass was to be occupied by the commando from Vrede; Van Reenen's Pass by the commandos from Harrismith and Winburg; and Tintwaspas by the commando from Kroonstad. Westwards, the burghers from Bethlehem were to guard Oliviershoekpas.

Commandant Steenekamp was very ill that night, and was unable to set out; he accordingly ordered me to take his place and to proceed forward with six hundred burghers.

Although I had only to cover six miles, it cost me considerable thought to arrange everything satisfactorily. This was due to the fact that real discipline did not exist among the burghers. As the war proceeded, however, a great improvement manifested itself in this matter, although as long as the struggle lasted our discipline was always far from perfect. I do not intend to imply that the burghers were unwilling or unruly; it was only that they were quite unaccustomed to being under orders. When I look back upon the campaign I realize how gigantic a task I performed in regulating everything in accordance with my wishes. It did not take me long to get everything arranged, and we made an early start.

It was impossible to say what might lie before us. In spite of the fact that I had visited the spot the day before, I had not been able to cross the frontier. The English might have been on the precipitous side of the mountains under the ridge without my being any the wiser. Perhaps on our arrival we should find them in possession of the pass, occupying good positions and quite prepared for our coming.

Everything went well with us, however, and no untoward incident occurred. When the sun rose the following morning the whole country, as far as the eye could reach, lay before us calm and peaceful.

Note Passes. Some with other names.

[17] A Vice-Commandant has no duties to fulfil so long as the Commandant is himself in camp and fit for work.

Crossing the border into Natal.[29] Note Laager in background.

I sent a full report of my doings to Commandant Steenekamp, and that evening he himself, although still far from well, appeared with the remaining part of the commando. He brought the news that war had started in grim earnest. General De la Rey had attacked and captured an armoured train at Kraaipan.

The armored train captured at Kraaipan.[30] [18]

Mountain guns captured at Kraaipan manned by Staats artillery.[31] [32] Note Heliograph at right.

[18] Kraaipan is a town between Mahikeng and Vryburg in the former Bechuanaland. After the declaration of war, on 12 October 1899, the Boers, led by General Koos de la Rey, looted an armoured train, the Mosquito, at Kraaipan. It was carrying two 7-pounder cannons, rifles, ammunition and supplies was derailed and after a five-hour fight the British surrendered the next morning. The cannons, rifles, ammunition, supplies and prisoners were taken. The Boer troops discovered British Mark IV ammunition (better known as dumdum) on the train. https://en.wikipedia.org/wiki/Battle_of_Kraaipan And https://af.wikipedia.org/wiki/Kraaipan

Another view of the apparently the same artillery unit. “Boer Artillery unit operating near Mafeking.”[19][33]

Some days after this a war council was held at Van Reenen’s Pass under Commander-in-Chief Marthinus Prinsloo. As Commandant Steenekamp, owing to his illness, was unable to be present, I attended the council in his place. It was decided that a force of two thousand burghers, under Commandant C.J. De Villiers, of Harrismith, as Vice-Vechtgeneraal,[20] should go down into Natal, and that the remaining forces should guard the passes on the Drakensberg.[21]

As Commandant Steenekamp was still prevented by his health from going to the front, I was ordered, as Vice-Commandant of the Heilbron commando, to proceed with five hundred men to Natal.

It soon became apparent that we had been sent to Natal with the object of cutting off the English [From the South] who were stationed [North] at Dundee and Elandslaagte. We were to be aided in our task by the Transvaalers who were coming from Volksrust and by a party of burghers from Vrede, all under the command of General Roch.

We did not arrive in time to be successful in this plan. That there had been some bungling was not open to question. Yet I am unable to assert to whom our failure was due—whether to the Commandants of the South African Republic, or to Commander-in-Chief Prinsloo, or to Vechtgeneraal De Villiers. For then I was merely a Vice-Commandant, who had not to give orders, but to obey them.

But whoever was to blame, it is certainly true that when, early in the morning of the 23rd of October, I cut the line near Dundee, I discovered that the English had retreated to Ladysmith. It was General Yule who had led them, and he gained great praise in British circles for the exploit.

Yule’s Retreat from Dundee after battle of Talana Hill.[34]

J. H. Yule.[35]

[19] “The guns which the Boers captured from us a Kraaipan have been used against gallant little Mafeking.” These two pictures appear to have been taken at the same time. Both of the above photographs are inscribed “Photo by L Botha” It does not appear that Louis Botha is in either photo and they may have been taken by him.

[20] Fighting general.

[21] Let me say, in parenthesis, that the laws of the Orange Free State make no allusion to the post of Vechtgeneraal. But shortly before the war began the Volksraad had given the President the power to appoint such an officer. At the same session the President was allowed the veto on all laws dealing with war.

If we had only reached our destination a little sooner we should have cut off their retreating troops and given them a very warm time. But now that they had joined their comrades at Ladysmith, we had to be prepared for an attack from their combined forces, and that before the Transvaalers, who were still at Dundee, could reinforce us.[22]

The British did not keep us long in anxiety.

At eight o'clock the following morning—the 24th of October—they came out of Ladysmith, and the battle of Modder Spruit[23] began. With the sole exception of the skirmish between the Harrismith burghers and the Carabineers at Bester Station on the 18th of October, when Jonson, a burgher of Harrismith, was killed—the earliest victim in our fight for freedom—this was the first fighting the Free-Staters had seen.

We occupied kopjes which formed a large semicircle to the west of the railway between Ladysmith and Dundee. Our only gun was placed on the side of a high kop on our western wing. Our men did not number more than a thousand—the other burghers had remained behind as a rear-guard at Bester Station.

With three batteries of guns the English marched to the attack, the troops leading the way, the guns some distance behind. A deafening cannonade was opened on us by the enemy's artillery, at a range of about 4,500 yards. Our gun fired a few shots in return, but was soon silenced, and we had to remove it from its position. Small arms were our only weapons for the remainder of the contest.

The English at once began as usual to attack our flanks, but they did not attempt to get round our wings. Their object appeared to be to keep us in small parties, so that we should be unable to concentrate a large force anywhere.

Meanwhile the troops which were making the attack pushed on closer and closer to us. The country was of such a nature that they were able to get quite near to us without coming under our fire, for small kloofs[24] and other inequalities of the ground afforded them excellent cover. But when they did show themselves they were met by such a frightful and unceasing fire that they could not approach nearer than two hundred paces from our lines.

The brunt of the attack was borne by the burghers from Kroonstad, who, under Commandant Nel, formed our western wing. More to the east, where I myself was, our men had less to endure. But every burgher, wherever he might be, fought with the greatest courage. Although there were some who fell killed or wounded, there was no sign of yielding throughout the whole battle, and every one of our positions we successfully held.

Till three o'clock in the afternoon we kept up our rifle fire on the English, and then we ceased, for the enemy, realizing the impossibility of driving us out of our positions, withdrew to Ladysmith. Shortly afterwards we were able to go over the battlefield. There were not many dead or wounded to be seen; but burghers who had been stationed on the high kop previously mentioned had seen the English remove their wounded during the engagement.[25]

We ourselves had eleven men killed and twenty-one wounded, of whom two subsequently died. This loss touched us deeply, yet it was encouraging to notice that it had not the effect of disheartening a single officer or burgher.

Casualties In Action	Killed		Wounded		Died of Wounds in SA Included in Wounded		Missing and Prisoners		Total Killed, Wounded, Missing and Prisoners	
	Officers	N.C.O.'s & Men	Officers	N.C.O.'s & Men	Officers	N.C.O.'s & Men	Officers	N.C.O.'s & Men	Officers	N.C.O.'s & Men
Rietfontein	1	11	6	98	-	4	-	2	7	111

British Casualties.[26] [27]

[22] 10/21/1899 General French, with about 2000 men, attacked a Boer force under General Kock at Elandslaagte.
10/22/1899 General Yule retired from Dundee on Ladysmith.

[23] Sometimes referred to as the battle of Rietfontein.

[24] Water-courses.

[25] 10/24/1899 Sir George White engaged Boers at Reitfontein [Modder Spruit] to cover Yule's Retreat from Dundee.

[26] Doyle, Arthur Conan. The Great Boer War: Seventh Impression 1900 (September) London, Smith, Elder & Co. Casualties from Outbreak of War until Saturday, September 8th, 1900 Pages 535 & 536.

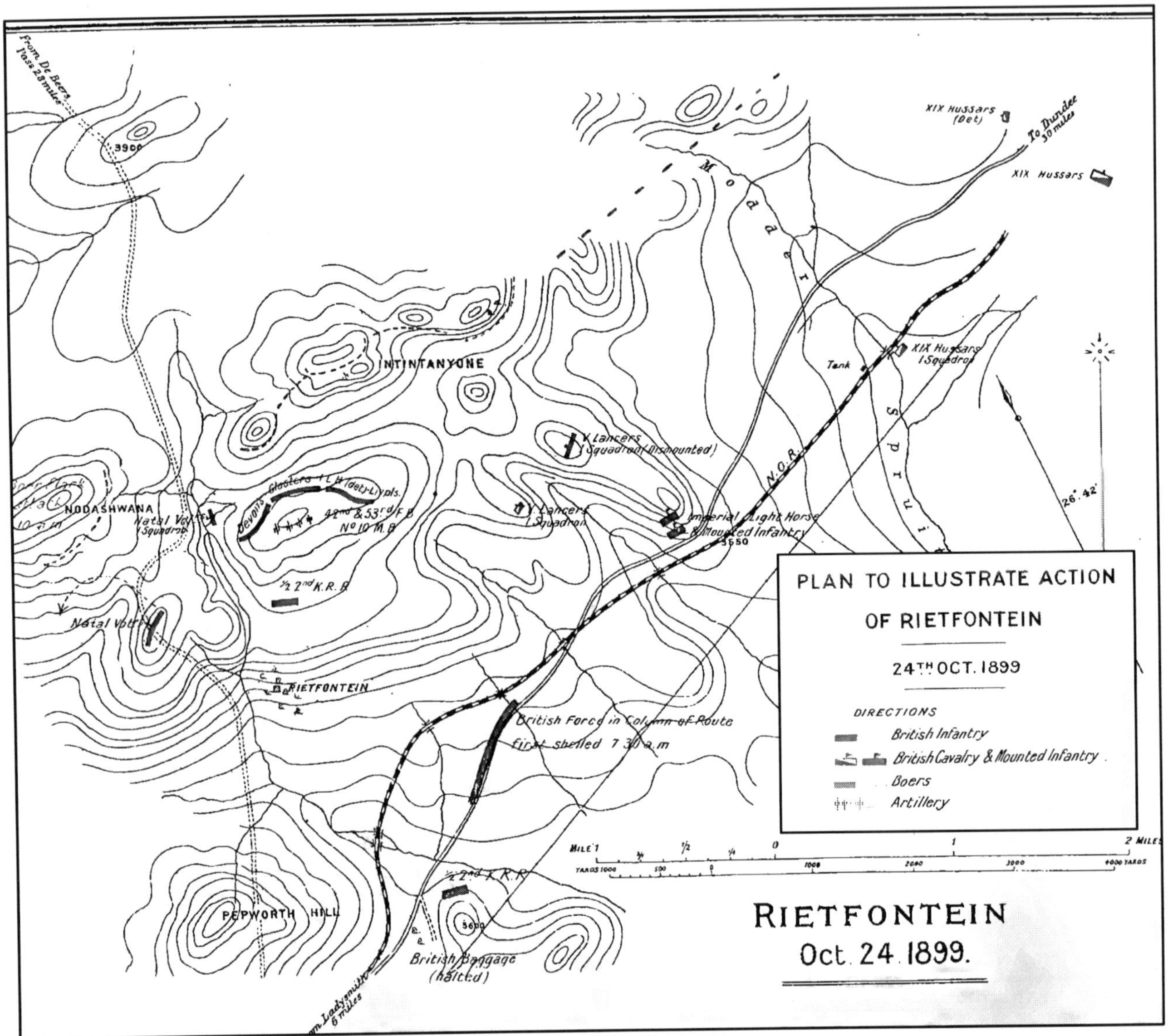

Rietfontein/ Modder River/Intintanyone, Oct 24th, 1899.[36]

Just as the battle began Mr. A.P. Cronje arrived on the scene. He had been nominated by the President [Steyn] as Vechtgeneraal, and had taken over the command from Vice-General C.J. De Villiers. He was most useful in this engagement. When it was over I agreed with him in thinking that our forces were too weak to pursue the retreating English troops. As soon as I was able to leave my position it gave me great pleasure to shake hands with him, for he was an old friend and fellow-member of the Volksraad. It was pleasant to greet him as Vechtgeneraal—he was the son of a valiant officer who had fought in the Basuto war of 1865 and 1866. He [De Villiers] had reached the age of sixty-six years, an age when it is very hard for a man to have to stand the strain which the duties of a Vechtgeneraal necessarily entail.

[27] On 21 October, British troops under Colonel Ian Hamilton and White's cavalry commander, Colonel John French, recaptured the station in the Battle of Elandslaagte. White was unable to follow up this success by relieving Penn-Symons's force (now commanded by Brigadier General J.H. Yule) as he feared that 6,000 Boers from the Orange Free State were about to attack Ladysmith. This decision highlighted Ladysmith's exposed position. https://en.wikipedia.org/wiki/Battle_of_Elandslaagte
Yule's force had to make an exhausting four-day march across the Biggarsberg hills before they could rejoin White in Ladysmith. On 24 October, White made a demonstration at Rietfontein against the Boers pressing forward down the railway line from Glencoe to Ladysmith, to distract them from attacking Yule. The British suffered 132 casualties, mainly from artillery fire.
At noon, after about four hours of engagement, White received a heliograph message from Yule's column that he had passed the last bottleneck and had a clear path in towards Ladysmith. White then retreated as well having accomplished his purpose to cover Yule's retreat.

2. Nicholson's Nek

Until the 29th of October we retained our positions at Rietfontein. On that date General Joubert joined us with a portion of the Transvaal commandos. On his arrival it was settled that the Transvaalers should proceed to the north of Ladysmith and occupy positions on the east of Nicholson's Nek, whilst the Free-Staters were to go to the west and north-west of that town.

General Joubert and staff at Modder Spruit.[37]

A party of burghers, under Commandant Nel, of Kroonstad, were ordered to station themselves on a kop with a flat top, called Swartbooiskop,[28] an hour and a half to the south of Nicholson's Nek. After the battle which was fought on the 30th of November (sic, 30th of October) this kop was christened by us Little Majuba.

Just after sunrise on the 30th of November (sic, 30th of October) the roaring of cannon came to our ears. The sound came from the extreme end of our position, where the Transvaalers were stationed. No sooner did we hear it than the order to off-saddle was given. I myself asked Commandant Steenekamp, who had arrived the previous day from Bezuidenhoutspas, to go to General Croup's laager, about two miles distant, and to request him to advance to where the firing was taking place. To this request General Croup acceded, and Commandant Steenekamp went there with three hundred men, of whom I was one. Our way led past the kop to the south of Nicholson's Nek. What a sight met our gaze on our arrival there!

The kop was occupied by the English.

This must be ascribed to the negligence of Commandant Nel, who had orders to guard the kop. He excused himself by assuring us that he had been under the impression that one of his Veldtcornets and a number of burghers were occupying the hill.

What could we do now?

Commandant Steenekamp and I decided that we must storm the hill with the three hundred men whom we had at our disposal. And this we did, and were sufficiently fortunate to capture the northern point of the kop.

On reaching the summit we discovered that the British troops occupied positions extending from the southern point to the middle of the mountain.

The enemy, the moment we appeared on the ridge, opened a heavy rifle fire upon us. We answered with as severe a fusillade as theirs. Whilst we were shooting, twenty of Commandant Nel's men joined us and helped us to hold our ground. When we had been engaged in this way for some time we saw that the only possible course was to fight our way from position to position towards the English lines.

[28] About nine miles: distance reckoned by average pace of ridden horse—six miles an hour.

I now observed that the mountain top was of an oblong shape, extending from north to south for about a thousand paces. At the northern end, where we were, the surface was smooth, but somewhat further south it became rough and stony, affording very good cover. In our present situation we were thus almost completely exposed to the enemy's fire. The English, on the other hand, had excellent positions. There were a number of ruined Kaffir kraals scattered about from the middle of the mountain to its southern end, and these the enemy had occupied, thus securing a great advantage.

Our bullets hailed on the English, and very shortly they retreated to the southernmost point of the mountain. This gave us the chance for which we had been waiting, for now we could take the splendid positions they had left.

Whilst this was going on an amusing incident occurred. A Jew came up to a burgher who was lying behind a stone, on a piece of ground where boulders were scarce.

"Sell me that stone for half-a-crown," whined the Jew.

"Loop!"[29] the Boer cried; "I want it myself."

"I will give you fifteen shillings," insisted the Jew.

Although the Boer had never before possessed anything that had risen in value with such surprising rapidity, at that moment he was anything but ready to drive a bargain with the Jew, and without any hesitation he positively declined to do business.

In the positions from which the English had retired we found several dead and wounded men, and succeeded in capturing some prisoners.

The enemy were now very strongly posted at the south end of the mountain, for there were in their neighbourhood many Kaffir kraals and huge boulders to protect them from our marksmen. Their fire on us became still more severe and unceasing, and their bullets whistled and sang above our heads, or flattened themselves against the stones. We gave at least as good as we got, and this was so little to their liking that very soon a few white flags appeared in the kraals on their left wing, and from that quarter the firing stopped suddenly.

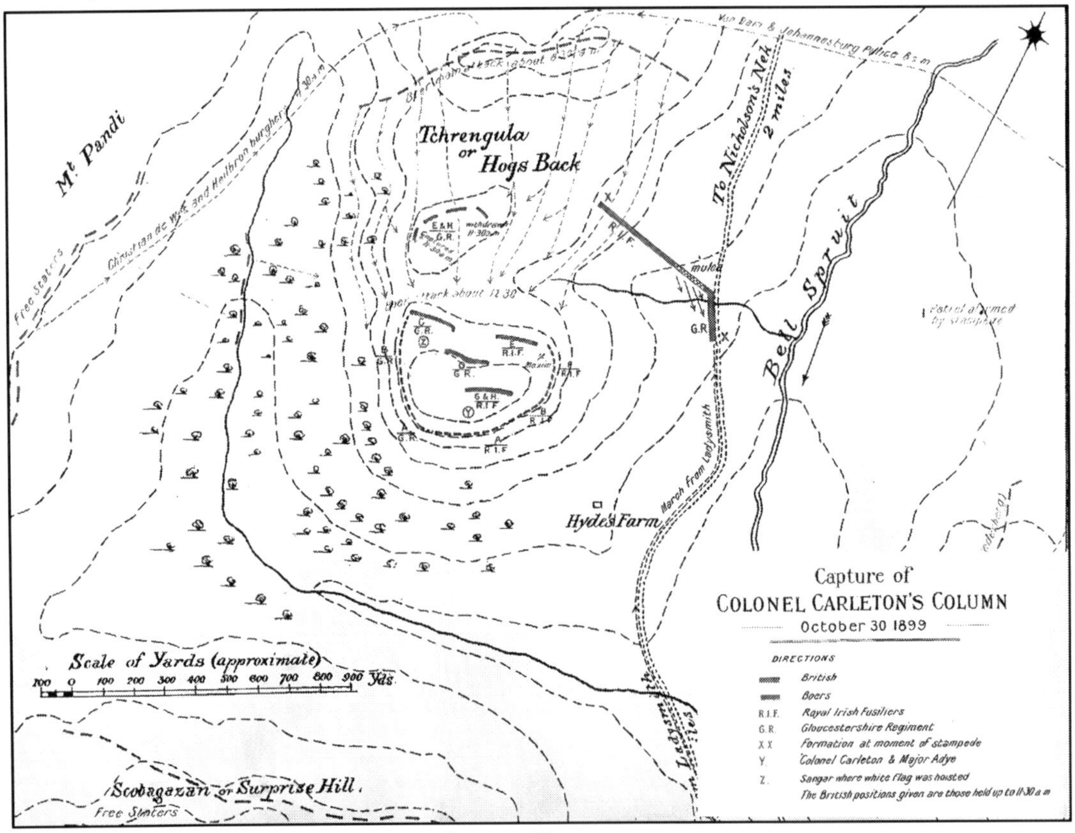

Capture of Carleton's Column, Oct 30th, 1899.[38]

[29] Clear off.

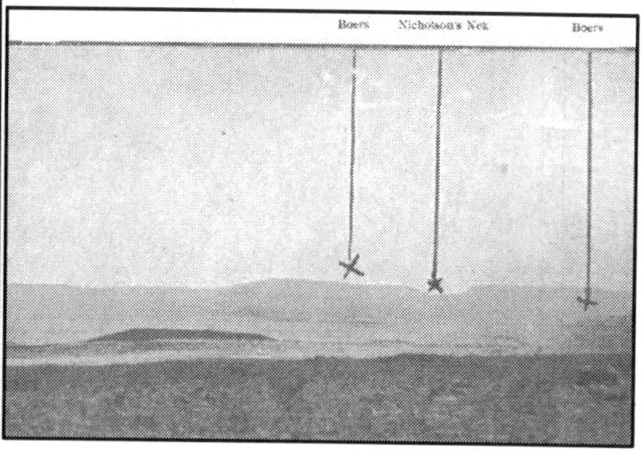

Nicholson's Nek, scene of De Wet's first victory.[39] Area of Lombard's Kop Battle on October 30th.[40] The 'Nek' is the dip in the middle of the ridge between the two Boer positions on either side.

I immediately gave the order to cease fire and to advance towards the enemy. All at once the English blazed away at us again. On our part, we replied with vigour. But that did not continue long. In a very short time white flags fluttered above every kraal—the victory was ours.

I have no wish to say that a misuse of the white flag had taken place. I was told when the battle was over that the firing had continued, because the men on our eastern wing had not observed what their comrades on their left had done. And this explanation I willingly accept.

Our force in this engagement consisted only of three hundred men from Heilbron, twenty from Kroonstad, and forty or fifty from the Johannesburg Police, these latter under Captain Van Dam. The Police had arrived on the battlefield during the fighting, and had behaved in a most praiseworthy manner.

But I overestimate our numbers, for it was not the whole of the Heilbron contingent that reached the firing line. We had to leave some of them behind with the horses at the foot of the kop, and there were others who remained at the first safe position they reached—a frequent occurrence at that period.

I took careful note of our numbers when the battle was over, and I can state with certainty that there were not more than two hundred burghers actually engaged.

Our losses amounted to four killed and five wounded. As to the losses of the English, I myself counted two hundred and three dead and wounded, and there may have been many whom I did not see. In regard to our prisoners, as they marched past me four deep I counted eight hundred and seventeen.

In addition to the prisoners we also captured two Maxim and two mountain guns. They, however, were out of order, and had not been used by the English. The prisoners told us that parts of their big guns had been lost in the night, owing to a stampede of the mules which carried them, and consequently that the guns were incomplete when they reached the mountain. Shortly afterwards we found the mules with the missing parts of the guns.

It was very lucky for us that the English were deprived of the use of their guns, for it placed them on the same footing as ourselves, as it compelled them to rely entirely on their rifles. Still they had the advantage of position, not to mention the fact that they out-numbered us by four to one.

The guns did not comprise the whole of our capture: we also seized a thousand Lee-Metford rifles, twenty cases of cartridges, and some baggage mules and horses.[30]

The fighting had continued without intermission from nine o'clock in the morning until two in the afternoon. The day was exceedingly hot, and as there was no water to be obtained nearer than a mile from the berg,[31] we suffered greatly from thirst. The condition of the wounded touched my heart deeply. It was pitiable to hear them cry, "Water! water!"

[30] 10/30/1899 Engagement at Lombard's Kop. George White sent out from Ladysmith to Nicholson's Nek a Mountain Battery, with the Irish Fusiliers and the Gloucesters, to turn the enemy's right flank. Mules, with guns and reserve ammunition, stampeded into enemy's lines. After gallantly defending their position for six hours, men's ammunition was exhausted, and about 800 were captured.

[31] Hill.

I ordered my burghers to carry these unfortunate creatures to some thorn-bushes, which afforded shelter from the scorching rays of the sun, and where their doctors could attend to them. Other burghers I told off to fetch water from our prisoners' canteens, to supply our own wounded.

As soon as the wounded were safe under the shelter of the trees I despatched a message to Sir George White asking him to send his ambulance to fetch them, and also to make arrangements for the burial of his dead. For some unexplained reason, the English ambulance did not arrive till the following morning.

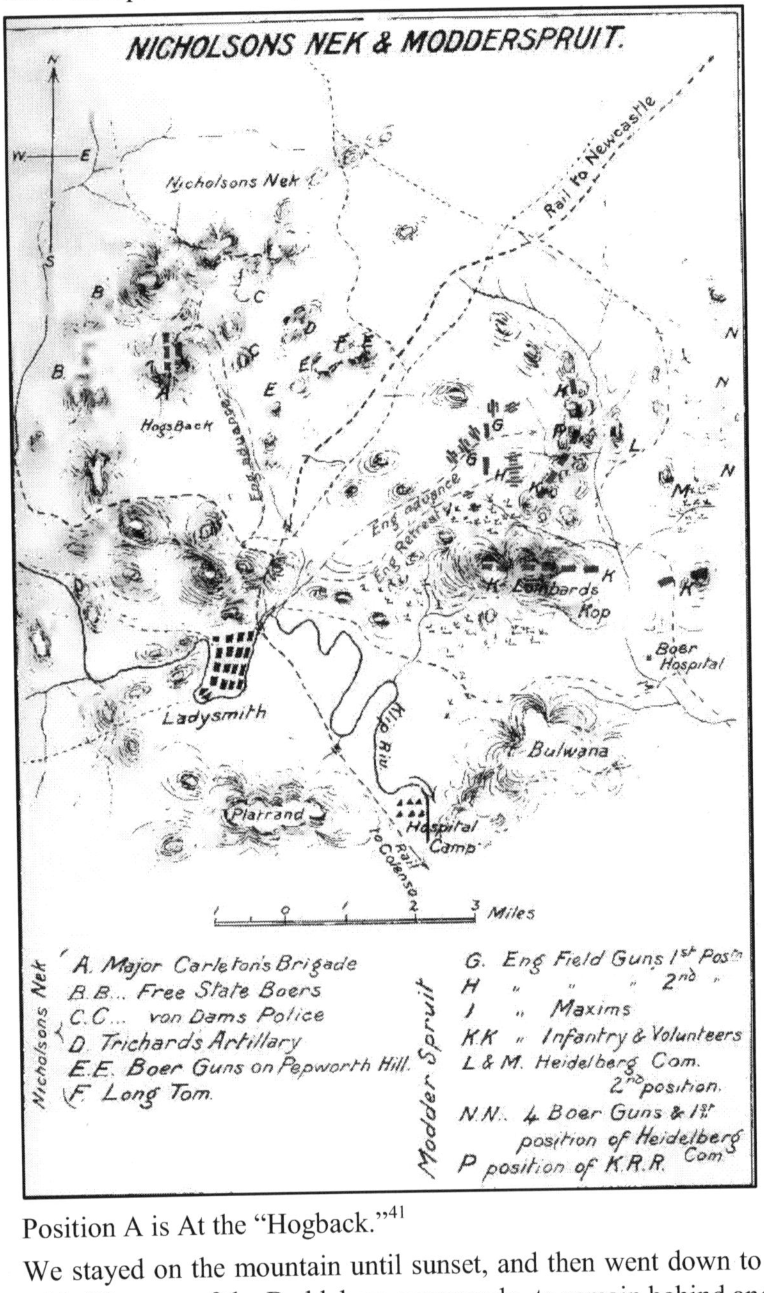

Position A is At the "Hogback."[41]

We stayed on the mountain until sunset, and then went down to the laager. I ordered my brother, Piet de Wet, with fifty men of the Bethlehem commando, to remain behind and guard the kop.

We reached camp at eight o'clock, and as the men had been without food during the whole day it can be imagined with what delight each watched his bout span frizzling on the spit. This, with a couple of stormjagers and a tin of coffee, made up the meal, and speedily restored them. They were exempted from sentry duty that night, and greatly enjoyed their well-earned rest.

To complete my narrative of the day's work, I have only to add that the Transvaal burghers were engaged at various points some eight miles from Nicholson's Nek, and succeeded in taking four hundred prisoners.

Battle of Ladysmith.

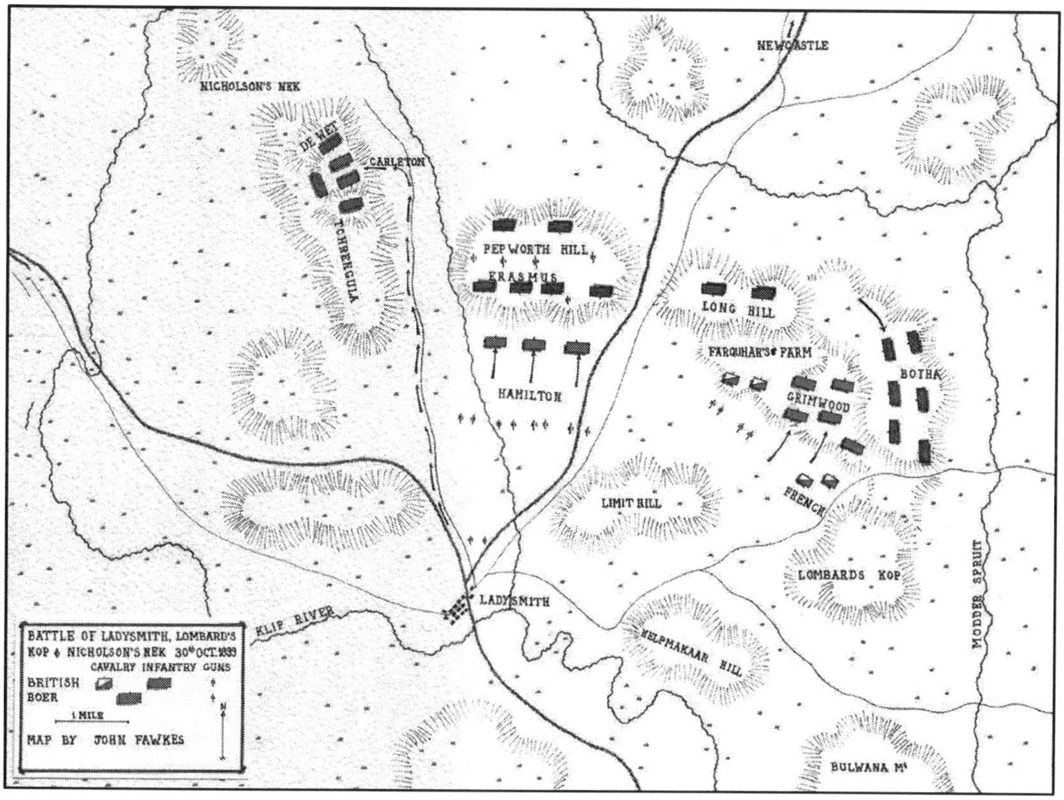

Ladysmith Battle Map.[42] Note Nicholson's Nek engagement upper left.[32]

British Mountain Battery.[43] Captured British 2.5. in. Mountain Gun.[44] One of the 10th Mountain Battery's Guns captured from the British at Nicholson's Nek on 30 October 1899.

We placed our sentries that evening with the greatest care. They were stationed not only at a distance from the camp, as Brandwachten,[33] but also close round the laager itself. We were especially careful, as it was rumoured that the English had armed the Zulus of Natal. Had this been true, it would have been necessary to exercise the utmost vigilance to guard against these barbarians.

[32] The Battle of Ladysmith/ Farquhar's Farm/ Lombard's Kop, took place simultaneously with that at Nicholson's Nek. Col. Carlton had been sent out in order to cut off any Boer retreat. The loss of his 'Mountain Guns' during the ascent of the kopje was a significant factor in his defeat as noted above by our narrator.

[33] Literally, watch-fire men. They were the furthest outposts, whose duty it was to signal by means of their fires.

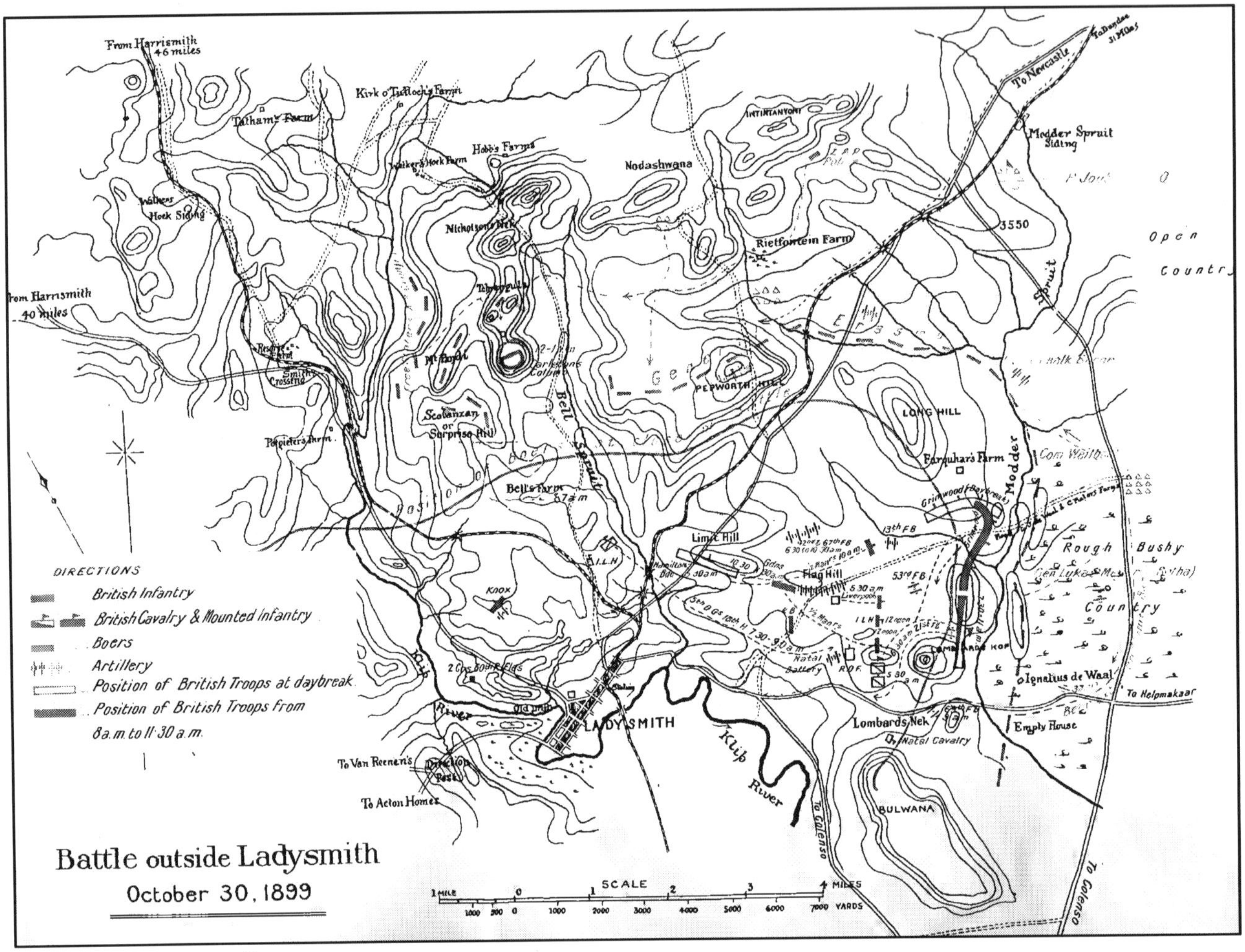

Battle of Ladysmith.[45]

Casualties In Action	Killed		Wounded		Missing and Prisoners			Forces Engaged		
	Officers	Men	Officers	Men	Officers	Men	Total	British	Boer	UK Loss
Nicholson's Nek and Ladysmith aka Farquhar's Farm	6	57	9	244	43	925	1,284	10,000	20,000	12.8%

Casualty Figures from, 'With the Flag to Pretoria.'[46 34]

[34] The main frontal attack was intended to capture Pepworth Hill. The column was led by Colonel Ian Hamilton and consisted of the 1st battalion, the Devonshire Regiment, 1st battalion, the Manchester Regiment, 2nd battalion, the Gordon Highlanders and 2nd battalion, the Rifle Brigade. This attack was to be supported by a column under Colonel Geoffrey Grimwood, consisting of the 1st and 2nd Battalions, the King's Royal Rifle Corps; 1st Battalion, the Leicestershire Regiment; 1st Battalion, the King's Liverpool Regiment; and the 2nd Battalion, the Royal Dublin Fusiliers, which would attack the supposed Boer left flank and capture Long Hill, about 1.5 miles (2.4 km) east of Pepworth Hill. The bulk of the mounted troops under Colonel John French (the 5th Lancers, 5th Dragoon Guards, 18th Hussars less a squadron lost at Talana Hill, 19th Hussars, several companies of Mounted infantry, the Natal Carbineers and the Imperial Light Horse were stationed in reserve or to Grimwood's right. Six batteries (the 13th, 21st, 42nd, 53rd, 67th and 69th) of 15-pounder guns of the Royal Artillery were to support the attacks.
Grimwood's brigade had deployed during the night around Lombard's Kop and Farquhar's Farm, and faced north towards Long Hill. As dawn broke, Grimwood found that half his brigade had straggled, and French's mounted troops had not reached their assigned position. Before this could be corrected, the British troops came under heavy rifle fire from their own right flank. The Boers in this sector were nominally commanded by the elderly Commandant Lucas Johannes Meyer, but were actually led by Louis Botha.
At the same time, the first shell from the Boers' "Long Tom" on Pepworth Hill landed in the town, causing consternation. The British field guns opened fire on Pepworth Hill and Long Hill, and temporarily silenced the Boer siege gun but Colonel Ian Hamilton's frontal attack on Pepworth Hill was called off as it became clear that Grimwood was in trouble, and Hamilton's attack would not be supported.

British forces retreating into Ladysmith after the battle.[47] Note Intact Clock Tower.

Since the very beginning of our existence as a nation—in 1836—our people had been acquainted with black races, and bitter had been their experience. All that our voortrekkers[35] had suffered was indelibly stamped on our memory. We well knew what the Zulus could do under cover of darkness—their sanguinary night attacks were not easily forgotten. Their name of "night-wolves" had been well earned. Also we Free-Staters had endured much from the Basutos, in the wars of 1865 and 1867.

History had thus taught us to place Brandwachten round our laagers at night, and to reconnoitre during the hours of darkness as well as in the day-time.

Perhaps I shall be able to give later on a fuller account in these pages—or, it may be, in another book—of the way we were accustomed to reconnoitre, and of the reasons why the scouting of the British so frequently ended in disaster. But I cannot resist saying here that the English only learnt the art of scouting during the latter part of the war, when they made use of the Boer deserters—the "Hands-uppers."

These deserters were our undoing. I shall have a good deal more to say about them before I finally lay down my pen, and I shall not hesitate to call them by their true name—the name with which they will be for ever branded before all the nations of the world.

Although the main body of South African Republic Boers under Commandant-General Piet Joubert were stationed behind Pepworth Hill, there were none on Long Hill, and fire on this hill was wasted. The British artillery itself came under accurate and effective fire from the Boers' field guns, which were fought as individual gun detachments, and were quickly moved between emplacements before British guns could find their range. The British guns were deployed as they had been drilled to do, in neat rows of six without using cover from artillery or even rifle fire. Their fire was not as effective as that of the Boer Staatsartillerie, and they incurred needless casualties.
After four hours, White had received no messages from Carleton, although firing could be heard from the position he was supposed to have occupied. With no prospect of a successful attack, he ordered his troops to "retire as opportunity offers." As they began to fall back into the open plain between Lombard's Kop and Ladysmith, they came under heavy fire from Botha's riflemen and the Boer gun on Pepworth Hill. Some of Grimwood's units (mainly those which had already suffered heavy casualties at Talana Hill, and were short of officers and tired) panicked and the retreat became a rout. Two batteries of field guns covered the retreat by withdrawing in stages, each battery providing covering fire in turn while the other limbered up and fell back. The British were also very fortunate that a detachment of naval guns (four 12-pounder and two 4.7-inch guns, under Captain Percy Scott of HMS Terrible) arrived by train in Ladysmith and went almost straight into action. Their first few rounds immediately found the range to Pepworth Hill and suppressed the Boers' "Long Tom."
The British fell back into Ladysmith, with units temporarily demoralised. The day was subsequently termed "Mournful Monday." Nevertheless, the Boers did not immediately take advantage of their victory. It is said that when Joubert was urged to follow up the British infantry, he replied, "When God stretches out a finger, do not take the whole hand."
https://en.wikipedia.org/wiki/Battle_of_Ladysmith

[35] Pioneers.

3. Ladysmith Besieged

The Orange Free State and the South African Republic held a joint council of war on the 1st of November, and it was then decided to lay siege to Ladysmith.

We also agreed to send out a horse-commando in the direction of Estcourt. This commando, under Vice-General Louis Botha, had several skirmishes with the enemy. On the 15th of November he engaged an armoured train, capturing a hundred of the British troops. This was General Botha's chief exploit, and shortly afterwards he returned to camp. But I must not anticipate.[36]

On the night of the council of war, General Piet Cronje was sent to occupy positions to the south and south-west of Ladysmith. He had with him the Heilbron burghers, a part of the commandos from Winburg and Harrismith, and two Krupp guns. On the following day a brush took place with the enemy, who, however, speedily fell back on Ladysmith. On the 3rd, [November] a few of their infantry regiments, with a thousand or fifteen hundred mounted troops, and two batteries of 15 and 12-pound Armstrong guns, marched out of the town in a south-westerly direction.

The English brought these two guns into position at such a distance from us that we could not reach them with the Mauser; nor would it have been safe for us to advance upon them, for between them and us lay an open plain, which would have afforded no cover. One of our guns, which was placed exactly in front of the enemy, did indeed begin to fire; but after a shot or two, it received so much attention from the English artillery that we were compelled—just as at Rietfontein—to desist.

The British infantry and cavalry did not show any excessive eagerness to tackle us; and we, on our side, were as disinclined to come to close quarters with them. Nevertheless, the enemy's infantry, backed up by the thunder of twelve guns, did make an attempt to reach us; but though they advanced repeatedly, they were for the most part careful to keep out of range of our rifles. When they neglected this precaution, they soon found themselves compelled to retire with loss.

Our second gun, which had been placed on a tafel-kop[37] to the east of the ground where the engagement was taking place, did excellent work. It effectually baulked the enemy's mounted troops in their repeated efforts to outflank us on that side, and also made it impossible for the English to bring their guns farther east, so as to command the tafel-kop. They did, indeed, make an attempt to place some guns between us and Platrand, which lay to the north of our eastern position, but it was unsuccessful, for our Krupp on the tafel-kop brought such a heavy fire to bear on the troops and gunners, that they were forced to retire.

We, on our part, as I have already said, found it equally impossible to storm the English positions. To advance would have been to expose ourselves to the fire of their heavy guns, whereas an attack to the south would have involved exposure to a cross-fire from the guns on Platrand.

Altogether it was a most unsatisfactory engagement for us both. Nothing decisive was effected; and, as is always the case in such battles, little was done except by the big guns, which kept up a perpetual roar from ten in the morning until five in the afternoon. At that hour the British fell back on Ladysmith.

Our loss was one killed and six wounded, among the latter being Veldtcornet Marthinus Els, of Heilbron. It was evident that the English did not escape without loss, but we were unable to ascertain its extent. My own opinion is that they did not lose very heavily.

From that day nothing of importance happened until I left Natal; though both the Transvaalers and Free State burghers had a few slight brushes with the enemy.

During the night of December the 7th, "Long Tom," the big Transvaal gun, which had been placed on Bulwana Hill, ['Puffing Billy' on map infra] had been so seriously damaged by dynamite, that it had to remain out of action for some time. We all admitted that the English on that occasion acted with great skill and prudence, and that the courage of their leaders deserved every praise. Yet, if we had only been on our guard, we might have beaten off the storming party; but they had caught us unawares. Nevertheless, the mishap taught us a useful

[36] 11/15/1899 Armoured train wrecked by Boers near Frere. Mr. Winston Churchill and a number of Dublin Fusiliers and Volunteers captured.

[37] A table-shaped mountain.

lesson: henceforth the Transvaal Commandants were more strict, and their increased severity had an excellent effect both on the burghers and gunners.[38]

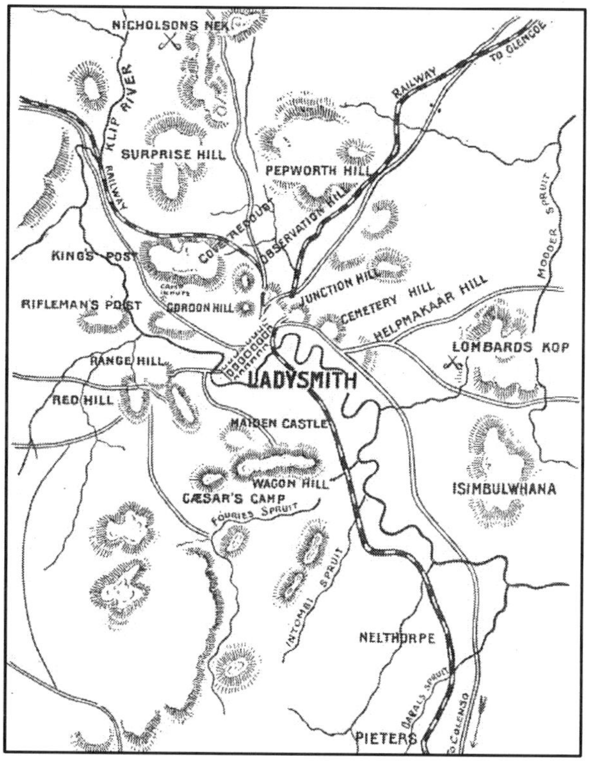

Ladysmith landmarks.[48]

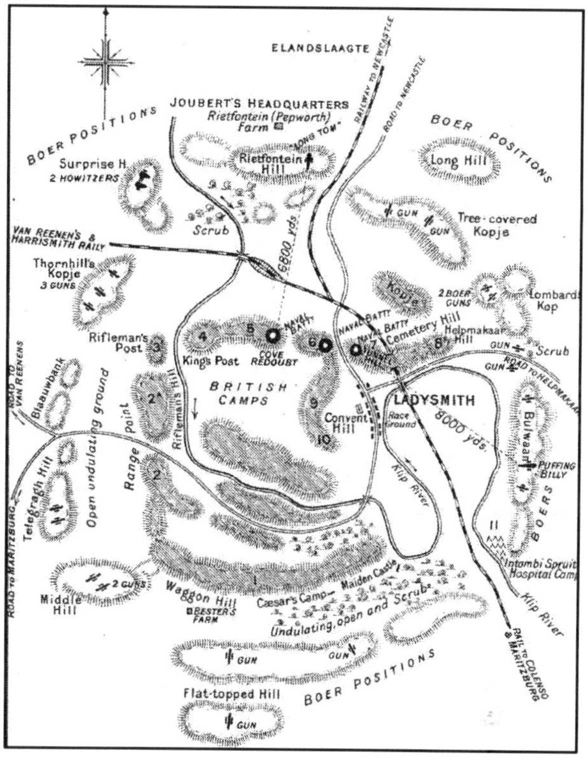

Surprise Hill to the North. Lombard's Kop to the East. Ladysmith Map.[49]

"Long Tom in hospital: A Boer gun under repair at Pretoria."[50] One of four Boer 'Long Tom' canons, damaged at Siege of Ladysmith, on its way to the Siege of Kimberley after it was repaired in Pretoria.[51]

General Sir Redvers Buller had landed at Cape Town early in November. We were now expecting every day to hear that he had assumed the chief command over the English army encamped between Estcourt and Colenso. The number of troops there was continually increasing owing to the reinforcements which kept pouring in from over the ocean.

Great things were expected of Sir Redvers Buller, to whom the Boers, by a play of words, had given a somewhat disrespectful nick-name. He had not been long in Natal before his chance came. I must, however, be silent about his successes and his failures, for, as I left Natal on the 9th of December [1899], I had no personal experience of his methods. But this I will say, that whatever his own people have to say to his discredit, Sir Redvers Buller had to operate against stronger positions than any other English general in South Africa.

[38] 12/8/1899 British sortie from Ladysmith, Lombard's Kop being carried. [South of Ladysmith] See Map.
12/9/1899 Lieutenant-Colonel Metcalfe, 2nd Rifle Brigade, with 500 men from Ladysmith, captured Surprise Hill, destroying a howitzer. [North of Ladysmith]

4. I Am Appointed Vechtgeneraal

Up to the 9th of December I had only been a Vice-Commandant, but on the morning of that day I received a telegram from States-President Steyn, asking me to go to the Western frontier as Vechtgeneraal.[39]

This came as a great surprise to me, and I telegraphed back to the President asking for time to think the matter over. To tell the truth, I should have much preferred to go through the campaign as a private burgher.

Almost immediately after this there came another telegram—this time from Mr. A. Fischer, a member of the Executive Council, and a man whom I respected greatly on account of his official position. He urged me not to decline the appointment, but to proceed at once to the Western borders. I did not know what to do. However, after deliberating for a short time, and with great difficulty overcoming my disinclination to leave my present associates, I decided to accept the post offered to me. Commandant Steenekamp was kind enough to allow me to take with me fourteen men, with whom I had been on especially friendly terms; and, after a few parting words to the Heilbron burghers, in which I thanked them for all the pleasant times I had passed in their company, I left the laager.

Abraham Fischer, Member of the Executive Raad of the Orange Free State.[52 40]

It was heart-breaking to tear myself away from my commando: that 9th of December was a day which I shall never forget.[41]

The following morning I arrived, with my staff, at Elandslaagte Station, on our way to Bloemfontein. A special train, provided by the Transvaal authorities, at the request of my Government, was waiting for us, and we started without a moment's delay. As we journeyed on, the conductor would sometimes ask me whether I should like to stop at such and such a station, but my answer was always:

"No! no! hurry on!"

But when we got as far as Viljoen's Drift, there was an end to my "special train!" In spite of the Government's orders that I was to be sent forward without delay, I had to wait six hours, and then be content to travel as an ordinary passenger. At Bloemfontein we found everything ready for us, and at once started on our journey of sixty or seventy miles to Magersfontein, where we arrived on December the 16th.

During the time I had spent in travelling, three important engagements had taken place, namely those of Colenso, Magersfontein and Stormberg. At Colenso, the English had suffered heavy losses, and ten guns had fallen into our hands. Magersfontein also had cost them dear, and there General Wauchope had met his fate;

[39] Vechtgeneraal = Combat general: the commanding officer of a division of the Boer army.

[40] Abraham Fischer (9 April 1850–16 November 1913) was a South African statesman. He was Prime Minister of the Orange River Colony in South Africa, and when that ceased to exist joined the cabinet of the newly formed Union of South Africa. https://en.wikipedia.org/wiki/Abraham_Fischer

[41] Also: 12/9/1899 Lieutenant-Colonel Metcalfe, 2nd Rifle Brigade, with 500 men from Ladysmith, captured Surprise Hill, destroying a howitzer.

while at Stormberg seven hundred of them had been taken prisoners, and three of their big guns had been captured by us.[42]

At Magersfontein were six or seven thousand Transvaal burghers under General Piet Cronje, with General De la Rey as second in command. Thus it fell to my lot to take over the command of the Free-Staters. The Commander-in-Chief of these Free State burghers, as well as of those who were camped round Kimberley, was Mr. C.J. Wessels;[43] Mr. E.R. Grobler commanded at Colesberg, and Mr. J.H. Olivier at Stormberg.

General E. R. Grobler [Grobelaar].[53 44 54]

John Hendrik Oliver.[55 45]

Ostrich Farm, South Africa.[56]

Ostrich Farm, South Africa.[57]

I spent my first few days at Magersfontein in organizing the Free State burghers. When this task had been accomplished, General De la Rey and I asked General Cronje's permission to take fifteen hundred men, and carry on operations in the [southerly] direction of Hopetown and De Aar with the intention of breaking Lord

[42] Also: 12/17/1899 General Gatacre advanced from Sterkstroom to Putters Kraal. General French established his headquarters at Arundel.

[43] This is not Cornelius Hermanus Wessels who is in the Netherlands as part of the diplomatic delegation to the Netherlands during the course of the conflict. https://en.wikipedia.org/wiki/Cornelius_Hermanus_Wessels

[44] Esaias Reinier Grobler (1861–1937) was born in 1861 near Philippolis in the Free State and followed a political life becoming a member of the Volksraad in 1886. At the start of the Boer War he commanded Boers Cape border. He was involved with General Olivier at Stormberg in December 1899. He also saw action at Colesberg and Paardeberg. He was wounded near Springfontein and captured at Winburg in May 1900. He re-entered politics after the war and became a Senator in the Union Government, and Administrator of Orange Free State Province. 11 March 1924 to March 1929 https://www.angloboerwar.com/name-search And https://en.wikipedia.org/wiki/List_of_administrators_of_former_South_African_provinces

[45] Jan Hendrik Olivier (1848–1930) In 1865 he became a Field Cornet and fought in the last Basuto War. He is rewarded with the farm Olifantsbeen where he becomes prosperous. He became a member of the Volksraad in 1883 for the Caledon River ward.
He was in command of the Rouxville and Thaba Nchu commandos, and moved with his mother through the Colesberg, Barkly East and Dordrecht areas. He was in command of the Boer forces at the Battle of Stormberg in December 1899. Later he worked with General De Wet in the eastern Free State and Brandwater areas. After getting into an argument with De Wet, he decided to join the Transvaal forces. On the way there, he was captured. He was exiled to Ceylon as Boer prisoners of war along with three of his sons. SA Wikipedia: J.H. Olivier - Wikipedia English Wikipedia: Jan Hendrik Olivier - Wikipedia

Methuen's railway communications. But Cronje would hear nothing of the scheme. Say what we would, there was no moving him. He absolutely refused to allow fifteen hundred of his men to leave their positions at Magersfontein, unless the Government found it impossible to procure that number of burghers from elsewhere. Thus our plan came to nothing.

Shortly afterwards De la Rey was sent to the commandos at Colesberg, and I succeeded him in the command of the Transvaalers at Magersfontein. The Government then put General Wessels in sole command at Kimberley, and gave General Cronje the chief command over the Free State burghers at Magersfontein. Thus it was that I, as Vechtgeneraal, had to receive my orders from Cronje. I had the following Commandants under me:

Du Preez, of Hoopstad; Grobler, of Fauresmith; D. Lubbe, of Jacobsdal; Piet Fourie,[46] of Bloemfontein; J. Kok and Jordaan, of Winburg; Ignatius Ferreira,[47] of Ladybrand; of Ficksburg; Du Plessis, and, subsequently, Commandant Diederiks, of Boshof.

The English had entrenched themselves at the Modder River, we at Magersfontein. There was little or nothing for us to do, and yet I never had a more troublesome time in my life. I had all the Transvaalers under my orders, in addition to the burghers of the Free State, and the positions which I had to inspect every day extended over a distance of fifteen miles from end to end. I had to listen to constant complaints; one of the officers would say that he could not hold out against an attack if it were delivered at such and such a point; another, that he had not sufficient troops with him, not to mention other remarks which were nonsensical in the extreme.

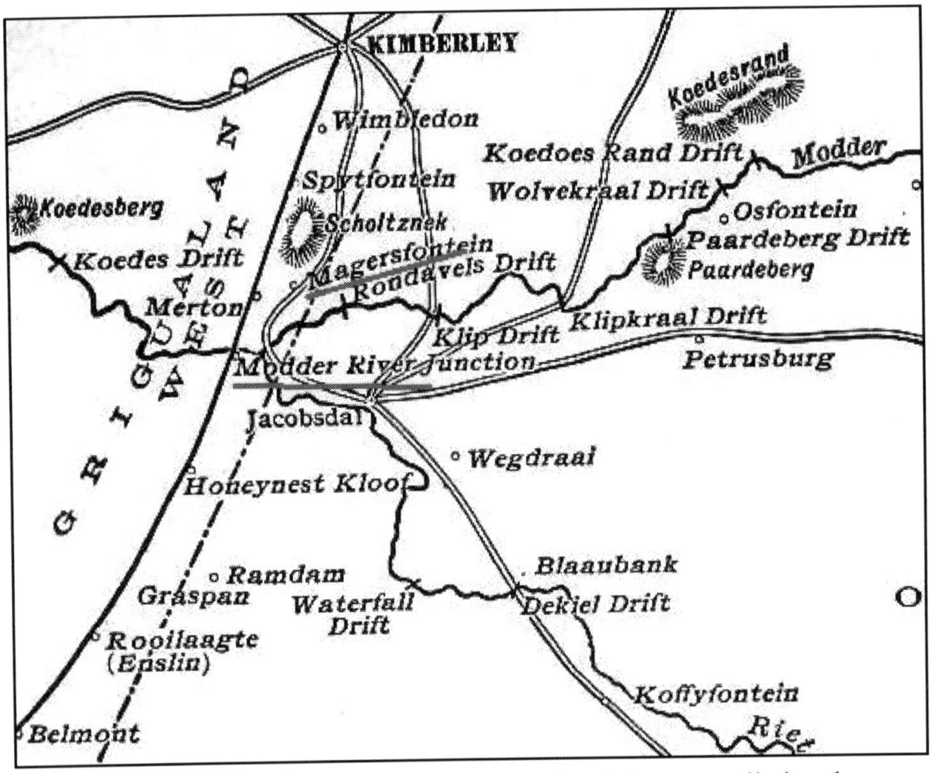

Modder River and Magersfontein are underlined and also generally in other maps in the book.

46 Fourie became the commander of the Bloemfontein Commando and fought all the major battles at the western front. After the capture of Bloemfontein by the British on 13 March 1900 Fourie participated in the Boer victory at Sanna's Post (31 March 1900), the failed attack on British positions at Jammersberg Drift (9 April 1900), the failure to stop the British advance in the Battle of Sand River southwest of Kroonstad (10 May 1900), and again a victory at Rooiwal (Roodewal) Station (7 June 1900), under the overall command of Christiaan de Wet. https://en.wikipedia.org/wiki/Piet_Fourie

47 Ignatius "(Oom) Naas" Stephanus Ferreira (21 March 1844–18 February 1900, Oom meaning Uncle in Afrikaans) was an Orange Free State Boer war general and Commander-in-Chief, who was fatally shot at the Battle of Paardeberg. Naas Ferreira should not be confused with his son and namesake Ignatius "Natie" Stephanus Ferreira (1875–1947), the gold miner Ignatius Philip Ferreira (1840–1921), and Boer general J. (Joachim) Ferreira (1835–1917).

Ferreira was succeeded as commander-in-chief of the Orange Free State troops by Christiaan de Wet on February 19, 1900. Ferreira's son and namesake "Natie" continued to serve as a secretary to Boer general Koos de la Rey. https://en.wikipedia.org/wiki/Naas_Ferreira

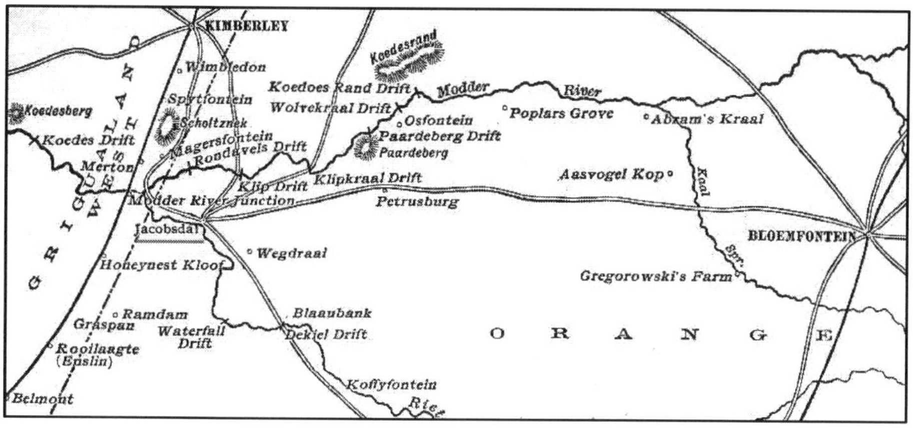

Modder and Reit Rivers.[58] Cronje's main laager is at Jacobsdal.

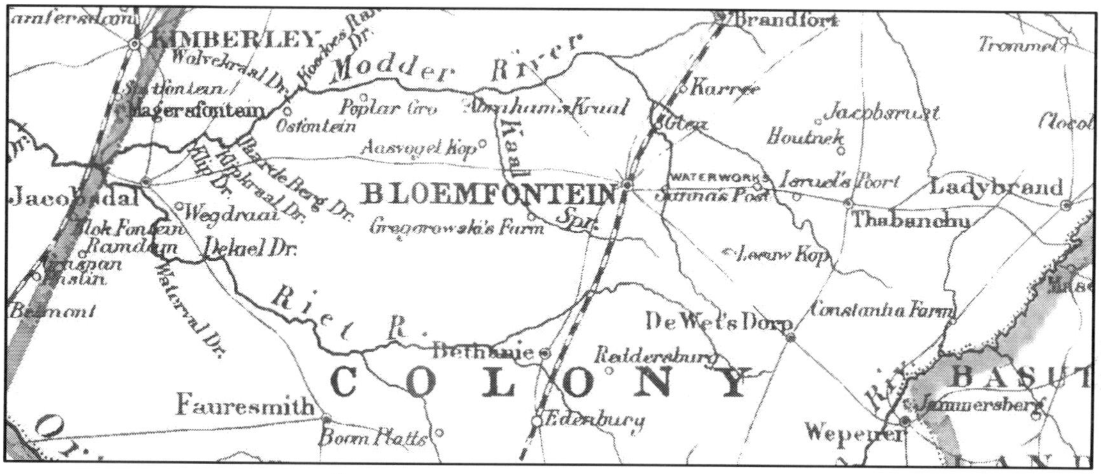

Orange River Colony South.[59]

Trenches at Magersfontein Battlefield.[60]

In the Orange River trenches holding back the Boers.[61]

Magersfontein.[62]

Graves of the Highlanders Killed at Magersfontein.[63]

In the meantime, the enemy was shelling our positions unceasingly. Not a day passed but two of their Lyddite guns dropped shells amongst us. Sometimes not more than four or five reached us in the twenty-four hours; at other times from fifty to two hundred, and once as many as four hundred and thirty-six.

In spite of this, we had but few mishaps. Indeed, I can only remember three instances of any one being hurt by the shells. A young burgher, while riding behind a ridge and thus quite hidden from the enemy, was hit by a bomb, and both he and his horse were blown to atoms. This youth was a son of Mr. Gideon van Tonder, a member of the Executive Council.

Another Lyddite shell so severely wounded two brothers, named Wolfaard, Potchefstroom burghers, that we almost despaired of their lives. Nevertheless, they recovered. I do not want to imply that the British Artillery were poor shots. Far from it. Their range was very good, and, as they had plenty of practice every day, shot after shot went home. I ascribe our comparative immunity to a Higher Power, which averted misfortune from us.

I had not been long at Magersfontein before I became convinced that Lord Methuen was most unlikely to make another attack on our extensive positions. I said nothing of this to any of the burghers, but on more than one occasion, I told General Cronje what I thought about the matter.

"The enemy," I repeated to him over and over again, "will not attack us here. He will flank us." But Cronje would not listen to me.

The presence of women in our laager was a great hindrance to me in my work. Indeed, I opened a correspondence with the Government on the matter, and begged them to forbid it. But here again my efforts were unavailing. Later on, we shall see in what a predicament the Republican laagers were placed through the toleration of this irregularity.

Meanwhile, the inevitable results of Cronje's policy became more and more apparent to me, and before long we had to suffer for his obstinacy in keeping us to our trenches and schanzes.[48]

Norvals Pont Railway Bridge.[64]

Modder River Bridge.[65]

Modder River Temporary Bridge.[66]

[48] A shelter-mound of earth and boulders.

5. The Overwhelming Forces of Lord Roberts

I speedily discovered the object which the English had in view in taking such advanced positions and in bombarding Magersfontein. They wished to give us the impression that they were able to attack us at any moment and so to keep us tied to our positions. In the meantime they were making preparations in another direction, for the movement which was really intended—namely, the advance of Lord Roberts with his overwhelming force.

The Commander-in-Chief, Piet de Wet (and before him Commandant H. Schorman), had plenty of work given them by the English. But General De la Rey had been so successful that he had prevented Lord Roberts, notwithstanding the enormous numbers he commanded, from crossing the Orange River at Norvalspont, and had thus forced him to take the Modder River route.

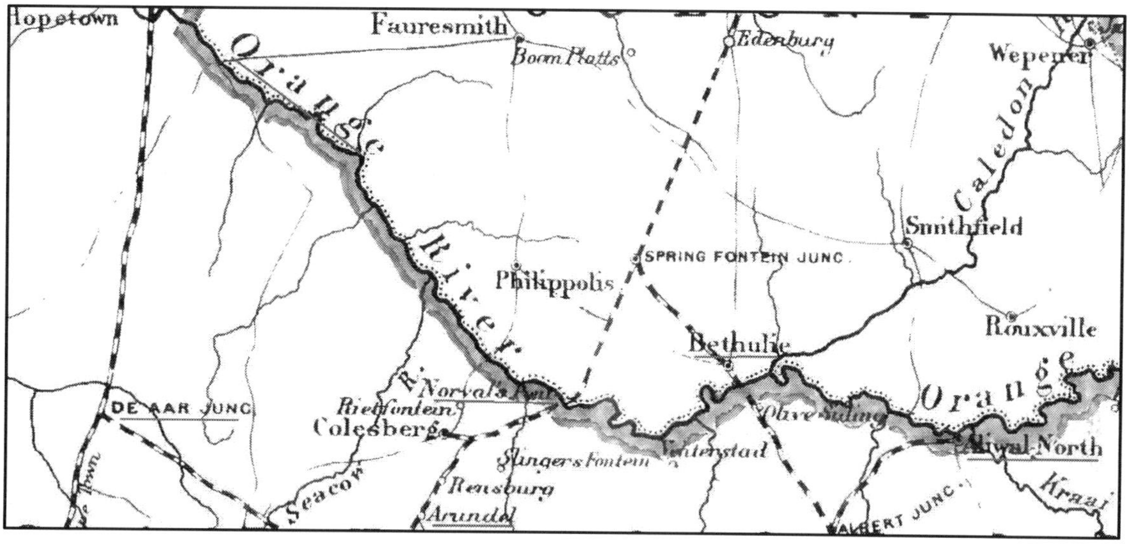

Orange River Colony South. Note Orange River and De Aar Junction Left, Norval's Point Lower Center above Arundel at river crossing and Bethulie as well as Aliwal North on right.[67]

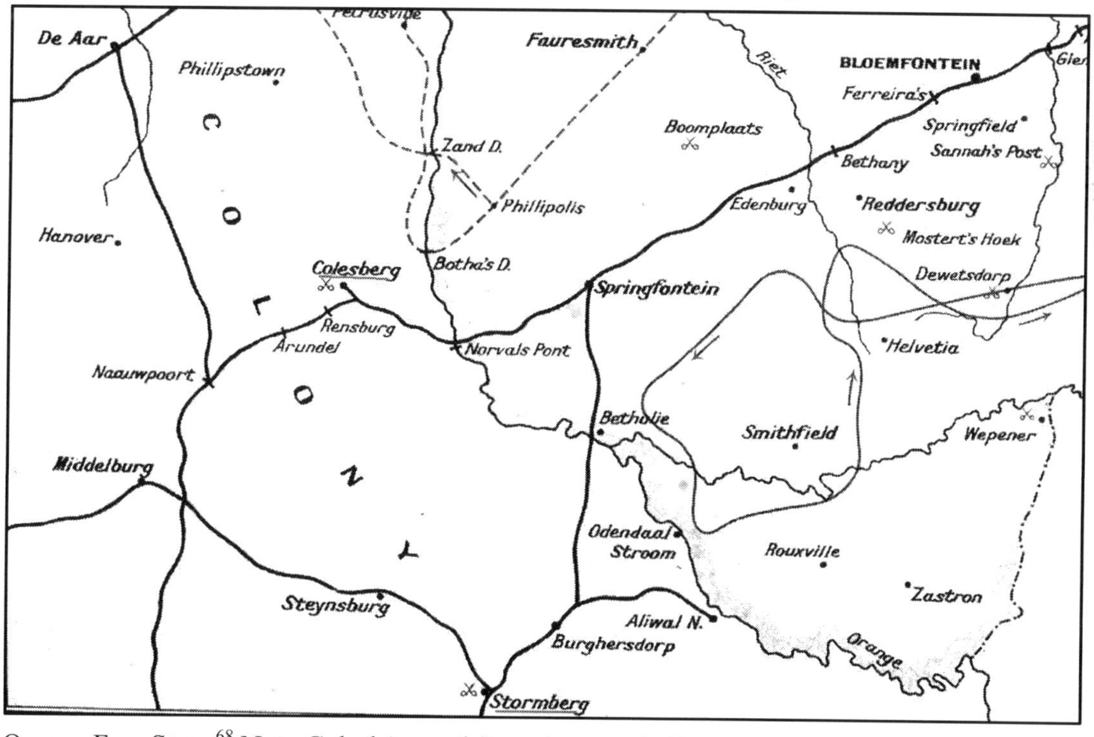

Orange Free State.[68] Note Colesberg and Stormberg underlined.

Lord Roberts would have found it more convenient to have crossed the Orange River, for the railway runs through Norvalspont. Yet had he attempted it, he would have fared as badly as Sir Redvers Buller did in Natal. Our positions at Colesberg, and to the north of the river, were exceedingly strong. He was wise, therefore, in his decision to march over the unbroken plains.

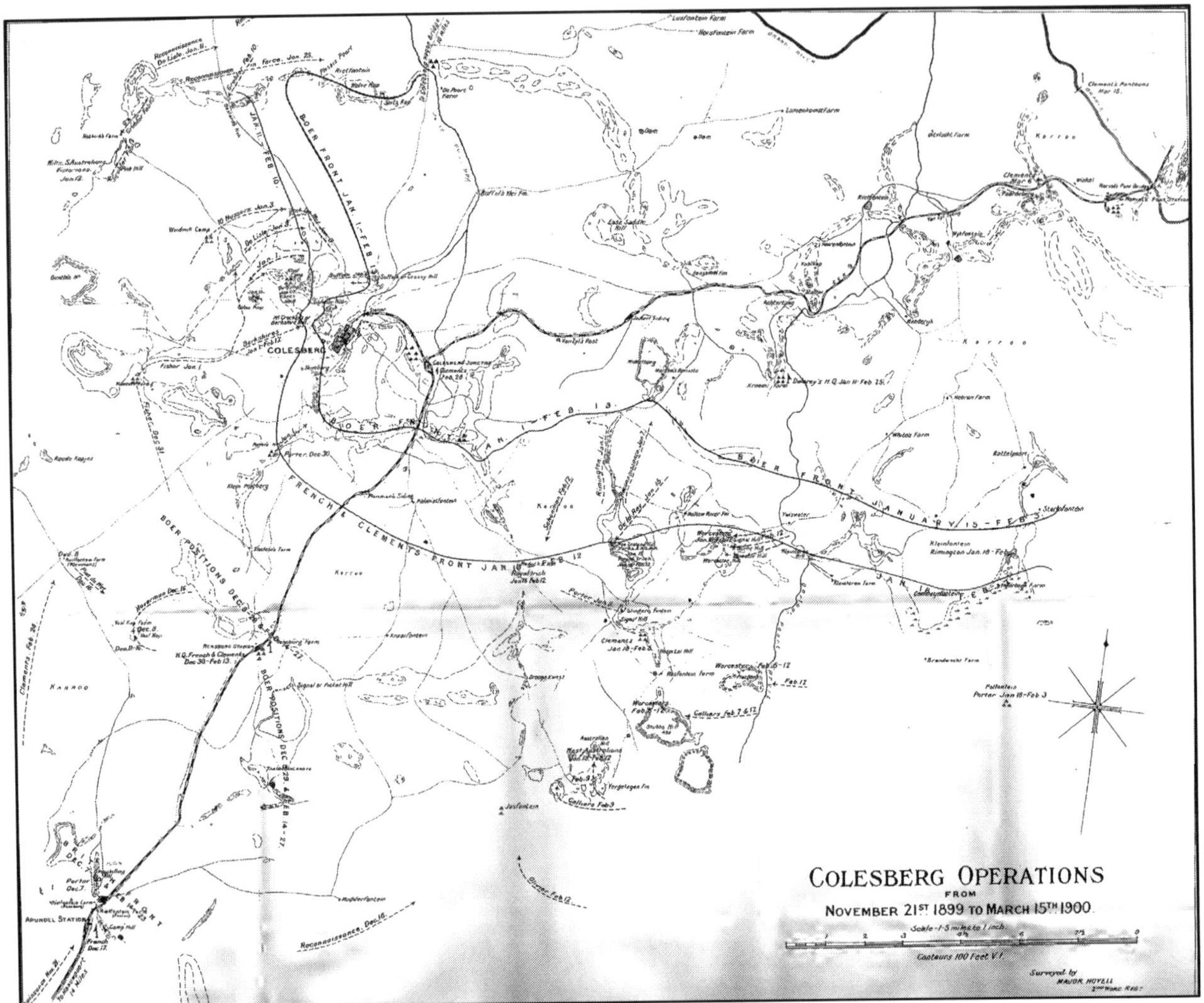

Colesburg Operations.[69] Note they occur from November 21st, 1899 through March 15th, 1900.

It was now, as I had foreseen, that the English renewed their flanking tactics.

On the 11th of February, 1900, a strong contingent of mounted troops, under General French, [Hector McDonald] issued from the camps at Modder River and Koedoesberg. This latter was a kop on the Riet River, about twelve miles to the east of their [British] main camp.

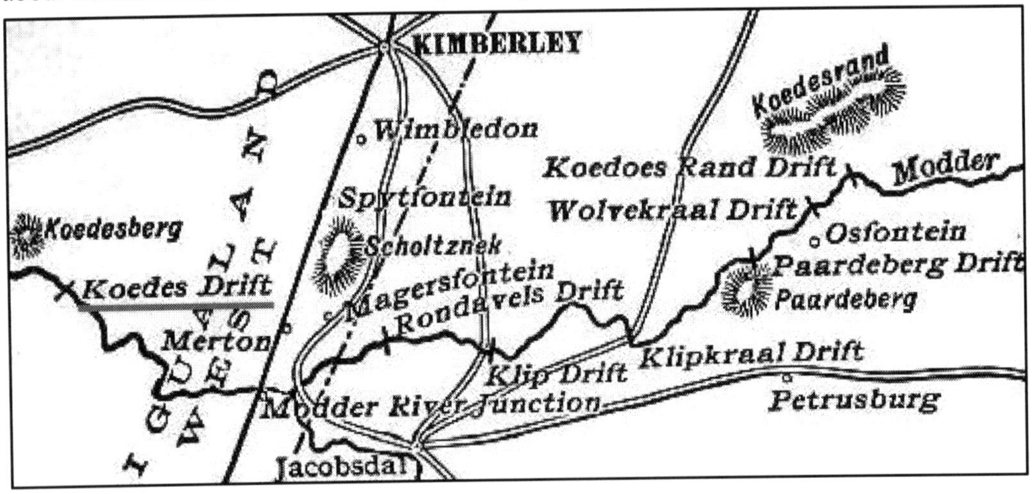

Koedes Drift, far left.

At ten o'clock in the morning, General French started. Immediately I received orders from General Cronje to proceed with three hundred and fifty men to check the advancing troops. As I stood on the ridges of Magersfontein, I was able to look down upon the English camps, and I saw that it would be sheer madness to pit three hundred and fifty men against General French's large force. Accordingly I asked that one hundred and fifty more burghers and two guns might be placed at my disposal. This request, however, was refused, and so I had to proceed without them.

When we arrived at Koedoesberg that afternoon, [11th] we found that the English had already taken possession of the hill. They were stationed at its southern end, and had nearly completed a stone wall across the hill from east to west. Their camp was situated on the Riet River, which flows beside the southern slopes of the berg. The enemy also held strong positions on hillocks to the east of the mountain, whilst on the west they occupied a ravine, which descended from the mountain to the river.

Commandant [Christoffel Cornelis] Froneman[49] and I determined to storm the berg without a moment's delay. We reached the foot of the mountain in safety, and here we were out of sight of the English. But it was impossible to remain in this situation, and I gave orders that my men should climb the mountain. We succeeded in reaching the summit, but were unable to get within seven hundred paces of the enemy, owing to the severity of their fire from behind the stone wall. And so we remained where we were until it became quite dark, and then very quietly went back to the spot where we had left our horses.[50]

As General French was in possession of the river, we had to ride about four miles before we could obtain any water.

Early the following morning [12th] we again occupied the positions we had held on the previous evening. Although under a severe rifle fire, we then rushed from position to position, and at last were only three hundred paces from the enemy. And now I was forced to rest content with the ground we had gained, for with only three hundred and fifty men I dare not risk a further advance, owing to the strength of the enemy's position.

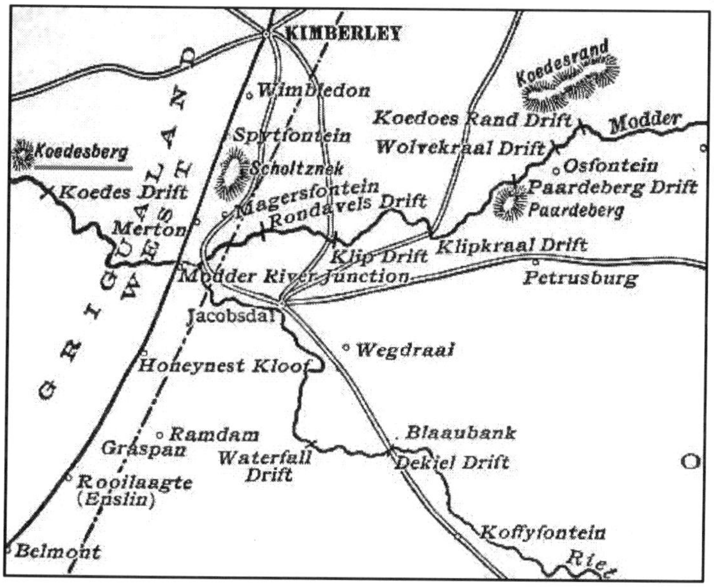

Modder and Reit Rivers; note Koedesberg, far left.

The previous day [11th] I had asked General Cronje to send me reinforcements, and I had to delay the advance until their arrival. In a very short time a small party of burghers made their appearance. They had two field-pieces with them, and were under the command of Major Albrecht. We placed the guns in position and trained them on the English.

With the second shot we had found our range, while the third found its mark in the wall, so that it was not long before the enemy had to abandon that shelter. To find safe cover they were forced to retreat some hundred paces. But we gained little by this, for the new positions of the English were quite as good as those from which we had driven them, and, moreover, were almost out of range of our guns. And we were unable to bring our field-pieces any nearer because our gunners would have been exposed to the enemy's rifle fire.

[49] Christoffel Cornelis Froneman, commonly known as Stoffel Froneman (Leliehoek, Winburg, 26 March 1846–Cypress, District Marquard, 12 March 1913), was veldkornet, general and Vice-Commander-in-Chief of the Orange Free State Boer forces during the Anglo-Boer War of 1899–1902. At the outbreak of the war in 1899 he was Commandant of the Ladybrand Commando. https://en.wikipedia.org/wiki/Christoffel_Cornelis_Froneman

[50] While Methuen's 1st Division demonstrated against the Boer entrenchments at Magersfontein and the Highland Brigade under Major General Hector MacDonald marched 20 miles (32 km) westward to Koedoesberg and fixed the Boers' attention to their right flank.

Our Krupps made good practice on the four English guns which had been stationed on the river bank to the south. Up till now these had kept up a terrific fire on our guns, but we soon drove them across the river, to seek protection behind the mountain. I despatched General Froneman to hold the river bank, and the sluit[51] which descended to the river from the north. While carrying out this order he was exposed to a heavy fire from the enemy's western wing, which was located in the above-mentioned ravine, but he succeeded in reaching the river under cover of the guns. Once there, the enemy's artillery made it impossible for him to move.

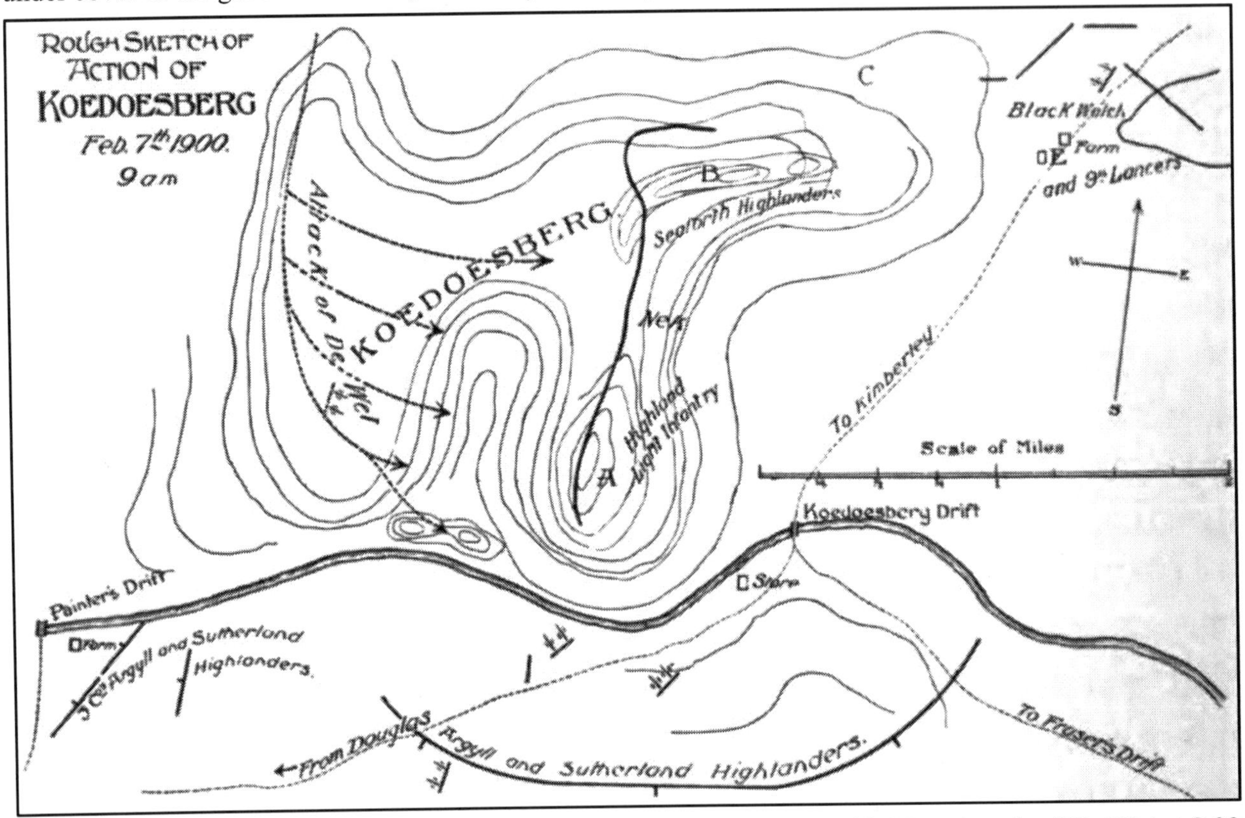

Cronje's main laager is at Jacobsdal at the road junction. Koedoesberg Drift. Note Attack of De Wet at 9:00 AM. No accounting for the date of the 7th but the action appears to be that described by de Wet.[70 52]

And now a curious incident occurred! A falcon, hovering over the heads of our burghers in the sluit, was hit by a bullet from one of the shrapnel shells and fell dead to the ground in the midst of the men. It was already half-past four, and we began to ask ourselves how the affair would end.

At this juncture I received a report from a burgher, whom I had placed on the eastern side of the mountain to watch the movements of the English at the Modder River. He told me that a mountain corps, eight hundred to a thousand men strong, was approaching us with two guns, with the intention, as it appeared, of outflanking us. I also learnt that eighty of my men had retreated. I had stationed them that morning on a hillock three miles to the east of the mountain, my object being to prevent General French from surrounding us.

It now became necessary to check the advance of this mountain corps. But how? There were only thirty-six men at my disposal. The other burghers were in positions closer to the enemy, and I could not withdraw them without exposing them too seriously to the bullets of the English. There was nothing for it, but that I with my thirty-six burghers should attack the force which threatened us.

We rushed down the mountain and jumping on our horses, galloped against the enemy. When we arrived at the precipice which falls sheer from the mountain, the English were already so near that our only course was to charge them.

In front of us there was a plain which extended for some twelve hundred paces to the foot of an abrupt rise in the ground. This we fortunately reached before the English, although we were exposed all the way to the fire of their guns. But even when we gained the rise we were little better off, as it was too low to give us cover. The

[51] A ravine or water-course.

[52] From British Chronology: 2/4/1900 General Macdonald occupied Koodoe's Drift.

English were scarcely more than four hundred paces from us. They dismounted and opened a heavy fire. For ten or fifteen minutes we successfully kept them back.

Then the sun went down! and to my great relief the enemy moved away in the direction of their comrades on the mountain. I ordered all my men from their positions, and withdrew to the spot where we had encamped the previous night. The burghers were exhausted by hunger and thirst, for they had had nothing to eat except the provisions which they had brought in their saddle-bags from the laager.

That evening Andreas Cronje—the General's brother—joined us with two hundred and fifty men and a Maxim-Nordenfeldt.

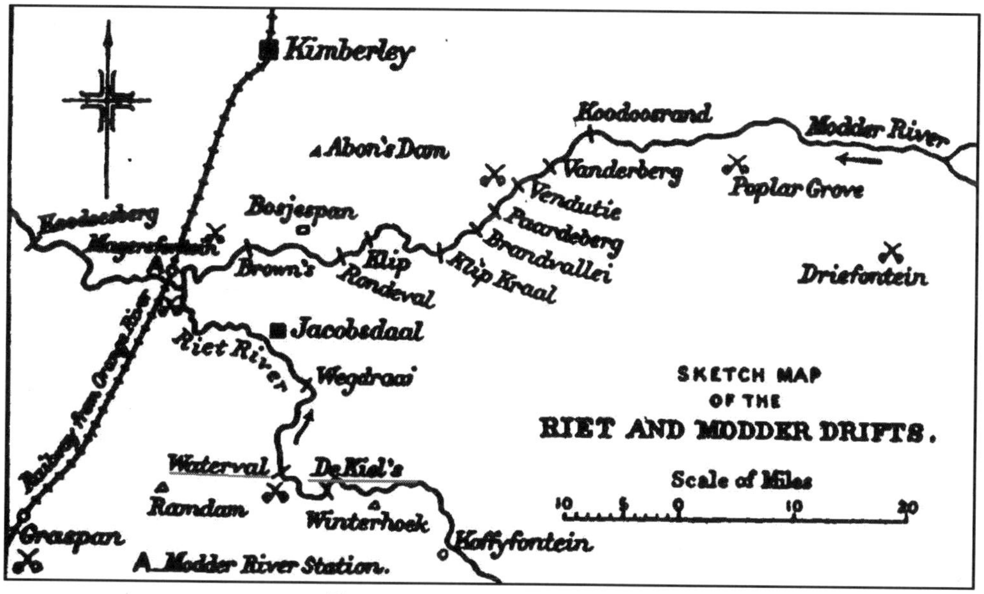

Reit & Modder River Drifts.[71] Dekiel's Drift is labeled De'Kid's Drift upstream of the waterfall.

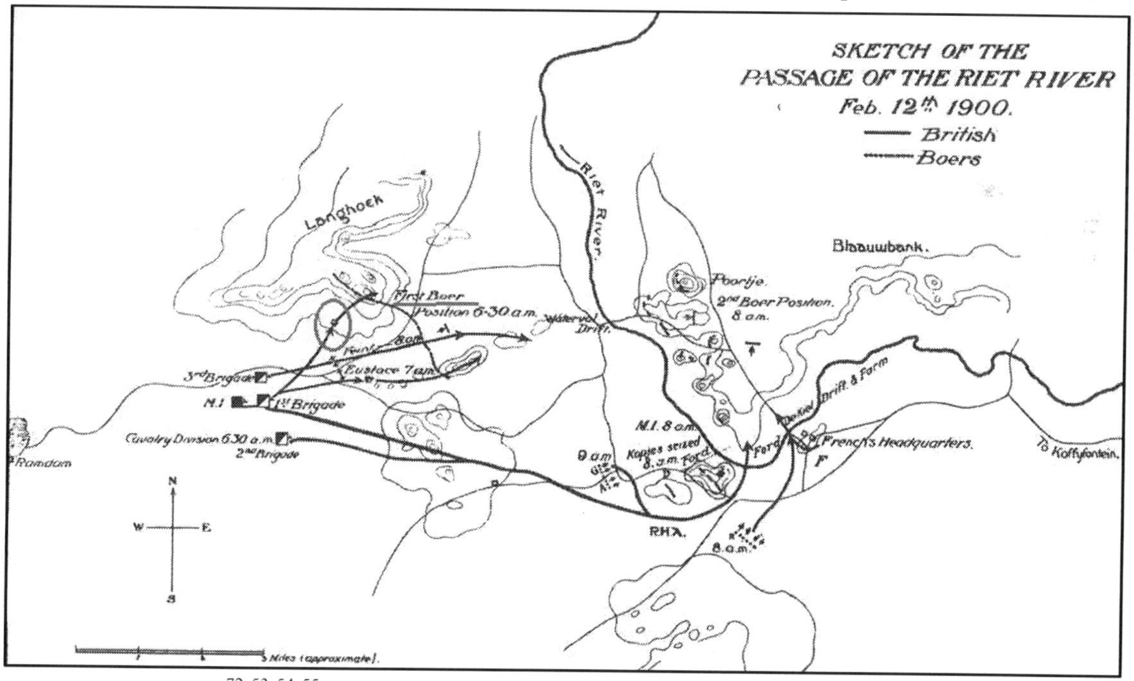

Reit River Passage.[72 53 54 55] You can see taking the 'First Boer Position' position at 6:30 AM and then also his check of the MI, mounted infantry sent up the kopje to contest the ground.

[53] Roberts's large force began marching east in secret, late on 11 February. By the evening of 12 February, his leading horsemen had secured fords across the first obstacle, the Riet River.

[54] 2/12/1900 General French with Cavalry Division, proceeding to the Relief of Kimberley, seized Dekiel's Drift.

When the sun rose on the following day, [13th] the veldt was clear of the enemy. General French had during the night retreated to headquarters. What losses he had suffered I am unable to say; ours amounted to seven wounded and two killed.[56]

Our task here was now ended, and so we returned [East] to Magersfontein.

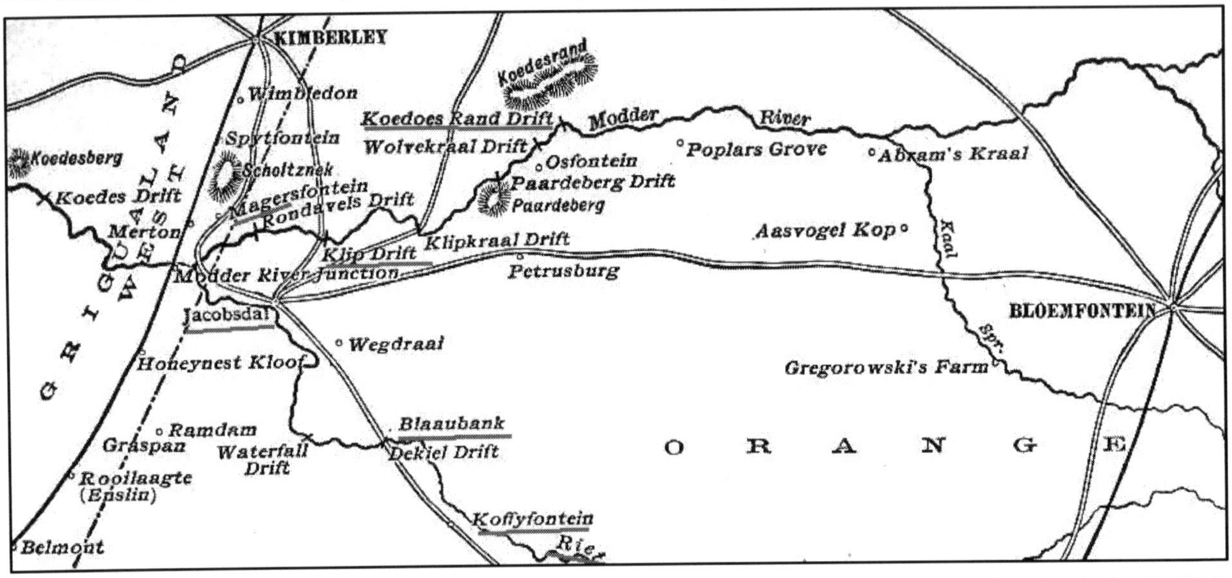

Modder and Reit Rivers.[73] Note Magersfontein, Koedesrand, Jacobsdal, Riet River, Blaaubank & Koffyfontein (bottom), Dekiel Drift, Klip Drift, etc.

The following morning [14th] a large force again left the English camp and took the direction of the Koffiefontein diamond mine. General Cronje immediately ordered me to take a force of four hundred and fifty men with a Krupp and a Maxim-Nordenfeldt, and to drive back the enemy. At my request, Commandants Andreas Cronje, Piet Fourie, Scholten and Lubbe joined me, and that evening we camped quite close to the spot where the English force was stationed!

Early the next day, [15th] before the enemy had made any movement, we started for Blauwbank,[57] and, having arrived there, we took up our positions. Shortly afterwards the fight began; it was confined entirely to the artillery.

We soon saw that we should have to deal with the whole of Lord Roberts' force, for there it was, advancing in the direction of Paardenberg's Drift. It was thus clear that Lord Roberts had not sent his troops to Koffiefontein with the intention of proceeding by that route to Bloemfontein, but that his object had been to divide our forces, so as to march via Paardenberg's Drift to the Capital.

I accordingly withdrew with three hundred and fifty of the burghers [South] in the direction of Koffiefontein, and then hid my commando as best I could. The remainder of the men—about a hundred in number—I placed under Commandant Lubbe, giving him orders to proceed in a direction parallel to the advance of the English, who now were nearing Paardenberg's Drift, and to keep a keen eye on their movements. It was a large force that Lubbe had to watch. It consisted chiefly of mounted troops; but there were also nine or ten batteries and a convoy of light mule waggons.

I thought that as General Cronje was opposing them in front, my duty was to keep myself in hiding and to reconnoitre.

[55] The next day, 13 February, the British mounted force made a grueling journey of 30 miles (48 km) North under a blazing sun to capture fords across the Modder. The effect of the heat was made worse when the dry grass of the veld caught fire from a carelessly discarded match. French's division had to wait at the fords (at Klip Drift) during the next day (16th) until the leading infantry reached them, after making an equally exhausting march. Luckily for the British, the move had taken the Boers by surprise and they did not move in strength to defend the fords or the hills nearby.

[56] 2/12/1900 General French, crossed Riet River, and next day, 2/13 with a strong force marched twenty-five miles into the Free State, seized Klip Drift on the Modder River, occupied the hills to the north, and captured three of the enemy's laagers, with supplies. 2/13/1900 (Kelly-Kenny's) Division on north bank of the Riet River at Waterfall Drift.

[57] In the district of Jacobsdal.

I wished to communicate with General Cronje before the English troops came up to him, and with this object I sent out a despatch rider. The man I chose for the mission was Commandant G.J. Scheepers—whose name later in the war was on every man's lips for his exploits in Cape Colony, but who then was only the head of our heliograph corps.

I informed General Cronje in my message that the English, who had been stationed at Blauwbank, had made a move in the direction of Paardenberg's Drift; and I advised him to get out of their road as quickly as he could, for they numbered, according to my computation, forty or fifty thousand men.[58]

I thought it wise to give Cronje this advice, on account of the women and children in our camps, who might easily prove the cause of disaster. When Scheepers returned he told me what reply General Cronje had made. It is from no lack of respect for the General, whom I hold in the highest honour as a hero incapable of fear, that I set down what he said. It is rather from a wish to give a proof of his undaunted courage that I quote his words.

"Are you afraid of things like that?" he asked, when Scheepers had given my message. "Just you go and shoot them down, and catch them when they run."

At Paardenberg's Drift there were some Free-Staters' camps that stood apart from the others. In these camps there were a class of burghers who were not much use in actual fighting. These men, called by us "water draggers," correspond to the English "non-combatants." I ordered these burghers to withdraw to a spot two hours' trek from there, where there was more grass. But before all had obeyed this order, a small camp, consisting of twenty or thirty waggons, was surprised and taken.[59] [60]

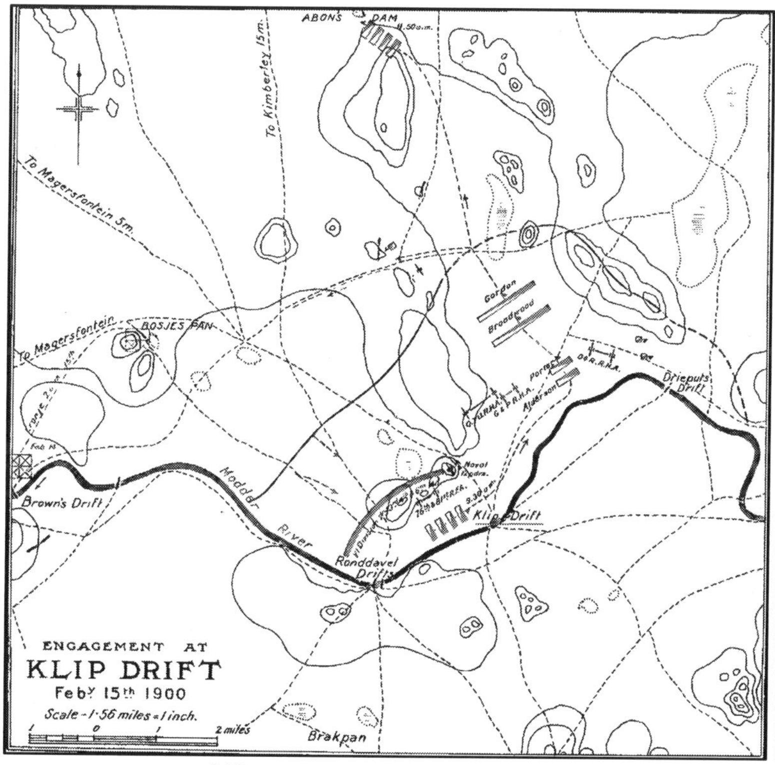

Klip Drift, Feb 15th.[74]

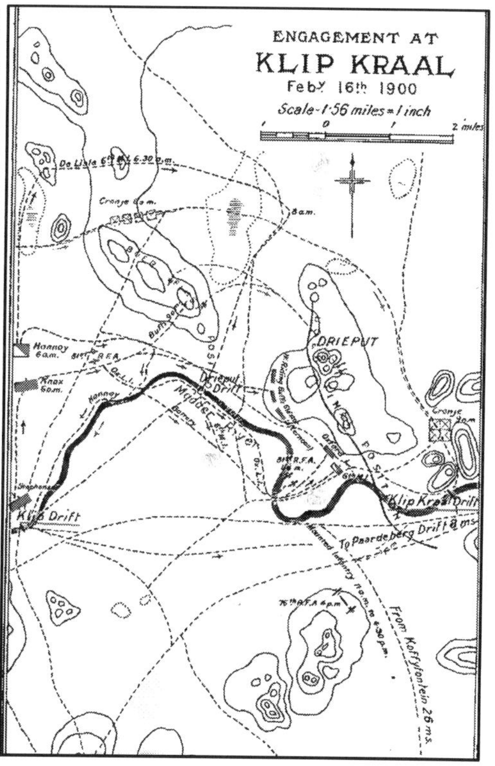

Klip Kraal Drift on Right, Feb. 16th.[75]

[58] 2/14/1900 Roberts advanced to Dekiel's Drift.

2/15/1900 General French reached and relieved Kimberley, The Boers evacuated Majersfontein and Spyfontein, retreating to Koodoosrand Drift.

[59] On 15 February, Cronjé's men, some 5,000 Transvaalers and Freestaters, finally evacuated their camp at Jacobsdal. Their position at Magersfontein was no longer relevant and they were in danger of being besieged in Jacobsdal by the British 7th Division under Lieutenant General Charles Tucker, which had turned west from Klip Drift.

[60] On the night of the 15th, the large convoy of Boer ox-wagons passed between the rear of French's division and the outposts of Lieutenant General Thomas Kelly-Kenny's 6th Division at the Modder River fords. Throughout the next day, the Boer mounted rearguards prevented the British 6th Division (with only one understrength mounted infantry unit) overtaking them.

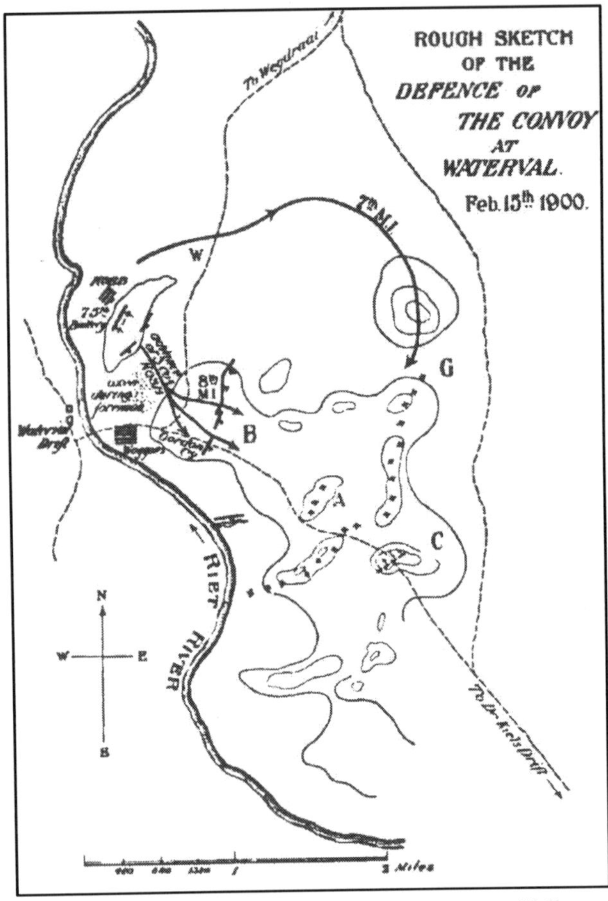

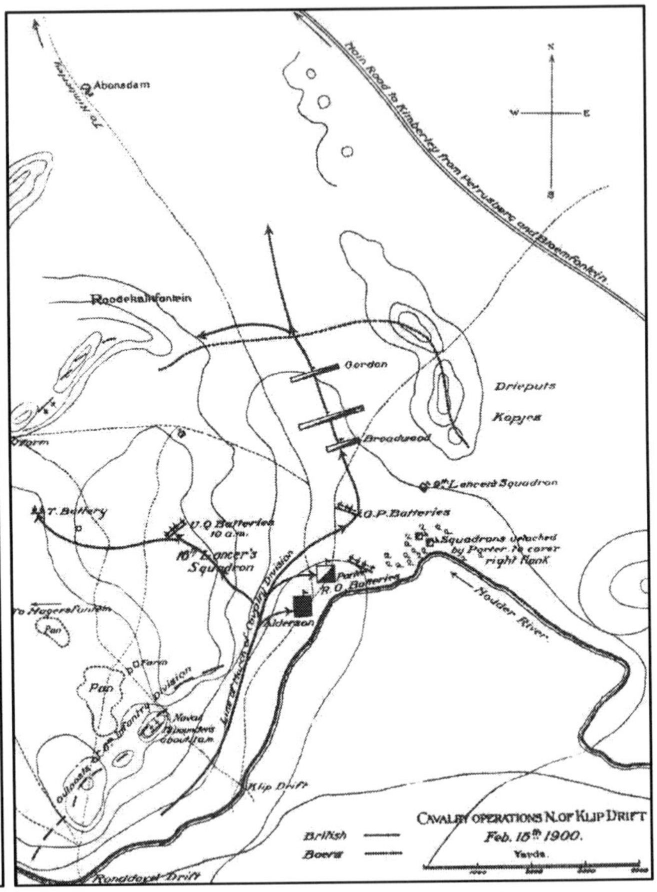

Defense of Waterval Drift on the Reit River.[76] [61] Cavalry Operations North of Klip Drift Feb 15th 1900 North of the Modder river.[77]

In the meantime, keeping my little commando entirely concealed, I spied out the enemy's movements.

On the 16th of February, I thought I saw a chance of dealing an effective blow at Lord Roberts. Some provision waggons, escorted by a large convoy, were passing by, following in the wake of the British troops. I asked myself whether it was possible for me to capture it then and there, and came to the conclusion that it was out of the question. With so many of the enemy's troops in the neighbourhood, the risk would have been too great. I, therefore, still kept in hiding with my three hundred and fifty burghers.[62]

I remained where I was throughout the next day; [17th] but in the evening I saw the convoy camping near Blauwbank, just to the west of the Riet River. I also observed that the greater part of the troops had gone forward with Lord Roberts.[63]

On the 18th I still kept hidden, for the English army had not yet moved out of camp. The troops, as I learnt afterwards, were awaiting the arrival of columns from Belmont Station.[64]

[61] And, 2/15/1900 General French reached and relieved Kimberley, The Boers evacuated Majersfontein and Spyfontein, retreating to Koodoosrand Drift.

[62] 2/16/1900 Roberts occupied Jacobsdal. Evacuation of Cronje's force and occupation of Majersfontein by the Guards. General Kelly-Kenny, in pursuit of Cronje retiring east with 10,000 men on Bloemfontein, captured 78 waggons with stores, 2 waggons with Mauser rifles, and 8 waggons with shell belonging to Cronje's column.

[63] On the 17th, the large convoy of Boer wagons reached the crossing of the Modder at Paardeberg Drift. They were starting to cross the river when a force of 1,500 British mounted troops, almost all of French's fit horses and men who had covered the 40 miles (64 km) from Kimberley in another desperately tiring march, opened fire on them unexpectedly from the north, causing confusion.
Cronjé then inadvisedly decided to form a laager and dig in on the banks of the Modder river. His reasons for doing so are unclear because the British had insufficient cavalry and it would therefore have been an easy matter for Cronjé to brush them aside and link up with other Boers east of the Modder. Boers under noted commander Christiaan De Wet were only 30 miles (48 km) away to the south-east and other forces under Chief Commandant Ignatius Ferreira were a similar distance to the north.

[64] At Paardeberg on 2/18/1900 Lieutenant General Kelly-Kenny, commanding the British 6th Division, planned to lay siege to Cronjé and bombard his force into surrender. Roberts was ill, and his chief of staff, General Herbert Kitchener, was now in overall command of the British force, and had different plans, he overruled Kelly-Kenny.

On the following day [19th] I attacked the convoy on the flank. The three or four hundred troops who were guarding it offered a stout resistance, although they were without any guns.[65]

After fighting for two hours the English received a reinforcement of cavalry, with four Armstrong guns, and redoubled their efforts to drive us from the positions we had taken up under cover of the mule waggons. As I knew that it would be a serious blow to Lord Roberts to lose the provisions he was expecting, I was firmly resolved to capture them, unless the force of numbers rendered the task quite impossible. I accordingly resisted the enemy's attack with all the power I could.

The battle raged until it became dark; and I think we were justified in being satisfied with what we had achieved. We had captured sixteen hundred oxen and forty prisoners; whilst General Fourie, whom I had ordered to attack the camp on the south, had taken several prisoners and a few water-carts.

We remained that night in our positions. The small number of burghers I had at my disposal made it impossible for me to surround the English camp.

To our great surprise, the following morning, [20th] we saw that the English had gone. About twenty soldiers had, however, remained behind; we found them hidden along the banks of the Riet River at a short distance from the convoy. We also discovered thirty-six Kaffirs on a ridge about three miles away. As to the enemy's camp, it was entirely deserted.[66]

Our booty was enormous, and consisted of two hundred heavily-laden waggons, and eleven or twelve water-carts and trollies. On some of the waggons we found klinkers,[67] jam, milk, sardines, salmon, cases of corned beef, and other such provisions in great variety. Other waggons were loaded with rum; and still others contained oats and horse provender pressed into bales. In addition to these stores, we took one field-piece, which the English had left behind. It was, indeed, a gigantic capture; the only question was what to do with it.

Our prisoners told us that columns from Belmont might be expected at any moment. Had these arrived we should have been unable to hold out against them.

By some means or other it was necessary to get the provisions away, not that we were then in any great need of them ourselves, but because we knew that Lord Roberts would be put in a grave difficulty if he lost all this food. I did not lose a moment's time, but at once ordered the burghers to load up the waggons as speedily as possible, and to inspan. It was necessary to reload the waggons, for the English troops had made use of the contents to build schanzes; and excellent ones the provisions had made.

The loading of the waggons was simple enough, but when it came to inspanning it was another matter. The Kaffir drivers alone knew where each span had to be placed, and there were only thirty-six Kaffirs left. But here the fact that every Boer is himself a handy conductor and driver of waggons told in our favour. Consequently we did not find it beyond our power to get the waggons on the move. It was, however, very tedious work, for how could any of us be sure that we were not placing the after-oxen in front and the fore-oxen behind? There was nothing left for it but to turn out the best spans of sixteen oxen that we could, and then to arrange them in the

On Sunday February 19th Kitchener proceeded to order his infantry and mounted troops into a series of uncoordinated frontal assaults against the Boer laager. This was despite the fact that the cost of frontal assaults against entrenched Boers had been demonstrated time and again the preceding months. It was no different this time. The land sloped down to the Boer position and lacked any cover for 800 metres (870 yd) or more. The Highland Brigade and the 2nd (Special Service) Battalion, The Royal Canadian Regiment of Infantry, led the attack.

The Boer soldiers withheld fire until the British soldiers were within 200 metres (220 yd). The British were pinned and the exchange of fire continued until nightfall when the British withdrew.

By nightfall some 24 officers and 279 men were killed and 59 officers and 847 men wounded. Judged by British casualties it was the most severe reverse of the war and became known as Bloody Sunday. the Canadian losses were 18 dead and 60 wounded. Bloody Sunday (1900) - Wikipedia

During the siege a further 2,000 Imperial soldiers died or were invalided at Paardeberg from illness, mostly due to drinking the water of the Modder River, downstream from where the Boer were throwing horse and cattle corpses killed by the artillery fire.

[65] 2/19/1900 Bombardment of Cronje's position began. Boer reinforcements driven back. Cronje asked for armistice, but Kitchener demanded his surrender; Cronje refused, and was then bombarded heavily. Reoccupation of Dordrecht by General Brabant.

[66] At Paardeberg, on Monday, 20th February, General Roberts arrived on the scene. He initially urged a resumption of the frontal assaults, but Cronjé requested a cease-fire to bury the dead. The British refused and Cronjé replied "If you are so uncharitable as to refuse me a truce as requested, then you may do as you please. I shall not surrender alive. Bombard as you will." The truce communications had taken up much of the day and there was no time for any more assaults.

[67] Biscuits.

way that struck us as being most suitable. It was all done in the most hurried manner, for our one idea was to be off as quickly as possible.

Even when we had started our troubles were not at an end. The waggons would have been a hard pull for sixteen oxen properly arranged; so that it is not surprising that our ill-sorted teams found the work almost beyond their strength. Thus it happened that we took a very long time to cover the first few miles, as we had constantly to be stopping to re-arrange the oxen. But under the supervision of Commandant Piet Fourie, whom I appointed Conductor-in-Chief, matters improved from hour to hour.

Drakensberg. Ox wagon with 16 oxen crossing a drift near Mont-Aux-Sources.[78] Ox wagon stuck in Crocodile River.[79]

Voorloopers and supply train.[80]

Oxen pulling wagons.[81]

After a short time I issued orders that the convoy should proceed over Koffiefontein to Edenberg. I then divided my burghers into two parties; the first, consisting of two hundred men with the Krupp gun, I ordered to proceed with the convoy; the second, consisting of a hundred and fifty men with the Maxim-Nordenfeldt, I took under my own command, and set out with them in the direction of Paardenberg's Drift.

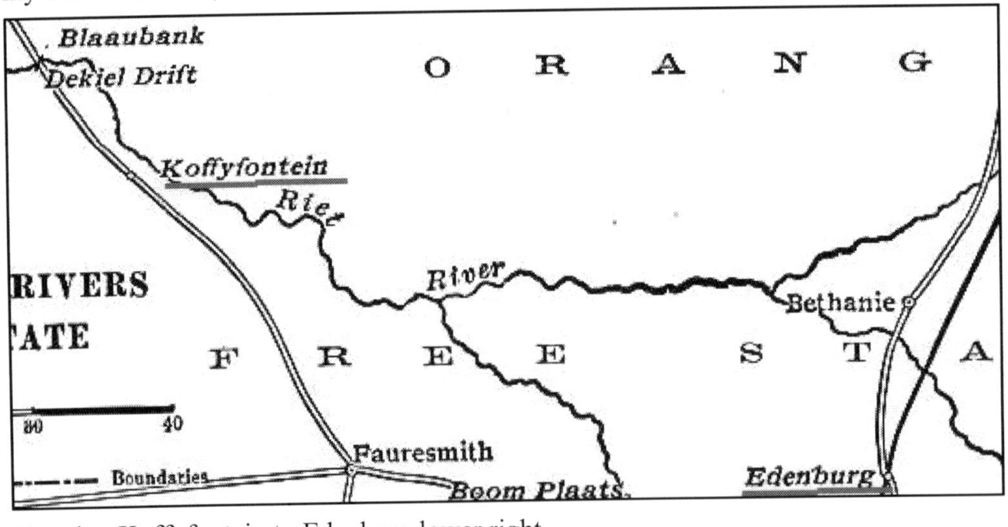

Showing Koffyfontein to Edenburg lower right.

My spies had informed me that there were some fifty or sixty English troops posted about eight miles from the spot where we had captured the convoy. We made our way towards them, and when we were at a distance of about three thousand yards, I sent a little note to their officer, asking him to surrender. It was impossible for his troops to escape, for they found themselves threatened on three sides.

The sun had just gone down when my despatch-rider reached the English camp; and the officer in command was not long in sending him his reply, accompanied by an orderly.

"Are you General De Wet?" the orderly asked me.

"I am," replied I.

"My officer in command," he said in a polite but determined voice, "wishes me to tell you that we are a good hundred men strong, that we are well provided with food and ammunition, and that we hold a strong position in some houses and kraals. Every moment we are expecting ten thousand men from Belmont, and we are waiting here with the sole purpose of conducting them to Lord Roberts."

I allowed him to speak without interrupting him; but when he had finished, I answered him in quite as determined a voice as he had used to me.

"I will give you just enough time to get back and to tell your officer in command that, if he does not surrender at once, I shall shell him and storm his position. He will be allowed exactly ten minutes to make up his mind—then the white flag must appear."

"But where is your gun?" the orderly asked. In reply I pointed to the Maxim-Nordenfeldt, which stood a few hundred paces behind us, surrounded by some burghers.

ZAR Staats Artillery using Maxim Pom Pom.[82]

ZAR Norden Maxim Mafeking.[83]

"Will you give us your word of honour," he asked me when he caught sight of the gun, "not to stir from your position till we have got ten miles away? That is the only condition on which we will abandon our positions."

I again allowed him to finish, although his demand filled me with the utmost astonishment. I asked myself what sort of men this English officer imagined the Boer Generals to be.

"I demand unconditional surrender," I then said. "I give you ten minutes from the moment you dismount on arriving at your camp; when those ten minutes have passed I fire."

He slung round, and galloped back to his camp, the stones flying from his horse's hoofs.

He had hardly dismounted before the white flag appeared. It did not take us long to reach the camp, and there we found fifty-eight mounted men. These prisoners I despatched that evening to join the convoy.

I then advanced with my commando another six miles, with the object of watching Lord Roberts' movements, in case he should send a force back to retake the convoy he could so ill spare.[68]

But the following [21st] day we saw nothing except a single scouting party coming from the direction of Paardenberg's Drift. This proved to consist of the hundred burghers whom I had sent with Commandant Lubbe to General Cronje's assistance. I heard from Lubbe that General French had broken through, and had in all

[68] 2/20/1900 Roberts defeated Boer reinforcements at Paardeberg.

probability relieved Kimberley;[69] and that General Cronje was retreating before Lord Roberts towards Paardeberg. I may say here that I was not at all pleased that Commandant Lubbe should have returned.

On account of Lubbe's information, I decided to advance at once in the direction of Paardenberg's Drift, and was on the point of doing so when I received a report from President Steyn. He informed me that I should find at a certain spot that evening, close to Koffiefontein, Mr. Philip Botha[70] with a reinforcement of one hundred and fifty men.

This report convinced me that the convoy I had captured would reach Edenberg Station without mishap, and accordingly I went after it to fetch back the gun which would no longer be needed. I found the convoy encamped about six miles from Koffiefontein. Immediately after my arrival, General Jacobs, of Fauresmith, and Commandant Hertzog,[71] of Philippolis, brought the news to me that troops were marching on us from Belmont Station. I told Jacobs and Hertzog to return with their men, two or three hundred in number to meet the approaching English.

We were so well supplied with forage that our horses got as much as they could eat. I had, therefore, no hesitation in ordering my men to up-saddle at midnight, and by half-past two we had joined Vice-Vechtgeneraal Philip Botha. I had sent him word to be ready to move, so that we were able to hasten at once to General Cronje's assistance. Our combined force amounted to three hundred men all told.

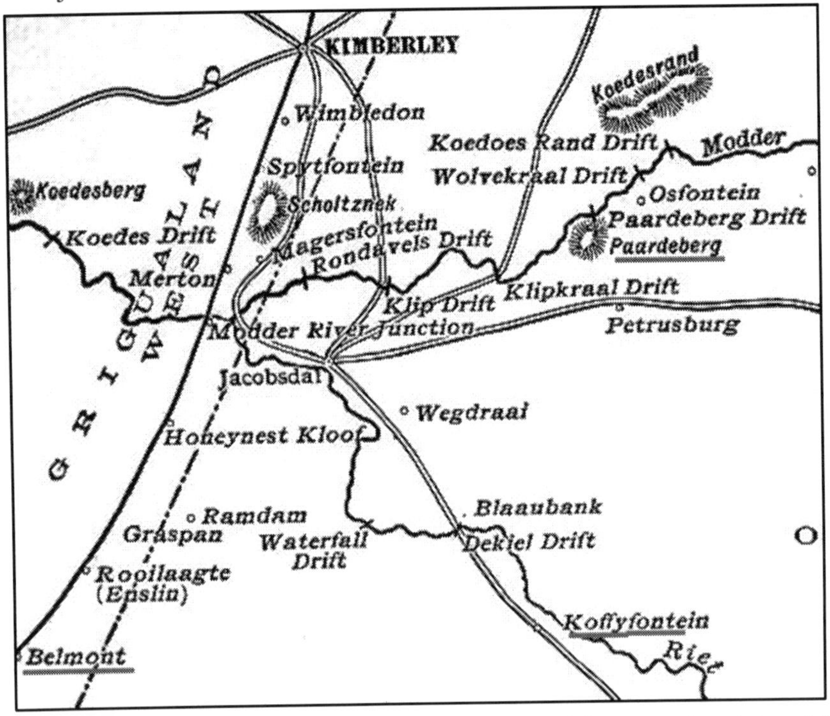

Belmont. De Wet is moving North towards Paardeberg on the night of the 21st. Philip Rudolph Botha.[84 72]

[69] Early on 15 February, French's division began the final march to relieve Kimberley. Only scattered and disorganised Boers opposed them, and the enormous mass of British horsemen broke through their thin line, concealed in the dust cloud they created. Late that evening they reached Kimberley, where they were greeted with cheering crowds. French should by rights have gone to the military commander of the besieged garrison, Lieutenant Colonel Kekewich. Instead he called first on Cecil Rhodes, the former prime minister of Cape Colony and foremost Imperialist, at the town's chief hotel.
The final day's ride had crippled most of French's division. Most of his British regular cavalry carried too much equipment and their un-acclimatised horses (and those of the seven batteries of horse artillery) were exhausted. His effective force was reduced to two regiments of New Zealand and Australian light horse, and two "brigades" (actually battalions) of mounted infantry.
French was to further tire his men on 16 February by futile attempts to intercept one of the Boers' Creusot 40-pounder siege guns (nicknamed "Long Tom") which was withdrawing to the north.

[70] Mr. Philip Botha had just been appointed Vice-Vechtgeneraal.

[71] Brother to Judge Hertzog.

[72] Philip Rudolph Botha (30 June 1851, Greytown, Colony of Natal–6 March 1901, Doornberg, Ventersburg, Orange Free State, South Africa) was a Second Boer War general, like his younger brothers Louis (1862–1919), Christiaan (1864–1902) and Theunis Jacobus (1867–1930). Philip Botha - Wikipedia

6. Paardeberg

An hour after sunrise [22nd] we off-saddled, and heard, from the direction of Paardeberg, the indescribable thunder of bombardment. That sound gave us all the more reason for haste. We allowed our horses the shortest possible time for rest, partook of the most hurried of breakfasts, and at once were again on the move, with the frightful roar of the guns always in our ears.

About half-past four that afternoon, we reached a point some six miles to the east of Paardeberg, and saw, on the right bank of the Modder River, four miles to the north-east of the mountain, General Cronje's laager. It was surrounded completely by the enemy, as a careful inspection through our field-glasses showed.

Immediately in front of us were the buildings and kraals of Stinkfontein, and there on the opposite bank of the river stood Paardeberg. To the left and to the right of it were khaki-coloured groups dotted everywhere about—General Cronje was hemmed in on all sides, he and his burghers—a mere handful compared with the encircling multitude.

What a spectacle we saw! All round the laager were the guns of the English, belching forth death and destruction, while from within it at every moment, as each successive shell tore up the ground, there rose a cloud—a dark red cloud of dust.

It was necessary to act—but how?

We decided to make an immediate attack upon the nearest of Lord Roberts' troops, those which were stationed in the vicinity of Stinkfontein, and to seize some ridges which lay about two and a half miles south-east of the laager. Stinkfontein was about a thousand paces to the north of these ridges, and perhaps a few hundred paces farther from where Cronje was stationed.

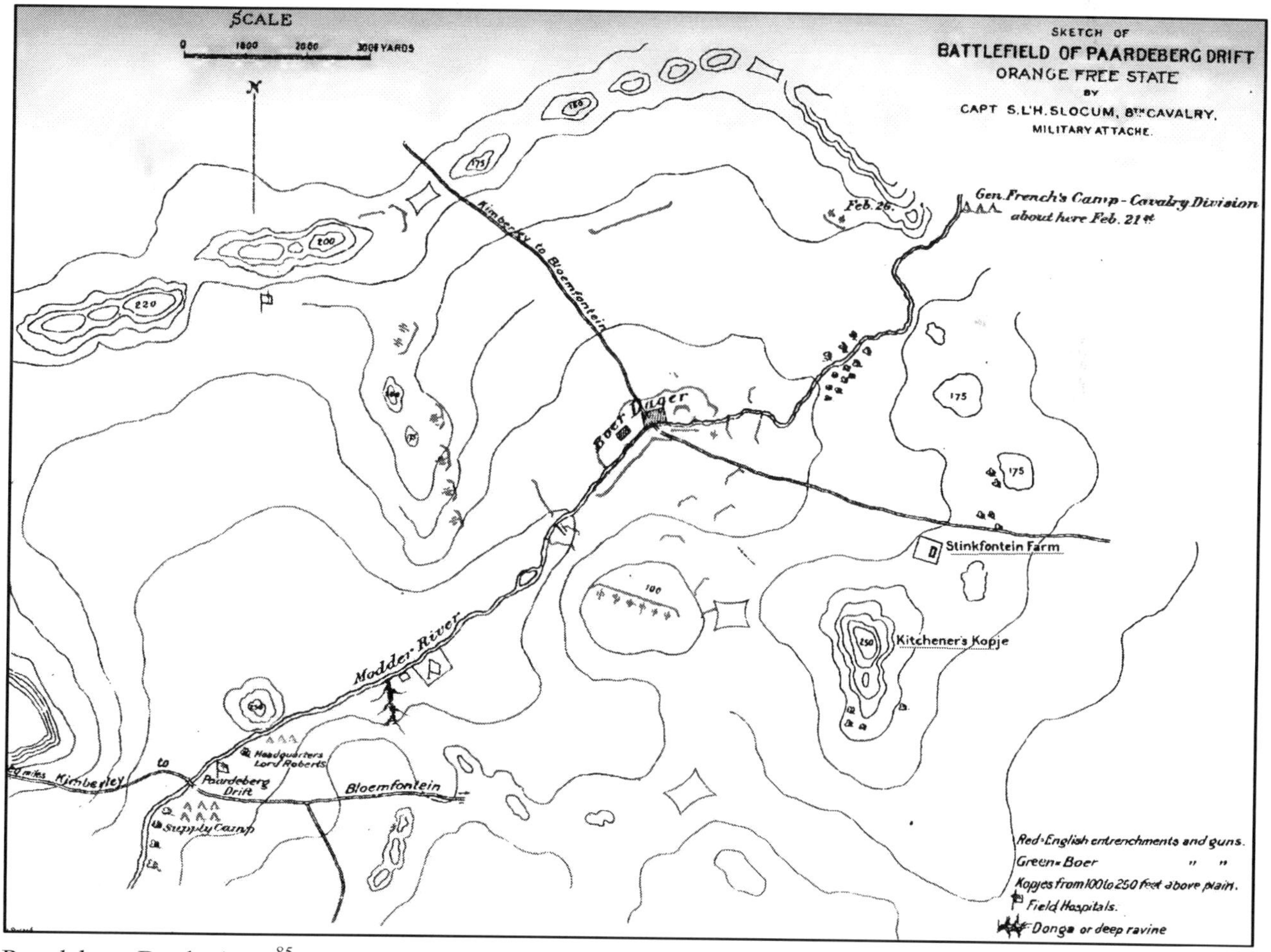

Paardeberg Battle Area.[85] Note Stinkfontein Farm and Kitchener's Kopje right side..

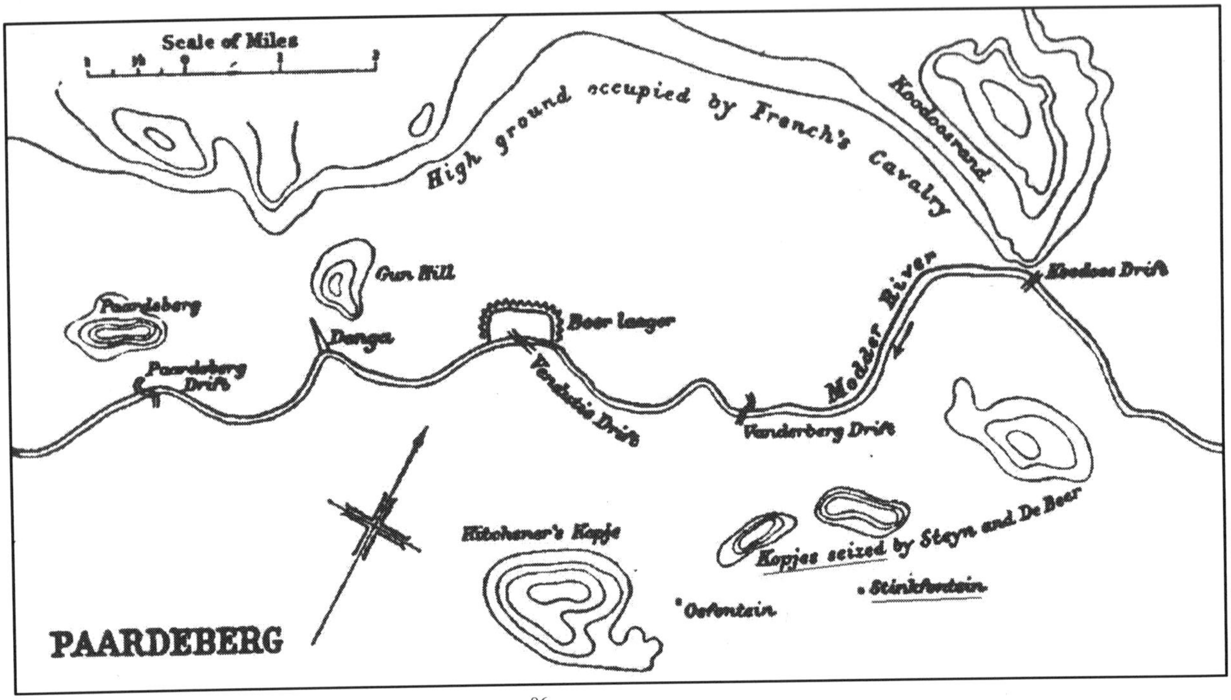

Another view with additional features noted. [86]

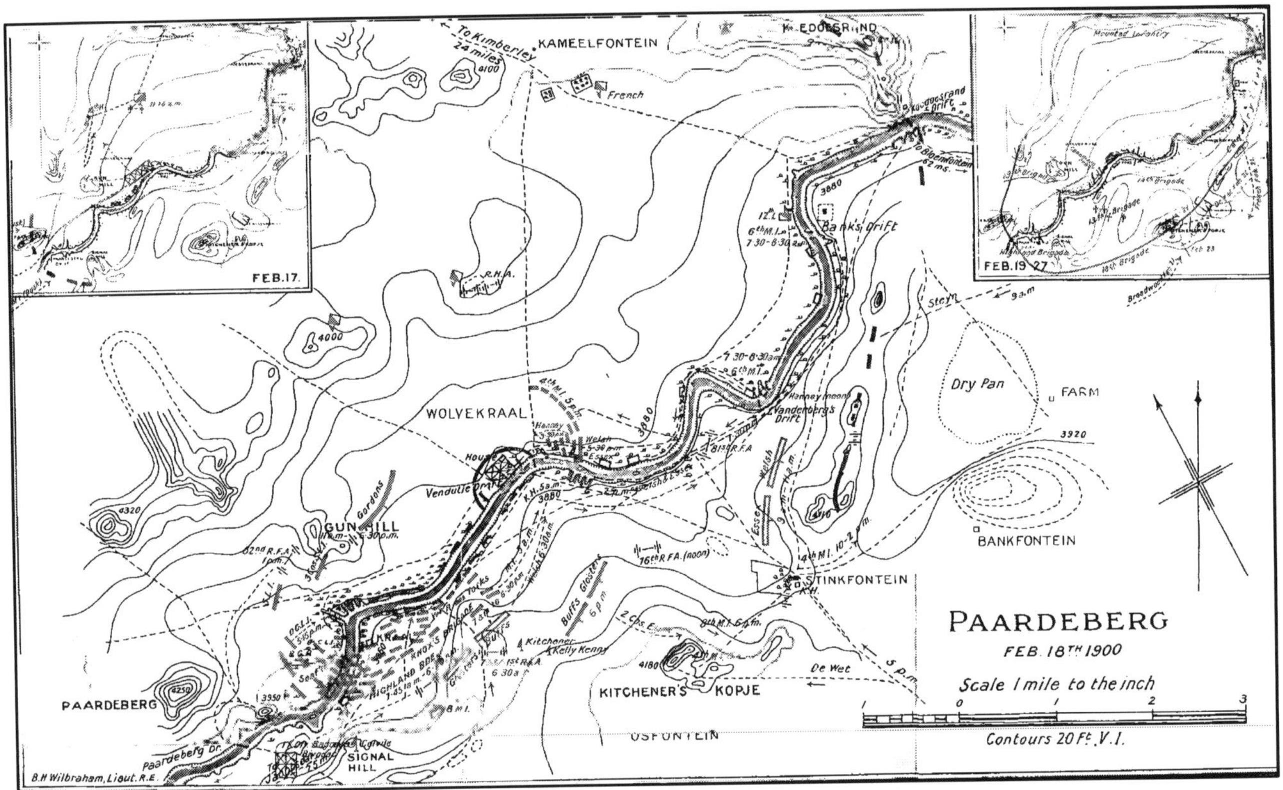

Times History Map of Paardeberg, February on 18th, 1900.[87]

We rode towards the ridges, and when we were from twelve to fourteen hundred paces from Stinkfontein, we saw that the place was occupied by a strong force of British troops.

General Botha and I then arranged that he should storm the houses, kraals and garden walls of Stinkfontein, whilst I charged the ridges. And this we did, nothing daunted by the tremendous rifle fire which burst upon us. Cronje's pitiable condition confronted us, and we had but one thought—could we relieve him?

We succeeded in driving the English out of Stinkfontein, and took sixty of them prisoners.

The enemy's fire played on us unceasingly, and notwithstanding the fact that we occupied good positions, we lost two men, and had several of our horses killed and wounded.[73]

We remained there for two and a half days—from the 22nd to the 25th of February—and then were forced to retire. While evacuating our positions, three of my burghers were killed, seven wounded, and fourteen taken prisoner.

But the reader will justly demand more details as to the surrender of Cronje, an event which forms one of the most important chapters in the history of the two Republics. I am able to give the following particulars.

After we had captured the positions referred to above, I gave orders that the Krupp and the Maxim-Nordenfeldt should be brought up. For with our hurried advance, the oxen attached to the big guns, as well as some of the burghers' horses, had become so fatigued, that the guns and a number of the burghers had been left behind. The ridges were so thickly strewn with boulders, that even on the arrival of the guns, it was impossible to place them in position until we had first cleared a path for them. I made up my mind to turn these boulders to account by using them to build schanzes, for I knew that a tremendous bombardment would be opened upon our poor Krupp and Maxim-Nordenfeldt as soon as they made themselves heard.

During the night we built these schanzes, and before the sun rose the following morning, the guns were placed in position.

In the schanzes.[88]

By daybreak the English had crept up to within a short distance of our lines. It was the Krupp and the Maxim-Nordenfeldt that gave our answer.

But we had to be very sparing of our ammunition, for it was almost exhausted, and it would take at least five days to get a fresh supply from Bloemfontein.

Our arrival on the previous day had made a way of escape for General Cronje. It is true that he would have been obliged to leave everything behind him, but he and his burghers would have got away in safety. The British had retreated before our advance, thus opening a road between us and the laager. That road was made yet wider by the fire from our guns.

But General Cronje would not move. Had he done so, his losses would not have been heavy. His determination to remain in that ill-fated laager cost him dearly.

The world will honour that great general and his brave burghers; and if I presume to criticize his conduct on this occasion, it is only because I believe that he ought to have sacrificed his own ideas for the good of the nation, and that he should have not been courageous at the expense of his country's independence, to which he was as fiercely attached as I.

[73] General Kelly-Kenny had warned Kitchener not to leave "Kitchener's Kopje" undefended. Possession of the kopje was essential to guard the south-east of the British position and prevent Cronjé's escape. But Kitchener, in his zeal for an all-out attack, had left the kopje defended by only a handful of "Kitchener's Horse" (volunteer British colonists). De Wet was therefore able to take the kopje. The strategic picture had now changed dramatically. De Wet could now make the British position on the south east bank of the Modder untenable, and the Boers now commanded a swathe of front stretching from the north east right through to the south east. As darkness fell, Kitchener ordered his troops to dig in where they were. Few received these orders and fewer still obeyed them. Desperately thirsty and exhausted, the surviving British trickled back into camp. Rescue for Cronjé now seemed the likely outcome.

Some of the burghers in the laager made their escape, for, on the second day, when our guns had cleared a wide path, Commandants Froneman and Potgieter (of Wolmaranstadt), with twenty men, came galloping out of the laager towards us.

Although we were only a few in number, the British had their work cut out to dislodge us. First they tried their favourite strategy of a flanking movement, sending out strong columns of cavalry, with heavy guns to surround us. It was necessary to prevent the fulfilment of this project. I, therefore, removed the Krupp and the Maxim-Nordenfeldt from their positions, and divided our little force into three portions. I ordered the first to remain in their position, the second was to proceed with the Krupp round our left wing, while I despatched the third party to hold back the left wing of the British. I had no wish to share General Cronje's unenviable position.

We succeeded in checking the advance of the enemy's wings; and when he saw that we were not to be outflanked he changed his tactics, and while still retaining his wings where they were, in order to keep our men occupied, he delivered at mid-day, on the 20th, (sic, 19th) an attack on our centre with a strong force of infantry.

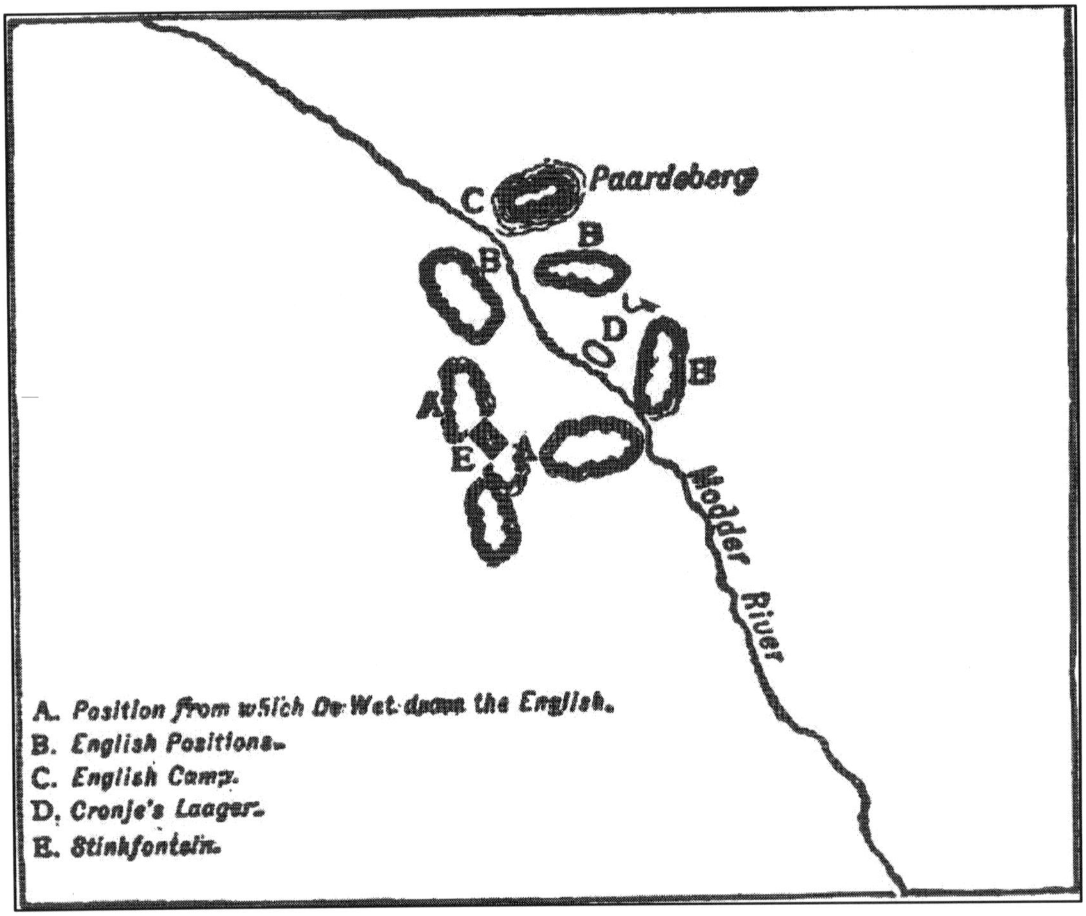

Original Map in De Wet's Book. "Paardeberg (Cronje's), From a Sketch by the Author."

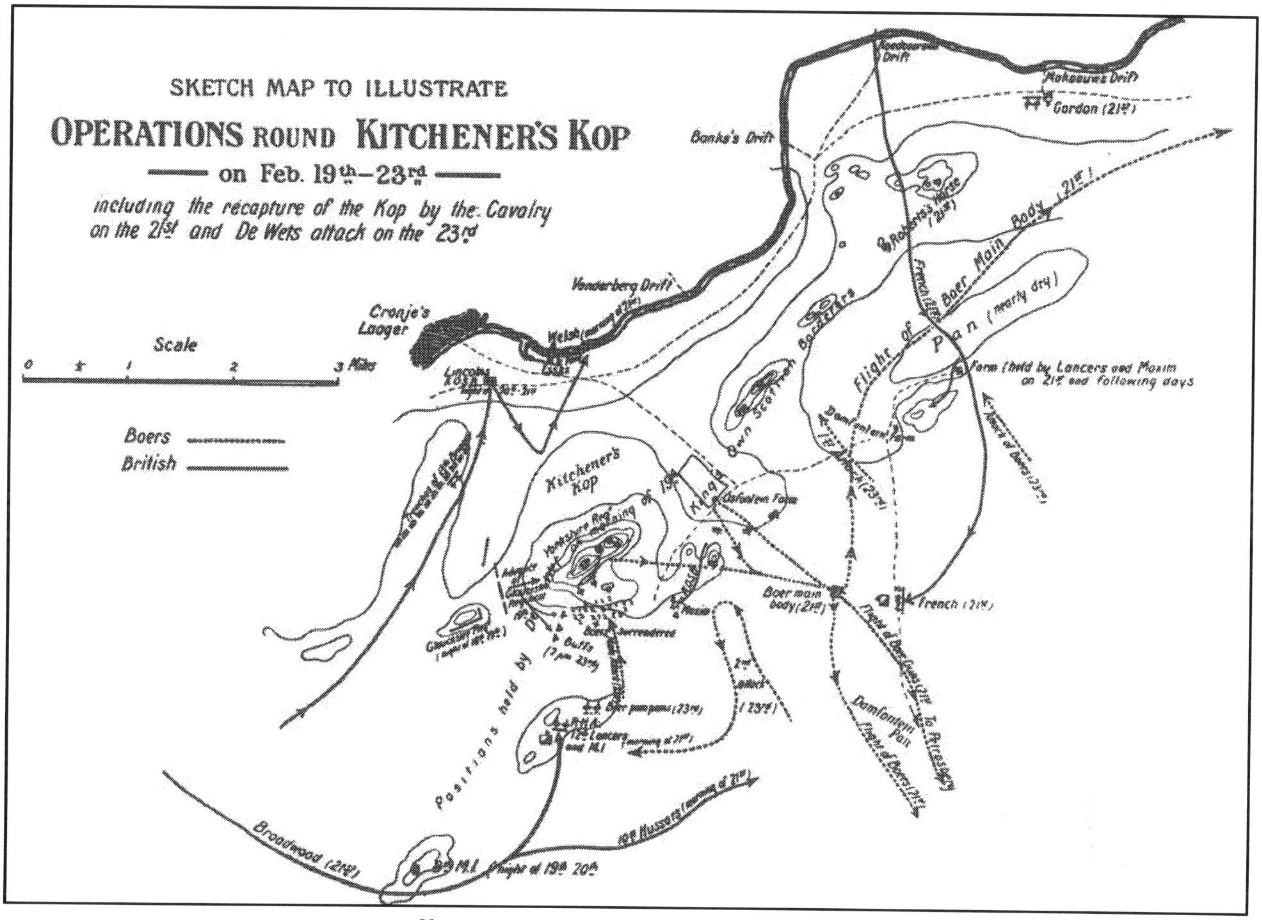

Operations around Kitchener's Kop.[89]

The result of this was that the British gained one of our positions, that, namely, which was held by Veldtcornet Meyer, an officer under Commandant Spruit. Meyer was entirely unable to beat off the attack, and, at nightfall, was compelled to retire about two or three hundred paces, to a little ridge, which he held effectively.

As the English took up the abandoned position, they raised a cheer, and Commandant Spruit, who was ignorant of its meaning, and believed that his men were still in possession, went there alone.

"Hoe gaat het?"[74] he called out.

"Hands up!" was the reply he received.

There was nothing left for the Commandant to do but to give himself up. The soldiers led him over a ridge, and struck a light to discover his identity. Finding papers in his pocket which showed that their prisoner was an important personage, they raised cheer upon cheer.[75]

I heard them cheering, and thought that the enemy were about to attempt another attack, and so gave orders that whatever happened our positions must be held, for they were the key to General Cronje's escape. However, no attack was delivered.

[74] "How is it with you?"

[75] Eleven or twelve days after, Commandant Spruit was again with us. When he appeared, he seemed to us like one risen from the dead. We all rejoiced, not only because he was a God-fearing man, but also because he was of a lovable disposition. I heard from his own mouth how he had escaped. He told me that the day after his capture, he was sent, under a strong escort, from Lord Roberts' Headquarters to the railway station at Modder River, and that he started from there, with a guard of six men on his road to Cape Town. During the night as they drew near De Aar, his guards fell asleep, and our brave Commandant prepared to leave the train. He seized a favourable opportunity when the engine was climbing a steep gradient and jumped off. But the pace was fast enough to throw him to the ground, though fortunately he only sustained slight injury. When daylight came he hid himself. Having made out his bearings he began to make his way back on the following night. He passed a house, but dared not seek admission, for he did not know who its occupants might be. As he had no food with him, his sufferings from hunger were great, but still he persevered, concealing himself during the day, and only walking during the hours of darkness. At last he reached the railway line to the north of Colesberg, and from there was carried to Bloemfontein, where he enjoyed a well-earned rest. In the second week of March he returned to his commando, to the great delight of everybody.

Nobody could have foreseen that two thousand infantry would give up the attack on positions which they had so nearly captured, and we all expected a sanguinary engagement on the following morning. We had made up our minds to stand firm, for we knew that if General Cronje failed to make his way out, it would be a real calamity to our great cause.

Fully expecting an attack, we remained all that night at our posts. Not a man of us slept, but just before dawn we heard this order from the English lines:

"Fall in."

"What can be the meaning of this?" we ask one another.

Lying, sitting or standing, each of us is now at his post, and staring out into the darkness, expecting an attack every moment. We hold our breath and listen. Is there no sound of approaching footsteps? And now the light increases. Is it possible? Yes, our eyes do not deceive us. The enemy is gone.

Surprise and joy are on every face. One hears on all sides the exclamation, "If only Cronje would make the attempt now." It was the morning of the 25th of February.

But the enemy were not to leave us alone for long. By nine o'clock they were advancing upon us again, with both right and left wing reinforced. I had only a few shots left for the Krupp, and thirty for the Maxim-Nordenfeldt, and this last ammunition must now be expended on the wings. One gun I despatched to the right, the other to the left, and the English were checked in their advance. I had ordered the gunners, as soon as they had fired their last round to bring their guns into safe positions in the direction of Petrusberg. Very soon I observed that this order was being executed, and thus learnt that the ammunition had run out.

The burghers who, with their rifles, had attempted to hold back the wings, now having no longer any support from the big guns, were unable to stand their ground against the overpowering forces of the enemy, and shortly after the guns were removed, I saw them retreat.

What was I to do? I was being bombarded incessantly, and since the morning had been severely harassed by small-arm fire. All this, however, I could have borne, but now the enemy began to surround me.[76] It was a hard thing to be thus forced to abandon the key to General Cronje's escape.

In all haste I ordered my men to retire. They had seen throughout that this was unavoidable, and had even said to me: "If we remain here, General, we shall be surrounded with General Cronje."

All made good their retreat, with the exception of Veldtcornet Speller, of Wepener, who, to my great regret, was taken prisoner there with fourteen men. That occurred owing to my adjutant forgetting, in the general confusion, to give them my orders to retreat. When Speller found that he, with his fourteen men, was left behind, he defended himself, as I heard later, with great valour, until at last he was captured by overpowering numbers. It cost the English a good many dead and wounded to get him out of his schanzes.

Although I had foreseen that our escape would be a very difficult and lengthy business, I had not thought that we should have been in such danger of being made prisoners. But the English had very speedily taken up positions to the right and left, with guns and Maxims, and for a good nine miles of our retreat we were under their fire. Notwithstanding the fact that during the whole of this time we were also harassed by small-arm fire, we lost—incredible as it may appear—not more than one killed and one wounded, and a few horses besides. The positions which we had abandoned the British now occupied, hemming in General Cronje so closely that he had not the slightest chance of breaking through their lines.

No sooner had we got out of range of the enemy's fire, than the first of the reinforcements, which we had expected from Bloemfontein, arrived, under the command of Vechtgeneraal Andreas Cronje. With him were Commandants Thewnissen, of Winburg, and Vilonel, of Senekal.

A council was at once held as to the best method of effecting the release of General Cronje. It was decided to recapture the positions which I had abandoned. But now the situation was so changed that there were three positions which it was necessary for us to take. We agreed that the attack should be made by three separate parties, that General Philip Botha, with Commandant Thewnissen, should retake the positions which we had

[76] Ignatius Ferreira's forces, to the North, which might have supported De Wet, had been left without direction after Ferreira was accidentally shot dead by one of his own sentries.

abandoned at Stinkfontein, General Froneman the position immediately to the north of these, and I, with General Andreas Cronje, others still further north.

The attack was made on the following morning. [26th] General Botha's attempt failed, chiefly owing to the fact that day dawned before he reached his position; a hot fight ensued, resulting in the capture of Commandant Thewnissen and about one hundred men. As I was so placed as to be unable to see how affairs were developing, it is difficult for me to hazard an opinion as to whether Commandant Thewnissen was lacking in caution, or whether he was insufficiently supported by General Botha. The burghers who were present at the engagement accused General Botha, while he declared that Thewnissen had been imprudent. However that may be, we had failed in our essay. The position had not been taken, and Commandant Thewnissen, with a hundred whom we could ill spare, were in the hands of the enemy, and to make matters still worse, our men were already seized with panic, arising from the now hopeless plight of General Cronje and his large force.

I, however, was not prepared to abandon all hope as yet. Danie Theron, that famous captain of despatch-riders, had arrived on the previous day with reinforcements. I asked him if he would take a verbal message to General Cronje—I dare not send a written one, lest it should fall into the hands of the English. Proud and distinct the answer came at once—the only answer which such a hero as Danie Theron could have given:

"Yes, General, I will go."

The risk which I was asking him to run could not have been surpassed throughout the whole of our sanguinary struggle.

I took him aside, and told him that he must go and tell General Cronje that our fate depended upon the escape of himself and of the thousands with him, and that, if he should fall into the enemy's hands, it would be the death-blow to all our hopes. Theron was to urge Cronje to abandon the laager, and everything contained in it, to fight his way out by night, and to meet me at two named places, where I would protect him from the pursuit of the English.

Danie Theron undertook to pass the enemy's lines, and to deliver my message. He started on his errand on the night of the 25th of February.

The following evening [26th] I went to the place of meeting, but to my great disappointment General Cronje did not appear.

Daniel (Danie) Theron.[90] [77]

Danie Theron's Scouts, at Dinner.[91]

[77] Theron Daniel (Danie) Johannes Stepha Commandant He was born in 9 May 1872 at Tulbagh in the Cape. Initially he worked as a teacher and later studied law. In 1895 he served in the Mmalebogo War. He rose to prominence for assaulting W F Moneypenny, editor of the Star, for allegedly insulting womanhood. He was fined for the assault but the money was paid by his sympathisers. When the Boer War started, he served a Captain in the Transvaal Cyclist Corps. He was present at Colenso. He then commanded a corps of scouts and became legendary. During the engagement at Paardeberg, he carried a message from General Christiaan de Wet to General Cronje. He later commanded Theron se Verkenningskorps with a strength of around 100. Theron was killed on 5 September 1900 at Elandsfontein near Potchefstroom, in action against Major General Hart's column. https://www.angloboerwar.com/name-search And https://en.wikipedia.org/wiki/Daniel_Theron

On the morning of the 27th of February Theron returned. He had performed an exploit unequalled in the war. Both in going and returning he had crawled past the British sentries, tearing his trousers to rags during the process. The blood was running from his knees, where the skin had been scraped off. He told me that he had seen the General, who had said that he did not think that the plan which I had proposed had any good chance of success.

At ten o'clock that day, General Cronje surrendered. Bitter was my disappointment. Alas! my last attempt had been all in vain. The stubborn General would not listen to good advice.[78]

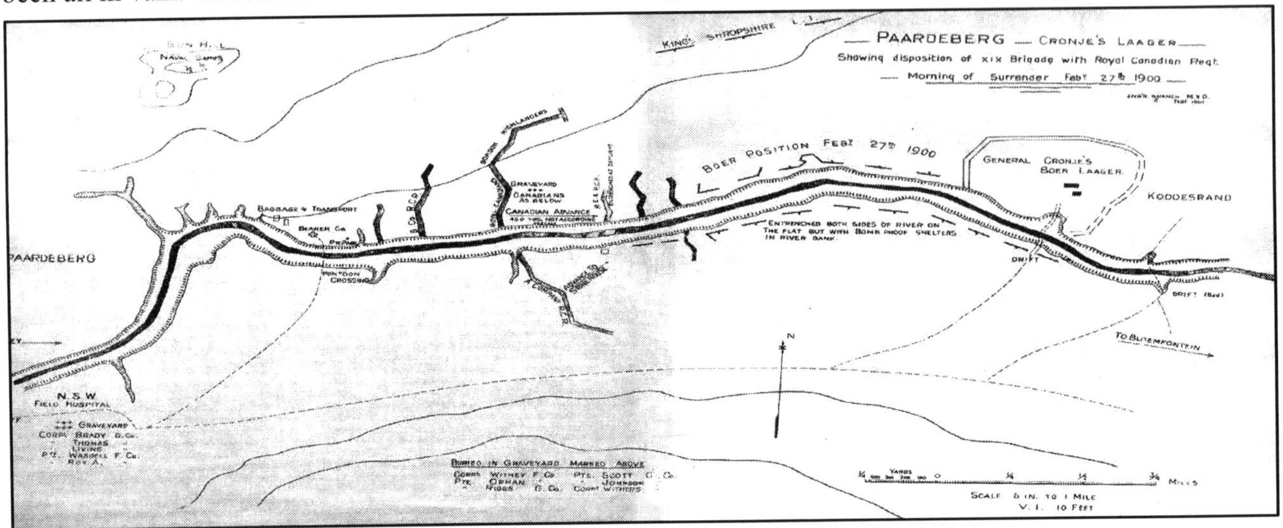

Positions on the Morning of Feb. 27th.[92]

I must repeat here what I have said before, that as far as my personal knowledge of General Cronje goes, it is evident to me that his obstinacy in maintaining his position must be ascribed to the fact that it was too much to ask him—intrepid hero that he was—to abandon the laager. His view was that he must stand or fall with it, nor did he consider the certain consequences of his capture. He never realized that it would be the cause of the death of many burghers, and of indescribable panic throughout not only all the laagers on the veldt, but even those of Colesberg, Stormberg and Ladysmith. If the famous Cronje were captured, how could any ordinary burgher be expected to continue his resistance?

It may be that it was the will of God, who rules the destinies of all nations, to fill thus to the brim the cup which we had to empty, but this consideration does not excuse General Cronje's conduct. Had he but taken my advice, and attempted a night attack, he might have avoided capture altogether.

I have heard men say that as the General's horses had all been killed, the attempt which I urged him to make must have failed—that at all events he would have been pursued and overtaken by Lord Roberts' forces. The answer to this is not far to seek. The English at that time did not employ as scouts Kaffirs and Hottentots, who could lead them by night as well as by day. Moreover, with the reinforcements I had received, I had about sixteen hundred men under me, and they would have been very useful in holding back the enemy, until Cronje had made his escape.

No words can describe my feelings when I saw that Cronje had surrendered, and noticed the result which this had on the burghers. Depression and discouragement were written on every face. The effects of this blow, it is not too much to say, made themselves apparent to the very end of the war.

[78] 2/27/1900 (on anniversary of Majuba, 1881) Cronje, with 44 commandants and other officers of all grades, and over 3,500 men, surrendered unconditionally to Roberts. Sir Redvers Buller's force captured the Boer position at Pieters. This action opened the road to Ladysmith. Boers retired north to Ladysmith.

Meeting of Lord Roberts & General Cronje.[93] February 27th, 1900. Cronje and British Officer.[94]

British troops at assembly point during probably surrender ceremony of General Piet Cronjé's units.[95] British soldier standing in the assembly line, possibly during the surrender of the Boer fighters ceremony.[96]

Paardeberg Battlefield.[97] Powder Barrels and Destroyed Wagons.[98] Remains of burnt wagons.[99]

Cronje's Chief Commandants after the surrender, Feb. 27th, at Paardeberg.[100] [101] A heap of Mausers with martini rifles surrendered by Cronje's army.[102]

Stacks of rifles taken from Cronje's Laager at Paardeberg.[103]

Captured Boer Rifles at Cape Town.[104]

The Boer prisoners (Cronje's men) resting on the road from Paardeberg to Modder River, S. Africa.[105]

		Killed		Wounded		Missing and Prisoners		
Date	Action	Officers	Men	Officers	Men	Officers	Men	Total
16-27-Feb	Paardeberg	17	239	74	1,136	7	62	1,535

British Casualties at Paardeberg.[106]

General Cronje and his officers on St. Helena.[107]

7. The Wild Flight from Poplar Grove

The surrender of General Cronje only made me all the more determined to continue the struggle, notwithstanding the fact that many of the burghers appeared to have quite lost heart. I had just been appointed Commander-in-Chief, and at once set my hand to the work before me.

Let me explain how this came about.

As I have already said, General C.J. Wessels had been appointed Commander-in-Chief at Kimberley. In the month of January he was succeeded by Mr. J. S. Ferreira, who at once proceeded to make Kimberley his headquarters. On the relief of that town, one part of the besieging force went to Viertienstroomen, another in the direction of Boshof, while a small party, in which was the Commander-in-Chief [Prinsloo] himself, set out towards Koedoesrand, above Paardeberg.

It was while I was engaged in my efforts to relieve Cronje, that a gun accident occurred in which General [J.S.] Ferreira was fatally wounded. Not only his own family, but the whole nation, lost in him a man whom they can never forget. I received the sad news the day after his death, and, although the place of his burial was not more than two hours' ride from my camp, I was too much occupied with my own affairs to be able to attend his funeral.

On the following day I received from President Steyn the appointment of Vice-Commander-in-Chief. I had no thought of declining it, but the work which it would involve seemed likely to prove anything but easy. To have the chief command, and at such a time as this! But I had to make the best of it.

I began by concentrating my commandos, to the best of my ability, at Modderrivierpoort (Poplar Grove), ten miles east of the scene of Cronje's surrender. I had plenty of time to effect this, for Lord Roberts remained inactive from the 24th of February to the 7th of March, in order to rest a little after the gigantic task he had performed in capturing Cronje's laager. His thoughts must have been busy during that period with even more serious matters than the care of his weary troops; for, if we had had two hundred killed and wounded, he must have lost as many thousands.

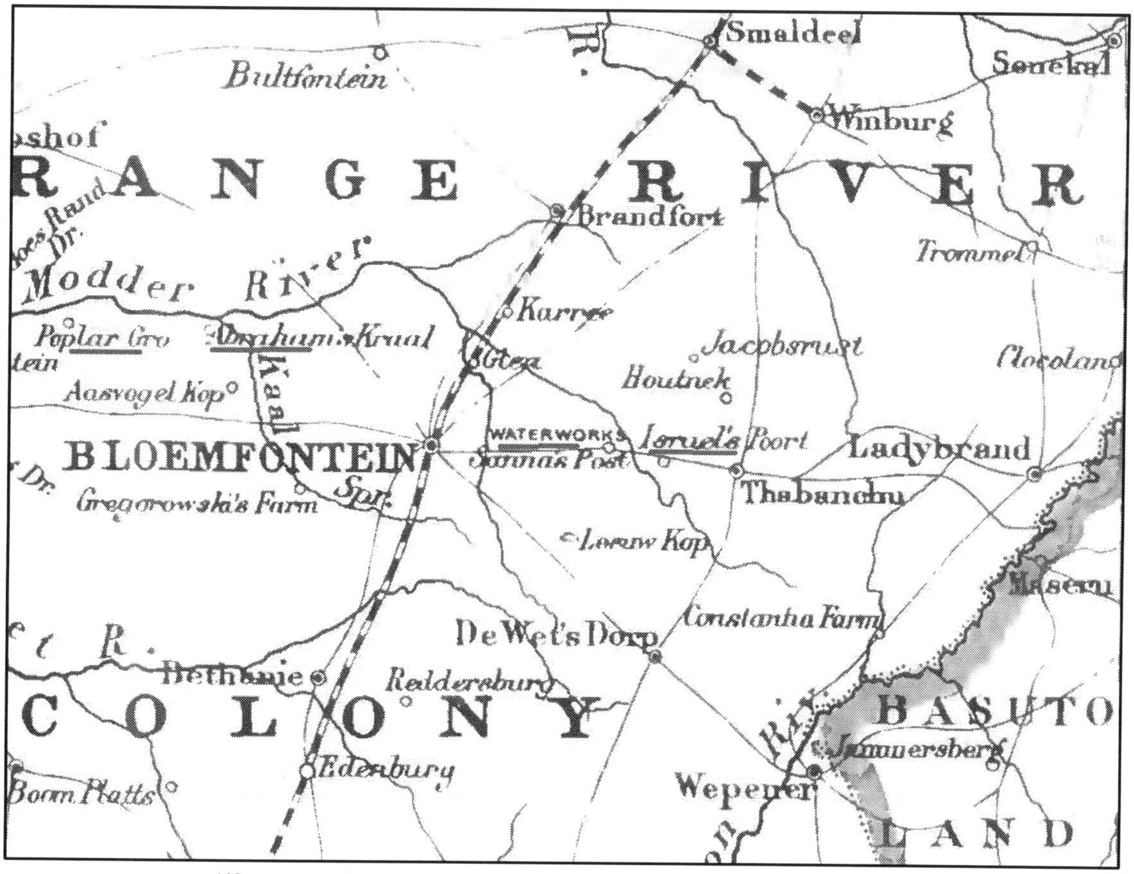

Orange Free State.[108] Note Poplar Grove, Abraham's Kraal, Waterworks and Sanna's Post and Israel's Poorte.

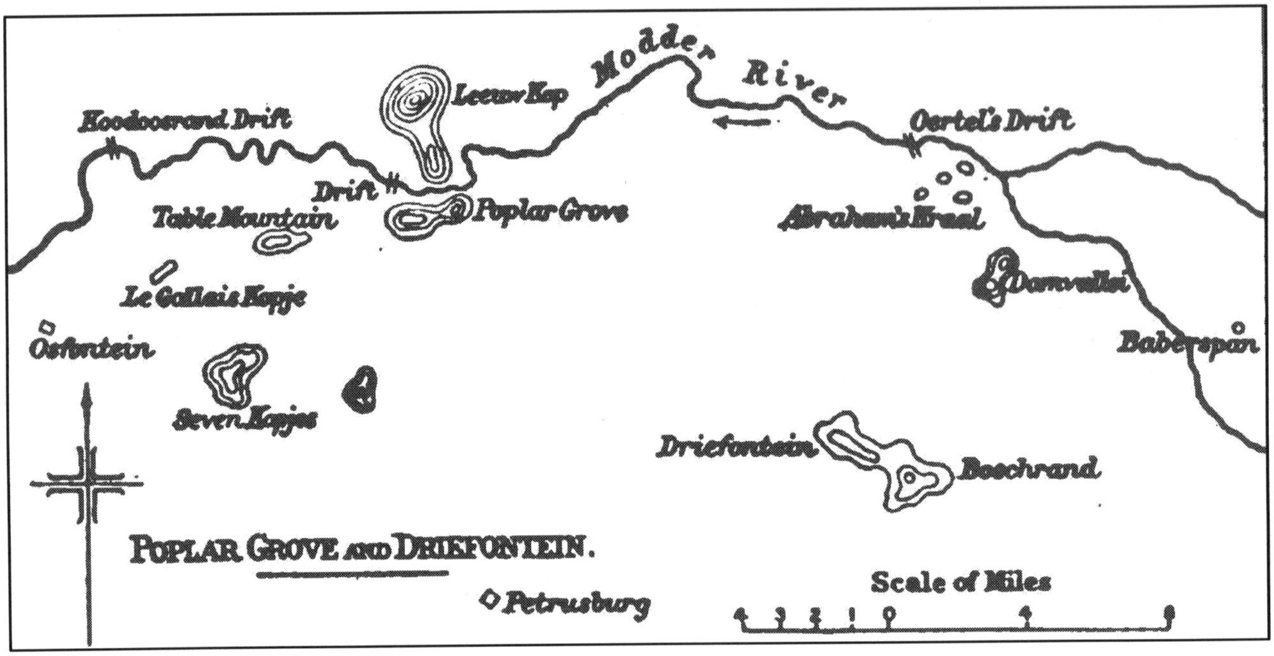

Poplar Grove & Driefontein Areas.[109]

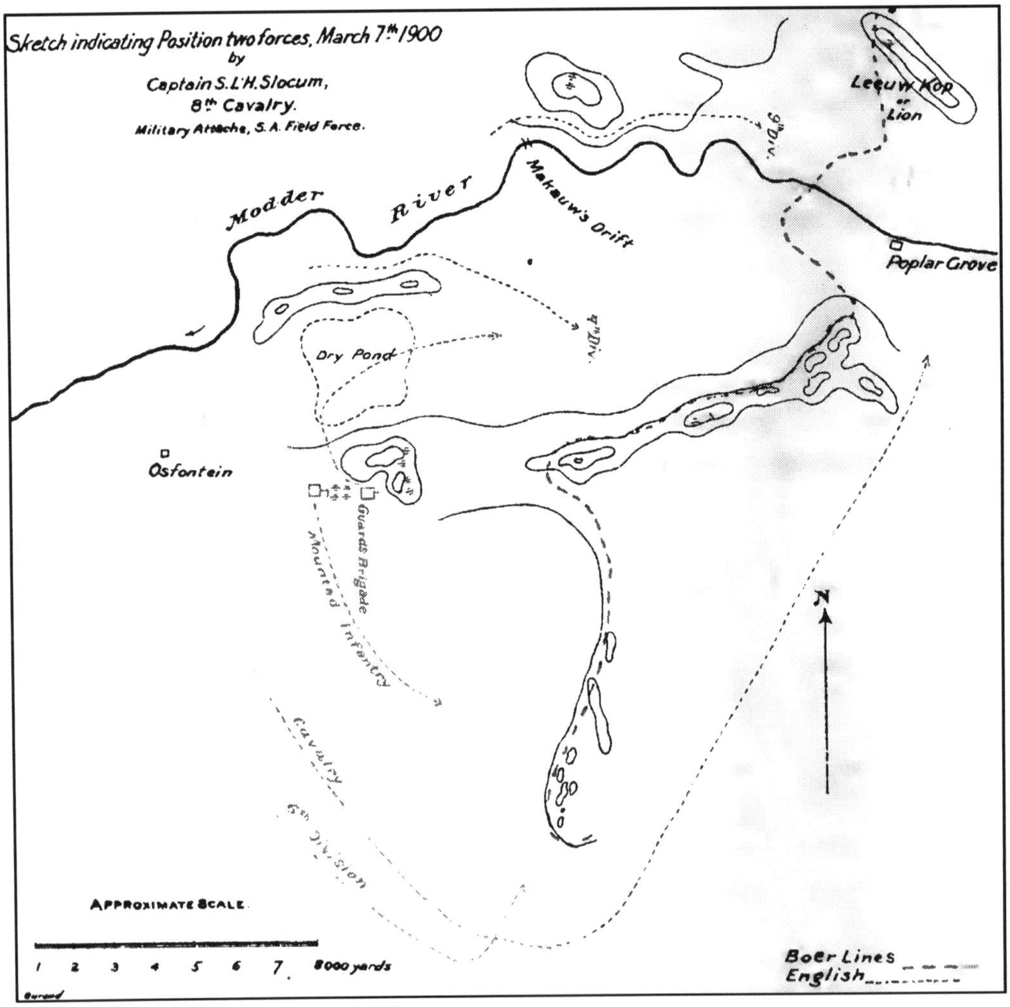

Positon of Forces on March 7th.[110]

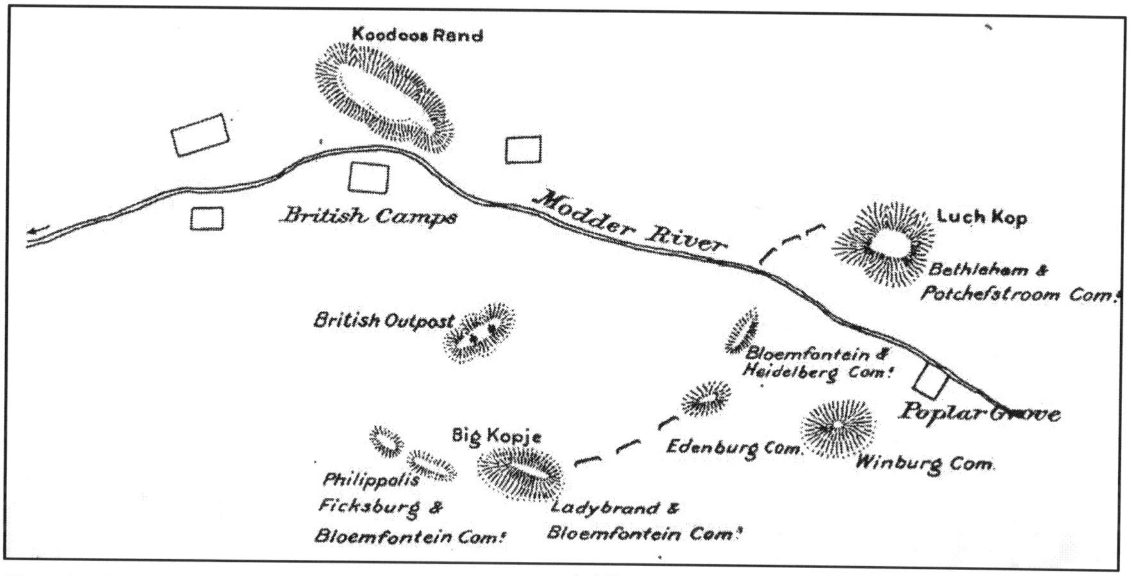

Sketch of Boer Position at Poplar Grove, March 7th.[111]

Those few days during which our enemy rested were also of advantage to me in enabling me to dispose of the reinforcements, which I was now receiving every day, and from almost every quarter.

While I was thus engaged, I heard that General Buller had relieved Ladysmith on the 1st of March, that General Gatacre had taken Stormberg on the 5th, and that General Brabant was driving the Boers before him.

These were the first results of General Cronje's surrender.

But that fatal surrender was not only the undoing of our burghers; it also reinforced the enemy, and gave him new courage. This was evident from the reply which Lord Salisbury made to the peace proposals made by our two Presidents on March 5th. But more of this anon.[79]

Our last day at Poplar Grove was signalized by a visit paid to us by President Kruger, the venerable chief of the South African Republic. He had travelled by rail from Pretoria to Bloemfontein; the remaining ninety-six miles of the journey had been accomplished in a horse-waggon—he, whom we all honoured so greatly, had been ready to undergo even this hardship in order to visit us.

The President's arrival was, however, at an unfortunate moment. It was March the 7th, and Lord Roberts was approaching. His force, extending over ten miles of ground, was now preparing to attack my burghers, whom I had posted at various points along some twelve miles of the bank of the Modder River. It did not seem possible for the old President even to outspan, for I had received information that the enemy's right wing was already threatening Petrusburg. But as the waggon had travelled that morning over twelve miles of a heavy rain-soaked road, it was absolutely necessary that the horses should be outspanned for rest. But hardly had the harness been taken off the tired animals when a telegram arrived, saying that Petrusburg was already in the hands of the English. President Kruger was thus compelled to return without a moment's delay. I saw him into his waggon, and then immediately mounted my horse, and rode to the positions where my burghers were stationed.

Again I was confronted with the baleful influence of Cronje's surrender. A panic had seized my men. Before the English had even got near enough to shell our positions to any purpose, the wild flight began. Soon every position was evacuated. There was not even an attempt to hold them, though some of them would have been almost impregnable. It was a flight such as I had never seen before, and shall never see again.

I did all that I could, but neither I nor my officers were able to prevent the burghers from following whither the waggons and guns had already preceded them. I tried every means. I had two of the best horses that a man could wish to possess, and I rode them till they dropped. All was in vain. It was fortunate for us that the advance of the English was not very rapid. Had it been so, everything must have fallen into their hands.

[79] 3/2/1900 Cronje and staff, moved to Simonstown under a guard and put on board H.M.S. Doris, and sent to St. Helena.
3/3/1900 General Buller formally entered Ladysmith. Skirmish near Osfontein.
3/4/1900 General Brabant advanced from Dordrecht against Labuschagne.
3/5/1900 General Gatacre occupied Stormberg without opposition. Ovetures of Peace made by Kruger and Steyn to British Govt.

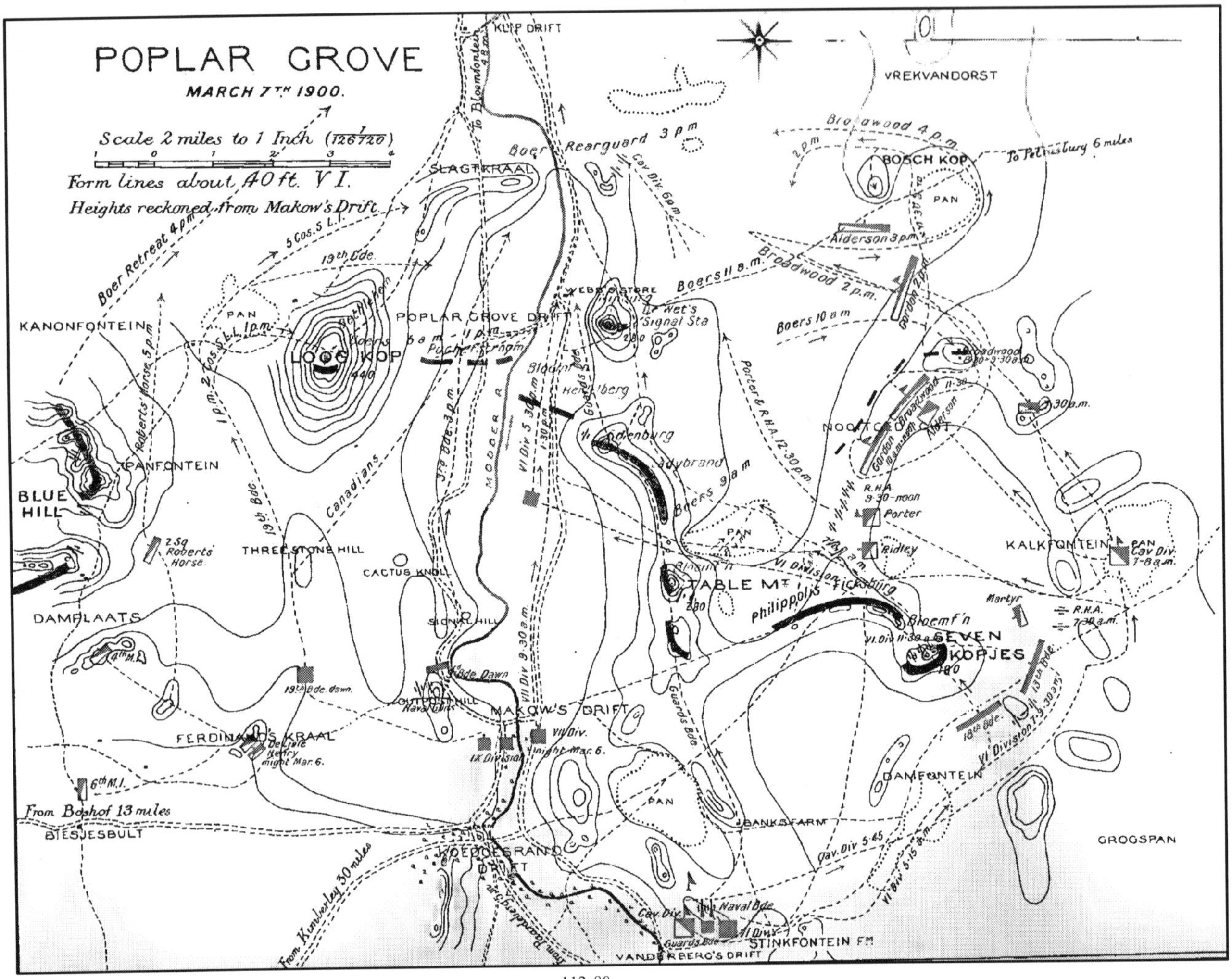

Times History Map of Poplar Grove Engagement.[112 80]

In the evening we came to Abraham's Kraal, a farm belonging to Mr. Charles Ortel, some eighteen miles from Poplar Grove. The enemy were encamped about an hour and a half's ride from us.

The next morning the burghers had but one desire, and that was to get away. It was only with the greatest difficulty that I succeeded in persuading them to go into position. I then hastened to Bloemfontein, in order to take counsel with the Government about our affairs generally, and especially to see what would be the most suitable positions to occupy for the defence of the capital. Judge Hertzog and I went out together to inspect the ground; we placed a hundred men in the forts, with Kaffirs to dig trenches and throw up earthworks.

I was back at Abraham's Kraal by nine o'clock on the morning of March the 18th. (sic 10th) I found that our forces had been placed in position by Generals De la Rey, Andreas Cronje, Philip Botha, Froneman and Piet de Wet, the last-named having arrived with his commandos from Colesberg a few days before the rout at Poplar Grove.

We had not long to wait before fighting began, fighting confined for the most part to the artillery. The English shells were at first directed against Abraham's Kraal, which was subjected to a terrific bombardment; later on they turned their guns upon Rietfontein, where the Transvaalers and a part of the Free State commandos, under General De la Rey, were posted. The attack upon these positions was fierce and determined; but De la Rey's burghers, though they lost heavily, repulsed it with splendid courage. I will not say more of this. It is understood that General De la Rey will himself describe what he and his men succeeded in accomplishing on that occasion.

[80] The Battle of Poplar Grove was an engagement on 7 March 1900 during the Second Boer War in South Africa. It followed on from the Relief of Kimberley (on February 15th) as the British Army moved East to take the Boer capital of Bloemfontein. General French's cavalry attacked the Boer force from the rear while mounted infantry and horse artillery attacked from the right flank. The Boers abandoned their positions. Battle of Poplar Grove - Wikipedia

From ten in the morning until sunset the fight continued, and still the burghers held their positions. They had offered a magnificent resistance. Their conduct had been beyond all praise, and it was hard to believe that these were the same men who had fled panic-stricken from Poplar Grove. But with the setting of the sun a change came over them. Once more panic seized them; leaving their positions, they retreated in all haste towards Bloemfontein. And now they were only a disorderly crowd of terrified men blindly flying before the enemy.

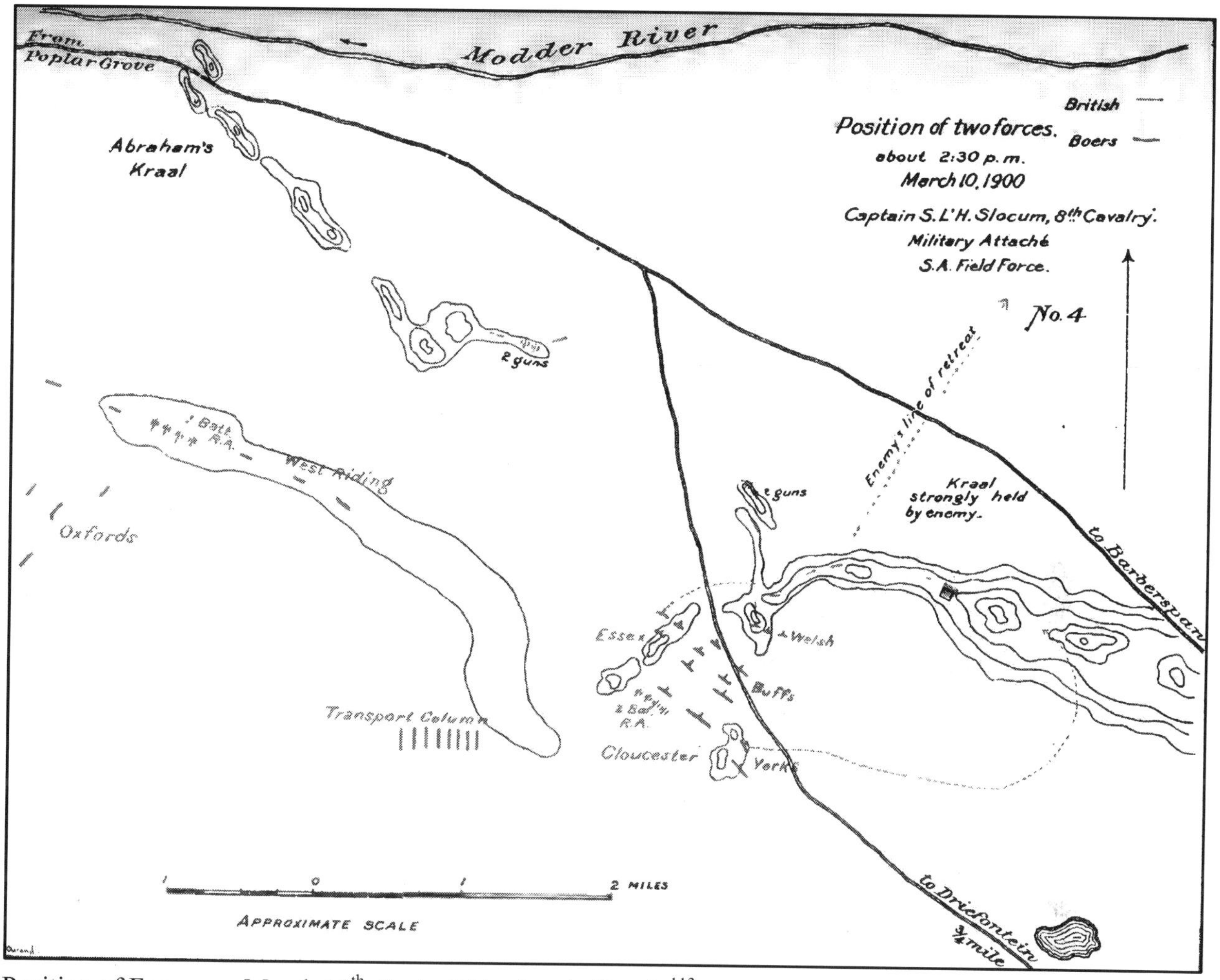

Position of Forces on March 10th, East of Abraham's Kraal .[113]

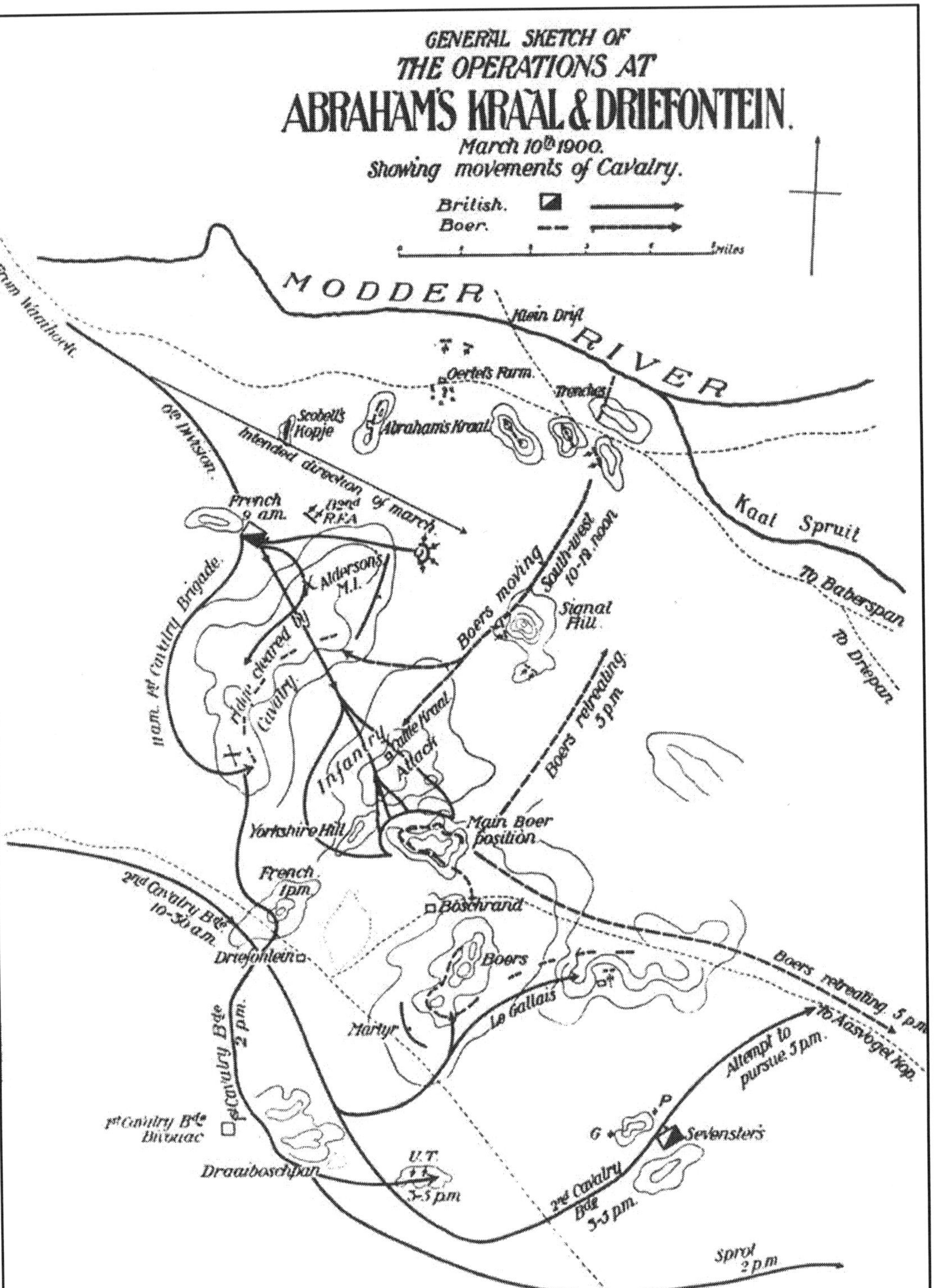

Abraham's Kraal & Driefontein.[114 81]

[81] The Battle of Driefontein on 10 March 1900 followed on the Battle of Poplar Grove in the Second Boer War between the British Empire and the Boer republics, in what is now South Africa. In the first half of 1900, the British made an offensive towards the two Boer republic capitals of Bloemfontein and Pretoria. The Boer forces under the command of Christiaan de Wet were holding a 7-mile (11 km) line covering the approach to Bloemfontein. Roberts subsequently ordered a division under Lieutenant General Thomas Kelly-Kenny to attack the position from the front, while Lieutenant General Charles Tucker's division moved against its left flank. The Boers were subsequently forced to withdraw losing 124 men killed and captured, while the British lost 82 killed and 342 wounded. Battle of Driefontein - Wikipedia
See Also: Battle of Driefontein, 10 March 1900 (historyofwar.org) The battle of Driefontein (Boer War) was the last attempt to prevent the British under Roberts from occupying Bloemfontein, the capital of the Orange Free State. The Boer position in the Orange Free State had collapsed very quickly. Only four weeks earlier their main army had been camped at Magersfontein, blocking the direct route to

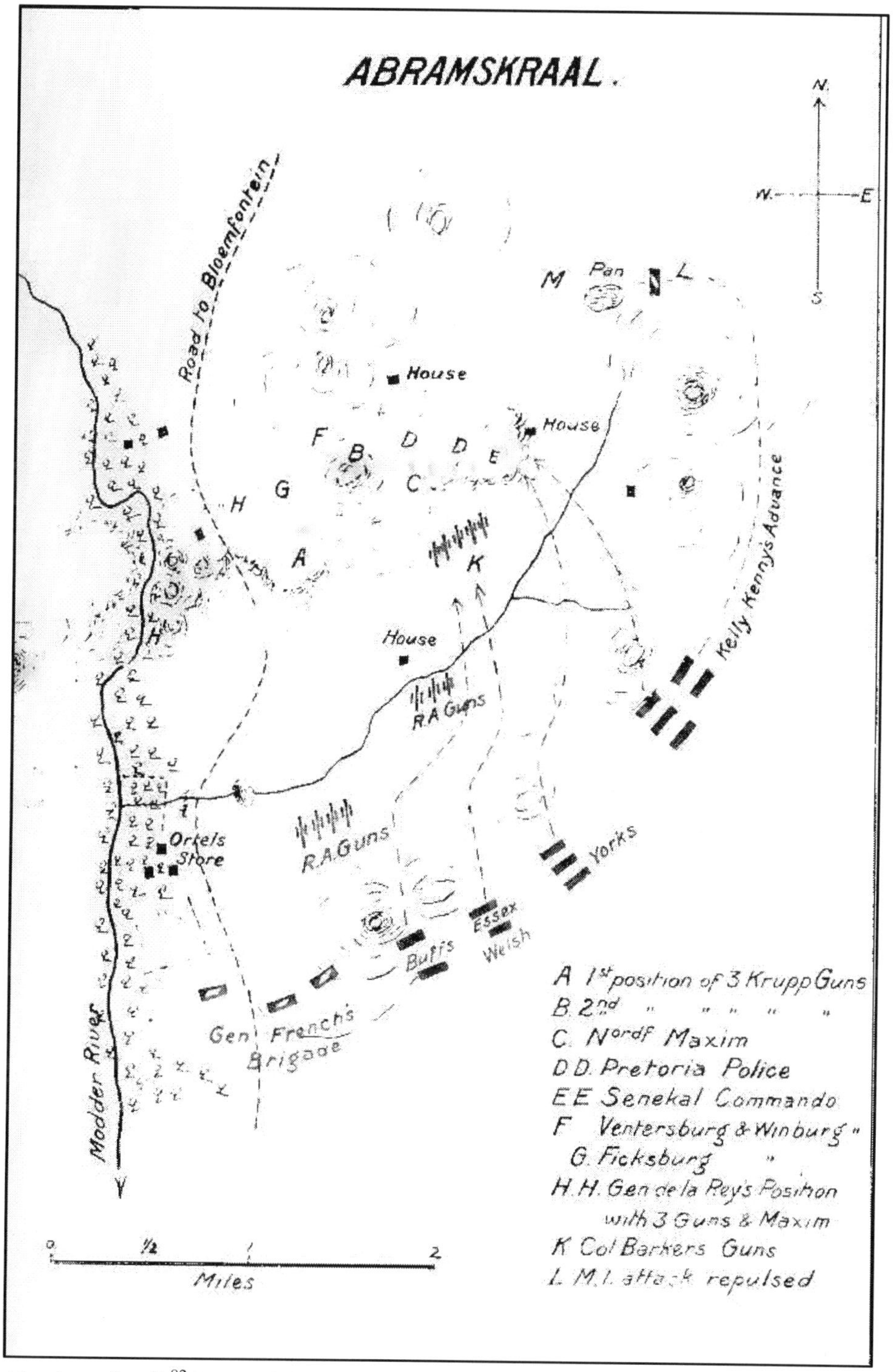

Abrahamskrall.[82] Map of the engagement with apologies for the quality.

Kimberley. On 11 February Roberts had begun his great flank march, which had seen Kimberley liberated on 15 March, and the army from Magersfontein captured at Paardeberg on 27 February. A first attempt to stop the British advance had failed with hardly any fighting at Poplar Grove (7 March). In the aftermath of the rout at Poplar Grove, President Kruger had made valiant attempts to rally the fleeing burghers, but only 1,500 of the 6,000 men from Poplar Grove could be convinced to stay.

Under the command of Christiaan De Wet, with assistance from De La Rey, the Boers organised a new defensive line at Dreifontein. This time they fought, and fought well. Roberts organised his men into three columns and attacked along the entire Boer line. Despite being heavily outnumbered, the Boers managed to hold on all day, but as darkness fell the British had seized the northern end of the Boer line, and were threatening the southern flanks.

Once again the burghers fled–De Wet described it as a panic. This time there was no stopping them. Three days later the British captured Bloemfontein without facing any more significant resistance.

British losses at Driefontein had been 82 dead and 342 wounded. Boer losses were at least 102 dead and 22 captured.

[82] Viljoen Ben, My Reminiscences, London: Hood, Douglas, & Howard, 11, Clifford's Inn, E.C. 1902. Maps from Drawings by P. Van Breda.

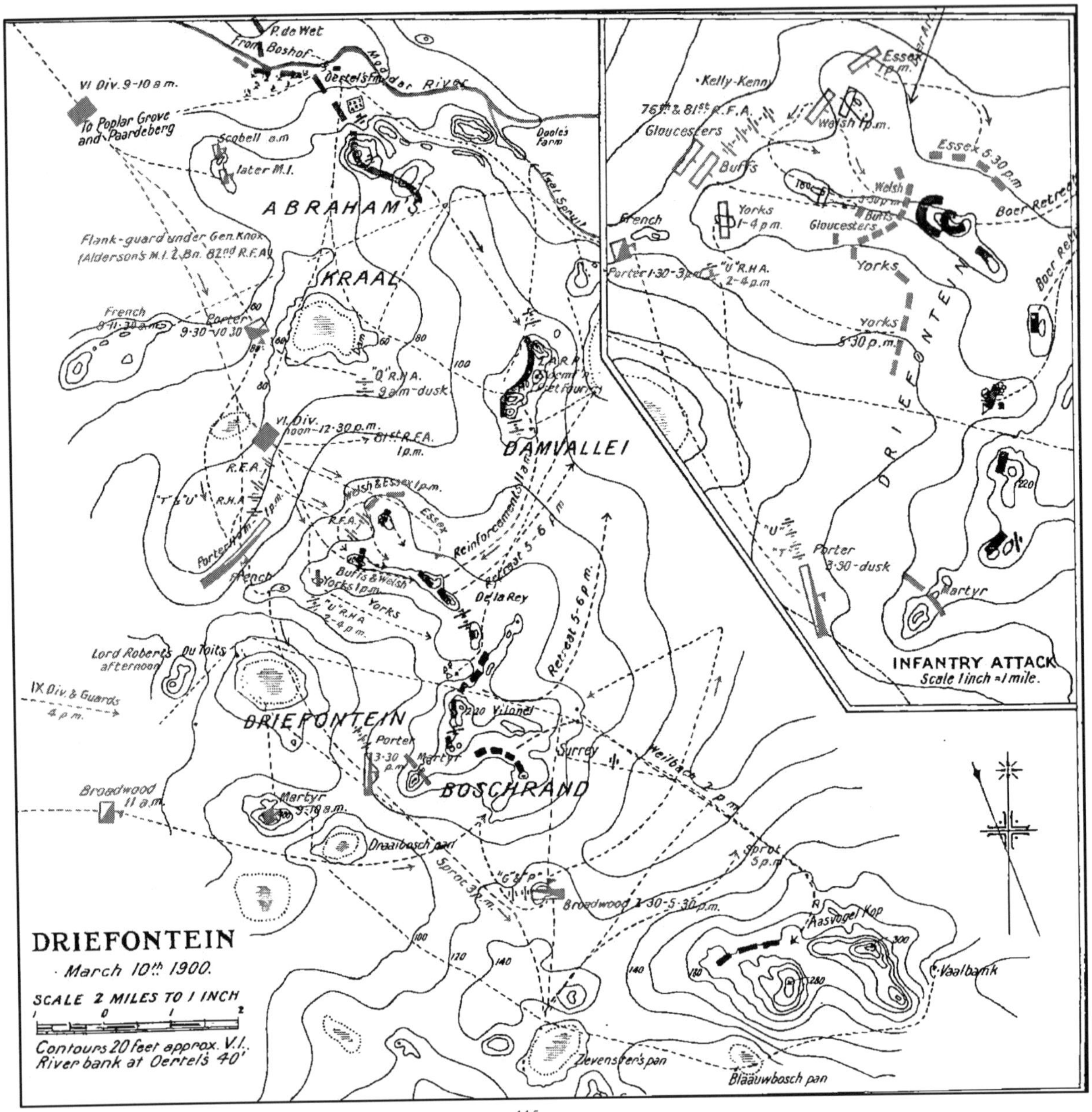

Times History Map of Driefontein Engagement.[115]

But it was Bloemfontein that lay before them, and the thought that his capital was in peril might well restore courage in the most disheartened of our burghers. I felt that this would be the case, and a picture arose before me of our men holding out, as they had never done before.

Before going further I must say a few words about the peace proposals which our Presidents made to the English Government on the 5th of March. They called God to witness that it was for the independence of the two Republics, and for that alone, that they fought, and suggested that negotiations might be opened with the recognition of that independence as their basis.

Lord Salisbury replied that the only terms he would accept were unconditional surrender. He asserted, as he did also on many subsequent occasions, that it was our ultimatum that had caused the war. We have always maintained that in making this assertion he misrepresented the facts, to use no stronger term.[83]

Naturally our Government would not consent to such terms, and so the war had to proceed.

[83] 3/11/1900 Presidents Kruger and Steyn received reply from the Prime Minister refusing to entertain their overtures for peace.

March 1900 Offer and Reply

In March of 1900 the Presidents of both Republics wrote directly to Robert Cecil[84] (The Prime Minister of the United Kingdom and known as the Marquess of Salisbury) in an attempt to negotiate an end to the conflict on honorable terms.[85] [86]

Kruger and Steyn's Letter to Lord Salisbury

From the States-President of the South African Republic and the Orange Free State

Bloemfontein, 5th March, 1900.

To His Excellency Lord Salisbury, London.

The blood and tears of the thousands who have suffered through this war, and the prospect of all the moral and material ruin which now threatens South Africa, render it necessary for both parties carrying on the war to ask themselves calmly, and in the faith of the Trinity, for what they are fighting and if the aims of both justify all this horrible misery and devastation.

On this account, and with an eye to the assertion of several English Statesmen that the war was begun and carried on with the determined end to undermine Her Majesty's authority in South Africa, and to establish in the whole of South Africa a Government independent of Her Majesty's Government, we consider it our duty to declare that this War was only commenced as a measure of defence and for the purpose of obtaining a guarantee for the threatened independence of the South African Republic, and was only continued in order to ensure the indisputable independence of both Republics as Sovereign International States, and to obtain the assurance that the subjects of Her Majesty who have taken part with us in the war will not suffer the least hurt either in their lives or their possessions.

On these conditions alone we demand, as in the past, to see peace restored in South Africa, and an end made to the wrong that now exists there.

But if Her Majesty's Government has decided upon destroying the independence of the Republic, nothing remains to us and our people but to persist to the bitter end on the road now taken, notwithstanding the overpowering might of the British Empire, trusting that God, who has lit the inextinguishable fire of the love of liberty in our hearts, and in the hearts of our fathers, will not abandon us, but will fulfil His work in us, and in our descendants.

We hesitated to lay this declaration earlier before Your Excellency, because we were afraid that as long as the advantage was on our side, and our Army had in their occupation positions of defence far into the British Colonies, such a declaration would have hurt the feelings of the English nation; but now that the prestige of the British Empire may be considered to be restored, through the capture of one of our armies, and we are compelled by this to sacrifice other positions which our armies occupied, this difficulty is removed, and we can no longer hesitate to tell you, in the face of the whole civilized world, why we are fighting, and on what conditions we are prepared to make peace.

[84] Robert Arthur Talbot Gascoyne-Cecil, 3rd Marquess of Salisbury (3 February 1830–22 August 1903), known as Lord Salisbury, was a British statesman and Conservative politician who served as Prime Minister of the United Kingdom three times for a total of over thirteen years. He was also Foreign Secretary before and during most of his tenure. He avoided international alignments or alliances, maintaining the policy of "splendid isolation." Robert Gascoyne-Cecil, 3rd Marquess of Salisbury - Wikipedia

[85] CDR Note: 'This correspondence will be found in Chapter 30.' [PF: It is in the Appendix with the correspondence. Chapter 30 is Steyn's Letter to Kitchener. It is reproduced here as well for its relevance to the story and the struggle.]

[86] In Dr. Conan Doyle's work, Cause and Conduct, these two communications are reproduced as well. The differences in language are notable, since each purports to be a full transcription.

Reply

From Lord Salisbury

To their Excellencies the States-Presidents of the South African Republic and Orange Free State.

London, 11th March, 1900.

I have the honour to acknowledge the receipt of your Honour's cable, dated 5th March, from Bloemfontein, of which the purport is principally whether Her Majesty's Government will acknowledge the indisputable independence of the South African Republic and Orange Free State and treat them as Sovereign International States, and will offer to conclude the war on these conditions.

In the beginning of October of this year, there was peace between the Queen and the two Republics, under the Convention which then held good. There was a discussion carried on during a few months between Her Majesty's Government and the South African Republic, of which the purport was the amendment of very serious grievances under which English inhabitants suffered in the South African Republic.

In the course of these negotiations, the South African Republic obtained the knowledge that Her Majesty's Government had made considerable preparations for war, and had taken steps to provide the necessary reinforcements for the English garrisons at Cape Colony and Natal. No inroad on the rights guaranteed by the Conventions had, until then, taken place on the English side.

Suddenly the South African Republic, after having two days previously issued an insulting ultimatum, declared War on Her Majesty; and the Orange Free State, with which there had been no disagreement, took a similar step.

Thereupon an inroad was made into Her Majesty's territory by the two Republics; three towns within the British frontier were besieged, a great part of the two Colonies was over-run, with great destruction of property and life, and the Republics claimed the right to treat the inhabitants of Her Majesty's territory as if this territory had been annexed by one of these States.

The Transvaal having these actions in view, had for years stored up, on an enormous scale, military provisions, which could only have been destined for use against England.

Your Excellencies made some remarks of a negative nature concerning the aim for which these preparations were made. I do not consider it necessary to discuss the question which you have thus raised, but the consequences of the preparations, made in great secrecy, have been that the British Empire has found itself forced to repel an inroad which has brought on a costly war, and caused the loss of thousands of valuable lives.

This great misfortune has been the punishment that Great Britain has had to undergo during the last few years for having suffered the two Republics to exist.

Keeping in sight the use which the two Republics have made of the position presented to them, and the misfortunes which their unprovoked attacks on Her Majesty's territory have brought, Her Majesty's Government can only reply to Your Honour's telegram by saying that they are not prepared to acknowledge the independence either of the South African Republic, or of the Orange Free State.

Robert Gascoyne-Cecil, 3rd Marquess of Salisbury.[116]

It was decided to send a deputation to Europe. This deputation, consisting of Abraham Fisher,[87] Cornelius H. Wessels,[88] and Daniel Wolmarans,[89] sailed from Delagoa Bay.[90]

The reader may ask the object which this deputation had in view. Was it that our Governments relied on foreign intervention? Emphatically, no! They never thought of such a thing. Neither in his harangue to the burghers at Poplar Grove, nor in any of his subsequent speeches, did President Steyn give any hint of such an intention. The deputation was sent in order that the whole world might know the state of affairs in South Africa. It fulfilled its purpose, and was justified by its results. It helped us to win the sympathy of the nations.

Diplomatic Delegation of the Boer Republics to Europe and America, 1900, l.t.r. front: A.D.W. Wolmarans, A. Fischer, C. H. Wessels, [Cornelius Hermanus Wessels] and back: Dr. W.J. Leyds and Dr. H.P.N. Muller.[117] [91]
To Hon. C. H. Wessels   See: Cornelius Hermanus Wessels.[92] [93]

But I must return to my narrative.

A few days before the flight from Poplar Grove, I had appointed Danie Theron captain of a scouting party. I now left him and his corps behind, with instructions to keep me informed of Lord Roberts' movements, and proceeded myself to Bloemfontein. There I disposed the available forces for defence, and kept them occupied in throwing up schanzes. These schanzes were erected to the west and south of the town, and at distances of from four to six miles from it.

[87] Member of the Free State Volksraad and Executive Council.

[88] Member of the Free State Volksraad and Executive Council, and also President of the Volksraad.

[89] Member of the first Volksraad of the South African Republic.

[90] This harbour, then the only harbour in South Africa open to us, was subsequently forbidden us by the Portuguese Government, whose officials even went so far as to arrest eight hundred of our burghers (who, for want of horses, had taken refuge in Portuguese territory), and to send them to Portugal. The ports of German West Africa cannot be counted among those which were available for us. Not only were they too far from us to be of any service, but also, in order to reach them, it would have been necessary to go through English territory, for they were separated from us by Griqualand West, Bechuanaland, and isolated portions of Cape Colony. We had, therefore, during the latter portion of the war, to depend for supplies upon what little we were able to capture from the enemy.

[91] Special Diplomatic Delegation of the Boer Republics to Europe and America, 1900, l.t.r. front: A.D.W. Wolmarans, A. Fischer, C. H. Wessels, Cornelius Hermanus Wessels and back: Dr. W.J. Leyds and Dr. H.P.N. Muller.

[92] File:CHWessels VA0963.jpg - Wikimedia Commons

[93] Sir Cornelius Hermanus Wessels (26 April 1851 -1 March 1924 was a South African statesman. Cornelius Hermanus Wessels Wessels stemmed from an important Orange Free State family clan, dedicated to farming and the Boer way of life. Mainly self-educated, he turned to politics in his early thirties. He developed his skills as a diplomat and mediator, and was involved in many of the important political decisions the Orange Free State had to make in the 1880s and 1890s. In 1897 he was appointed president of the Volksraad. During the South African War (1899–1902), Wessels was first a member of the war cabinet, and later a member of the Joint Diplomatic Delegation of the Boer Republics sent to Europe and the United States to muster support for the Boer cause. After the war Wessels sided with De Wet and Hertzog and became a cabinet minister in the Free State government. In 1915, the colonial government appointed Wessels as administrator of the Orange Free State. He was awarded a knighthood for his services to the state in the 1920 New Year Honours.

On the evening of the 12th of March, Lord Roberts appeared, and a few skirmishes ensued south of the town, but no engagement of any importance took place. We awaited the morrow with various forebodings.

For myself, I believed that that 13th of March should see a fight to the finish, cost what it might! for if Bloemfontein was to be taken, it would only be over our dead bodies.

With this before my eyes, I made all necessary arrangements, riding at nightfall from position to position, and speaking both to the officers and to the private burghers. They must play the man, I told them, and save the capital at any cost. An excellent spirit prevailed amongst them; on every face one could read the determination to conquer or to die.

But when, about an hour before midnight, I reached the southern positions, I heard a very different story. They told me there that Commandant Weilbach had deserted his post early in the evening. What was I to do? It was impossible to search for him during the night, and I was compelled to take burghers away from other commandos, and to place them in the abandoned positions. On their arrival there, they discovered that no sooner had Weilbach failed us than the enemy had seized his post—the key to Bloemfontein! We did all that we could, but our situation had been rendered hopeless by the action of a Commandant who ought to have been dismissed out of hand for his conduct at Poplar Grove.

That night I did not close an eye.

The morning of the 13th of March dawned.

Hardly had the sun risen, when the English in the entrenchments which Commandant Weilbach had deserted, opened a flank fire on our nearest positions.

First one position and then another was abandoned by our burghers, who followed one another's example like sheep; few made any attempt to defend their posts, and in spite of my efforts and those of the officers under me, they retreated to the north.

Thus, without a single shot being fired, Bloemfontein fell into the hands of the English.

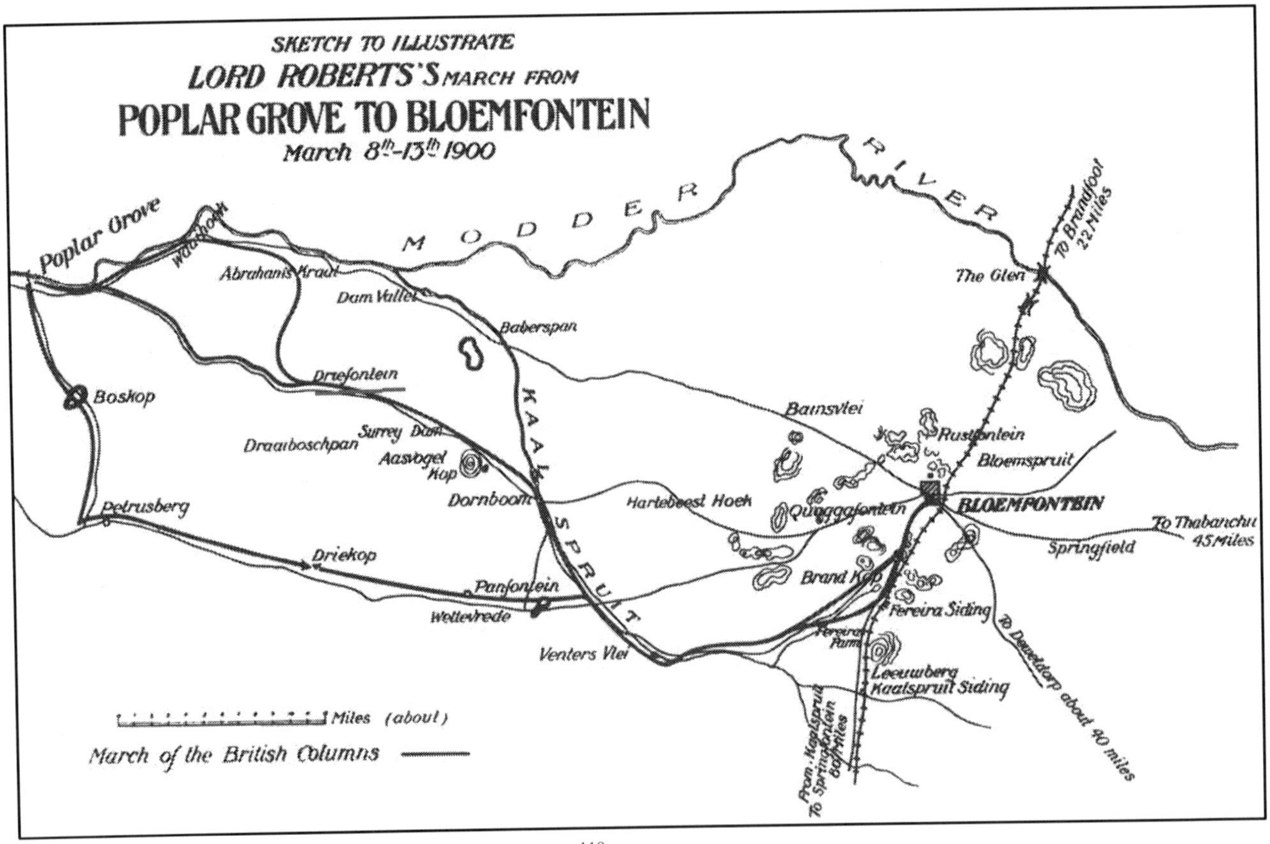

Poplar Grove to Bloemfontein. Note Driefontein.[118]

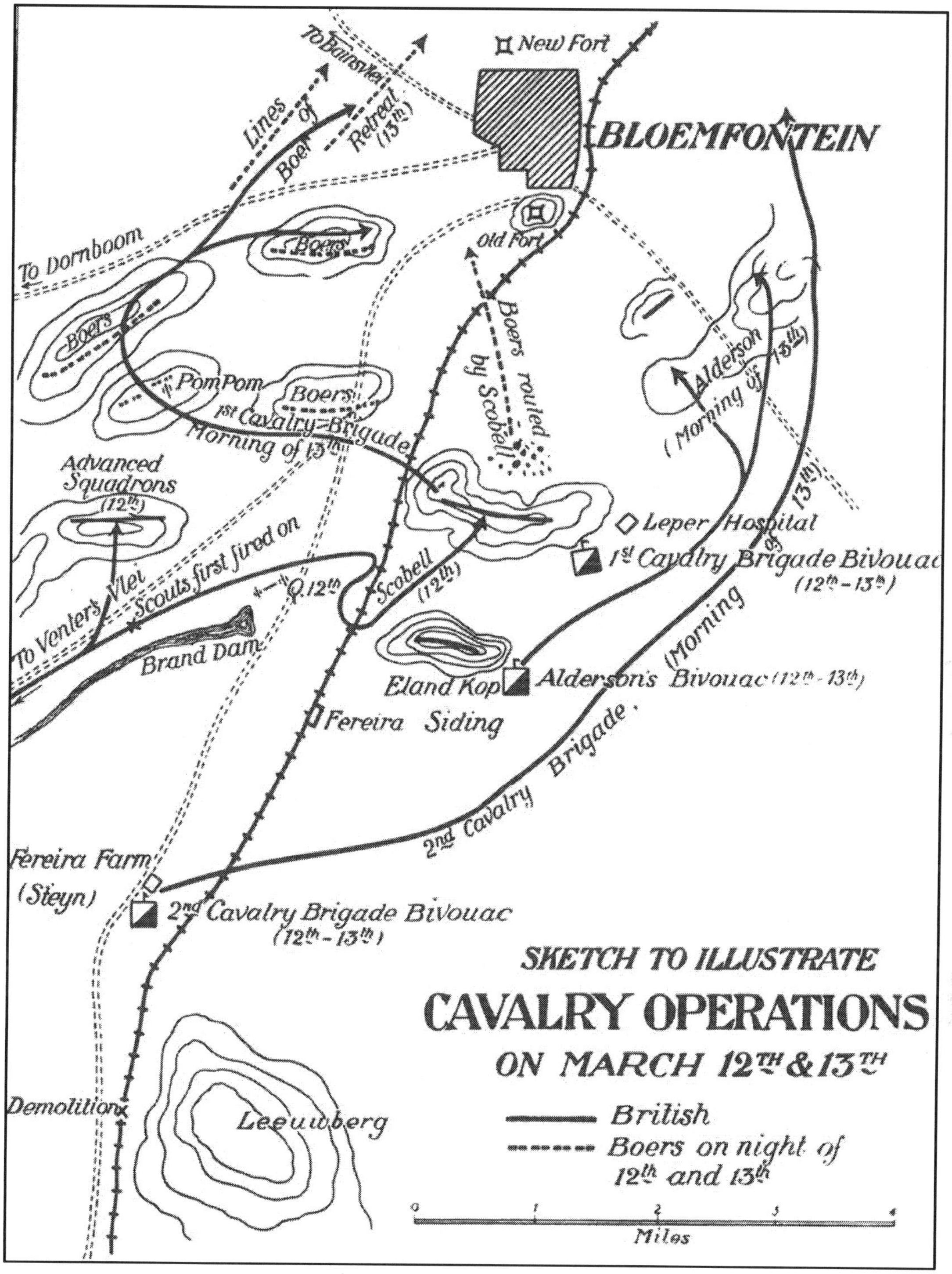

Cavalry Operations On March 12th and 13th, 1900.[119] [94]

[94] 3/12/1900 General French (with cavalry, R.H.A., and Mounted Infantry) arrived before Bloemfontein, captured the railway. 3/13/1900 Roberts occupied Bloemfontein. His despatch ran:—"The British flag now flies over the Presidency vacated last evening by Mr. Steyn, late President of the Orange Free State. The inhabitants gave the troops a cordial welcome."

General view of Bloemfontein.[120]

The Guards Brigade Entering Bloemfontein.[121]

Changing of the Guard at Presidency, Bloemfontein.[122]

8. The Burghers Receive Permission to Return to Their Homes

Thus Bloemfontein had fallen into the hands of the English; but whatever valuables it contained were spared by the enemy. I did not myself consider the place much superior to any other town, and I would not have thought it a matter of any great importance if it had been destroyed. Still, I felt it to be very regrettable that the town should have been surrendered without a shot.

How can I describe my feelings when I saw Bloemfontein in the hands of the English? It was enough to break the heart of the bravest man amongst us. Even worse than the fall of our capital was the fact that, as was only to be expected, the burghers had become entirely disheartened; and it seemed as if they were incapable now of offering any further resistance. The commandos were completely demoralized. Indeed! the burghers from Fauresmith and Jacobsdal had already returned home from Poplar Grove without asking for permission to do so; and now all the others were hurrying back in the greatest disorder to their own districts.

I felt sure that Lord Roberts' troops would remain for some time in the capital, in order to obtain the rest they must have sorely needed. And I now asked myself what I could do whilst the English were remaining inactive. For notwithstanding all that had happened, I had not for a single moment the thought of surrender. It seemed to me that my best course was to allow the burghers, who had now been away from their families for six months, an opportunity to take breath![95]

After everything had been arranged I went to Brandfort and thence to Kroonstad, at which place I was to meet President Steyn, who had left Bloemfontein the evening before it fell.

On my road to Kroonstad I fell in with General P. J. Joubert, who had come to the Free State, hoping to be able to discover some method for checking the advance of Lord Roberts. He was anything but pleased to hear that I had given my men permission to remain at home till the 25th of March.

"Do you mean to tell me," he asked, "that you are going to give the English a free hand, whilst your men take their holidays?"

"I cannot catch a hare, General, with unwilling dogs," I made reply.

But this did not satisfy the old warrior at all. At last I said:

"You know the Afrikanders as well as I do, General. It is not our fault that they don't know what discipline means. Whatever I had said or done, the burghers would have gone home; but I'll give you my word that those who come back will fight with renewed courage."

I knew very well that there were some who would not return, but I preferred to command ten men who were willing to fight, rather than a hundred who shirked their duties.

Meanwhile President Steyn had proclaimed Kroonstad as the seat of the Government, so that in future all matters were to be settled there.[96]

On March 20th, 1900, a war council was held, which was attended by from fifty to sixty officers. President Steyn presided; and there sat beside him that simple statesman, grown grey in his country's service—President Kruger.

The chief officers at this council were Commandant General Joubert, Generals De la Rey, Philip Botha, Froneman, C.P. Cronje, J.B. Wessels, and myself. A number of the members of both Governments also put in an appearance at this meeting.

[95] The men I still had with me belonged to commandos from Bloemfontein, Ladybrand, Wepener, Ficksburg, Bethlehem and Winburg. They were respectively under Commandants Piet Fourie, Crowther, Fouche, De Villiers, Michal Prinsloo and Vilonel; and these Commandants took orders from Vechtgeneraals J.B. Wessels, A.P. Cronje, C.C. Froneman, W. Kolbe and Philip Botha.

The Colesberg and Stormberg commandos had received the order to go northwards in the direction of Thaba'Nchu and Ladybrand. These commandos also had been panic-stricken since General Cronje's surrender.

The Kroonstad, Heilbron, Harrismith and Vrede burghers, under Commander-in-Chief Prinsloo, were directed to remain where they were, and guard the Drakensberg.

General De la Rey followed my example, and gave his men permission to return home for some time.

[96] 3/15/1900 General Gatacre occupied Bethulie. Boers attacked Colonel Plumer's camp and were repulsed.

3/16/1900 Fighting at Fourteen Streams.

3/19/1900 Kitchener occupied Prieska.

3/20/1900 Rouxville occupied by Major Cumming.

3/21/1900 Smithfield occupied by British troops.

Do not let it be imagined that the object we had in view was to come to an agreement on any peace proposal made by the English. Nothing could have been further from our minds than this. Lord Salisbury's letter to our two Presidents, demanding unconditional surrender, had rendered any thought of peace impossible. On the contrary, we were concerned to discover the best method of continuing the war. We knew, I need scarcely say, that humanly speaking ultimate victory for us was out of the question—that had been clear from the very beginning. For how could our diminutive army hope to stand against the overwhelming numbers at the enemy's command? Yet we had always felt that no one is worthy of the name of man who is not ready to vindicate the right, be the odds what they may. We knew also, that the Afrikanders, although devoid of all military discipline, had the idea of independence deeply rooted in their hearts, and that they were worthy to exist as a Free Nation under a Republican form of Government.

I shall not enter upon all that happened at that meeting. I shall merely note here that besides deciding to continue the war more energetically than ever, we agreed unanimously that the great waggon-camps should be done away with, and that henceforth only horse-commandos should be employed. The sad experience we had gained from six months' warfare, and more especially the great misfortune that had overtaken the big waggon-camp of General Cronje, were our reasons for this new regulation.[97]

I left the meeting firmly determined that, come what might, I should never allow another waggon-camp. But, as the reader will see before he has concluded the perusal of these pages, it was not until many months had elapsed that the waggons were finally suppressed. All the mischief that they were destined to bring upon the African Nation was not yet completed.

One of the effects of this council was to produce an unusually good spirit among the officers and burghers. There was only one thought in my mind, and only one word on every tongue: "FORWARD!"

I proceeded from Kroonstad to the railway bridge at Zand River, and remained there until the 25th of March, when the commandos reassembled. What I had foreseen occurred. The burghers were different men altogether, and returned with renewed courage to the fight. They streamed in such large numbers on this and the following days, that my highest hopes were surpassed.

It is true that certain burghers had remained behind. Such was the case with the men from Fauresmith and Jacobsdal, and with a large proportion of the commandos from Philippolis, Smith field, Wepener, and Bloemfontein. But with these burghers I was unable to deal on account of Lord Roberts' Proclamations, which made it impossible for me to compel the burghers to join the commando; and I decided that I had better wait until I had done some good work with the men I had, before I made any attempt to bring the others back to the commando.

On the 25th of March we went to Brandfort. The arrival of the burghers at the village doubled and even trebled its population. I was forced to close the hotels, as I discovered that my men were being supplied with drink. From this I do not wish the reader to infer that the Afrikanders are drunkards, for this is far from being the case. On the contrary, when compared with other nations, they are remarkable for their sobriety, and it is considered by them a disgrace for a man to be drunk.

[97] This council also enacted that officers should be very chary in accepting doctors' certificates. The old law had laid it down that if a burgher produced a medical certificate, declaring him unfit for duty, he should be exempted from service. That there had been a grave abuse of this was the experience of almost every officer. There were several very dubious cases; and it was curious to note how many sudden attacks of heart disease occurred—if one were to credit the medical certificates. I remember myself that on the 7th of March, when the burghers fled from Poplar Grove, I had thrust upon me suddenly eight separate certificates, which had all been issued that morning, each declaring that some burgher or other was suffering from disease of the heart. When the eighth was presented to me, and I found that it also alleged the same complaint, I lost all patience, and let the doctor know that was quite enough for one day. When this question of certificates was discussed at the council, I suggested in joke that no certificate should be accepted unless it was signed by three old women, as a guarantee of good faith. The system had indeed been carried to such lengths, and certificates had been issued right and left in such a lavish manner, that one almost suspected that the English must have had a hand in it!

9. Sanna's Post

On the 28th of March a council of war was held. The first business transacted referred to disciplinary matters; the council then proceeded to lay down the conditions under which the commandos were to operate. It was decided that General De la Rey with his Transvaalers should remain at Brandfort with certain Free State commandos under General Philip Botha, and that the remaining troops, under my command, should withdraw in the evening.

Great was the curiosity of the officers and burghers concerning our movements, but no man learnt anything from me. I was determined that in future my plans should be kept entirely secret. Experience had taught me that whenever a commanding officer allows his intentions to become public something is sure to go wrong, and I made up my mind to hold the reins of discipline with a firmer hand.

It is, of course, true that scarcely anything could be done without the free co-operation of the burghers. They joined the commando when they wished, or, if they preferred it, stayed away. But now I intended that the men who joined the commando should be under a far stricter discipline than formerly, and success rewarded my efforts.

We left Brandfort on the same evening. My object was to surprise the little garrison at Sanna's Post, which guarded the Bloemfontein Water Works, and thus to cut off the supply of water from that town.

I started [North] in the direction of Winburg, so as to throw every one off the scent. On all sides one heard the question, "Where are we really going? What can we have to do at Winburg?"

The following day [29th] I concealed my commando, and that evening some spies, on whom I could rely, and who were aware of my secret intentions, brought me all the information I required.

At this point I had a great deal of trouble with Commandant Vilonel. It appeared that, notwithstanding the express interdiction of the council of war, there were some thirty waggons, belonging to burghers from Winburg who were under his orders. I reminded him of the decision to which the council had come; but he replied that he did not wish his burghers to have to undergo the hardship of travelling without waggons. We started that evening, and, sure enough, there he was with his lumber following behind us.

I gave him notice in writing the next morning that he must send back the waggons that very night when we were on the march. This provoked from him a written request that a war council should be summoned to revise the decision come to at Kroonstad. I answered that I absolutely declined to do any such thing.

In the course of that day I received a number of reports. I was informed that General Olivier was driving General Broadwood from Ladybrand towards Thaba'Nchu. A little later I heard from General Froneman and Commandant Fourie how matters stood at Sanna's Post. I had disclosed my plan to them, and sent them out to reconnoitre. There were—so they told me—according to their estimation, about two hundred English troops which were stationed in such and such positions.

I at once summoned Generals A.P. Cronje, J.B. Wessels, C.C. Froneman, and Piet de Wet, and took council with them, telling them of my plans and enjoining strict secrecy. I then gave orders that Commandant P. Fourie and C. Nel, with their burghers, three hundred and fifty in number, should proceed under my command to Koorn Spruit, and be there before break of day.

We settled that Generals Cronje, Wessels, Froneman, and Piet De Wet should proceed with the remaining burghers, numbering eleven hundred and fifty, to the ridges east of the Modder River, right opposite Sanna's Post. They were to take with them the guns, of which we had four or five, and bombard Sanna's Post as soon as it was light.

The English, I expected, would retreat to Bloemfontein, and then from my position in Koorn Spruit I should be able to decimate them as they passed that ravine. I had sent a large number of burghers with the four generals so that our force might be sufficiently strong to turn General Broadwood, in case he should hear that there was fighting at Sanna's Post and come up to reinforce the garrison.

Here again I had trouble with Commandant Vilonel. I had little time to argue—the sun was already setting, and we had to be off at once. I had declined to allow a single waggon to go with me, but the Commandant declared that he would not abide by the decision of the council of war. He also refused to allow his burghers to go into positions which he himself had not reconnoitred. He asked that the attack should be postponed until he had examined Sanna's Post through his telescope.

My patience was now at an end. I told Commandant Vilonel that he must obey my orders, and that if he did not do so I should dismiss him, unless he himself resigned. He preferred to resign. My secretary procured paper, and the Commandant wrote out his resignation. I at once gave him his dismissal, and felt that a weight had been taken off my shoulders now that I was free from so wrong-headed an officer.

There was no time now for the burghers to elect a new Commandant in the usual way. I therefore assembled the Winburg commando, and told them that Vilonel had resigned, that an opportunity of choosing a substitute should be given to them later on, but that in the meanwhile I should appoint Veldtcornet Gert Van der Merve. Nobody had anything to say against "Gerie," who was a courageous and amiable man; and, after he had given orders that the waggons should be sent home, we continued our march.

Electing a Field-Cornet.[123]

I met some of my spies at a rendezvous which I had given them on the road to the Water Works, and learnt from them that the force under General Broadwood had come that evening from the direction of Ladybrand and now occupied Thaba'Nchu.[98]

I had ordered my generals to take up positions opposite Sanna's Post and east of the Modder River. I now left them and rode on to Koorn Spruit, not knowing that General Broadwood had left Thaba'Nchu after nightfall and had proceeded to the Water Works. My advance was made as quietly as possible, and as soon as we reached Koorn Spruit I hid my burghers in the ravine, placing some to the right and some to the left of the drift[99] on the road from Thaba'Nchu and Sanna's Post to Bloemfontein.

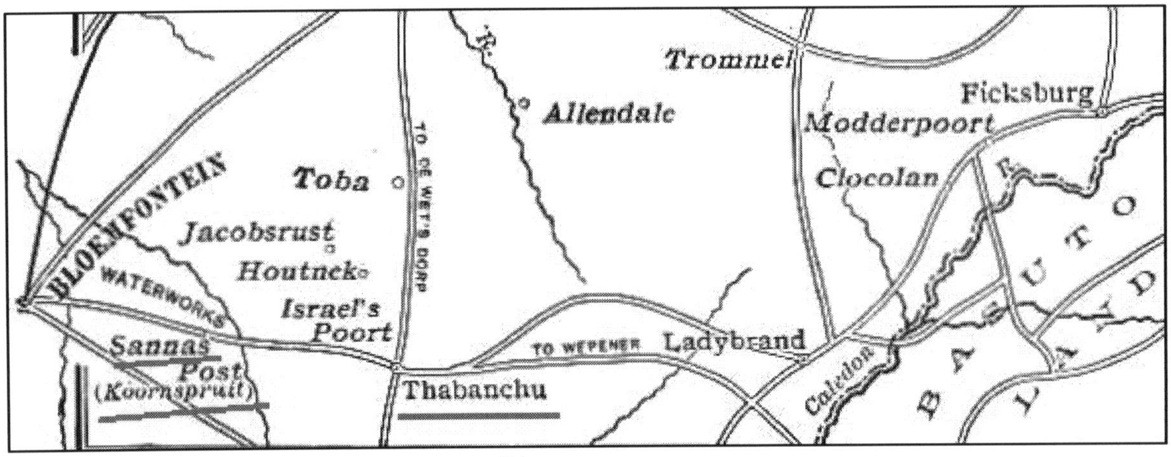

Sanna's post (Koornspruit and Thabanchu).[124]

[98] 3/27/1900 General Clements occupied Fauresmith, and arrested the landrost. Death of General Joubert.
3/29/1900 Action at Karree Siding. Boer position taken. Wepener occupied by Brabant's Horse under Colonel Dalgety.
3/30/1900 Colonel Broadwood with Cavalry Brigade and two batteries Royal Horse Artillery at Thabanchu retired on waterworks pressed by the enemy.

[99] Ford.

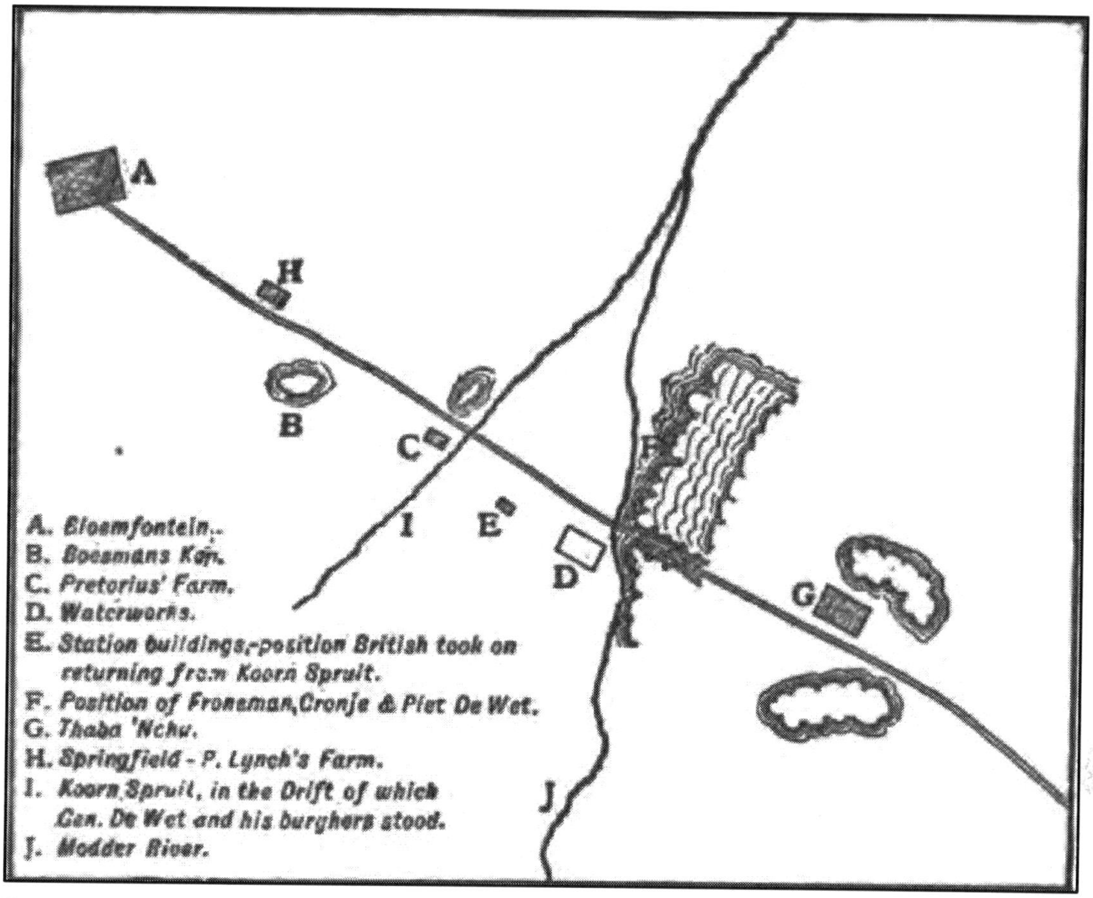

Sanna's Post. From a Sketch by the Author. This is the only map that shows the marshy ground or impoundment for the city water.

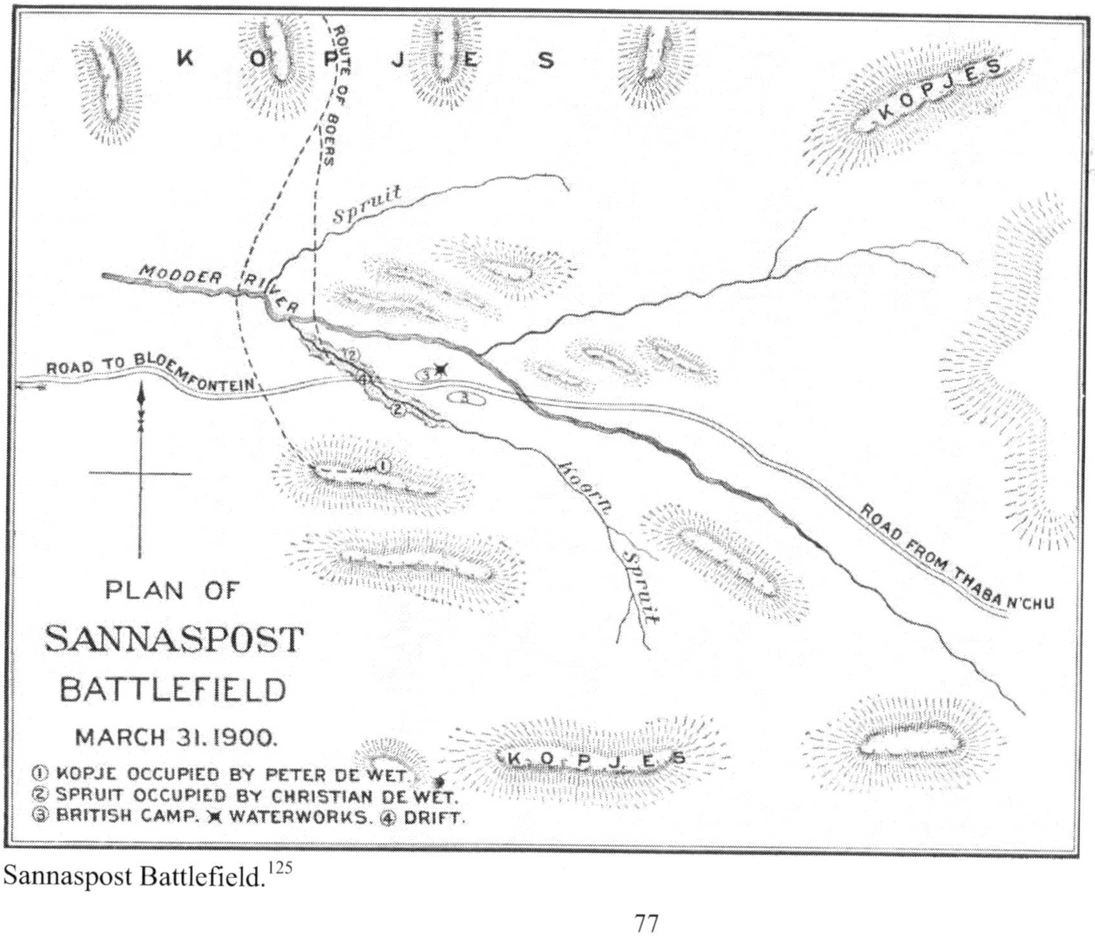

Sannaspost Battlefield.[125]

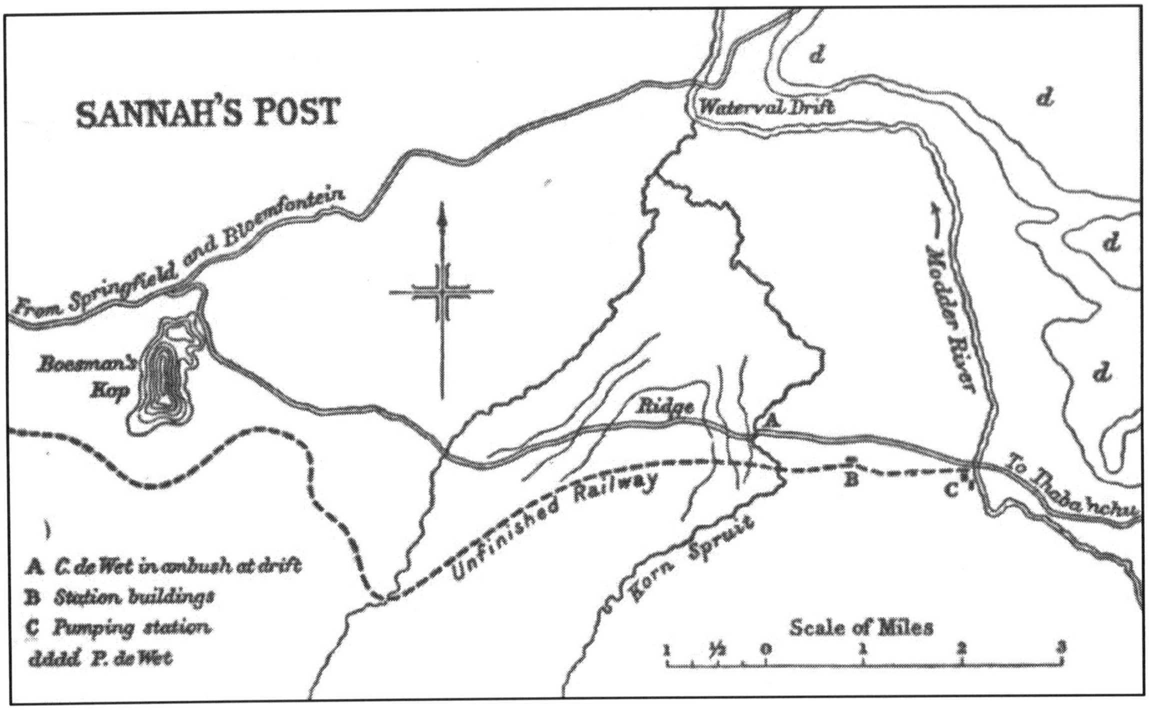

Sannah's Post.[126]

As soon as it became light enough to see anything we discovered that just above the spruit[100] stood a waggon, with some Kaffirs and a number of sheep and cattle beside it. The Kaffirs told us that the waggon belonged to one of the "hands-uppers" from Thaba'Nchu, and that they had been ordered to get it down to Bloemfontein as quickly as possible and to sell it to the English. The owner of the sheep and of the cattle, they said, was with General Broadwood, whose troops had just arrived at Sanna's Post.

The light grew brighter, and there, three thousand paces from us, was Broadwood's huge force.

I had only three hundred and fifty men with me; the other generals, to the east of the Modder River, had not more than eleven hundred and fifty between them.

The numbers against us were overwhelming, but I resolved to stand my ground; and, fortunately, the positions which I had chosen were much to our advantage—there would be no difficulty in concealing my burghers and their horses.

I ordered that every one should still remain hidden, even when our party to the east of the Modder River began to shoot, and that not a round was to be fired until I gave the command.

General Broadwood was preparing to strike camp. It was then that I told my men to allow the British troops to get to close quarters and "hands-up" them, without wasting a single bullet.

Then our guns began to fire.

The result was a scene of confusion. Towards us, over the brow of the hill, came the waggons pell-mell, with a few carts moving rapidly in front. When the first of these reached the spruit its occupants—a man with a woman beside him—became aware that something was wrong.

I was standing at the top of the drift with Commandants Fourie and Nel. I immediately ordered two of my adjutants to mount the cart and to sit at the driver's side.

The other carts came one after the other into the drift, and I ordered them to follow close behind the first cart, at the same time warning the occupants that if they gave any signal to the enemy, they would be shot.

The carts were filled with English from Thaba'Nchu. I was very glad that the women and children should thus reach a place of safety, before the fighting began.

[100] Water-course or ravine.

So speedily did the carts follow each other that the English had no suspicion of what was occurring, and very shortly the soldiers began to pour into the drift in the greatest disorder. As soon as they reached the stream they were met by the cry of "Hands up!"

Directly they heard the words, a forest of hands rose in the air.

More troops quickly followed, and we had disarmed two hundred of them before they had time to know what was happening. The discipline among the burghers was fairly satisfactory until the disarming work began. If my men had only been able to think for themselves, they would have thrown the rifles on the bank as they came into their hands, and so would have disarmed far more of the English than they succeeded in doing. But, as it was, the burghers kept on asking:

"Where shall I put this rifle, General? What have I to do with this horse?"

That the work should be delayed by this sort of thing sorely tried my hasty temper.

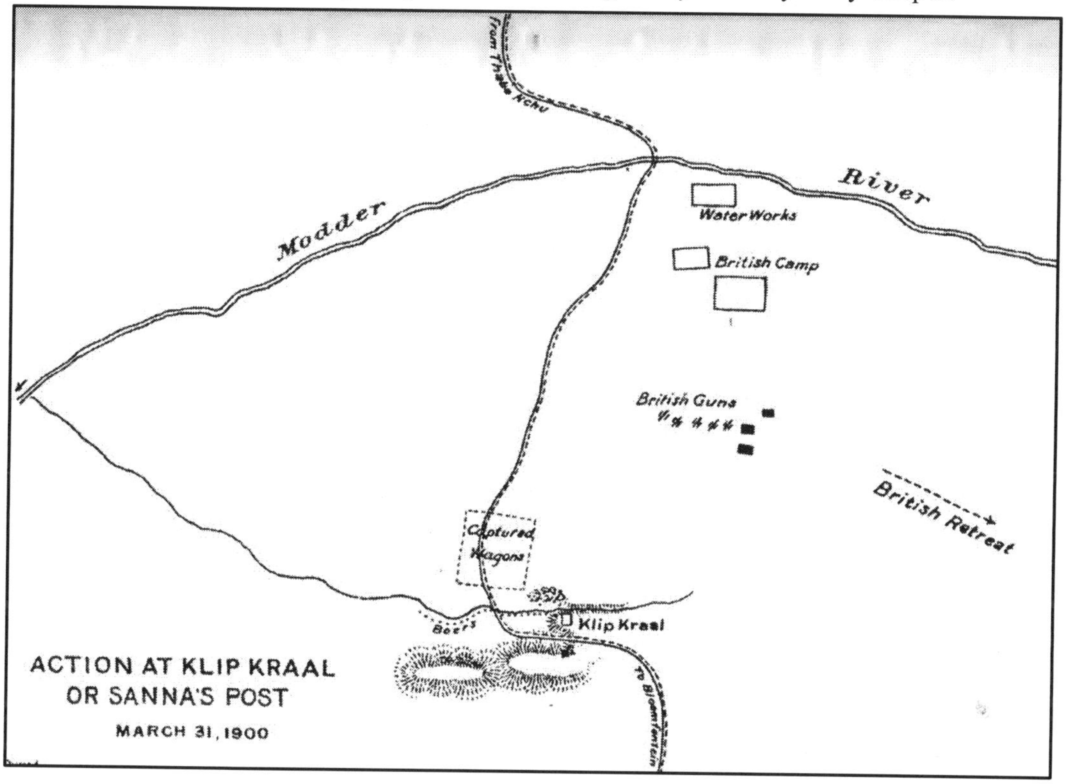

Action at Klip Kraal, Koorn Spruit, or Sanna's Post. March 31st.[127]

Very soon the enemy in the rear discovered that there was something wrong in the drift, for one of their officers suddenly gave orders that the troops should fall back. But in the meantime, as I have already stated, we had disarmed two hundred men; while, about a hundred paces from us on the banks of the spruit stood five of their guns, and more than a hundred of their waggons, in one confused mass. A little further off—two or three hundred paces, perhaps—two more of the enemy's guns had halted.

The English fell back some thirteen hundred yards, to the station on the Dewetsdorp-Bloemfontein railway. I need scarcely say that we opened a terrific fire on them as they retreated. When they reached the station, however, the buildings there gave them considerable protection. I little knew when I voted in the Volksraad for the construction of this line, that I was voting for the building of a station which our enemies would one day use against us.

An attempt was made by the English to save the five guns, but it was far beyond their powers to do so. They did succeed, however, in getting the other two guns away, and in placing them behind the station buildings. From there they severely bombarded us with shrapnel shell.

While the English troops were running to find cover in the buildings, they suffered very heavily from our fire, and the ground between the station and the spruit was soon strewn with their dead and wounded, lying in heaps. But having arrived at the railway they rallied, and posting themselves to the right and left of the station, they fired sharply on us.

The eleven hundred and fifty burghers who were to the east of the Modder River now hurried up to my assistance. But unfortunately, when they attempted to cross the river, they found that the Water-Works dam had made it too deep to ford. So they proceeded up stream over some very rough ground, being much inconvenienced by the dongas which they had to cross. When they had covered three miles of this they were again stopped, for an impassable donga blocked the way. They had therefore to retrace their steps to the place whence they had started. Ultimately they crossed the river below the dam, in the neighbourhood of the waggon-drift.

This delay gave General Broadwood a good three hours in which to tackle us. And had it not been for the excellent positions we had taken on the banks of the spruit, we would have been in a very awkward predicament. But, as it was, only two of my men were hit during the whole of that time.

As soon as our reinforcements had crossed the river, General Broadwood was forced to retire; and his troops came hurrying through Koorn Spruit both on the right and on the left of our position. We fired at them as they passed us, and took several more prisoners. Had I but commanded a larger force, I could have captured every man of them. But it was impossible, with my three hundred and fifty men, to surround two thousand.

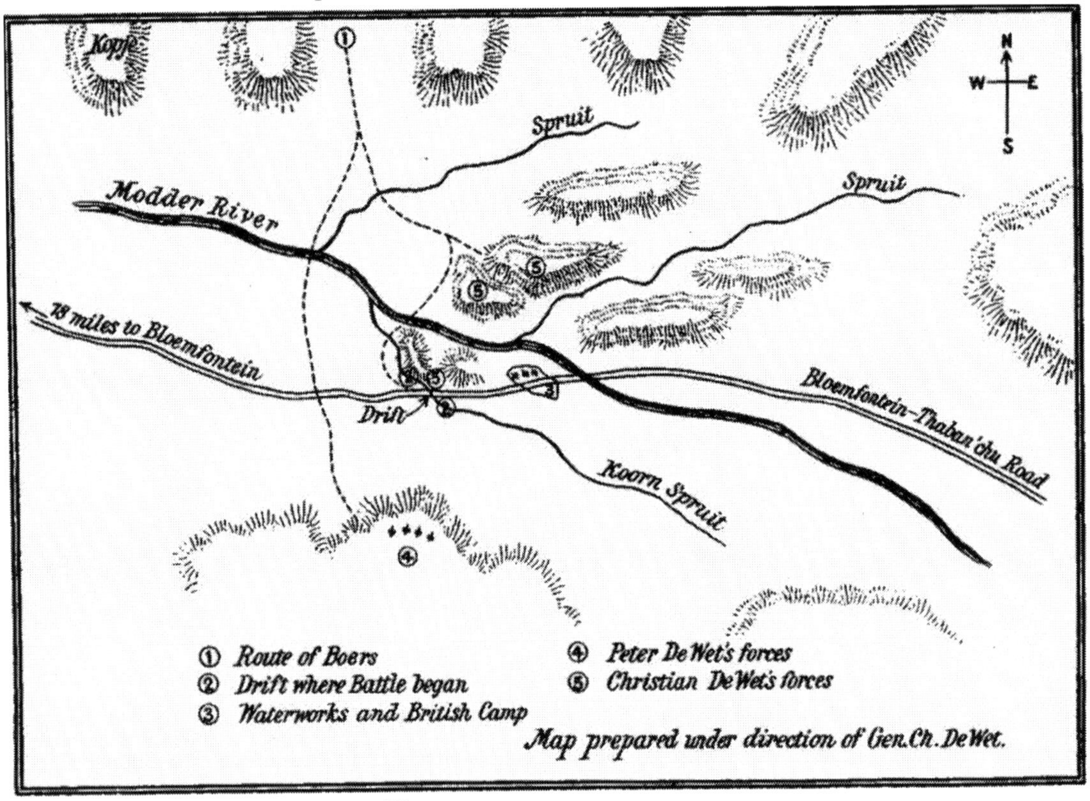

Map of Sannaspost engagement.[128] Note Koorn Spruit.

Our men on the Modder River now attacked the enemy with the greatest energy, and succeeded in putting them to flight, thus bringing the battle to an end.

The conduct of my burghers had been beyond praise. I had never seen them more intrepid. Calm and determined, they stood their ground, when the enemy streamed down upon them like a mighty river. Calm and determined they awaited their arrival, and disarmed them as they came. It was a fresh proof to me of the courage of the Afrikander, who indeed, in my judgment, is in that quality surpassed by no one.

Our loss was three killed and five wounded. Among the latter was Commandant General Van der Merve, who, although very seriously injured, fortunately recovered. I had no time myself to note the enemy's losses, but, from their own report, it amounted to three hundred and fifty dead and wounded. We captured four hundred and eighty prisoners, seven guns, and one hundred and seventeen waggons.

		Killed		Wounded		Missing and Prisoners		
Date	Action	Officers	Men	Officers	Men	Officers	Men	Total
31-Mar	Sanna's Post	3	15	16	121	18	408	581

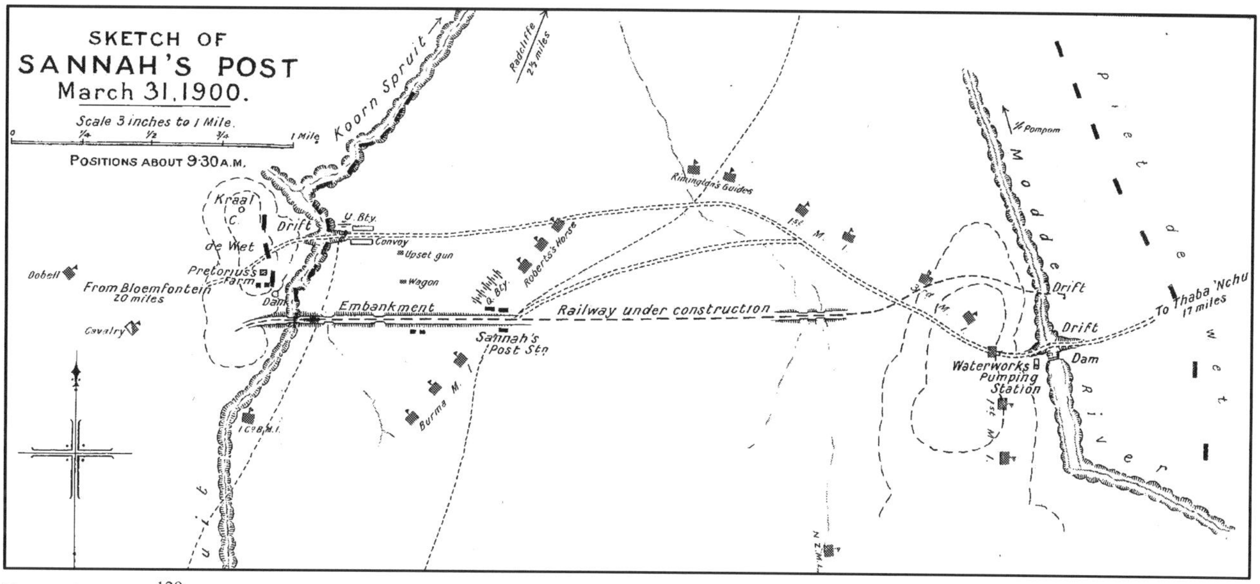

Sanna's Post.[129]

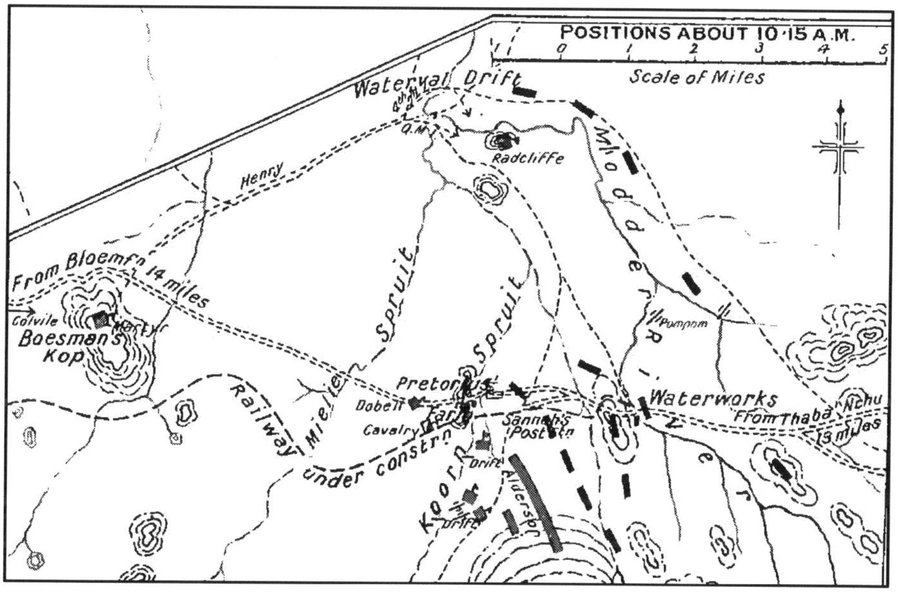

Sanna's Post.[130]

Here again I had the greatest trouble in unravelling the medley. Many of the horses, mules and oxen had been killed, whilst some of the waggons were broken. Everything was in a state of indescribable confusion, and at any moment a force might arrive from Bloemfontein.

But, fortunately, no reinforcement appeared. Our burghers who had pursued the retreating English, saw, at about twelve o'clock, a body of mounted troops approaching from Bloemfontein. But this force at once came to a halt, remaining at the spot where we had first seen it.[101]

When everything was over a party of troops from General Olivier's commando arrived on the scene of the recent operations. They had been following General Broadwood, and on hearing the firing that morning, had

[101] I may note here that it seemed very strange to me and to all whose opinion I asked, that Lord Roberts, with his sixty thousand men, sent no reinforcements from Bloemfontein. The battle had taken place not more than seventeen miles from the capital, and it had lasted for four hours; so that there had been ample time to send help. The English cannot urge in excuse that, owing to our having cut the telegraph wire, Lord Roberts could know nothing of General Broadwood's position. The booming of the guns must have been distinctly heard at Bloemfontein, as it was a still morning. In addition to this plain warning, the English had an outpost at Borsmanskop, between Koorn Spruit and Bloemfontein. I do not mention these things with the object of throwing an unfavourable light upon Lord Roberts' conduct, but merely to show that even in the great English Army, incomprehensible irregularities were not unknown, and irregularities of such a character as to quite put in the shade the bungles we were sometimes guilty of. But the Republics, young though they were, never thought of boasting about the order, organization, or discipline of their armies; on the contrary they were perhaps a little inclined to take too lenient a view when irregularities occurred.

hastened in our direction, maintaining on their arrival, that it was quite impossible for them to have come any sooner.

Calling for volunteers to man cannon captured at Sannaspost.[131]

Corps of the State Artillery, Orange Free State. Telegraphists, heliographers and artillerymen commanded by Lieut. Baay. Commando of General Olivier outside Colesberg.[132]

Sanna's Post prisoners.[102]

British prisoners of war en route for Pretoria[103]

[102] A view taken looking down the train. The old burgher in the front, was one of those in charge. The irrepressible 'Thomas' enlivened the proceedings by rendering 'Rule Britannia' and other national songs along the way. Anglo Boer War - Images from 'On the Transvaal border' published by W H Gill, 1900 - Page 2 - Boer War Forum Credit Dr. David Biggins

10. Four Hundred and Seventy English Taken Prisoner at Reddersburg

In the evening of the day on which the events described in the last chapter occurred, (March 31st) I handed over the command to Generals Piet de Wet and A.P. Cronje, and taking with me three of my staff, rode to Donkerpoort, in the direction of Dewetsdorp, on a reconnoitring expedition.

Andries Petrus Johannes Cronjé.[133]

General Peter De Wet.[134]

Early the following morning I came to a farm called Sterkfontein, where, at noon, I received the news that a party of English, coming from Smithfield, had occupied Dewetsdorp. [De Wet's 'town']

It was thirty miles from Sterkfontein to my commando, but, notwithstanding this, I sent an order that 1,500 men, under Generals J.B. Wessels, C.C. Froneman and De Villiers, should come up with all haste and bring three guns with them.

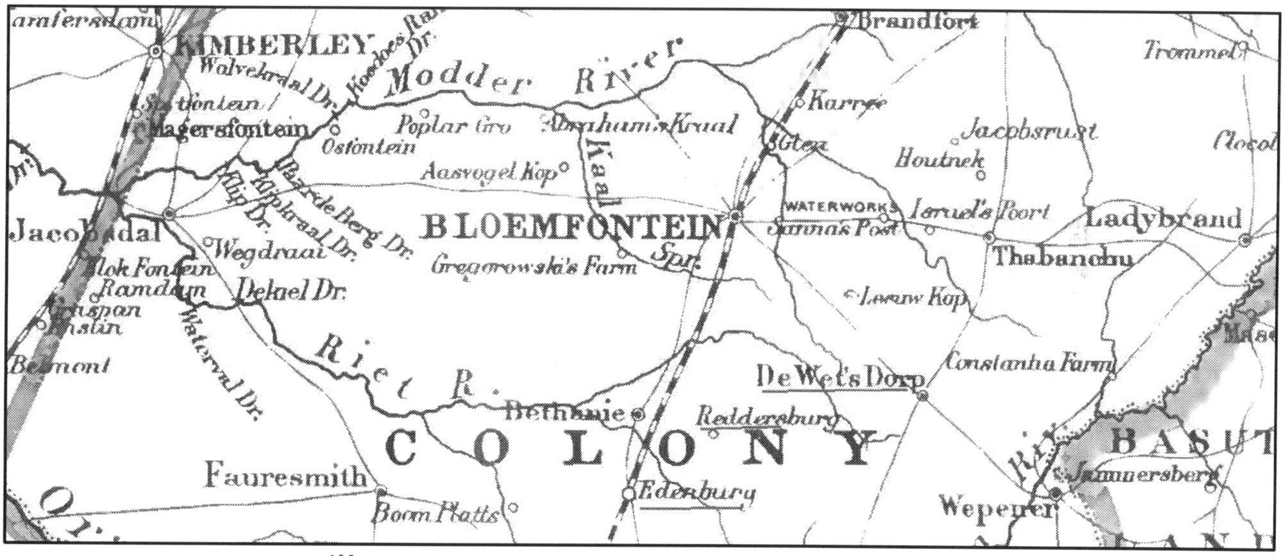

Orange River Colony South.[135] Dew Wet's Dorp middle right. Reddersburg above Edenburg.

103 The men on this train were captured at Sanna's Post. The Boer guards allowed gifts of tobacco, cigarettes etc. to be given to them. Conversations with the men were also permitted. Anglo Boer War - Images from 'On the Transvaal border' published by W H Gill, 1900 - Page 2 - Boer War Forum Credit Dr. David Biggins

During the time that must necessarily elapse before the arrival of this force, I sent men out to visit the farms of those burghers who had gone home after the fall of Bloemfontein, with orders to bring them back to the front.

By the evening of the 1st of April I had all the men of the district together; but it was then too late to make a start.

A CASE OF ROOT HOG OR DIE.
From the Minneapolis Journal, February 13, '01

A Case of Root Hog or Die. General Christiaan De Wet prods the boar 'The Cape Boer' with a bayonet..[136] [104]

Oath of Neutrality-Informal

In the absence of any official forms, hand-written "oaths" were penned and signed by surrendering Boer combatants. This example was, signed at Bloemfontein on 29 March 1900, two weeks after the capital fell to the British forces.

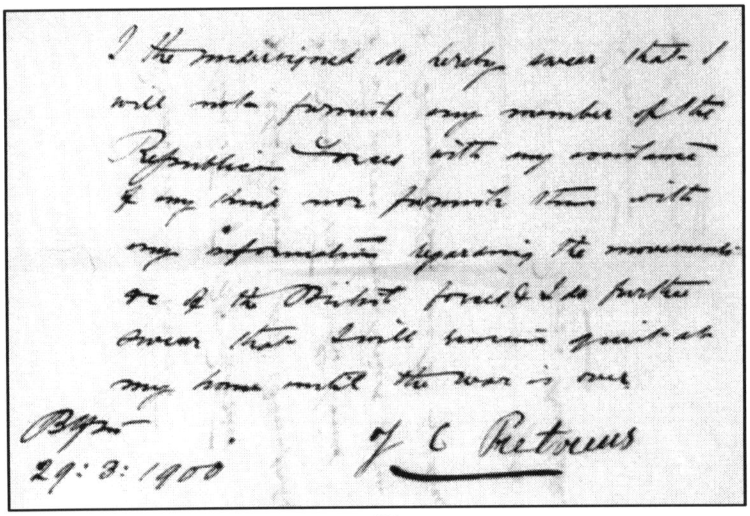

29: 3: 1900
J C Pretorius

"I the undersigned do hereby swear that I will not furnish any member of the Republican Forces with my assistance of any kind nor furnish them with any information regarding the movement etc. of the British Forces and I do further swear that I will remain quiet at my home until the war is over." J. C. Pretorius.

Handwritten Oath of Neutrality.[137]

Revocation of Oath

I, ________________, the undersigned burgher of the South African Republic, declare under oath that the oath of neutrality taken by me, in the hands of the enemy, was taken by me without the sanction of my military officers, and I consider the same null and void."[138]

[104] This cartoon is from later when De Wet was attacking the Cape Colony of British South Africa in hopes of provoking a general uprising of the Boers resident there. This from the Minneapolis Journal in the USA on Feb, 13th, 1901.

At ten o'clock the following morning the English left Dewetsdorp, and marched [South] towards Reddersburg. Directly I received news of this, I sent word to the Generals, that they must hasten to Reddersburg; while I, with the men who had rejoined, made my way to the north, so as to take up a position on the enemy's flank. I had with me one hundred and ten men in all. Many of them were without rifles, having given up their arms at Bloemfontein.

Others were provided with serviceable achterlaaiers,[105] but had little or no ammunition, because they had already fired off their cartridges in mere wantonness in the belief that they might have to give up their rifles any day. My handful of burghers were thus as good as unarmed.

During our march I kept the English continually under surveillance. They were unable to advance very rapidly, as the bulk of their force was made up of infantry. But they were too far ahead for the commandos whom I had sent in pursuit to be able to get at them; and for me, with the handful of almost unarmed burghers which I commanded, to have attempted an attack would have been worse than folly.

On the evening of the 2nd of April, the English encamped on the hill to the west of a farm called Oollogspoort; whilst we off-saddled to the north of them, on Mr. Van der Walt's farm. The enemy, however, was not aware of the position of our laager.

The following morning, at four o'clock, I sent a third report to the commandos. They had been some way on the road to Dewetsdorp, and thus, far out of the course to Reddersburg, when my second report reached them; and now my despatch rider met only Generals Froneman and De Villiers with seven hundred men and three guns, and was too late to prevent General Wessels from going on to Dewetsdorp.

Shortly after sunrise General Froneman received my report. He had been riding all night through without stopping, and many of his horses were already tired out. But as my order was that the Generals were to leave behind those who were unable to proceed, and to hasten on at once without so much as off-saddling, he did not wait to be told twice, but pushing forward with all speed, arrived on the 3rd of April at Schwarskopjes on the Kaffir River. He had left Sanna's Post on the afternoon of the previous day.

Those who consider that he was marching with seven hundred men and three Krupp guns, and that his horses were so exhausted that some of them had to be left behind, will agree with me that he did a good day's work in those twenty-four hours.

Fortunately for us, it was not at that time the habit of the English to start on their march before the sun had risen. And, by another lucky chance, our opponents were off their guard, and quite unsuspicious of attack, although they must, undoubtedly, have heard something of what had happened at Sanna's Post.

General Froneman gave me to understand that it was necessary to off-saddle the horses, and to give them a long rest, as he had been riding without any break since the previous evening.

"However necessary it may be," I replied, "it is impossible;" and I pointed out to him that if we were to delay, the English would occupy the ridge between Muishondsfontein and Mostertshoek, and thus obtain the best position. I, therefore, ordered the men to proceed with all speed, and to leave behind those who could not go on. The General did not appear to be "links"[106] at this, but called out with his loud voice, "Come on, burghers!"

We were fortunate in being able to keep up with the enemy by riding along a little plain, which was hidden from them by an intervening hill. Our course ran in a direction parallel to their line of march, and at a distance of about six miles from it. But unluckily, the English were the first to reach the ridge. When we appeared at the point where the hill which had concealed us from them came to an end, their vanguard had just passed the eastern end of the ridge at which we were both aiming; and we had still some four or five miles to go before we could reach it.

I saw that the enemy was not strong enough to occupy the whole ridge, so I at once gave orders to General De Villiers to advance, and to seize the western end at a point just above the farmstead of Mostertshoek. The enemy, observing this manoeuvre, took up their position on the eastern extremity of the ridge. Whereupon I divided the remaining burghers into small companies, with orders to occupy kopjes from six to seven hundred paces still further to the east; leaving to myself and Commandant Nel the task of seizing a small ridge which lay south-east of the English lines.

[105] As previously noted: Literally Backloaders as in breech loading rifles as opposed to a muzzle loader

[106] Vexed.

All these positions would have to be taken under fire, and before making the attempt I sent the following note to the British Commanding Officer:—

"Sir,—

"I am here with five hundred men, and am every moment expecting reinforcements with three Krupps, against which you will not be able to hold out. I therefore advise you, in order to prevent bloodshed, to surrender."

I sent this note post haste, and then rested a little while awaiting the return of the despatch rider.

And now a shameful incident occurred. The messenger had received the answer to my letter, and had covered about a hundred paces on his way back, when the enemy opened so heavy a fire upon him that it is inexplicable how he managed to come through unscathed.

The answer which he brought from the officer was in the following terms:—

"I'm d——d if I surrender!"

I at once ordered my men to rush the positions which I had already pointed out to them; and notwithstanding the fierce opposition of the enemy, they succeeded in carrying out my orders.

But although we had thus gained very good positions, those which the English held were quite as good, and perhaps even better, except for the fact that they were cut off from the water. However, when they had first become aware of our presence—that is, while they were at Muishondsfontein—they had taken the precaution of filling their water-bottles.

Our guns did not arrive until so late in the afternoon that only a few shots could be fired before it became dark.

Acting upon my orders, the burghers kept such good watch during the night that escape was impossible for the English. I also sent a strong guard to a point near Reddersburg, for I had heard that a reinforcement of from thirteen hundred to two thousand British troops had come from the direction of Bothathanie railway station, and were now encamped at Reddersburg.

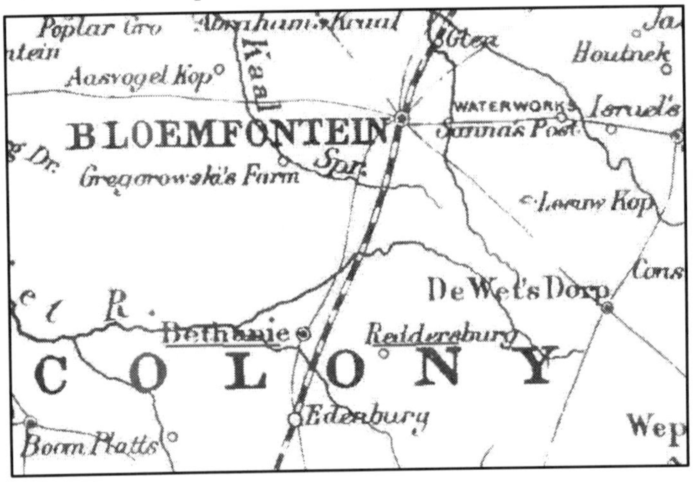

Bethanie and Reddersburg.

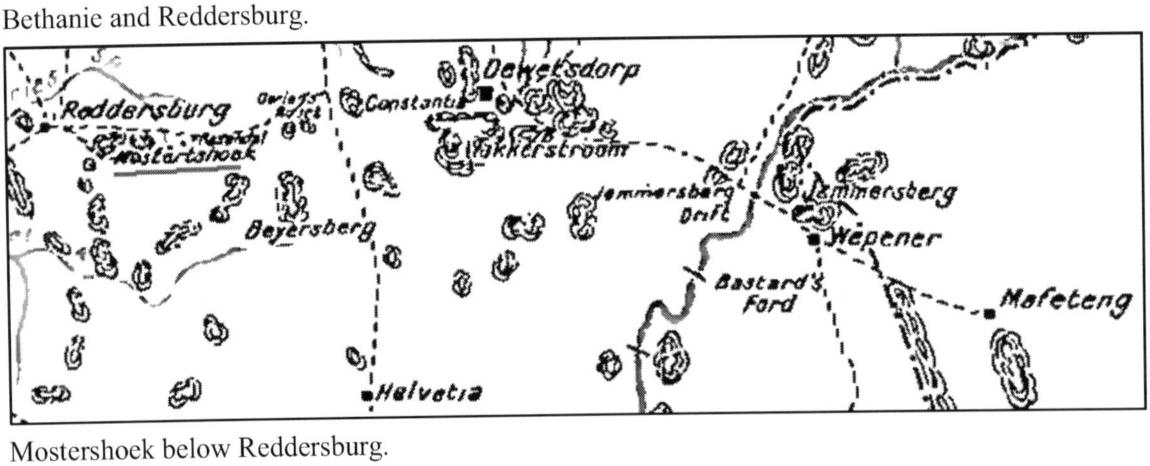

Mostershoek below Reddersburg.

I had begun operations with only four hundred men under me, but before the sun rose on the following day my force had been doubled by the addition of those who had been compelled to remain behind and rest their tired horses.

On the previous evening it had seemed to me highly improbable that we should be able to storm the ridge in the morning. I had expected that the force at Reddersburg—which lay only about four or five miles from Mostertshoek—would have seen the fight in progress, or heard the cannonading, and would have hastened to the assistance of their comrades.[107] Nevertheless, I had given orders that as soon as it was daylight, every one must do his utmost to force the English to surrender.

It was now rapidly growing lighter, and I ordered the gunners to keep up a continuous fire with our three Krupps. This they did from half-past five until eleven o'clock, and then the enemy hoisted the white flag.

My men and I galloped towards the English, and our other two parties did the same. But before we reached them, they again began to shoot, killing Veldtcornet Du Plessis, of Kroonstad. This treacherous act enraged our burghers, who at once commenced to fire with deadly effect.

Soon the white flag appeared above almost every stone behind which an Englishman lay, but our men did not at once cease firing. Indeed! I had the greatest difficulty in calming them, and in inducing them to stop, for they were, as may well be imagined, furious at the misuse of the white flag.

Strewn everywhere about on the ground lay the English killed and wounded. According to the official statement, they had a hundred casualties, the commanding officer himself being amongst the killed.

We took four hundred and seventy prisoners of war, all of them belonging to the Royal Irish Rifles and the Mounted Infantry. But I cared nothing to what regiment they belonged or what was the rank of the officer in command. Throughout the whole war I never troubled myself about such matters.

Our loss, in addition to Veldtcornet Du Plessis, whose death I have just described, was only six wounded.

7.5 cm Krupp Field Gun.[139]

Front View of 7.5 cm Krupp Gun.[140]

[107] I have never been able to understand why the great force, stationed at Reddersburg, made no attempt to come to the aid of the unfortunate victims at Mostertshoek. Their conduct seems to me to have been even more blameworthy than the similar negligence which occurred at Sanna's Post. They were not more than five miles off, and could watch the whole engagement—and yet they never stirred a foot to come and help their comrades. And it was fortunate for us that it was so, for we should have stood no chance at all against a large force.

To oppose successfully such bodies of men as our burghers had to meet during this war demanded rapidity of action more than anything else. We had to be quick at fighting, quick at reconnoitring, quick (if it became necessary) at flying! This was exactly what I myself aimed at, and had not so many of our burghers proved false to their own colours, England—as the great Bismarck foretold—would have found her grave in South Africa.

Loading Drill.[141]

Firing Drill.[142]

I had no longer any need to fear a reinforcement from Reddersburg, but nevertheless there was no time to be lost, for I had just heard from a prisoner of war that a telegram had been sent from Dewetsdorp to the garrison at Smithfield, bidding them consult their own safety by withdrawing to Aliwal North. I made up my mind to capture that garrison before it could decamp. I waited until I saw that the English ambulances were busy with their wounded, and then with all speed rode off.

As the direct road might prove to be held by Lord Roberts, I caused the prisoners of war to be marched to Winburg via Thaba'Nchu. From thence they were to be sent forward by rail to Pretoria.

		Killed		Wounded		Missing and Prisoners		
Date	Action	Officers	Men	Officers	Men	Officers	Men	Total
3-4-Apr	Reddersburg	2	10	2	33	8	397	452

PF: This is one instance where Dr. Conan Doyle's official summary is at odds with the Boer estimates.[108]

11. An Unsuccessful Siege

My object now was to reach Smithfield. We set out at once and late in the evening I divided my commandos into two parties. The first, some five hundred men in all, consisted chiefly of Smithfield burghers under Commandant Swanepoel, of Yzervarkfontein, but there were also some Wepener men amongst them. I gave General Froneman the command over this party, and ordered him to proceed without delay and attack the small English garrison at Smithfield. With the second party I rode off to join the burghers who were under General J.B. Wessels.

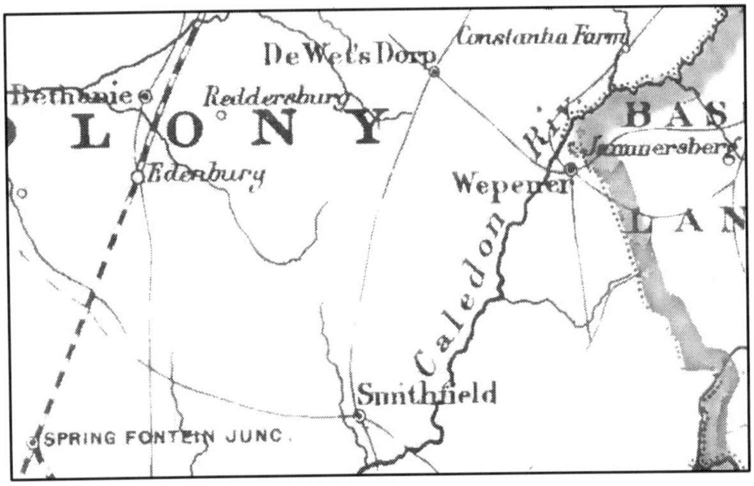

Smithfield.

[108] 4/4/1900 Troops led by General Christiaan de Wet attacked a convoy of 600 of the Royal Irish Rifles, and obtained their surrender the next day. General Gatacre is later relieved of command for not breaking up the siege in time.
4/5/1900 British occupation of Reddersburg. General Villebois killed near Boshof. General Clements received the submission of 4,000 rebels.

I came up with Wessels' division on the 6th of April at Badenhorst, on the road from Dewetsdorp to Wepener. Badenhorst lies at a distance of some ten miles from a ford on the Caledon River, called Tammersbergs drift, where Colonel Dalgety, with the highly renowned C.M.R.[109] and Brabant's Horse were at that time stationed. I call them "highly renowned" to be in the fashion, for I must honestly avow that I never could see for what they were renowned.

During the fight at Mostertshoek on the previous day I had kept them under observation, with the result that I learnt that they had entrenched themselves strongly, and that they numbered about sixteen hundred men, though this latter fact was a matter of indifference to me. The history of Ladysmith, Mafeking, and Kimberley, however, served me as a warning, and I asked myself whether it would be better to besiege the wolf or to wait and see if he would not come out of his lair.

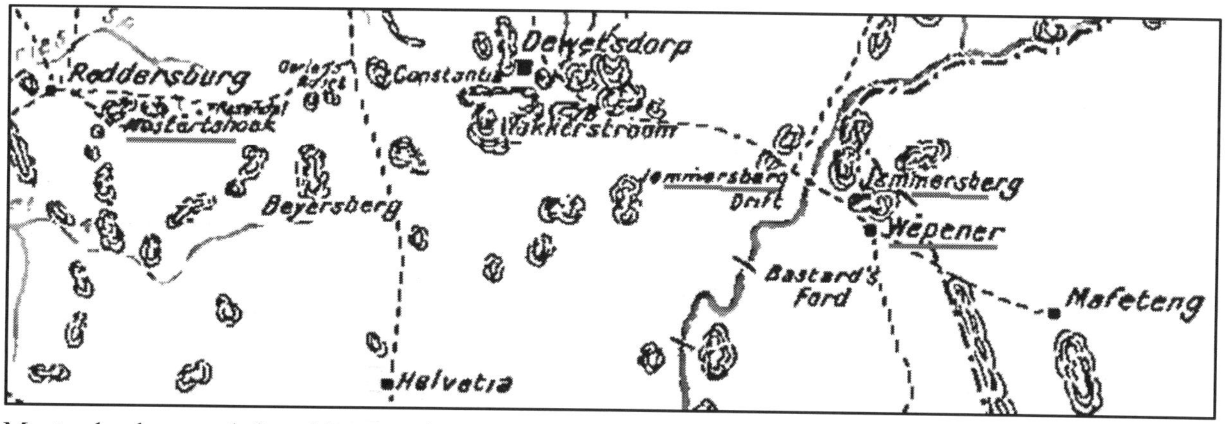

Mostershoek upper left and Jammersberg Drift on the way to Jammersberg and Wepner.

But the wolf, on this occasion, was not to be enticed out on any pretext; and moreover it was probable that Lord Roberts would be able to send a relieving force from Bloemfontein; so I decided to attack at once. First, however, I despatched some of my best scouts in the direction of Bloemfontein and Reddersburg, while I ordered the commandos under Generals Piet de Wet and A.P. Cronje to take up positions to the east and south-east of the capital.

Early in the morning of the 7th of April I made an attack on two points: one to the south-west, the other to the south-east of Dalgety's fortifications, opening fire on his troops at distances of from five to fifteen hundred paces. I dare not approach any nearer for lack of suitable cover. The place was so strongly fortified that many valuable lives must have been sacrificed, had I been less cautious than I was.

After a few days I received reinforcements, and was thus enabled to surround the English completely. But their various positions were so placed that it was impossible for me to shell any of them from both sides, and thus to compel their occupants to surrender.

Day succeeded to day, and still the siege continued.[110] [111]

Before long we had captured some eight hundred of the trek-oxen, and many of the horses of the enemy. Things were not going so badly for us after all; and we plucked up our courage, and began to talk of the probability of a speedy surrender on the part of the English.

To tell the truth, there was not a man amongst us who would have asked better than to make prisoners of the Cape Mounted Rifles and of Brabant's Horse. They were Afrikanders, and as Afrikanders, although neither Free-Staters nor Transvaalers, they ought, in our opinion, to have been ashamed to fight against us.

The English, we admitted, had a perfect right to hire such sweepings, and to use them against us, but we utterly despised them for allowing themselves to be hired. We felt that their motive was not to obtain the franchise of the Uitlanders, but—five shillings a day! And if it should by any chance happen that any one of them should find his grave there—well, the generation to come would not be very proud of that grave. No! it would be

[109] Cape Mounted Rifles.

[110] See: Wepener Account and Medal Roll by David and Alexander Biggins. Anglo Boer War - Wepener Account and Medal Roll Published by Token Publishing: Coin, Banknote and Medal Collector's Magazines. Token Publishing–Numismatic Interest

[111] 4/9/1900 Colonial Division attacked at Wepener. In April 1900, 2000 British soldiers were besieged by Gen Christiaan de Wet's forces for 17 days until reinforcements arrived on 25 April 1900. The British forces in the village were led by Colonel E.H. Dalgety.

regarded with horror as the grave of an Afrikander who had helped to bring his brother Afrikanders to their downfall.

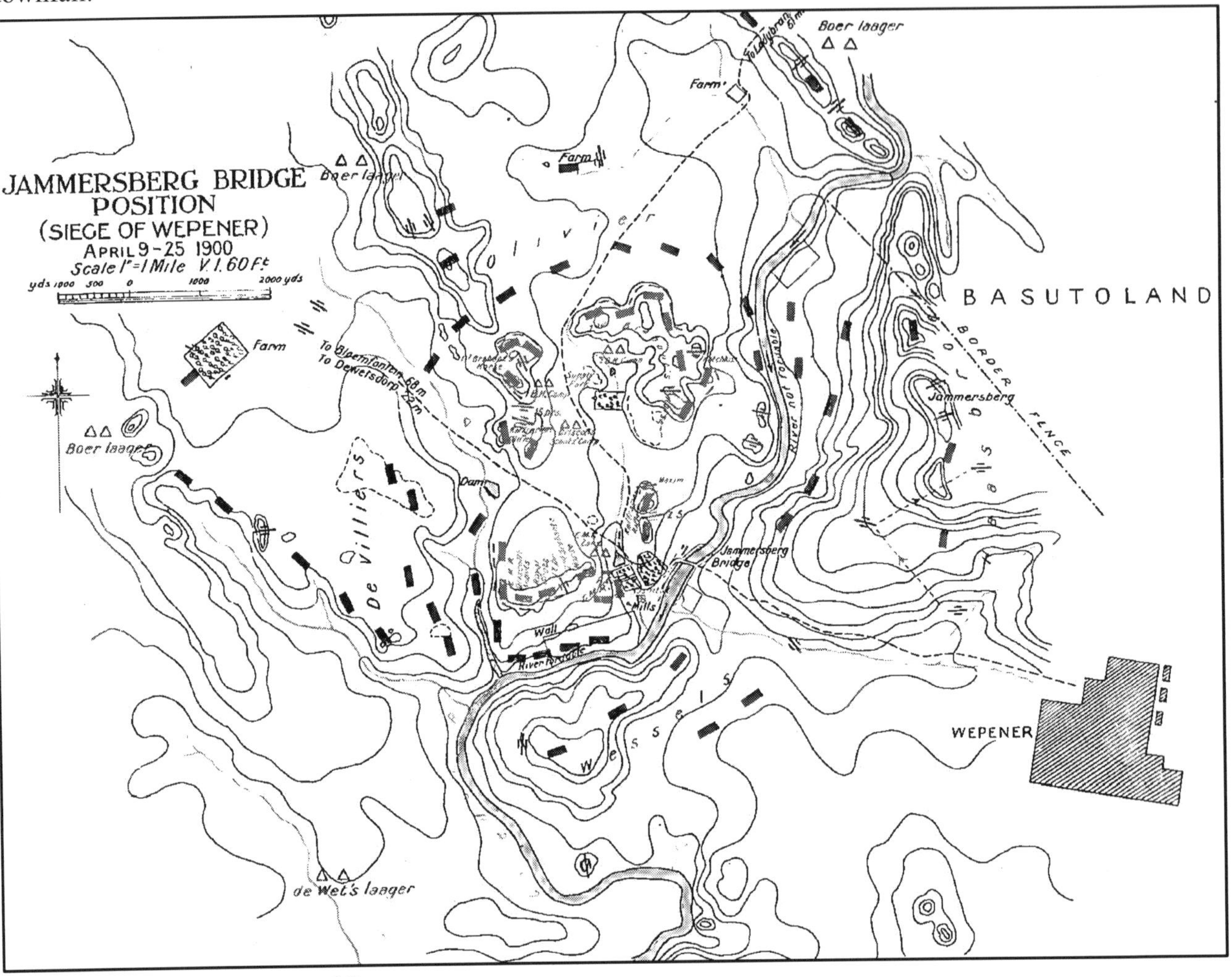

Jammersberg Bridge Position.[143]

Although I never took it amiss if a colonist of Natal or of Cape Colony was unwilling to fight with us against England, yet I admit that it vexed me greatly to think that some of these colonists, for the sake of a paltry five shillings a day, should be ready to shoot down their fellow-countrymen. Such men, alas! there have always been, since, in the first days of the human race, Cain killed his brother Abel. But Cain had not long to wait for his reward!

Brabant's Horse.[144]

Edward Yewd Brabant.[112]

[112] Major-General Sir Edward Yewd Brabant, KCB, CMG (31 May 1839–13 December 1914) was a British military commander in colonial South Africa. During the Second Boer War (1899–1902), he commanded the Colonial Division in 1900, and the Colonial

Whilst we were besieging these Afrikanders, news came that large columns from Reddersburg and Bloemfontein were drawing near. So overwhelming were their numbers that the commandos of Generals A.P. Cronje and Piet de Wet were far too weak to hold them in check, and I had to despatch two reinforcing parties, the first under Commandant Fourie, the second under General J.B. Wessels.

General Froneman had now returned from Smithfield, whither I had sent him to attack the garrison. He told me that he had been unable to carry out my orders, for, on his arrival at Smithfield, he had discovered that the garrison—which had only consisted of some two or three hundred men—had just departed. He learnt, however, that it was still possible to overtake it before it reached Aliwal North. Unfortunately, he was unable to persuade Commandant Swanepoel, who was in command of the burghers, to pursue the retreating troops. He therefore had to content himself with the fifteen men he had with him. He came in sight of the enemy at Branziektekraal, two hours from Aliwal North; but with the mere handful of men, which was all that he had at his command, an attack upon them was not to be thought of, and he had to turn back.

His expedition, however, had not been without good result, for he returned with about five hundred of those burghers who had gone home after our commandos had left Stormberg.

A Boer Commando Entering Aliwal, North.[145]

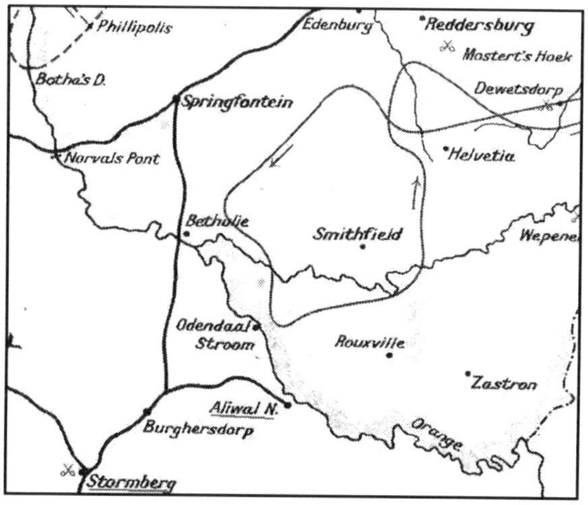

Note Stormberg and Aliwal N. on map.

We had to thank Lord Roberts for this welcome addition to our forces. The terms of the proclamation in which Lord Roberts had guaranteed the property and personal liberty of the non-combatant burghers had not been abided by. In the neighbourhood of Bloemfontein, Reddersburg, and Dewetsdorp, and at every other place where it was possible, his troops had made prisoners of burghers who had remained quietly on their farms.[113] The same course of action had been pursued by the column which fell into our hands at Mostertshoek—I myself had liberated David Strauss and four other citizens whom I had found there. While peacefully occupied on their farms they had been taken prisoners by the English column, which was then on its way from Dewetsdorp to Reddersburg.

Defence Force of Cape Colony in 1901. As a Brigadier General of the Eastern Cape troops, his command included: Cape Mounted Riflemen, the 79th Battery, RFA, the Kaffrarian Rifles, the Queenstown Volunteers, part of the 1st Battalion, Royal Scots, and Brabant's Horse. He occupied Jamestown and the Herschel district in February 1900. His units operated round the Queenstown/Dordrecht area and moved north to hold the Jammersburg Drift at Wepener, which they did under appalling rain and cold against a superior Boer force led by Christiaan De Wet. Edward Brabant - Wikipedia

See Also: He was born in 1839, and had had a long and distinguished career in politics and arms, He entered the 2nd Derby Militia as Ensign in 1855, and joined the Cape Mounted Rifles with similar rank in 1855, from which he retired on half pay with Captain's rank in 1870. He entered the arena of polities as Member of the Legislative Assembly for East London in 1873, and was re-elected in the following year. In 1875 he was appointed Field Commandant of the Cape Colonial Forces; became Colonel of the 1st Cape Yeomanry in 1879; was made CMG in 1880; was re-elected member for East London in 1882, and again in 1885. He was a member of the Defence Commission in 1896, and in 1897 was President of the South African League. General Brabant served through the Boer War, at first in command of the Colonial Division and subsequently as Inspector-General of the Colonial Defence Force, until the end of 1901, when he retired ......He resumed his duties in the Cape Parliament, and soon after seceded from his old political leader, Sir Gordon Sprigg, and joined the new Progressive party under Dr Smartt, with whom he was associated in connection with the Suspension movement. He resigned his seat in Parliament on his reappointment in Dec, 1902, to the command of the Cape Colonial Forces, from which He retired in 1904. Subsequently he re-entered the Cape Assembly as member for East London. He was a keen sympathiser with the loyalists who suffered from the effects of the war, and marked his departure from England after the Coronation by the public declaration that **"Loyalty does not pay."** Source: List of KCB recipients. Anglo Boer War - Name search

[113] See Proclamations infra.

This disregard of his proclamations did not increase the respect which the burghers felt for Lord Roberts. They felt that the word of the English was not to be trusted, and, fearing for their own safety, they returned to their commandos. I sent President Steyn a telegram, informing him that our burghers were rejoining, and adding that Lord Roberts was the best recruiting sergeant I had ever had!

General Froneman and the men whom he had collected soon found work to do. The enemy was expecting a reinforcement from Aliwal North, and I sent the General, with six hundred troops, to oppose it. He came into touch with it at Boesmanskop, and a slight skirmish took place.

In the meanwhile I received a report from General Piet de Wet, who was at Dewetsdorp, notifying me that the English forces outnumbered his own so enormously that he could not withstand their advance. He suggested that I ought at once to relinquish the siege and proceed in the direction of Thaba'Nchu.[114]

I also received discouraging news from General Piet Fourie, who had had a short but severe engagement with the troops that were coming from Bloemfontein, and had been compelled to give way before their superior forces.

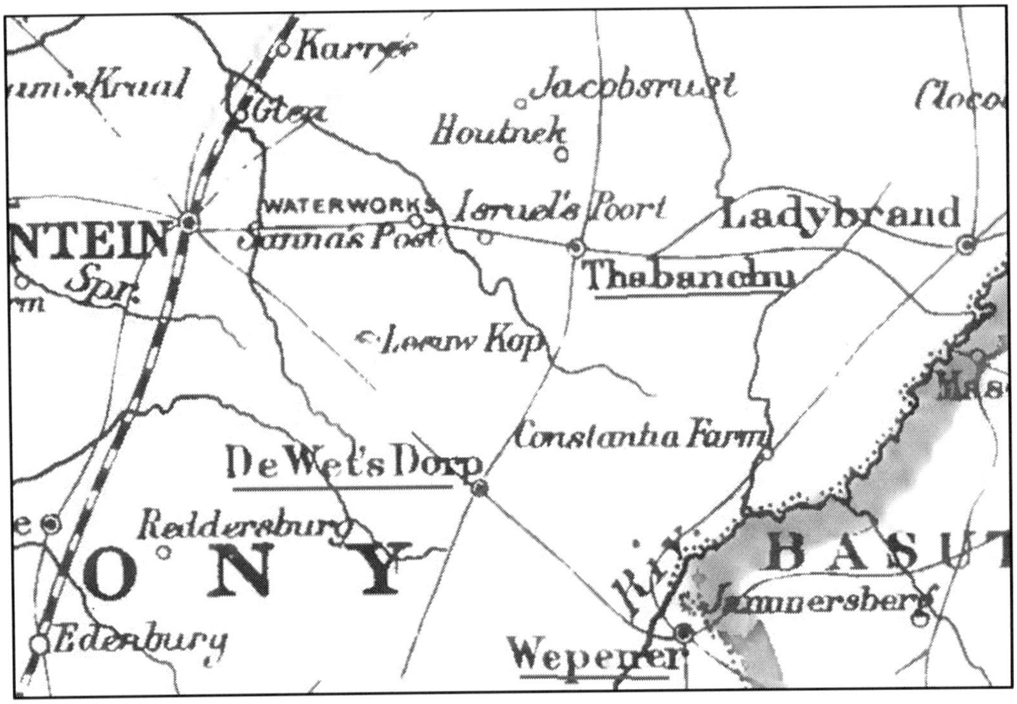

De Wet's Dorp, Wepener and Thabanchu.

Piet de Wet's advice appealed to me all the more strongly since reinforcements were pouring in upon the enemy from all sides. But I was of opinion that I ought to go with a strong force after the enemy in the direction of Norvalspont, as I was convinced that it was no longer possible to check their advance. But General Piet de Wet differed from me on this point, and held that we ought to keep in front of the English, and I was at last compelled to give in to him.

Accordingly I issued orders to General Froneman to desist from any further attack upon the reinforcement with which he had been engaged, and to join me. When he arrived I fell back on Thaba'Nchu.

My siege of Colonel Dalgety,[115] with his Brabant's Horse and Cape Mounted Rifles,[116] had lasted for sixteen days. Our total loss was only five killed and thirteen wounded. The English, as I learnt from prisoners, had suffered rather severely.[117]

[114] 4/20/1900 Boer positions attacked at Dewetsdorp.

[115] Dalgety Edmund Henry Colonel (Capt 1st Cape Yeory); CGHGSM (1) Basutoland (Capt C M Yeo); QSA (4) (Colonel CB CMR); KSA (2) Source: Wepener Siege Account and Medal Roll.

[116] See Article on Cape Mounted Rifles. https://www.angloboerwar.com/?option=com_content&view=article&id=316

12. The English Swarm Over Our Country

On April 25th we arrived at Alexandrie, six miles from Thaba'Nchu. The latter place was already occupied by English outposts. General Philip Botha now joined me; he had been engaging the enemy in the triangle formed by Brandfort, Bloemfontein and Thaba'Nchu. My commandos numbered some four thousand men, and I decided that it was time to concentrate my forces.[118]

'Village and Mountain the Thaba N'Chu.'[146]

Lord Roberts was about to carry out the plans which he had formed at Bloemfontein, namely, to outflank us with large bodies of mounted troops. He attempted to do this to the north-east of Thaba'Nchu, but at first was not successful. On a second attempt, however, he managed, after a fierce fight, to break through our lines. It was during this action that Commandant Lubbe was shot in the leg, and had the misfortune to be taken prisoner. At Frankfort also, Lord Roberts met with success, and General De la Rey was forced to retreat northwards.

I was now firmly convinced, although I kept the belief to myself, that the English would march to Kroonstad; and I could see, more clearly than ever, the necessity of operating in their rear. I had suggested to President Steyn when he had visited us at Alexandrie, that I should proceed to Norvalspont, or even into Cape Colony, but he was against any such project. This, however, was not because he disapproved of my suggestion in itself, but because he feared that the Transvaalers might say that the Free-Staters, now that their own country was in the enemy's hands, were going to leave them in the lurch.

On 2nd April Colonel Dalgety left that town to occupy Wepener and to command Jammersberg Drift and bridge over the Caledon River. Colonel Dalgety arrived at Wepener on the night of the 4th, and on the 5th the troops under his command were one company Royal Scots Mounted Infantry; CMR, 427; 1st and 2nd Brabant's Horse, 804; the Kaffrarian Rifles, 393; Driscoll's Scouts, 58; with a few Royal Engineers and the Artillery Detachment—93—of the CMR, with two 15-pounders, two naval guns, two 2'5 guns, and one Hotchkiss. A position, about six miles in circumference, was taken up. On the 6th it was seen Wepener was to be isolated, and defensive works were pushed on. On the morning of the 9th the enemy opened with artillery. In his report Colonel Dalgety said: "The weakest part of the position was on the extreme left rear, which was held by the Cape Mounted Rifles, and it was here that the heaviest casualties took place, the CMR losing 21 killed and 75 wounded, out of a total of 33 killed and 133 wounded." Colonel Dalgety gives an account in his report of the fierce attacks made by the enemy, especially on the position occupied by the CMR. He said that "it was found to be impossible to contract our lines or to give up any-portion of the position held, so that I had no reserve available for relief, and consequently for sixteen days and nights the whole force was constantly in the trenches, and in the case of the Royal Scots, CMR, and Scouts, they had nothing but cold food and water during the whole sixteen days, while for three days the trenches were flooded by rain. I cannot speak too highly of the behaviour of the whole force during all this time; all did their work cheerfully and well, although the continued strain was telling on all ranks." He brought forward the names of a number of officers, non-commissioned officers, and men of this corps [Report of Colonel Dalgety, dated 20th April 1900]. Wepener was relieved on the 24th April. During the siege Major Springer, Lieutenant Taplin, and Sergeant Major Court, and about 20 non-commissioned officers and men were killed; 7 officers and about 70 non-commissioned officers and men wounded.

[117] 4/9/1900 Colonial Division attacked at Wepener. In April 1900, 2000 British soldiers were besieged by Gen Christiaan de Wet's forces for 17 days until reinforcements arrived on 25 April 1900. The British forces in the village were led by Colonel E.H. Dalgety.

[118] 4/23/1900 General Carrington arrived at Beira.

4/25/1900 General Chermside occupied Dewetsdorp. Bloemfontein Waterworks recaptured. Wepener siege raised.

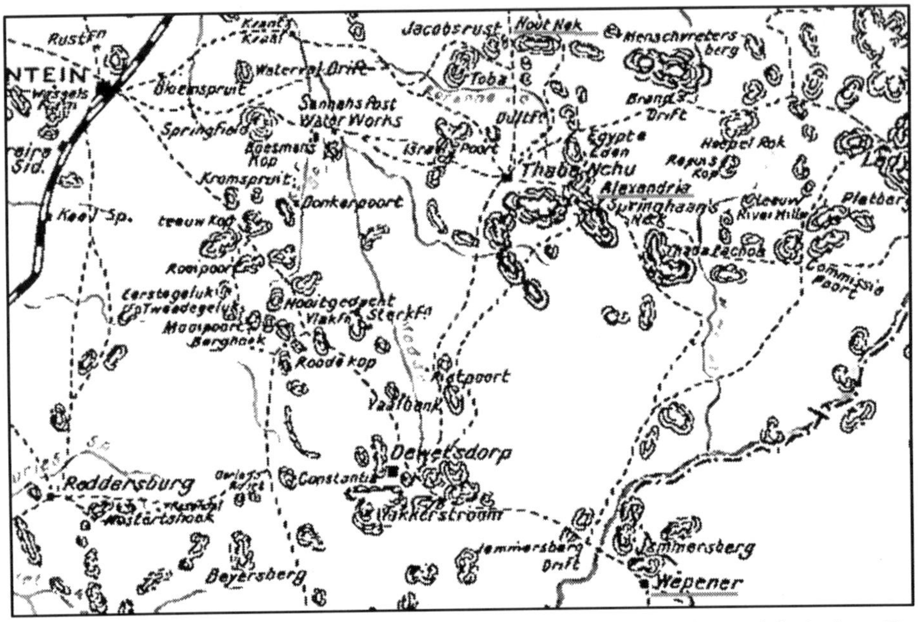

Wepner at the bottom right and Thaba'Nchu middle top. Alexandria is just East of it. Jacobsrust and Hout Nek at top center.

'President Kruger and President Steijn Addressing the Boers in the Market Hall at Kroonstad after the occupation of Bloemfontein.'[147]

Yet in spite of his opposition, I had ultimately to carry out my own ideas, for, even if I was misunderstood, I had to act as I thought best. I can only say that each man of us who remained true to our great cause acted up to the best of his convictions. If the results proved disastrous, one had best be silent about them. There is no use crying over spilt milk.

We now pushed our commandos forward to Zand River. At Tabaksberg General Philip Botha had a short but severe engagement with Lord Roberts' advanced columns. I was the last of the Generals to leave Thaba'Nchu.[119]

I was very anxious to prevent the "granary"[120] of the Orange Free State from falling into the hands of the English; with this object in view, I left behind me at Korannaberg General De Villiers, with Commandants De

[119] 4/27/1900 Town of Thabanchu occupied.
4/28/1900 Fighting near Thabanchu Mountain.

Villiers, of Ficksburg, Crowther, of Ladybrand, Roux, of Wepener, and Potgieter, of Smithfield, and ordered the General [De Villiers] to carry on operations in the south-eastern districts of the Free State.

This valiant General did some fine work, and fought splendidly at Gouveneurskop and Wonderkop, inflicting very serious losses upon the English. But nevertheless he had to yield to the superior numbers of the enemy, who ultimately gained possession of the "granary" districts. But he made them pay for it dearly.

General followed the English to Senekal and Lindley, and at Biddulphsberg, near the first named village, he again engaged them successfully, killing and wounding many of them. But a grave misfortune overtook us here, for the General received a dangerous wound on the head.

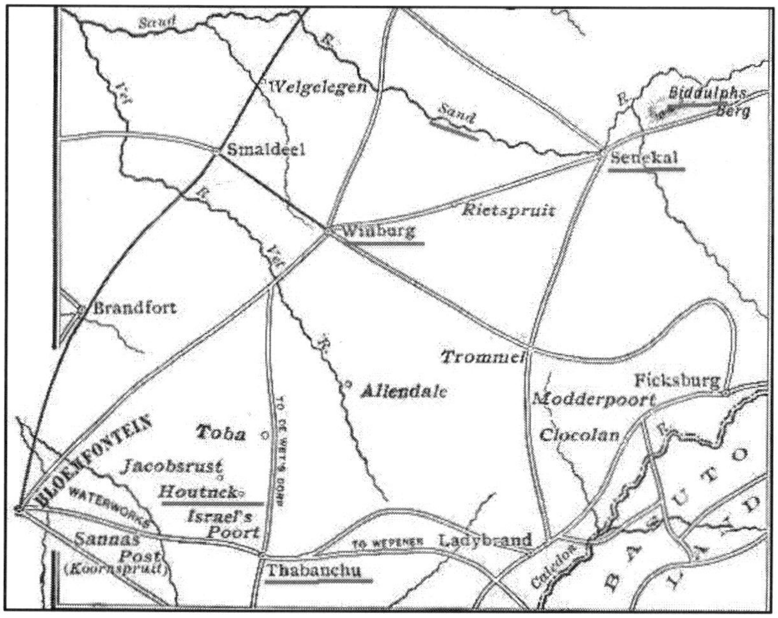

Thabanchu south, Houtnek, Windburg, Sand River, Senekal and Biddulphsberg.

'Hout Nek' is just east of Jacobsrust.

[120] This "granary" lay in the Ladybrand, Ficksburg and Bethlehem districts, and not only supplied the Free State, but also the greater part of the Transvaal. If the districts of Wepener, Rouxville, Bloemfontein, and Thaba'Nchu be included, this "granary" was the source of a very large yield of corn, and there had been an especially rich harvest that year. As the men were away on commando, the Kaffirs reaped the corn under the supervision of the Boer women; and where Kaffirs were not obtainable the women did the work with their own hands, and were assisted by their little sons and daughters. The women had provided such a large supply, that had not the English burnt the corn by the thousand sacks, the war could have been continued. It was hard indeed for them to watch the soldiers flinging the corn on the ground before their horses' hoofs. Still harder was it to see that which had cost them so much labour thrown into the flames.

In spite of the fact that the English, in order to destroy our crops, had let their horses and draught oxen loose upon the land, there was still an abundant harvest—perhaps the best that we had ever seen. And so it happened that whilst the men were at the front, the housewives could feed the horses in the stable. But Lord Roberts, acting on the advice of unfaithful burghers, laid his hand upon the housewives' work, and burnt the grain that they had stored.

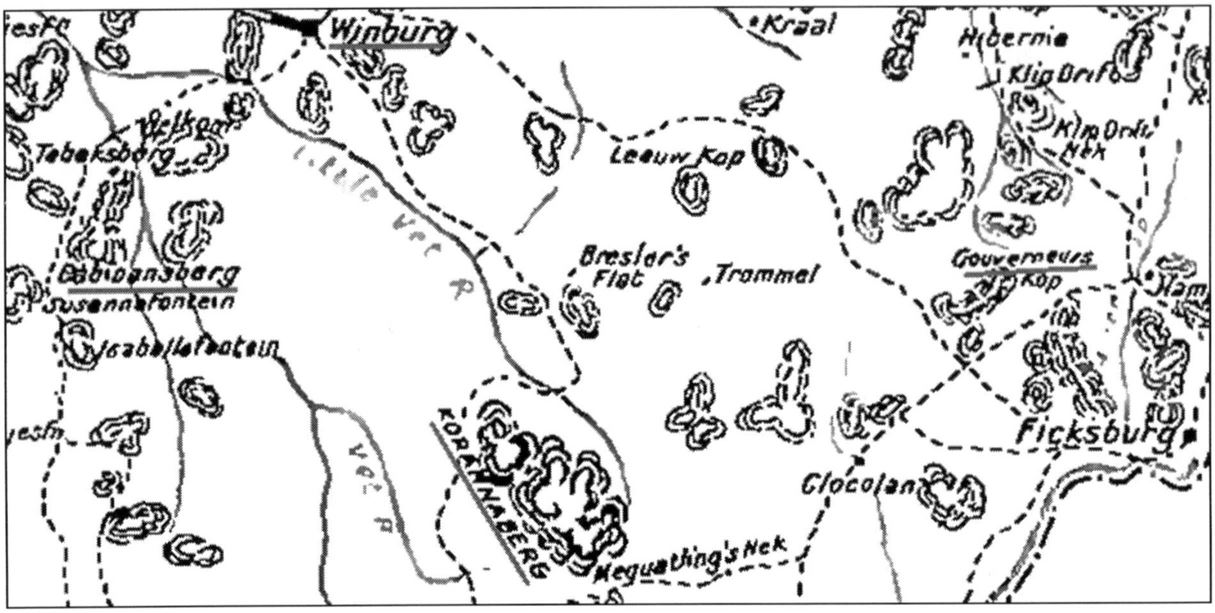

'Bididansberg' left center, Winburg, Korannaberg center bottom and Gouvernoeurs Kop center right.[121]

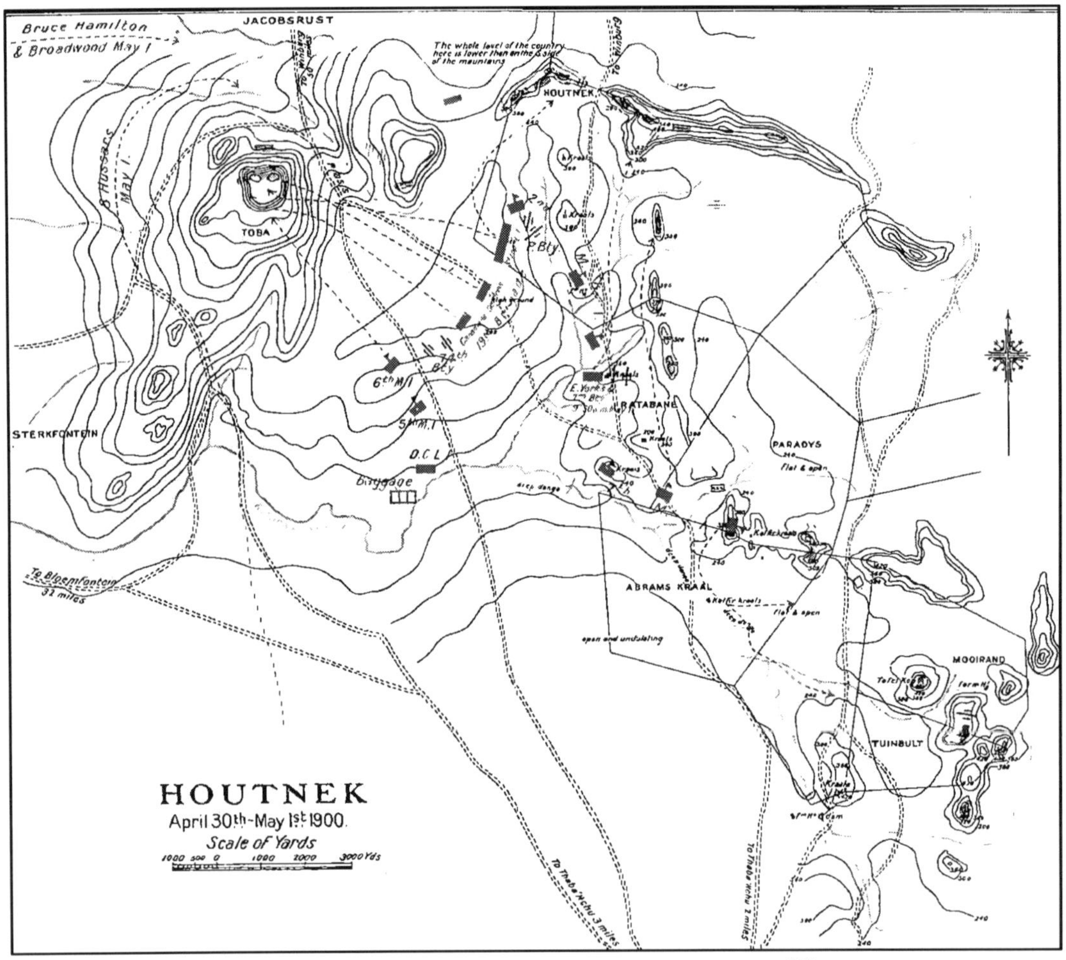

Houtnek April 30th to May 1st 1900. Houtnek is North of Thabanchu.[148]

[121] 5/1/1900 General Hamilton captured Houtnek.
5/3/1900 Hamilton's force reach "Jacobsrust (Isabellasfontein)" per WSC.
5/5/1900 British occupation of Brandfort. Roberts's further advance to the Vet River.
5/6/1900 The Vet River (Tributary of the Vaal River) passed and Smaldeel occupied.
5/7/1900 General Hunter occupied Fourteen Streams.
5/8/1900 Ladybrand deserted by the Boers.
5/9/1900 Capture of Welgelegen. Mafeking Relief Force reached Vryburg.

There was still another most deplorable occurrence. In some way or other the grass caught fire; and as it was very dry, and a high wind was blowing, the flames ran along the ground to where many of the English wounded were lying. There was no time to rescue them; and thus in this terrible manner many a poor fellow lost his life.

General De Villiers' wound was so serious, that the only course open was to ask the commanding officer of the Senekal garrison to let him have the benefit of the English doctors' skill. This request was willingly granted, and De Villiers was placed under the care of the English ambulance. Sad to say, he died of his wound.

C. J. De Villiers.[149]

Some time later I was informed that the man who had carried the request into Senekal was ex-Commandant Vilonel, who was then serving as a private burgher. A few days later he surrendered, so that one naturally inferred that he had arranged it all during his visit to Senekal.

Shortly after he had given up his arms, he sent a letter to one of the Veldtcornets, asking him to come to such and such a spot on a certain evening, to meet an English officer and himself. The letter never reached the hands of the person to whom Vilonel had addressed it; and instead of the Veldtcornet, it was Captain Pretorius with a few burghers, who went to the appointed place. The night was so dark that it was impossible to recognize anybody.

"Where is Veldtcornet—?" asked Mr. Vilonel.

"You are my prisoner," was Captain Pretorius' reply, as he took Vilonel's horse by the bridle.

"Treason! treason!" cried poor Vilonel.

They brought him back to the camp, and sent him thence to Bethlehem. A court-martial[122] was shortly afterwards held at that town, and he was condemned to a long term of imprisonment.

In the place of General De Villiers I appointed Deacon Paul Roux as Vechtgeneraal. He was a man in whom I placed absolute confidence. As a minister of religion he had done good service among the commandos, and in the fiercest battles he looked after the wounded with undaunted courage. His advice to the officers on matters of war had also been excellent, so that he was in every way a most admirable man. But his fighting career unfortunately soon came to an end, for he was taken prisoner in a most curious way near Naauwpoort, when Prinsloo surrendered.

I must now retrace my steps, and give some account of what I myself had been doing during this time.

I proceeded to the west of Doornberg, and only halted when I reached the Zand River. What memories does the name of that river bring back to me! It was on its banks that in 1852 the English Government concluded a Convention with the Transvaal—only to break it when Sir Theophilus Shepstone annexed that country on the 12th of April, 1877. But this Convention was re-established by Gladstone—greatest and noblest of English statesmen—when he acknowledged the independence of the South African Republic.

Here on the banks of this river, which was so pregnant with meaning, we should stand, so I thought, and hold the English at bay. But alas! the name with all its memories did not check the enemy's advance.

[122] This Court was not composed of officers, but consisted of three persons, one of whom was a lawyer.

On the 10th of May Lord Roberts attacked us with his united forces; and although his losses were heavy, he succeeded in breaking through our lines near Ventersburg, at two points which were held by General Froneman. And thus the English were free to advance on Kroonstad.[123]

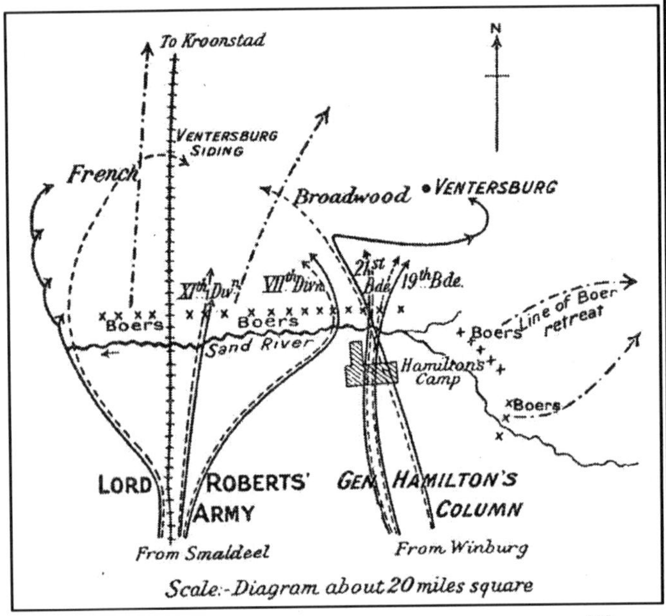

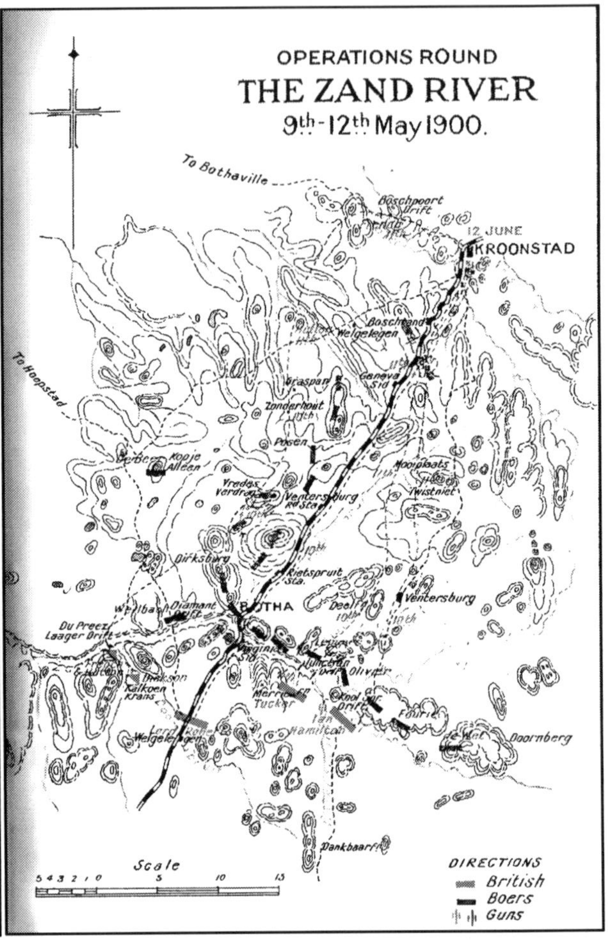

Diagram to Explain the Passage of the Sand River,[150] May 10th, 1900. The dotted lines show what was proposed, the continuous lines show what was done. The crosses indicate the Boers..[151] Operations around the Zand River.[152]

I gave orders to my commando to move on to Doornkop, which lies to the east of Kroonstad. I myself, with Commandant Nel and some of his adjutants, followed them when the sun had set. We rode the whole of that night, and reached the township on the following morning. We immediately arranged that the Government should withdraw from Kroonstad, and that very day it was removed to Heilbron. President Steyn, however, did not go to Heilbron, but paid a visit to General Philip Botha, whose commando had held back the English outposts some six miles from Kroonstad.

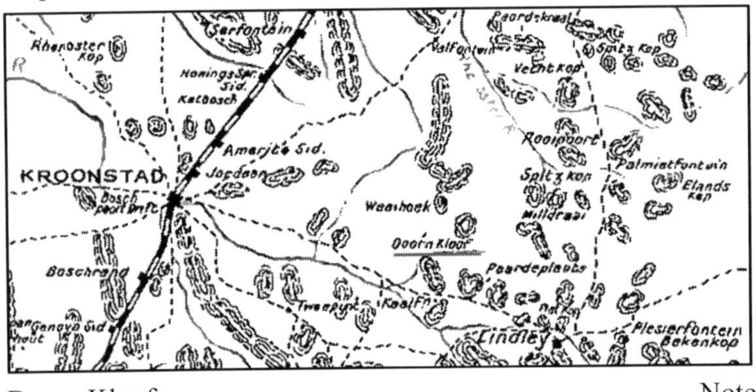

Doorn Kloof.

Note Valsch River.

[123] 5/10/1900 Battle of Zand (Sand) River. Occupation of Ventersburg.

The President, before leaving the town, had stationed police on the banks of the Valsch River with orders to prevent burghers from entering the dorp;[124] he had only just crossed the drift before my arrival. I came upon some burghers who, as they had been ordered, had off-saddled at the south side of the river, and I asked them if they had seen the President. As they were Transvaalers, they answered my question in the negative.

"But has nobody on horseback crossed here?" I said.

"Oh, yes! the Big Constable[125] crossed," one of them replied. "And he told us not to pass over the drift."

"What was he like?" I inquired.

"He was a man with a long red beard."

I knew now who the "Big Constable" had been; and when I afterwards told the President for whom he had been taken, he was greatly amused.

President Steyn [center] on his Way to the Sand River Battle.[153] Von Lossburg is on the right with the bandage on his head under his hat. Also, that looks like De Wet on the left. Pres. Marthinus Theunis Steyn.[154]

General Philip Botha discussed the state of affairs with me, and we both came to the conclusion that if Lord Roberts attacked us with his united forces, his superior numbers would render it impossible for us to hold our disadvantageous positions round Kroonstad. We had also to take into consideration the fact that my commando could not reach the town before the following day. Whilst we were still talking, news arrived that there was a strong force of cavalry on the banks of the Valsch River, six miles from Kroonstad, and that it was rapidly approaching the town.

On hearing this, I hastened back to the south of the township, where a body of Kroonstad burghers had off-saddled, and I ordered them to get into their saddles immediately, and ride with me to meet the enemy. In less time than it takes to describe it, we were off. As we drew near to the English we saw they had taken up a very good position. The sun had already set, and nothing could be done save to exchange a few shots with the enemy. So, after I had ordered my men to post themselves on the enemy's front till the following morning, I rode back to Kroonstad.

When I arrived there, I found that the last of the Transvaal commandos had already retreated through the town and made for the north. I at once sent orders to the burghers, whom I had just left, to abandon their positions, and to prepare themselves to depart by train to Rhenosterriviersbrug.

At Kroonstad there was not a single burgher left. Only the inhabitants of the township remained, and they were but too ready to "hands-up."

One of these, however, was of a different mould. I refer to Veldtcornet Thring, who had arrived with me at Kroonstad that morning, but who had suddenly fallen ill. On the day following he was a prisoner in the hands of the English.

124 Township.

125 Police Agent.

Thring was an honourable man in every way. Although an Englishman by birth, he was at heart an Afrikander, for he had accepted the Orange Free State as his second fatherland. Like many another Englishman, he had become a fellow-citizen of ours, and had enjoyed the fat of the land. But now, trusty burgher that he was, he had drawn his sword to defend the burghers' rights.

His earliest experiences were with the Kroonstad burghers, who went down into Natal; later on he fought under me at Sanna's Post and Mostertshoek, and took part in the siege of Colonel Dalgety at Jammersbergs drift. He had stood at my side at Thaba'Nchu and on the banks of the Zand River. I had always found him the most willing and reliable of officers, and he had won the respect and trust of every man who knew him.

Veldcornet Thring's Commandos near Ladysmith, Natal, South Africa, 1900.[155] [126]

He was faithful to the end. Although he might well have joined our enemies, he preferred to set the seal of fidelity upon his life by his imprisonment. Long may he live to enjoy the trust of the Afrikander people!

I remained late that evening in the town. It was somewhat risky to do so, as the place was full of English inhabitants, and of Afrikanders who did not favour our cause. In fact, I was surrounded by men who would have been only too pleased to do me an injury.

I said farewell to Kroonstad at ten o'clock that night, and was carried to Rhenosterriviersbrug, thirty-four miles from Kroonstad, by the last train that left the town. But before I departed, I took care that the bridge over the Valsch River should be destroyed by dynamite.[127]

Bridge across the Valsch River at Kroonstadt, blown up by the Boers.[156] Convoy of the 11th Division "outspanned" on the march to Kroonstadt.[157]

In the meantime, those portions of the Heilbron and Kroonstad commandos which had gone into Natal at the beginning of the war, received orders to leave the Drakensberg. Obeying these orders they joined me, and, with

[126] Thring Alfred Lester Veldkornet Awarded the DTD (Dekoratie voor Trouwe Dienst) and received the Wound Ribbon (Lint voor Verwonding). Kroonstad Commando https://www.angloboerwar.com/?option=com_content&view=article&id=3027 250 Records.

[127] 5/12/1900 Roberts occupied Kroonstad without resistance.

my other troops, had occupied splendid positions on either side of the railway line. Commandant General Louis Botha was also there with his Transvaal burghers, having arrived in the Free State a few days previously. Captain Danie Theron was still with me as my trustworthy scout, and he constantly kept me informed of Lord Roberts' movements.[128]

Field Marshal Frederick Roberts riding with his escort through the valley of the Zand River.[158] Reporters and photographers await the arrival of Roberts in Kroonstad on Saturday, May 12, 1900.[159]

Kroonstadt Camp.[160]

Arrival of the English Mail at Kroonstadt.[161]

For a few days Lord Roberts remained at Kroonstad, but about the 18th of May he again began to move his enormous forces. He sent out four divisions. The first he despatched from Kroonstad to Heilbron; the second from Lindley to the same destination; the third from Kroonstad to Vredefort and Parijs, and the fourth from Kroonstad along the railway line.

The two Governments had agreed that Commandant General Louis Botha should cross the Vaal River, [North into the Transvaal] and that we Free-Staters should remain behind in our own country. And this was carried out, with our full approval.

The Governments had also decided that even if the English entered the Transvaal, the Free State commandos were not to follow them. I had long ago wished that something of this nature should be arranged, so that we might not only have forces in front of the enemy, but also in their rear. Thus the orders of the Governments exactly coincided with my desires.

[128] 5/13/1900 General Buller advanced towards the Biggarsberg.
5/14/1900 Occupation of Dundee.
5/15/1900 Occupation of Glencoe.
5/15/1900 Christiana occupied.
5/17/1900 General Ian Hamilton occupied Lindley. Colonel Mahon, at the head of the relief force, entered Mafeking. Methuen entered Hoopstad.
5/18/1900 Occupation of Newcastle.

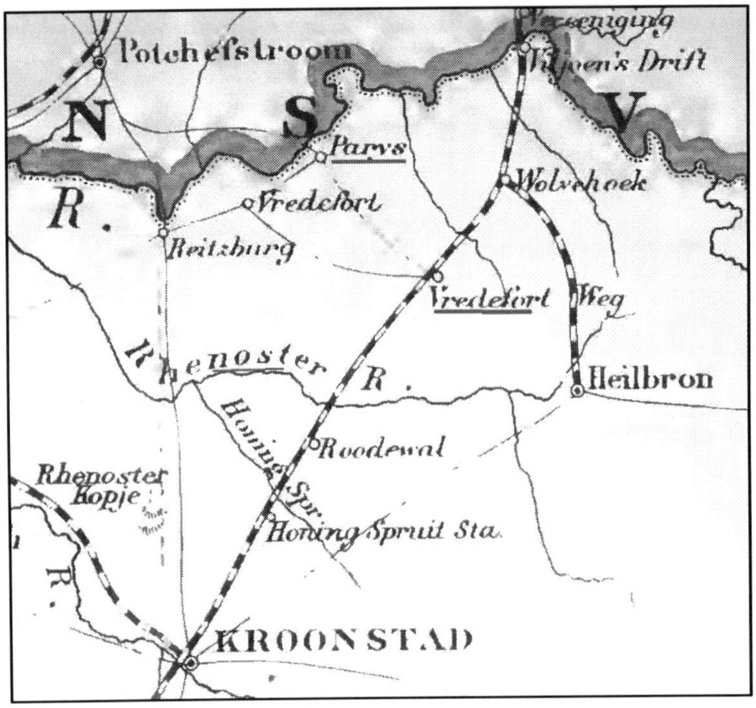

Rhenoster River north of Kroonstad and Vredefort.

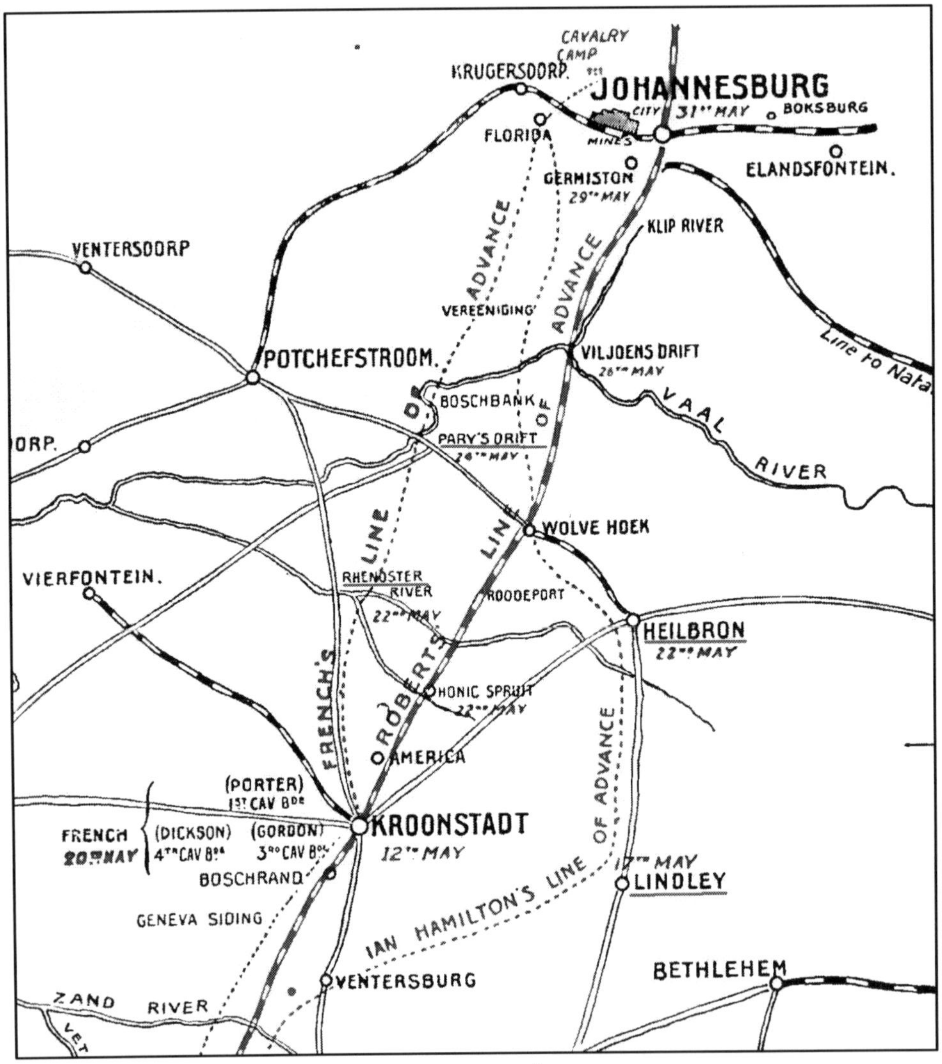

Roberts' Advance. Note Pary's Drift, May 24th and the area on the Rhenoster River between Heilborn and Lindley to the East.

Lest any one should think that the Transvaalers and the Free-Staters separated here on account of a squabble, or because they found that they could not work harmoniously together, let me state that this decision was arrived at for purely strategic reasons. We had now been reduced to a third of the original number of forty-five thousand burghers with which we had started the campaign. This reduction was due partly to Cronje's surrender, and partly to the fact that many of our men had returned to their farms. How, then, could we think of making a stand, with our tiny forces, against two hundred and forty thousand men, with three or four hundred guns? All we could do was to make the best of every little chance we got of hampering the enemy. If fortune should desert us, it only remained to flee.

To flee—what could be more bitter than that? Ah! many a time when I was forced to yield to the enemy, I felt so degraded that I could scarcely look a child in the face! Did I call myself a man? I asked myself, and if so, why did I run away? No one can guess the horror which overcame me when I had to retreat, or to order others to do so—there! I have poured out my whole soul. If I did fly, it was only because one man cannot stand against twelve.

Ambulance Train blown up at Kroonstad.[162]

The Red Cross hospital train wrecked.[163]

After the Transvaalers had crossed the Vaal River, I took twelve hundred men to Heilbron, where there was already a party of my burghers. General Roux with other Free-Staters was stationed east of Senekal, and the remainder of our forces lay near Lindley. But the commandos from Vrede and Harrismith, with part of the Bethlehem commando, still remained as watchers on the Drakensberg.

First camp inside the Transvaal.[164]

When I arrived at Heilbron, late at night, I received a report that fighting was taking place on the Rhenoster River, between Heilbron and Lindley, and that General J.B. Wessels and Commandant Steenekamp had been driven back. But on the following morning, when the outposts came in, they stated that they had seen nothing of this engagement. I immediately sent out scouts, but hardly had they gone, before one of them came galloping back with the news that the enemy had approached quite close to the town. It was impossible for me to oppose a force of five or six thousand men on the open plain; and I could not move to suitable positions, for that would involve having the women and children behind me when the enemy were bombarding me. I had therefore to be off without a moment's delay. I had not even time to send my wife and my children into a place of safety.

General Christian De Wet.[165] Mrs. Christian De Wet and children.[166] PF: The three sons who accompany him on campaign are: Kootie, Isaac, and Christiaan, see supra.

Our whole stock of ammunition was on the rail at Wolvehoek. I had given orders to Mr. Sarel Wessels, who had charge of the ammunition, to hold himself in readiness to proceed with it by rail, through the Transvaal, to Greylingstad as soon as he received orders to do so.

But now the ammunition could not remain there, as Sir Redvers Buller was gaining ground day by day towards the veldt on the Natal frontier and the am munition would thus be in danger of being taken. Therefore there was nothing left for me but to get it through by way of Greylingstad Station. It had to be done, and,—I had no carriages by which I could convey it, as I had not sufficient hands to take carriages from the trucks.[129] There was only one way (course) open; the commandos from Smithfield, Wepener and Bethulie still had, contrary to the Kroonstad resolution, carriages with them at Frankfort; I hastened to that village and sent the necessary number of these carriages under a strong escort, to fetch the ammunition from Greylingstad.

In order to do this responsible work I required a man whom I could trust. Captain Danie Theron was no longer with me, because he, being a Transvaaler, had gone with General Louis Botha. But there was another: Gideon J. Scheepers.[130]To him I entrusted the task of reconnoitring the British, so that the carriages which were going to fetch the ammunition could do in safety what they were required to do, and I knew that he would do it.

Loading Up Ammunition For The Transvaal Artillery.[167] De Amunite Trein [The Ammunition Train][168]

[129] Railway trucks.

[130] Everyone will know him, this brave man of pure Afrikander blood, subsequently a famous Commander, a martyr. I appointed him Captain of Scouts, and from the moment that he commenced his work I saw that a man had come forward. It was sad to think in what manner such a man was deprived of his life. I shall speak more of him later on, for, as our proverb says, "I had eaten too much salt" to pass over his career unnoticed.

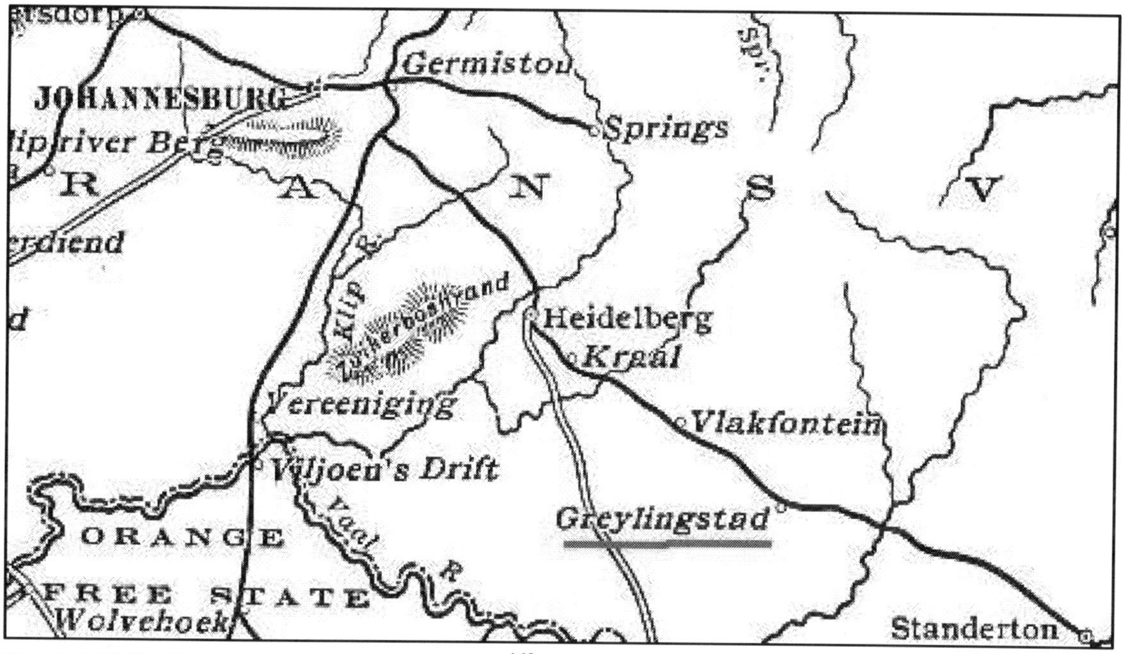

South Middle Transvaal. Note Greylingstad.[169]

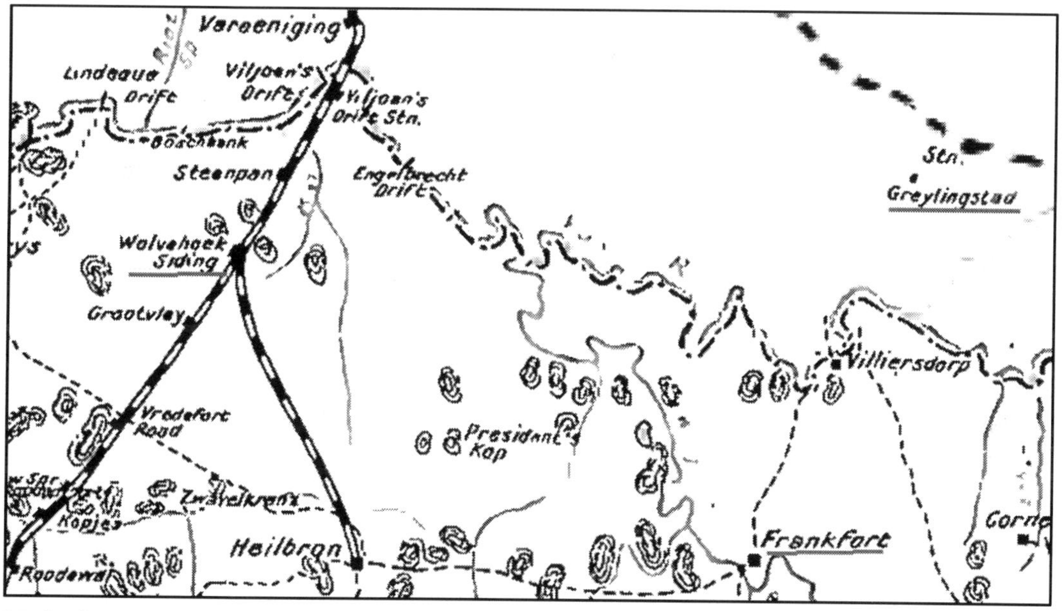

Wolvehoek Siding, Frankfort and Greylingstad in the Transvaal.

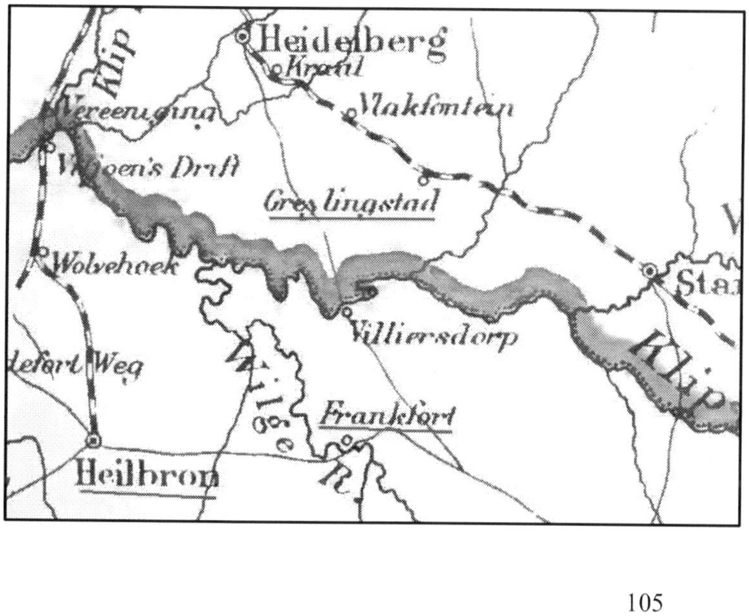

13. Our Position at the End of May, 1900

Once more it became necessary that the seat of Government should be changed, and towards the latter part of May our administrative headquarters were established at a **place between Frankfort and Heilbron**. The object of our Government in choosing this position was to be able to keep up telegraphic communication with the Transvaal. And their choice was soon to be justified, for after Johannesburg had been taken on May 31st and Pretoria on July (sic June) 5th, the only telegraphic connexion between the Free State and the South African Republic was via Frankfort, Greylingstad and Middlesburg. The terminus, at the Transvaal end, was situated not far from Pretoria.[131]

But, for the moment, it looked as if fortune were again going to smile on us, after our long spell of ill luck. On May the 31st Lindley and its garrison of Yeomanry fell into the hands of General Piet de Wet. The Yeomanry lost heavily, and five hundred of them, including, as I was told, several noblemen, were taken prisoner.[132]

These were the last prisoners of war that we were able to send into the South African Republic. Soon afterwards, when Pretoria was on the point of falling into the enemy's hands, the prisoners there had to be sent further east, but—owing either to the stupidity of the Transvaal Government, or to the treachery of the guards—a great many of them were left behind for Lord Roberts to release and re-arm against us. Our burghers grumbled much at this, and blamed the negligence of the Transvaalers.

Three Thousand British Soldiers Released from Prison at Pretoria Being Re-Armed in the Public Square.[170]

Before we had had time to get the captured Yeomanry through into the Transvaal, Sir Redvers Buller had forced his way over the Natal frontier, crossing the Drakensberg between Botha's Pass and Laing's Nek. This event, which happened on June the 17th, caused yet another panic among our commandos.[133]

"We are now," they said, "surrounded on all sides. Resistance and escape are equally impossible for us."

Never during the whole course of the war were President Steyn and I so full of care and anxiety as at this time. With Buller across our frontier, and the enemy within the walls of Johannesburg and Pretoria, it was as much as we could do to continue the contest at all. However brave and determined many of our burghers and officers might be, and, in fact, were, our numerical weakness was a fact that was not to be got over, and might prove an insuperable obstacle to our success. Moreover, the same thing was now going on in the Transvaal after the capture of Pretoria, as we had witnessed in the Free State after the fall of Bloemfontein—nearly all the burghers

[131] 5/29/1900 Battle of Doornkop: Boers defeated. https://en.wikipedia.org/wiki/Battle_of_Doornkop

5/30/1900 Occupation of Utrecht by General Hildyard. Charles Warren defeated the enemy near Douglas.

[132] 5/31/1900 Battalion of Irish Yeomanry captured at Lindley. Gen. Piet de Wet defeated the 13th Battalion Imperial Yeomanry, killing and wounding 80 soldiers and capturing 530 British soldiers.

[133] 6/12/1900 Almond's Nek having been forced the previous day, the Boers evacuated Laing's Nek and Majuba at nightfall, and General Buller encamped four miles north of Volksrust. The battle of Diamond Hill. Roberts defeated Botha 15 miles east of Pretoria. The Boers retreated in the night farther east.

were leaving their commandos and going back to their farms. Plenty of officers, but no troops! This was the pass to which we were come.

It was only the remembrance of how the tide had turned in the Free State that gave us the strength to hold out any longer.

President Steyn and I sent telegram after telegram to the Government and to the chief officers, encouraging them to stand fast. Meanwhile the two Generals, De la Rey and Louis Botha, were giving us all a splendid example of fortitude. Gazing into the future unmoved, and facing it as it were with clenched teeth, they prosecuted the war with invincible determination.

That the reader may the better appreciate the actual condition of our affairs at this time, I think it well to make a short statement as to the various districts of the Orange Free State, and the number of men in each on whom we could still rely!

Botha's Pass and Laings Nek.

Alleman's Nek Between the passes.

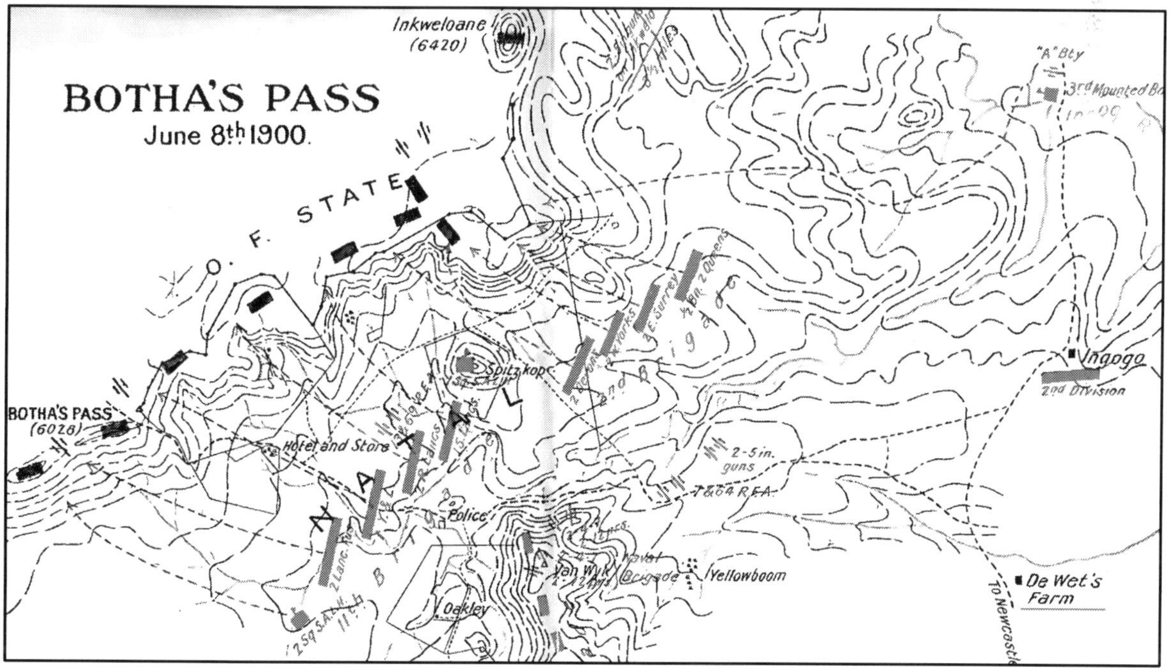

Action at Botha's Pass, June 8th, 1900.[171] Note 'De Wet's Farm' lower right.

The burghers of Philippolis and Kaapstad had surrendered en masse to the English. In the first named of these districts, only Gordon Fraser and Norval, in the second only Cornelius du Preez and another, whose name has escaped my memory, remained loyal to our cause. I mention these men here, because their faithfulness redounds to their everlasting honour.

In the district of Boshof, we could still reckon on Veldtcornet Badenhorst,[134] and twenty-seven men.

Jacobsdal was represented by Commandant Pretorius (who had succeeded Commandant Lubbe, after the latter had been wounded and taken prisoner at Tabaksberg), and forty men.

In the district of Fauresmith, Commandant Visser and some seventy men had remained faithful.

In Bethulie, Commandant Du Plooij, with nearly a hundred men, were still in arms.

Bloemfontein was represented by Commandant Piet Fourie and two hundred burghers.

The commandos of Rouxville, Smithfield, Wepener and Ladybrand, fell far short of their full complement of men, as a great number had remained behind at home.

Of the burghers from Winburg, Kroonstad and Heilbron, many had already laid down their arms, and the drain upon our troops in these districts was still continuing.

None of the burghers belonging to the districts of Ficksburg, Bethlehem,[135] Harrismith and Vrede had yet surrendered—their turn was to come.

All told, we were 8,000 burghers.

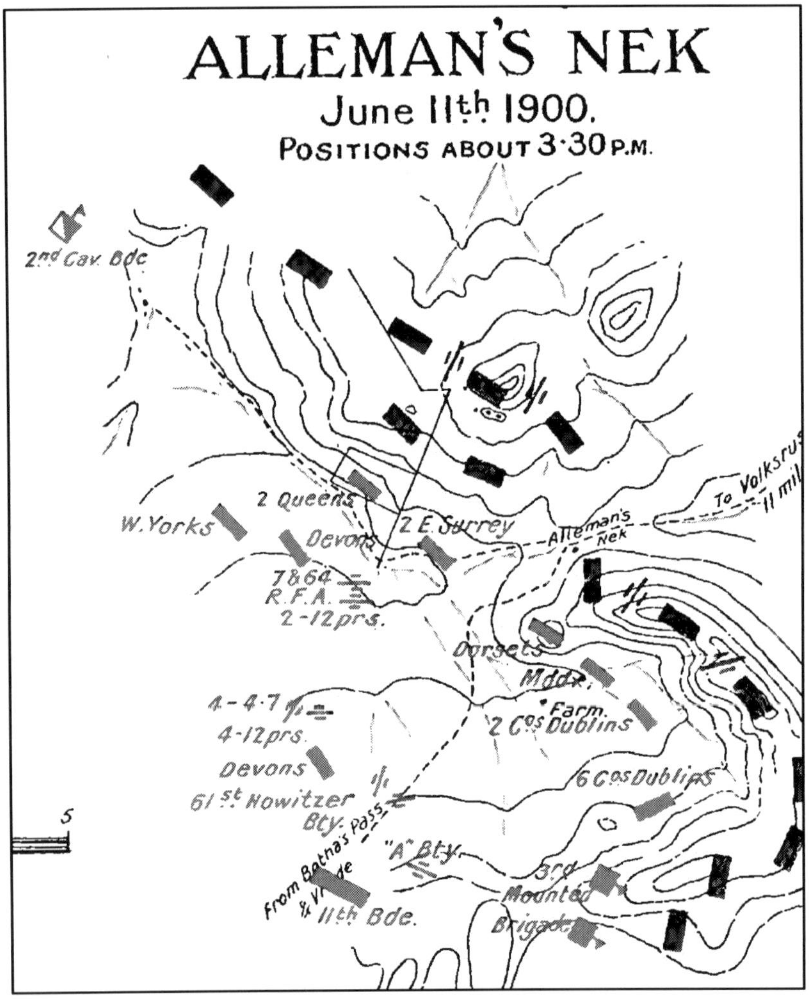

Alleman's Nek.[172]

[134] Afterwards Commandant, and, still later, Assistant Commander-in-Chief.

[135] At the conclusion of peace it was the Bethlehem commando which had the greatest number of burghers under arms.

After my men had gone northwards, those burghers of Hoopstad, Jacobsdal, Fauresmith, Philippolis, Bethulie, Smithfield, Rouxville, Wepener, Bloemfontein and the southern part of Ladybrand, who had laid down their arms and remained at home between the beginning of March and the end of May, were left undisturbed by Lord Roberts—so far as their private liberty was concerned.

I was now camped at Frankfort, waiting for the ammunition, which ought to have already arrived from Greylingstad Station. It was about this time that the Government decided, on the recommendation of some of the officers, that the rank of Vechtgeneraal should be abolished. In consequence of this decision all the officers of that rank resigned. I did not approve of this course of action, and obtained from the Government the rank of Assistant Commander-in-Chief. I was thus able to re-appoint the old Vechtgeneraals, Piet de Wet, C.C. Froneman, Philip Botha and Paul Roux, and I at once proceeded to do so.

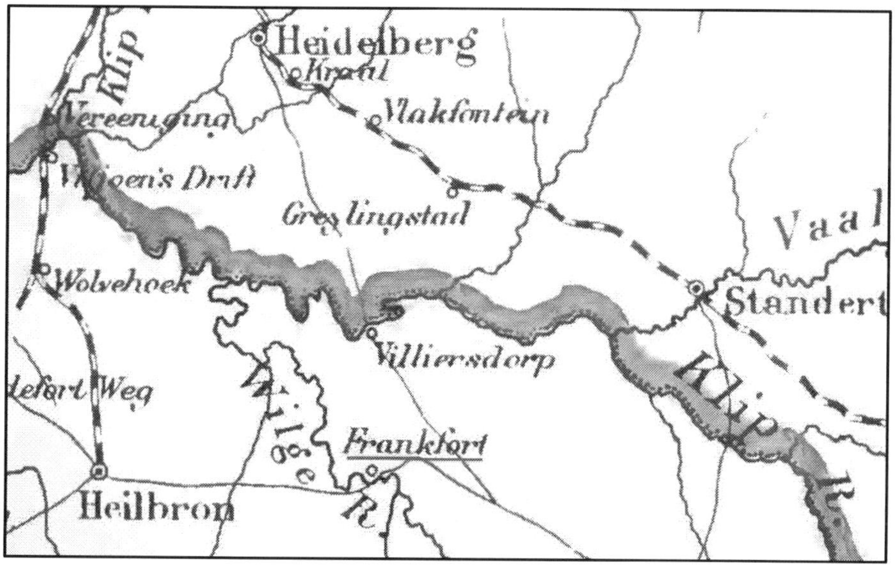

Frankfort, lower center on Wilge River.

14. Roodewal

The ammunition arrived safely, and towards the end of May I made my way to a certain hill, some twelve miles from Heilbron, to which we had given the name of Presidents kopje, and where Commandants Steenekamp and J.H. Olivier were posted.

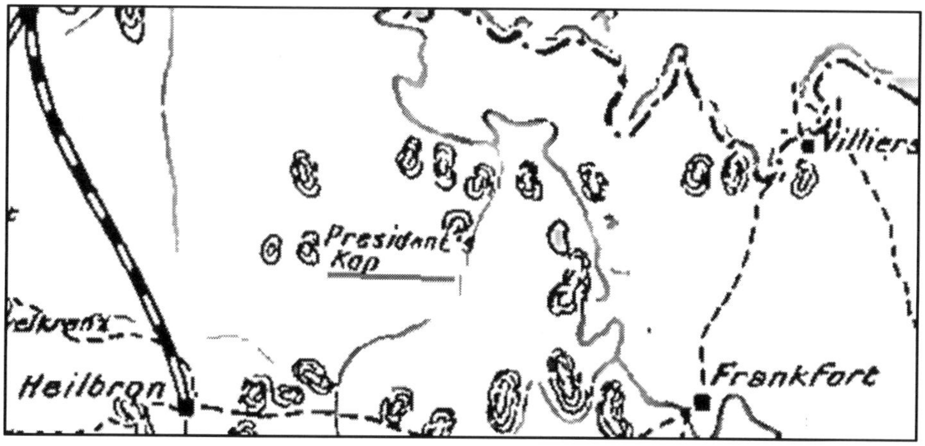

President's Kop.

Here I left the greater part of my commandos. But I myself, on the 2nd of June, set out in the direction of Roodewal Station, taking with me six hundred burghers, mounted on the best horses that were to be obtained. I reached the farm of Leeuwfontein the same night, and found it an excellent place in which to hide my men out of sight of the Heilbron garrison. The farm stood about nine miles to the south of that town.

The following evening we moved on as far as Smiths drift, which is a drift on the road from Heilbron to Kroonstad. There again I concealed my men.

On the afternoon of the next day, June the 4th, news was brought me that a convoy was on its way to Heilbron from Rhenoster River. This convoy encamped that evening at the distance of a mile from the farm of Zwavelkrans; the spot chosen was about five hundred paces from the Rhenoster River, and quite unprotected.

Before sunrise I sent a party of burghers down to the river, some five hundred paces from where the convoy was encamped, and by daybreak we had entirely surrounded the enemy.

No sooner had the sun appeared than I despatched a burgher with a white flag to the English officer in command. I ordered my messenger to inform the officer that he was surrounded, that escape was out of the question, and that if he wished to avoid unnecessary bloodshed, his only course was to surrender.

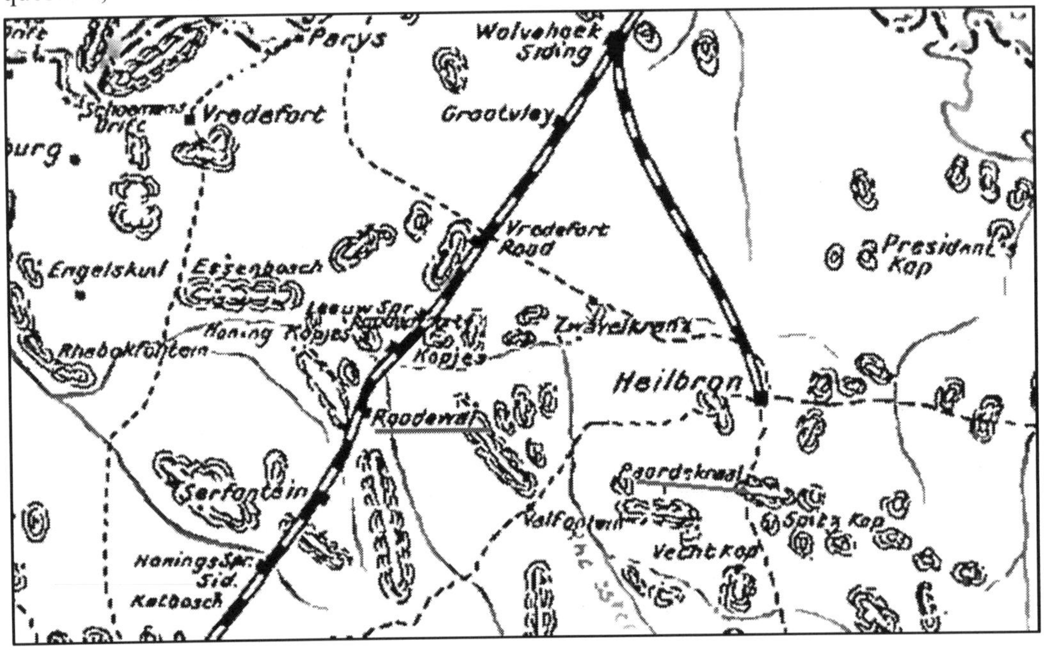

Roodenwal and Pardskraal.

On hearing this one of their men came to me with the object of demanding certain conditions. It goes without saying that my answer was—"Unconditional surrender!"

He asked for time to communicate this to the officer in command. I granted this request, and he returned to the convoy.

We were not left in suspense for long. The white flag was hoisted almost immediately, and two hundred Bergschotten,[136] with fifty-six heavily laden waggons, fell into our hands.

Fortunately, all this occurred out of sight of Roodewal Station and Heilbron, and, as not a single shot had been fired, I had no reason so far to fear that there was any obstacle in the way of my main project—the capture of the valuable booty at Roodewal.

I at once returned with my capture to the spot where we had been the previous night. General Philip Botha conducted the prisoners and the booty to the President's camp, returning to our laager on the following morning.

On the evening of the 6th of June I started on my road to Roodewal. At Walfontein I divided my troops into three parties. The first party, consisting of three hundred men with one Krupp, I despatched under Commandant Steenekamp to Vredefort Road Station, with orders to attack it the following day at sunrise. General Froneman, with Commandants Nel and Du Plooij, were in command of the second party, which consisted of three hundred burghers, with two Krupps and one quick-firing gun. My orders were that, at daybreak, they were to attack an English camp which was lying a mile to the north of the railway station at Rhenoster River, and close to some brick- coloured ridges. The third party I commanded myself. It consisted of Commandant Fourie and eighty burghers, with one Krupp; and with this force I pushed on to Roodewal Station.

At Doorndraai I left behind me a few waggons, with twenty men to guard them. I had previously stationed a hundred burghers there, with the object of keeping in touch with the enemy.

[136] Highlanders.

The information which Captain Scheepers had gained while scouting was amply sufficient to show me how the land lay.

Although I had heard that there were not more than fifty of the enemy at Vredefort Road Station, I had nevertheless sent three hundred burghers there. This was because I was aware that the main English force lay to the north of the station, so that these fifty men might be reinforced at the shortest possible notice. The numbers which General Froneman had to encounter were much greater, and the enemy held safe positions. But as General Froneman was himself able to take quite as good positions, I only gave him the same number of troops as I had assigned to Commandant Steenekamp. I also gave orders that two guns should proceed with him.

I was informed that there were only one hundred of the English at Roodewal, but that these hundred were very securely entrenched. My information was, however, at fault, for I discovered later on that there were at least double that number.

I arrived at Roodewal very early in the morning of the 7th of June.[137] I brought my men up to within eight hundred paces of the station, and ordered them to unharness the horses which were attached to the Krupp, and to place it in position.

But listen! There is the crack of rifles in the distance! That must be the sound of the enemy's fire on General Froneman. Again, and yet again, the sound meets my ears. Then all is quiet once more.

It was still two hours before the sun would rise, and I took full advantage of the opportunities which the darkness gave me. I ordered four of my burghers to approach as close to the station as was possible, and to find out everything they could about the enemy's position. Following my directions, they crept with extreme caution towards the English lines, until only a hundred paces separated them from the station. They returned before it was light, and brought back word that unless the enemy had thrown up unusually high schanzes, there must be an untold quantity of provisions piled up there. Everything had been very quiet, and they had seen no one stirring.

The day now began to dawn, and as soon as it was light I sent a message to the enemy demanding their surrender. The answer came back at once. On the back of my note these words had been written:

"We refuse to surrender."

I instantly opened a hot fire upon them, bringing the Krupp as well as the Mausers into action. But the reply of the enemy was no less severe.

We had no cover. There was only a shallow pan[138]—so shallow that it scarcely afforded protection to the horses' hoofs! A thousand paces to the north-west of the railway I had observed a deep pan where the horses would have had better cover, but even there our men would have been just as exposed as they now were. I had decided against taking up my position in this pan, because I should have been obliged to cross the line to reach it, and in doing so should have run the risk of being observed by the English.

Thus it was that the burghers were compelled to lie flat down in order to afford as little mark as possible to the enemy. But the men who served the Krupp were naturally unable to do this; and, seeing that the gun must be moved, I gave this order: "Inspan the gun, gallop it three thousand paces back; then blaze away again as fiercely as you can!"

Under a hail of bullets the horses were attached to the gun. Whilst this was being done, I ordered my men to fire upon the English entrenchments with redoubled energy, and thus, if possible, prevent the enemy from taking careful aim.

Incredible though it may appear, Captain Muller got the gun away without a single man or horse being hit. When he had covered three thousand paces, he halted, and turning the Krupp on the enemy, he shelled them with good effect.

At about ten o'clock, General Froneman succeeded in forcing the English troops which he had attacked to surrender. I therefore ordered the two Krupps which he had with him to be brought up with the utmost despatch. At half-past seven they arrived, and immediately opened fire on the English.

[137] 6/5/1900 The British flag hoisted in Pretoria.
6/7/1900 The 4th Battalion Derbyshire Regiment (Sherwood Foresters) captured by the enemy at Roodeval.
[138] A pond which only contains water during "the rains."

When the enemy had been under the fire of three guns and eighty Mausers for an hour, they thought it best to hoist the white flag. We accordingly ceased firing, and I rode out towards the station. Before I had reached it, I was met by two of the officers. They told me that they were willing to surrender, on condition that they were allowed to retain their private property and the mail bags, for it appeared that there were two English mails under their charge.

I replied that so far as their private belongings were concerned, they were welcome to keep them, as I never allowed the personal property of my prisoners to be tampered with in my presence.[139] But I told them that the letters were a different matter, and that I could not allow them to reach their destination—unless they were directed to a bonfire!

There was nothing left for the officers to do, except to agree to my terms then and there; for had they hesitated even for a moment, I should certainly have stormed the station.

But they wisely surrendered.

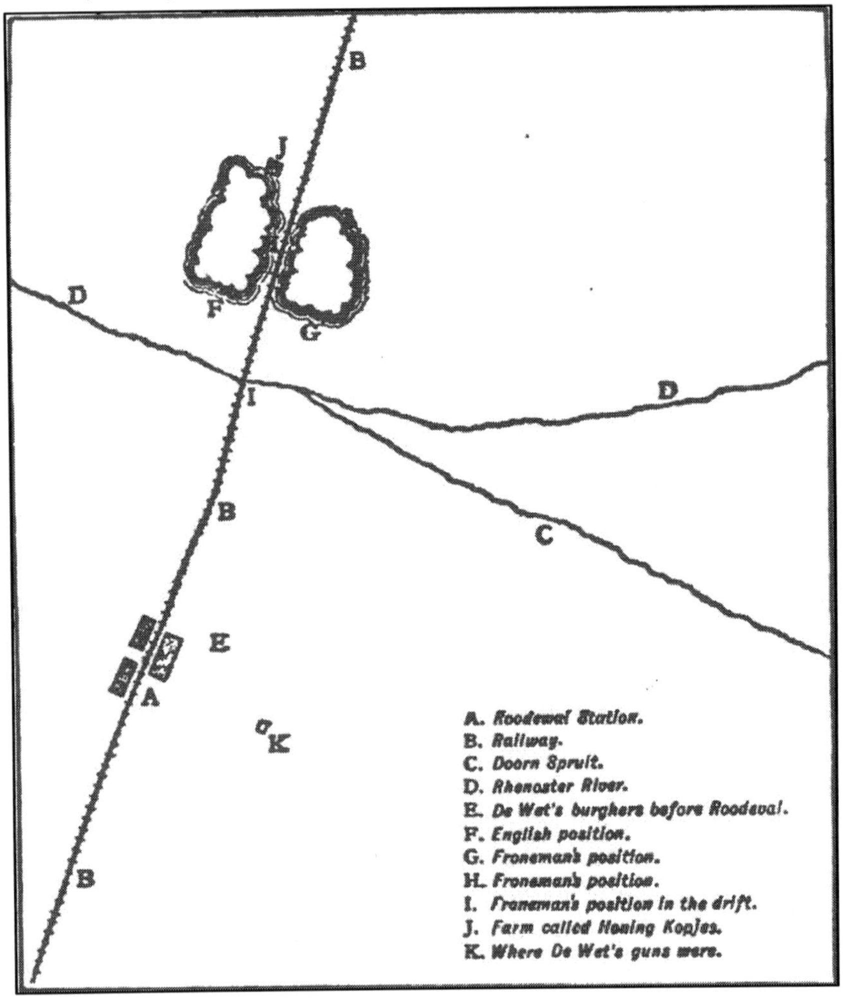

Roodewal. From a Sketch by the Author.

On our arrival at the station, we were all filled with wonder at the splendid entrenchments the English had constructed from bales of cotton, blankets and post-bags. These entrenchments had been so effectual that the enemy's loss was only twenty-seven killed and wounded—a remarkably small number, when it is remembered that we took two hundred of them prisoners.[140]

[139] The Uitschudden (stripping) of the enemy had not become necessary at that date. I can say for myself that when, at a later period, it came into practice, I never witnessed it with any satisfaction. Yet what could the burghers do but help themselves to the prisoners' clothing, when England had put a stop to our imports, and cut off all our supplies?

[140] Roodewal Station. A force of Gen. Christiaan de Wet, together with Commander Piet Fourie, attacked the British garrison, who were guarding the fully loaded trains.. They refused surrender, whereupon De Wet began his attack. The battle lasted for five hours and swung in De Wet's favor when General Froneman arrived with two guns. The British surrendered immediately as the bombs began to fall between them. The Boers looted more than 500 crates full of ammunition. https://af.wikipedia.org/wiki/Slag_van_Roodewalstasie

I had expected that our booty would be large, and my expectations were more than realized. To begin with, there were the bales of clothing that the English had used as entrenchments. Then there were hundreds of cases of necessaries of every description. Of ammunition, also, there was no lack, and amongst it there were projectiles for the Naval guns, with which Lord Roberts had intended to bombard Pretoria.

Some of the burghers attempted to lift these gigantic shells, but it took more than one man to move them.

I read in the newspapers afterwards that I had inflicted a loss of three quarters of a million sterling on the English Government—let that give the extent of my capture.

But at that moment we did not realize how much harm we had done to them. We had little time for anything which did not directly forward our cause. I was, however, very sorry that I could not carry away with me the blankets and boots which we found in large quantities, for they would have been most valuable for winter use. But there was no time for this, as the English held the railway and could at any moment bring up reinforcements from Bloemfontein, from Kroonstad, or from Pretoria. So, as I could not take the booty away with me, I was obliged to consign it to the flames.

But before I did this I gave the burghers permission to open the post-bags, and to take what they liked out of them. For in these bags there were useful articles of every description, such as underclothing, stockings, cigars and cigarettes.

Very soon every one was busy with the post-bags—as if each burgher had been suddenly transformed into a most zealous postmaster!

Whilst my men were thus pleasantly occupied, two prisoners asked me if I would not allow them also to open the post-bags, and to investigate their contents. I told them to take just what they fancied, for everything that was left would be burnt.

It was a very amusing sight to see the soldiers thus robbing their own mail! They had such a large choice that they soon became too dainty to consider even a plum-pudding worth looking at!

Although I had ordered my men to wreck the bridges both to the north and to the south of us, I still did not feel secure—any delay on our part was fraught with danger, and the sooner we were off the better.

But before we could start, I had to find some method of removing the ammunition which I wished to take with me. Since I possessed no waggons available for this purpose, my only course was to order my burghers to carry away the quantity required. But my burghers were busily engaged in looting.

Those who have had any experience of our commandos will not need to be told that it was a difficult task to get any men to help me in the work. I did succeed, however, in dragging a few of the burghers away from the post-bags. But the spirit of loot was upon them, and I was almost powerless. Even when I had induced a burgher to work, he was off to the post-bags again the instant my back was turned, and I had to go and hunt him up, or else to find some other man to do the work. Yet, in spite of this, I succeeded in removing the gun and Lee-Metford ammunition. We carried away some six hundred cases of this ammunition,[141] and hid it at a spot about three hundred paces from the station.

When the sun set, [On June 7th] the burghers were again on the march. But what a curious spectacle they presented!

Each man had loaded his horse so heavily with goods that there was no room for himself on the saddle; he had, therefore, to walk, and lead his horse by the bridle. And how could it be otherwise? For the burghers had come from a shop where no money was demanded, and none paid!

But the most amusing thing of all was to watch the "Tommies" when I gave them the order to march. The poor Veldtcornet, who was entrusted with the task of conducting them to our camp, had his hands full when he tried to get them away from the booty; and when at last he succeeded, the soldiers carried such enormous loads, that one could almost fancy that every man of them was going to open a store. But they could not carry such burdens for long, and soon they were obliged to diminish their bulk, thus leaving a trail of parcels to mark the road they had taken!

The British losses were 45 killed, 123 wounded, and 645 captured. The Boers lost 8 civilians and 24 were wounded.

[141] At this time the burghers were beginning to use the rifles which they had taken from the enemy.

And now it was time for the fire to do its work, and I ordered fifteen men to set the great heap of booty alight. The flames burst out everywhere simultaneously—our task was completed.

In an instant we had mounted our horses and were off.

When we had covered fifteen hundred paces, we heard the explosion of the first shells, and wheeled round to view the conflagration. The night was very dark, and this rendered the sight that met our eyes still more imposing. It was the most beautiful display of fireworks that I have ever seen.

One could hear, between the thunder of the big bombs, the dull report of exploding cordite. Meanwhile the dark sky was resplendent with the red glow of the flames.

Destruction of a train of supplies near Greylingstad, February 13th, 1901.[173] Mail Bags at the 'Post Office.'[174] [142]

I must now give some description of General Froneman's engagement to the north of Rhenoster-riviersbrug.[143]

The firing we had heard before sunrise came from the English outposts, as they were retreating to their camp. The burghers and the English had both seized positions on small hills and in abandoned Kaffir kraals.[144]

Although the English had very good positions, and out-numbered our men by two to one, they found it impossible to hold out against our fire. They had no guns, whilst we possessed, as the reader knows, two Krupps and a quick-firing gun, which latter had the same effect as a Maxim-Nordenfeldt. Thus the enemy was forced to surrender; and five hundred of them were taken prisoner, among whom were Captain Wyndham Knight and several other officers. Their casualties amounted to the large total of one hundred and seventy killed and wounded, Colonel Douglas being one of the killed.

Commandant Steenekamp had also met with success, for he had captured the English camp at Vredefortweg Station, and taken thirty prisoners, without firing a shot.

Thus we had made eight hundred of the enemy our prisoners, and destroyed an enormous amount of their ammunition, and this with scarcely any loss on our side. At Roodewal only two of my men had been wounded, whilst General Froneman had lost but one killed—a burgher named Myringen—and two slightly wounded.

It had been a wonderful day for us—a day not easily forgotten.

We were deeply thankful for our success. Our only regret was that it had been impossible for us to keep more of the clothing and ammunition. But although we had not been able to retain it, neither had the enemy. It was winter, and we had managed to burn their warm clothing. The English would certainly feel the want of it; and some time must elapse before they could receive a fresh supply from Europe.

Undoubtedly Lord Roberts would be very angry with me; but I consoled myself with the thought that his anger would soon blow over. I felt sure that after calm consideration he would acknowledge that I had been altogether within my rights, and that he had been rather unwise in heaping together at one place so large a quantity of insufficiently protected stores. He should have kept his supplies at Kroonstad, or, better still, at Bloemfontein,

142 "This engagement was fought on De Wet's own farm, Rooival, ...the yield in spoils was large, including two trains, fifteen hundred bags of accumulated mail for Pretoria, and the entire winter outfit and uniforms for the invading army, including 13,000 suits of underclothing." "There was a train load of ammunition for the twelve, fifteen and 240 pound guns, and $325,000 in specie. The enormous supply of stores captured caused De Wet to facetiously remark that it was a better crop than he had ever raised on his farm at any previous season." Richard Harding Davis With Both Armies in South Africa. Charles Scribner's Sons, New York 1900. Chapter 17.

143 Rhenoster River bridge.

144 These dated back to the time of Moselekatze (Umzilygazi).

until he had reconstructed all the railway bridges which we had blown up on the line to Pretoria. Lord Roberts had already begun to trust the Free-Staters too much; and he had forgotten that, whatever else we may have been thinking about, never for a single moment had we thought of surrendering our country.

I received a report the following day [June 8th] that thirty English troops had been seen eight miles to the west of Roodewal, and moving in the direction of Kroonstad. I despatched General Froneman with thirty of the burghers to fetch them in.

The next day, which was the 9th of June, I went with our prisoners to within three miles of the railway, and left them there under Veldtcornet De Vos,[145] ordering him to conduct them the rest of the way.

It was now my duty to bring away the ammunition which I had left at Roodewal and to put it into some safe place. With this in view, I sent the Commandants, when night had fallen, to Roodewal, each with two waggons, and ordered them to bring it to my farm at Roodepoort, which was three miles [East] away from the railway bridge over the Rhenoster River.

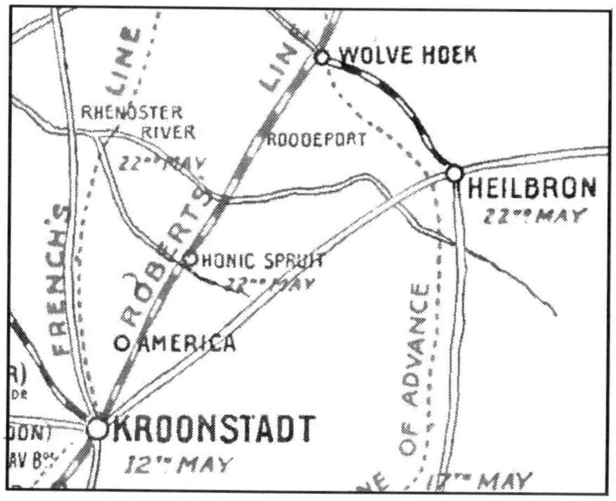

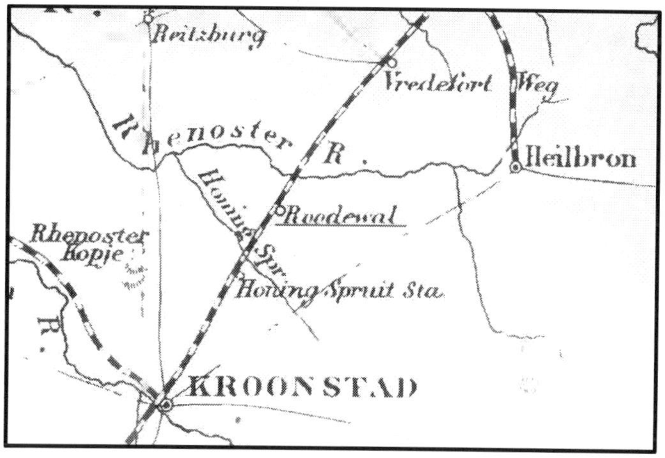

Roodenport in the first map and Roodewal in the second.

There was a ford near my farm with sandy banks; and I told the Commandants to bury the ammunition in this sand, on the south side of the river, and to obliterate all traces of what they had done by crossing and re-crossing the spot with the waggons. I found out subsequently that the Commandants had left some of the ammunition behind at Roodewal.

Before I conclude this chapter I have to record an event which filled me with disgust.

Veldtcornet Hans Smith, of Rouxville, contrived to have a conversation with Captain Wyndham Knight, who, as I have already stated, was one of our prisoners. The Veldtcornet obtained from him a "free pass" to Kroonstad through the English lines, and also a written request to the British authorities there to allow him and twenty burghers to proceed without hindrance to Rouxville. Alas! that any Free State officer should be capable of such conduct!

Captain Wyndham Knight will be held in high esteem by all who truly serve their country, for he was a man who never deserted the cause of his fatherland, no matter what dangers he encountered.

Veldtcornet Hans Smith with his twenty burghers decamped on the night of the 10th of June, but some days had passed before I discovered the mean trick he had played.

It was far easier to fight against the great English army than against this treachery among my own people, and an iron will was required to fight against both at once. But, even though one possessed an iron will, such events caused many bitter moments; they were trials which, as an African proverb says, no single man's back was broad enough to carry.[146]

[145] He was afterwards appointed Commandant.

[146] Literally the proverb runs as follows: "There are some trials which will not sit in one man's clothes."

15. I Make Lord Kitchener's Acquaintance

On the morning of June the 10th my anticipations were realized by the approach of a large English force from Vredefortweg and Heilbron. Commanded by Lord Kitchener, and numbering, as I estimated, from twelve to fifteen thousand men, this force was intended to drive us from the railway line.

Military reconnaissance balloon with an escort. It took four carts to carry the gas tubes and one to carry the equipment, each cart was drawn by four horses. Two extra carts carried the baggage.[175]

Long column of British troops, both mounted and infantry, crossing a river.[176] Note the Band with the drum.

Column of the Royal Artillery on the move.[177]

Column of the Royal Artillery on the move.[178]

I gave orders that the few waggons which we had with us should proceed in the direction of Kroonstad, to the west of the line; once out of sight, they were to turn sharply to the west, and continue in that direction. This manoeuvre, I hoped, would serve to mislead the enemy, who was on the look-out for us.

So much for the waggons. For the rest, I felt that it would never do for us to withdraw without having fired a shot, and I therefore got my men into position on some kopjes (where Captain Wyndham Knight had been four days previously, and which lay to the north of Rhenosterriviersbrug) on my farm Roodepoort,[147] and on the Honingkopjes.

The English, with their well known predilection for a flank attack on every possible opportunity, halted for an hour, and shelled our positions with Lyddite and other guns. This did not have the desired effect of inspiring terror in the burghers who were under my command at Honingkopjes.

Then the enemy began to move. I saw masses of their cavalry making for a piece of rising ground to the north of Roodepoort. As the burghers there were hidden from me, I was unable to observe from where I stood the effect of this flank movement. Knowing that if they were able to give way and to retreat along the river we should have no means of discovering the fact until it was too late and we were surrounded, I came to the conclusion that it was essential for me to go to Roodepoort to assure myself that the cavalry had not yet got round. But it was most important that no suspicion of the danger which threatened us should be aroused in the burghers—anything calculated to weaken their resistance was to be avoided on such an occasion. Accordingly I merely told them that I was going to see how affairs were progressing at Roodepoort, and that in the meantime they must hold their position.

I rode off, and discovered that the English were already so close to our troops at Roodepoort that fighting with small arms had begun. I had just reached an eminence between Roodepoort and the Honingkopjes when I saw that the burghers in the position furthest towards the north-west were beginning to flee. This was exactly what I had feared would happen. Immediately afterwards the men in the centre position, and therefore the nearest to me, followed their comrades' example. I watched them loosening their horses, which had been tethered behind a little hill; they were wild to get away from the guns of the English and from the advance of this mighty force.

Boer Horses During Action Under Crest of Hill Occupied by Burghers.[179]

[147] To deter further Boer resistance British Commander-in-Chief Roberts dynamited De Wet's farm Roodepoort on 16 June [1900]. From: https://en.wikipedia.org/wiki/Piet_Fourie

It was impossible for me now to go and tell the burghers on the Honingkopjes that the time had come when they too must retreat. My only course was to order the men near me not to effect their escape along the well protected banks of the river, but to the south, right across the stream, by a route which would be visible to burghers on the Honingkopjes. They obeyed my orders, and rode out under a heavy gun and rifle fire, without, however, losing a single man. The men on the Honingkopjes saw them in flight, and were thus able to leave their position before the enemy had a chance of driving them into the river or of cutting them off from the drift.

Unfortunately, seven burghers from Heilbron were at a short distance from the others, having taken up their position in a kliphok.[148] Fighting hard as they were, under a deafening gun-fire from the enemy, who had approached to within a few paces of them, they did not observe that their comrades had left their positions. Shortly afterwards, despairing of holding the kliphok any longer, they ran down to the foot of the hill for their horses, and saw that the rest of the burghers were already fleeing some eight or nine hundred paces in front of them,[149] and that their own horses had joined in the flight.

There was now only one course open to them—to surrender to the English.[150]

I ordered the burghers to retreat in the direction of Kroonstad, for by now they had all fled from Roodepoort and Honingkopjes—a name which, since that day, has never sounded very sweet to me.[151]

During the morning I received a report informing me that there were large stores at Kroonstad belonging to the English Commissariat, and that there was only a handful of troops to protect them. I had no thought, however, of attempting to destroy the provisions there, for I felt sure that the British troops, who had but just now put us to flight, would make for Kroonstad. They would know that the stores stood in need of a stronger guard, and moreover they would naturally think that we should be very likely to make an attack at a point where the defence was so weak.

Obviously, under these circumstances, it would never do for us to go to Kroonstad.

Accordingly, as soon as darkness came on, I turned suddenly to the west, and arrived at Wonderheuve late at night. I found there Veldtcornet De Vos with the prisoners of war.

Meanwhile, as I had anticipated, the vast English army marched up along thirty-four miles of railway to Kroonstad. Lord Kitchener, as I heard later on, arrived there shortly after noon on the following day.

We left Wonderheuve early in the morning, and advanced along Rietspruit until we reached the farm of Vaalbank, where we remained until the evening of the next day, June the 13th. That night I saw clearly that it was necessary for us to cross the line if we wanted to keep ourselves and our prisoners out of the clutches of Lord Kitchener; he had failed to find us at Kroonstad, and would be certain to look for us in the country to the west of the line.

I also felt myself bound to wreck this line, for it was the only railway which Lord Roberts could now utilize for forwarding the enormous quantities of stores which his vast forces required.[152] I resolved therefore to cross it at Leeuwspruit, north of Rhenoster River bridge (which the English had recently repaired), and then, in the morning, to attack the English garrisons which had again occupied Roodewal and Rhenoster River bridge.

[148] i.e. the ruins of Kaffir stone huts, built in the time of Moselekatze.

[149] Among these seven burghers were Willie Steyn, Attie Van Niekerk, and a certain young Botha. It was Steyn and Botha, with two men of the name of Steytler, and two other Free-Staters whose names I have forgotten, who managed to escape from the ship that lay anchored in the harbour of Ceylon. They swam a distance of several miles to a Russian ship, by which they were carried to one of the Russian ports, where they received every hospitality. I shall always be grateful to the Russians for this. They then travelled through Germany into Holland, being subsequently conveyed in a German ship to German West Africa. Thence they made their way through Boesmansland to Cape Colony, and, after many adventures, joined General Hermanus Maritz's commando. Botha, unfortunately, was killed in a skirmish some time later. What will the world say of these young burghers? Surely, that more valiant and faithful men than they have never lived. I regret that I do not remember the names of all Willie Steyn's comrades. I travelled with him by train from the Free State to Cape Town, where I had to join General Louis Botha and J.H. De la Rey, so as to accompany them to Europe on my nation's behalf. He promised then to give me all the particulars of his escape, but I suppose there has been some obstacle in the way.

[150] 6/11/1900 De Wet engages then retreats during engagement with Meuthen/Kitchener near his home base in Roodepoort and Honingkopjes. His farm: "at Roodepoort, ... three miles away from the railway bridge over the Rhenoster River." .

[151] The word honing means honey.

[152] At that time the Natal and Delagoa Bay railways were still in our possession.

Rhenoster bridge midway through reconstruction.[180]

I had given orders that all the cattle along the railway line should be removed; General Louis Botha had made the same regulation in regard to the country round Pretoria and Johannesburg. If only our orders had been carried out a little more strictly, and if only the most elementary rules of strategy had been observed in our efforts to break the English lines of communication, Lord Roberts and his thousands of troops in Pretoria would have found themselves in the same plight as the Samaritans in Samaria—they would have perished of hunger. It was not their Commander-in-Chief's skill that saved them, not his habit of taking into account all possible eventualities—no, they had to thank the disobedience of our burghers for the fact that they were not all starved to death in Pretoria.

I arranged with General Froneman that he should cross the line at the point I had already selected, that is to say, north of Rhenoster River bridge, and that in the morning he should attack, from the eastern side, the English who were posted at Leeuwspruit Bridge. I, in the meanwhile, would make my way with a Krupp to the west side of the line, and having found a place of concealment near Roodepoort, would be ready to fall upon the English as soon as I heard that the other party had opened fire on them from the east.

But my plan was to come to nothing. For when, during the night, Froneman reached the line, a skirmish took place then and there with the English outposts at Leeuwspruit railway bridge. At the same time a train arrived from the south, on which the burghers opened such a fierce fire that it was speedily brought to a standstill. General Froneman at once gave orders to storm the train, but his men did not carry out his orders.

Had they done so, Lord Kitchener would have fallen into our hands!

Nobody knew that he was in the train, and it was only later that we heard how, when the train stopped, he got a horse out of one of the waggons, mounted it, and disappeared into the darkness of the night.

Shortly afterwards the train moved on again, and our great opportunity was gone!

General Froneman succeeded in overpowering the garrison at the railway bridge, and took fifty-eight prisoners. He then set fire to the bridge, which was a temporary wooden structure, having been built to replace another similar one, which had been blown up with gunpowder.

Three hundred Kaffirs were also made prisoners on this occasion. They protested that they had no arms, and had only been employed in work upon the railway line. This absence of rifles was their saving. Possibly they had really been in possession of arms, and had thrown them away under cover of the darkness; but the burghers could not know this, and therefore acted upon the principle that it is better to let ten culprits escape than to condemn an innocent man to death.

Two men employed by the British Army seated beside a tree. Their brass arm band(s) denote military service.[181]

Almost complete.[182] [Note cattle in background.] The bridge is now well on towards completion. A large number of men, with plenty of native labour are busy making a deep cutting near the river bank, the rails are all ready to be laid, and in a short time trains will be able to cross. Trial train crossing.[183] In this last picture, the first engine is seen going slowly over to test the bridge, which is now finished. This means that railway communication was once more restored from Cape Town to Johannesburg and Pretoria, the line being intact beyond this point.

General Froneman went on towards the east of Doorndraai. He was very well satisfied with his bridge-burning and his capture of prisoners, and in his satisfaction he never gave thought to me.

I waited in my hiding-place, expecting that, as we had agreed, the firing would begin from the east, but nothing happened. I did not care to make an attack on my own account from the west, for my positions were not practicable for the purpose, and being short of men, I feared that such an attempt might end in disaster.

It was now ten o'clock.

A few English scouts appeared on the scene, and four of my men attacked them. One of the enemy was shot, and the rest taken prisoners. And still I did not hear anything from General Froneman.

At last I came to the conclusion that he must have misunderstood my instructions. If that were the case, I must do the best I could myself. Accordingly I opened fire on the English with my Krupp.

Still no news of General Froneman!

Then I ordered my burghers to advance. Our first movement was over the nearest rise to the north-west; we halted for a moment, and then made a dash for Leeuwspruit Bridge—but we found nothing there.

Late in the evening I met General Froneman, and heard from him the narrative which I have given above.

The following day I sent well on to twelve hundred prisoners of war—including Kaffirs—to the President's camp, which lay east of Heilbron. We then advanced to a point on the Rhenoster River, near Slootkraal, remaining in concealment there until the night of the 16th of June. The following morning we occupied some ridges at Elandslaagte, on the look-out for a large English force which was marching from Vredefortweg to Heilbron.

My intention was to give them battle at Elandslaagte, and to hold on to our positions there as long as possible; and then, if we could not beat them off, to retire. If only the burghers had carried out my orders strictly, we should certainly have inflicted heavy losses on the English, even if we had not won a complete victory.

The English had not sent out their scouts sufficiently far in advance, and came riding on, suspecting nothing. We occupied positions on the right and left of the road along which they were advancing, and my orders were that the burghers should let the troops get right between our ridges, which were about three hundred paces from each other, and then fire on them from both sides at once.

Instead of doing this, however, the burghers began to fire when the English were five hundred paces from them—before, that is to say, they had got anywhere near the door of the trap which I had set for them.

The enemy wheeled round, and galloped back for about fifteen hundred paces. They then dismounted, and fired on us. But, having no sort of cover, they were soon compelled to mount their horses again and retire to their guns, which were about three thousand yards from us. These guns now opened a heavy fire upon our ridges; we replied with our three Krupps, with which we made such good practice that we might have been able to hold out there indefinitely, had not a Lyddite and an Armstrong gun happened just then to arrive from Heilbron, which lay about ten miles behind us. Thus attacked both in front and rear, there was nothing to do but retire. Fortunately, we had not lost a single man.

First we rode in a southerly direction, but as soon as we got into cover we struck off to the east, setting our faces towards Heilbron.

Then, to our immense relief, the sun went down. How often during our long struggle for independence had not the setting of the sun seemed to lift a leaden weight from my shoulders! If, on a few occasions, the approach of night has been to our disadvantage, yet over and over again it has been nothing less than our salvation.

We got back safely, under cover of the darkness, to our little camp near Slootkraal, and there remained in hiding until the following day. It was there that Commandant Nel handed in his resignation. In his place the burghers of Kroonstad chose Mr. Frans Van Aard as their Commandant.

That night we set out for Paardenkraal, twenty miles to the north-east of Kroonstad, staying there until the evening of the 19th.[153] [154]

The time for my attack on the railway line having now come, I divided my men into three parties for that purpose. I sent on Commandant J.H. Olivier, who had joined me at Paardenkraal, to Honingspruit Station, General Froneman to America Siding, while I myself made my way to Serfontein Siding.

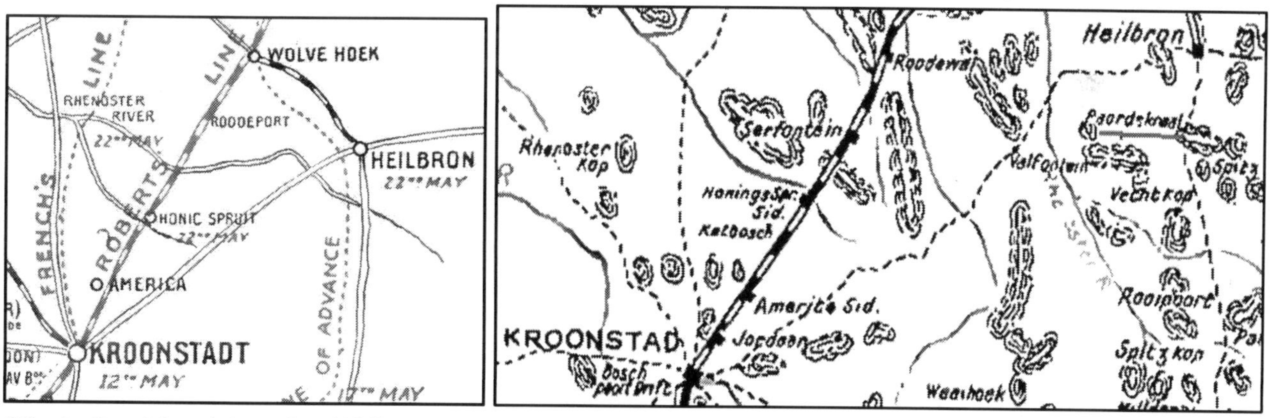

'Honic Spruit' and America (Siding). Note Paardenkraal upper right below Heilbron.

[153] This is a distinct location from the Paardekraal Monument which is located in Krugersdorp outside Johannesburg. Paardekraal Monument - Wikipedia The term itself simply means "Horse Corral" in Afrikaans.

[154] 6/13/1900 Continued actions on the Senekal-Ficksburg line. The Senekal-Winburg telegraph line was damaged. General Lyttelton occupied Wakkerstroom.

6/14/1900 Rustenburg occupied by General Baden Powell. Botha's rearguard surprised and "thoroughly routed" by General Ian Hamilton's Mounted Infantry. Position on Zand River attacked by 800 Boers with three guns. Enemy driven off by General Knox.

6/18/1900 General Baden-Powell arrived at Pretoria. General Hunter occupied Krugersdorp.

At daybreak General Froneman wrecked the line near America Siding, and I did the same at other places, also destroying the telegraph poles. Each pole was first shot through with the Mauser, and then pulled until it snapped at the point where the bullet had pierced it.

Things did not go so well with Commandant Olivier. He attacked the station, but, unfortunately, not so early as had been arranged. Consequently he was not able to bring his gun into action before the enemy had observed him. When I came up to him there was a strong English reinforcement from Kroonstad close at hand. We had too few men with us to be able to offer resistance, and had to retreat, returning to Paardenkraal at nightfall.

16. Bethlehem is Captured by the English

It was at this time that I decided to make my way to Lindley, which had been retaken by the English a few days after General Piet de Wet had captured the Yeomanry in that town. The object of my journey was to discover if it were not possible to again seize the place. On the 21st of June I covered half the distance to Lindley, and the following day I arrived within ten miles of the town.[155]

I rode round the town with Piet de Wet the next day, in order to find out our best method of attacking it.

Commandant Olivier had been sent by me that morning in the direction of Kroonstad to oppose a strong English column, which I had been informed was approaching. But my plan must have leaked out in some way or other, for the enemy carefully chose so well protected a route that they gave Commandant Olivier no chance of attacking them. Thus the following morning the English arrived safely at Lindley, and now there was no possibility of capturing the town.

In the meantime President Steyn's laager had moved from the east of Heilbron and joined us. He himself, with the members of the Government, had gone to Bethlehem. General Marthinus Prinsloo was there too; he had resigned his post of Commander-in-Chief of the commandos which guarded the Drakensberg. Commandant Hattingh of Vrede had been chosen in his place, and he also was at Bethlehem.

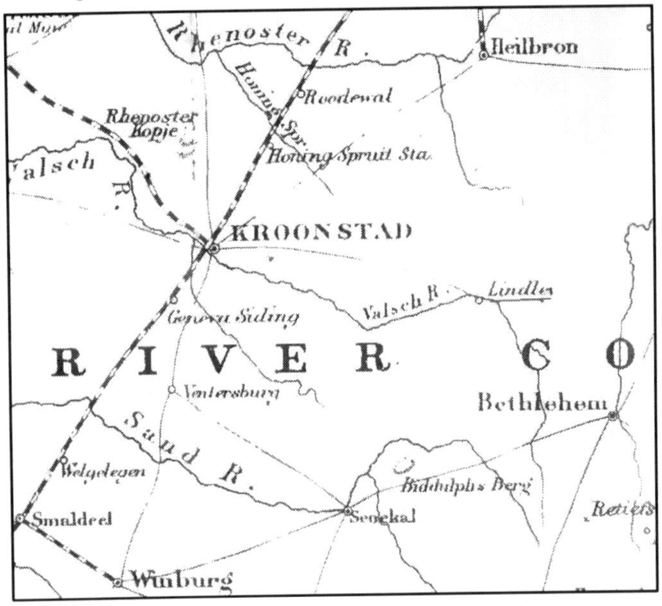

Lindley North of Bethlehem.

A difficulty now arose as to Prinsloo's position. The President declared that Prinsloo was nothing more than a private burgher; but Commandant Olivier was not satisfied with this, and asked that there might be an election of a Commander-in-Chief. This request, however, the President refused to grant.

I did not wish the office of Commander-in-Chief to devolve upon myself, for I knew that I did not possess the confidence of the officers. And as some eight miles to the east of Lindley there was telegraphic communication

[155] 6/19/1900 De Wet, Froneman and Oliver wreck train line and telegraphs near America Siding, Honing Spruit station and Serfontein siding before retreating to Paardenkrall from reinforcements under Muethen from Kroonstadt.
6/20/1900 Surrender of De Villiers. Extinction of rebellion in Cape Colony.

with Bethlehem, I was able to hold a conversation with the President over the wires. I accordingly again asked him to permit an election. But it was all in vain; the President declined to allow an election to take place.

I now took matters into my own hands. I collected the officers together with the object of holding a secret election. Thus I should discover what their opinion of me might be as chief of the Free State forces. I was firmly resolved that should the majority of the officers be against me, and the President should still refuse his consent to an election, that I would send in my resignation, and no longer continue to hold the post of Commander-in-Chief.

Commander-in-Chief Hattingh, Vechtgeneraal Roux, and all the oldest commandants of the Free State, were present at this meeting. The voting was by ballot; and the result was that there were two votes for General Marthinus Prinsloo, one for General Piet de Wet, and twenty-seven for myself.

I at once wired to the President, and told him what had occurred. He was ready to abide by the decision, and I was satisfied now that I knew exactly where I stood. Mr. Marthinus Prinsloo was also contented with the turn events had taken. And I must say this of him, that it was not he who had insisted on an election.

It soon became apparent that the enemy's object was the capture of Bethlehem. The English forces round Senekal advanced towards Lindley, and having been joined by the troops stationed there, had proceeded in the direction of Bethlehem; consequently a very large British force was marching on that town.[156]

We on our part now numbered over five thousand men, for General Roux had joined us with some[157] of his burghers.

The English were unopposed until they reached Elandsfontein, but there a battle took place in which big guns played the main role, although there was also some heavy fighting with small arms.

In this engagement Commandant Michal Prinsloo did a brave deed. I arrived at his position just after the burghers had succeeded in shooting down the men who served three of the enemy's guns. With a hundred men he now stormed the guns, hoping to be able to bring them back with him to our lines. Whilst he charged, I cannonaded the enemy, with a Krupp and fifteen pound Armstrong, to such good effect that they were forced to retreat behind a ridge.

A British Ordnance BL 15 pounder gun at camp near Boshof 1900.[184]

In this way Commandant Prinsloo reached the guns safely, but he had no horses with him to drag them back to us. He could do nothing but make the attempt to get them away by the help of his burghers, and this he tried to accomplish under a fierce fire from the English. But he would still have succeeded in the endeavour, had not unfortunately a large force of the enemy appeared on the scene, and attacked him and his hundred burghers. I was unable to keep the English back, for both my guns had been disabled. The nipple of the Armstrong had been blown away, and—for the first time—the lock of the Krupp had become jammed. Had it not been for this mishap, Commandant Prinsloo would certainly have been able to remove the guns to the other side of a ridge, whither teams of our horses were already approaching. But, as it was, he had to hurry away as fast as possible, and leave the guns behind.

[156] 6/22/1900 Dundonald occupied Standerton.
6/24/1900 General Clements defeated the Boers at Winburg. General Ian Hamilton occupied Heidelburg.
[157] He had left the remainder of his burghers at Witnek and at Houtnek, near Ficksburg.

When the enemy arrived they had outflanked us so far to the north, that we had nothing open to us but again to abandon our positions. We therefore retired to Blauwkop, and on the following day to Bethlehem.[158]

In the meantime I had once more become encumbered with a large waggon camp, which proved a source of great danger. During the last few weeks waggons had been accumulating round me without attracting my attention. The reason that the burghers were so anxious to bring their waggons with them, was to be found in the fact that the English, whenever they arrived at one of our farms, always took the waggons and oxen. The Boers felt it very hard to be robbed in this way of their property; and they hoped to be able to save their waggons and carts by taking them to the commando.

A typical Boer camp.[185]

Burghers Crossing a river on a pontoon ferry.[186]

It was natural for them to wish to save all they could; but I was convinced that the waggons could only be saved at the expense of our great cause. But nobody could see it in that light. And as I could only appeal to the free will of my burghers, I dare not attempt to get rid of the waggons by force. If I had made any such attempt, serious consequences would certainly have followed, even if a revolt had not ensued. The great fault of the burghers was disobedience, and this came especially to the fore when their possessions were in jeopardy.

I now made up my mind to defend the town of Bethlehem. The following morning I went with the Generals and Commandants to reconnoitre the country, so that I might be able to point out to each of them the position that I wished him to occupy.

Our line of defence began at the south of Wolhuterskop (a kop to the south-west of Bethlehem), and extended from there to the north-west of the town.

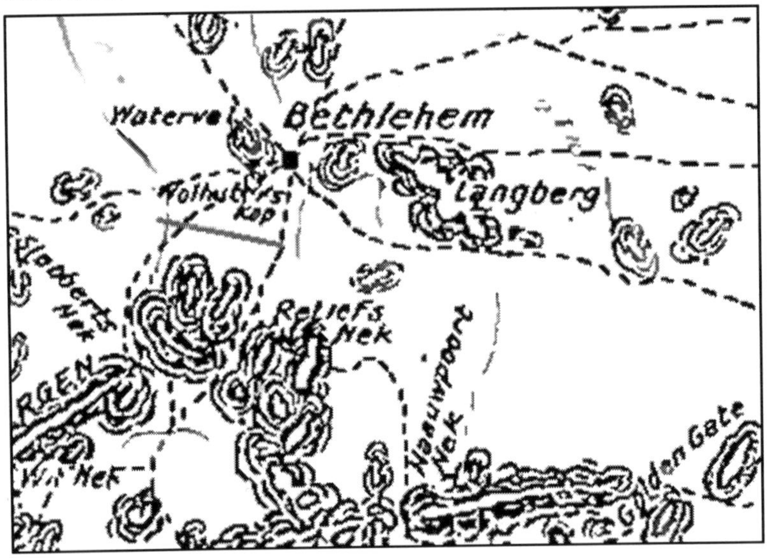

Wolhuters kop.

[158] 6/26/1900 Boer attack repulsed near Senekal, and enemy's laager burned.
6/27/1900 Attack on British at Roodeval Spruit. Boers beaten off.

When I had given my instructions to the officers, they returned to their commandos, which were stationed behind the first ridges to the south of Bethlehem, and brought them to the positions I had assigned to them.

So many of the horses were exhausted, that a large number of the burghers had to go on foot. Such of these Voetgangers[159] as were not required to attend to the waggons, I placed at Wolhuterskop.

When I had done this I gave notice to the inhabitants of Bethlehem, that as the dorp would be defended, I must insist on the women and children leaving it at once. It was not long before a number of women and children, and even a few men, started out on their way [further East] to Fouriesburg. The prisoner Vilonel, also, was conducted to this town.

Boer family with a wagon on a farm in South Africa circa 1900.[187] Titled: Father and son going to the front.

Family of Boer refugees.[188]

A Boer refugee family peering out from their covered wagon.[189]

At four o'clock that afternoon the advance guards of the enemy approached; and fifteen of their scouts made their appearance on the ridge to the north of the town. The burghers reserved their fire until these men were almost upon them. Then they let their Mausers speak, and in a moment there were nine riderless horses. The other six English made their escape, although they must have had wounds to show for their rashness.[160]

Only a few moments had passed before the roar of guns was mingled with the crack of rifles, and the whole air was filled with the thunder of battle.

[159] Infantry.

[160] As I have already stated, I intend to write on another occasion a book dealing with the art of scouting; and the above incident will there form a striking proof of how foolishly the English scouts did their work.

Everywhere the burghers fought with the utmost valour; the Voetgangers on Wolhuterskop were perhaps the bravest of them all. Whenever the enemy approached our positions, they were met by a torrent of bullets. And thus the day came to a close.

But the next day a large force of English appeared from the direction of Reitz. This had come from the Transvaal, and, if I remember rightly, was commanded by General Sir Hector Macdonald. He had come up and joined Generals Clements, Hunter, Broadwood and Paget, with the object of once and for all making an end of the Free-Staters.

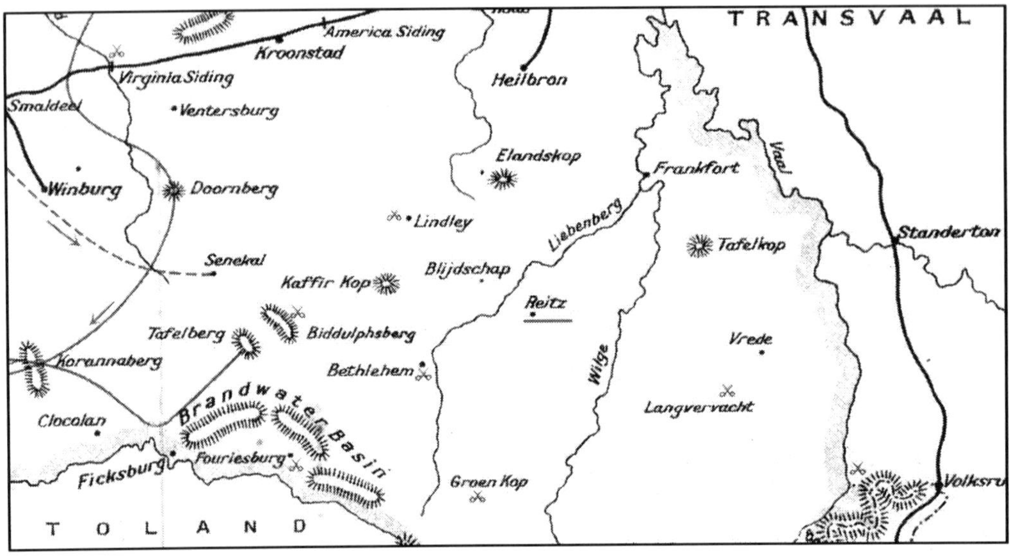

Reitz (Center) and Brandwater Area.

Our positions were now exposed to a most terrific bombardment, but fortunately without any serious consequences. I must describe here the fearful havoc that one lyddite shell wrought. It fell into the position held by Commandant Steenekamp, to the north-west of Bethlehem, and struck a rock behind which twenty-five of our horses were standing. Without a single exception every horse was killed!

The attack was pressed with the greatest vigour on the positions held by Commandants Van Aard and Piet Fourie. It became impossible for these officers to maintain their ground; and, at about twelve o'clock, before I was able to send them any reinforcements, they were compelled to give way.

Thus retreat became inevitable, and the enemy entered Bethlehem.[161]

One of our guns we were unable to remove; but before we withdrew it was thrown down the krans[162] of the mountain, and broken to pieces.

I knew at the time the number the English had lost, but now it had slipped my memory. I obtained the information from a man named Bland, who acted as our telegraphist. He had tapped the telegraph wire at Zwingkrans, and before General Clements had detected that he was not communicating with Senekal, he had received from that General a full list of the English killed and wounded.

We withdrew our commandos in a southerly direction to Retiefsnek, whither President Steyn and the Government had already preceded us.

[161] 7/1/1900 Generals Hunter and MacDonald joined hands at Frankfort.
7/4/1900 General Paget drove the enemy from strong positions towards Bethlehem.
7/7/1900 Bethlehem captured by Generals Clements and Paget.
[162] Precipice.

17. The Surrender of Prinsloo

The English, now that they had taken Bethlehem, were in need of rest; and this was especially the case with General Macdonald, who had come up by forced marches from the far-off Transvaal. A short breathing space was also a great benefit to us, for we had many preparations to make in view of probable events in the near future. I did not deceive myself as to the meaning of the present situation; now that all of us, except two small parties at Commandonek and Witnek, had retreated behind the lofty Roodebergen, I could see that, in all probability, we must before long be annihilated by the immense forces of the enemy.

The Roodebergen, which now separated us from the English, is a vast chain of mountains, extending from the Caledon River on the Basuto frontier to Slabbertsnek, then stretching away to Witzeshoek, where it again touches Basutoland. The passes over this wild mountain range are Commandonek, Witnek, Slabbertsnek, Retiefsnek, Naauwpoort and Witzeshoek. These are almost the only places where the mountains can be crossed by vehicles or horses; and, moreover, there are long stretches where they are impassable even to pedestrians.

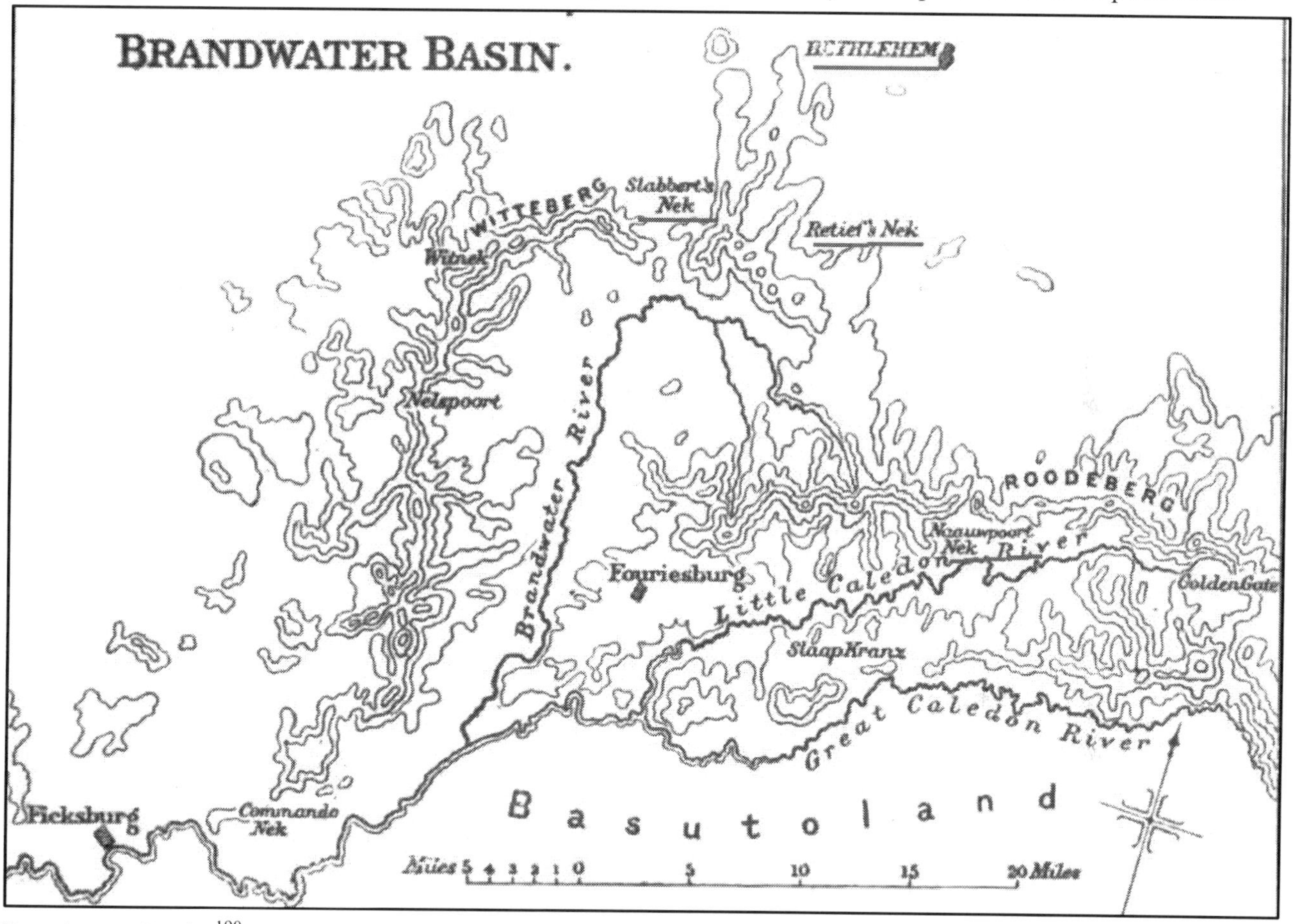

Brandwater Basin.[190] Note Retief's Nek and Slabbert's Nek (top center), south of Bethlehem, and Naauwpoort Nek.

It is plain enough, therefore, that nothing would have pleased the English more than for us to have remained behind the Roodebergen. If those Free-Staters—they must have been thinking—try to make a stand there, it will be the last stand they will ever make.

And the English would have been quite right in their anticipations. To have stayed where we then were would, without doubt, have been the end of us. Therefore, when the proposal was made that we should take positions in the mountains, I opposed it as emphatically as I could, alleging incontrovertible arguments against it. It was then decided that all our forces, with the exception of a small watch, should issue forth from behind the mountains.

We also arranged to divide the whole of the commandos[163] we had with us into three parts:—

[163] The Harrismith and Vrede commandos had also received orders to join us.

I was in supreme command of the first division, which was to march under the orders of General Botha. It consisted of burghers from Heilbron, under Commandant Steenekamp, and of Kroonstad men, under Commandant Van Aard. Besides these, there were also five hundred men from Bethlehem, under Commandant Michal Prinsloo; the burghers from Boshof, under Veldtcornet Badenhorst; a small number of Colonials from Griqualand, under Vice-Commandant Van Zyl; and some Potchefstroom burghers, who happened to be with us. Further, I took with me, for scouting purposes, Danie Theron and his corps of eighty men, recruited from almost every nation on the face of the earth; Captain Scheepers and his men also served me in the same capacity.

The Government and its officials were placed under my protection; and I was to set out, on July the 15th, in the direction of Kroonstad-Heilbron.

Witteberge at Lundean's Nek.[191] [164]

The second division was entrusted to Assistant Commander-in-Chief Paul Roux,[165] with P.J. Fourie and C.C. Froneman as Vechtgeneraals. It was composed of burghers from Fauresmith, under Commandant Visser; from Bloemfontein, under Commandant Du Plooij; from Wepener, under Commandant Roux; from Smithfield, under Commandant Potgieter; from Thaba'Nchu, under Commandant J.H. Olivier; from Jacobsdal, under Commandant H. Pretorius; and of the Deetje Bloemfontein commando, under Commandant Kolbe.

This force was to wait until the day after my departure, that is, until the 16th, and then proceed in the evening in the direction of Bloemfontein. From the capital it was to go south, and during its advance it was to bring back to the commandos all those burghers in the southern districts who had remained behind.

General Crowther[166] was given the command over the third division, which consisted of the burghers from Ficksburg, under Commandant P. De Villiers; from Ladybrand, under Commandant Ferreira; from Winburg, under Commandant Sarel Harebroek; and from Senekal, under Commandant Van der Merve.

This division was to start on the 16th, and marching to the north of Bethlehem, was to continue advancing in that direction until it fell in with the commandos from Harrismith and Vrede under Commander-in-Chief Hattingh. It would then operate, under his directions, in the north-eastern districts.

The remainder of Commandant Michal Prinsloo's Bethlehem men—that is to say, the burghers of Wittebergen—were to stay behind as a watch, and to take orders from Mr. Marthinus Prinsloo. This watch was

164 These hills are known as the Roodebergen, Wittebergen, Korannaberg, Viervoet, &c., and rise to nearly 7000 ft. See: Witteberg - Wikipedia

165 Paul Hendrik Roux (28 September 1862–8 June 1911) was a Second Boer War general and a Protestant pastor in the Dutch Reformed Church in South Africa (NGK). After the outbreak of the Second Boer War Roux served the troops from his home town of Senekal first as an almoner in the Colony of Natal, but later he was appointed general by Orange Free State president M. T. Steyn after general de Villiers was fatally wounded. https://en.wikipedia.org/wiki/Paul_Hendrik_Roux

166 Jonathan Crowther (also Jan Crowther, 20 October 1853–24 November 1926) was a Boer war general. https://en.wikipedia.org/wiki/Jonathan_Crowther_(general)

divided into three sections: the first to occupy a position at Slabbertsnek, the second at Retiefsnek, and the third at Naauwpoort. They were forbidden to use waggons; thus if the enemy should appear in overwhelming numbers, it would always be possible for them to escape across the mountains.

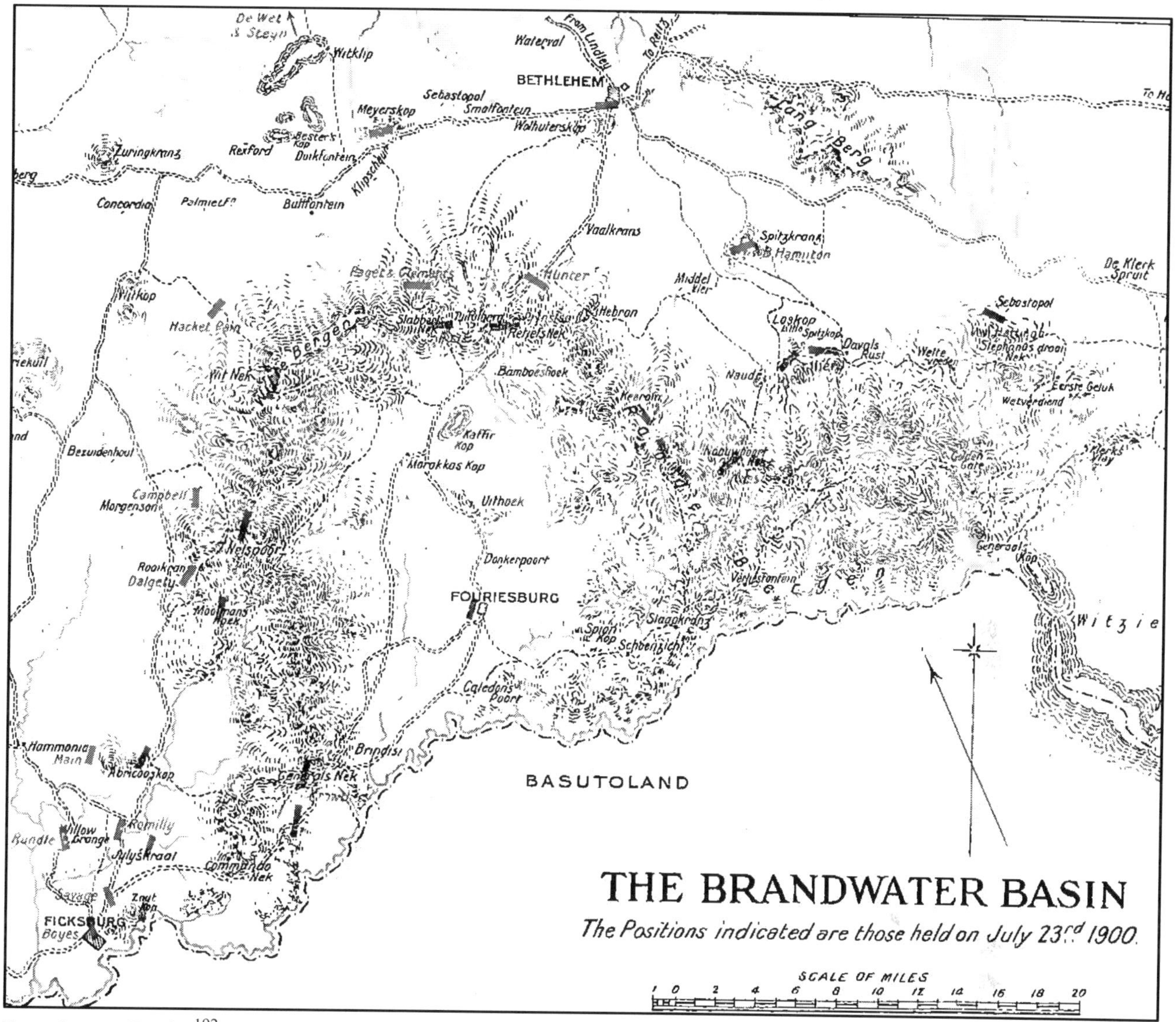

Brandwater Basin.[192] Note Topography.

My reason for selecting these men in preference to others, was that they belonged to the district, and thus were well acquainted with every foot of this rough and difficult country. Their duties were simply to protect the large numbers of cattle which we had driven on to the mountains, and I anticipated that there would be no difficulty about this, for now that all our commandos had left those parts, the English would not think it worth while to send a large force against a mere handful of watchers.

Thus everything was settled, and on the 15th of July I set out through Slabbertsnek, expecting that the other generals would follow me, conformably to my orders and the known wishes of the Government.

But what really happened?

Immediately after my departure, some of the officers, displeased that Assistant Commander-in-Chief Roux should have been entrusted with the command, expressed the wish that another meeting should be held and a new Assistant Commander-in-Chief elected. This would have been absolutely illegal, for the Volksraad had decreed that the President should be empowered to alter all the commando-laws. But even then, all would have gone well if Roux had only stood firm.

Unfortunately, however, he yielded, and on July the 17th a meeting was called together at which Mr. Marthinus Prinsloo was chosen Assistant Commander-in-Chief. He had a bare majority even at the actual meeting, and several officers, who had been unable to be present, had still to record their votes.

Not only, therefore, had Prinsloo been elected irregularly, but his election, such as it was, could only be considered as provisional. Nevertheless, for the moment, power was in his hands. How did he use it?

He surrendered unconditionally to the English.

On the 17th and 18th of July the enemy had broken through at Slabbertsnek and Retiefsnek, causing the greatest confusion among our forces.

Many of the officers and burghers were for an immediate surrender, as appears from the fact that the same assembly which, in defiance of the law, elected Mr. Prinsloo as Commander-in-Chief, also decided, by seventeen votes to thirteen, to give up their forces to the enemy. But this decision was at once rescinded—an act of policy on the part of the officers—and it was agreed to ask for an armistice of six days, to enable them to take counsel with the Government.

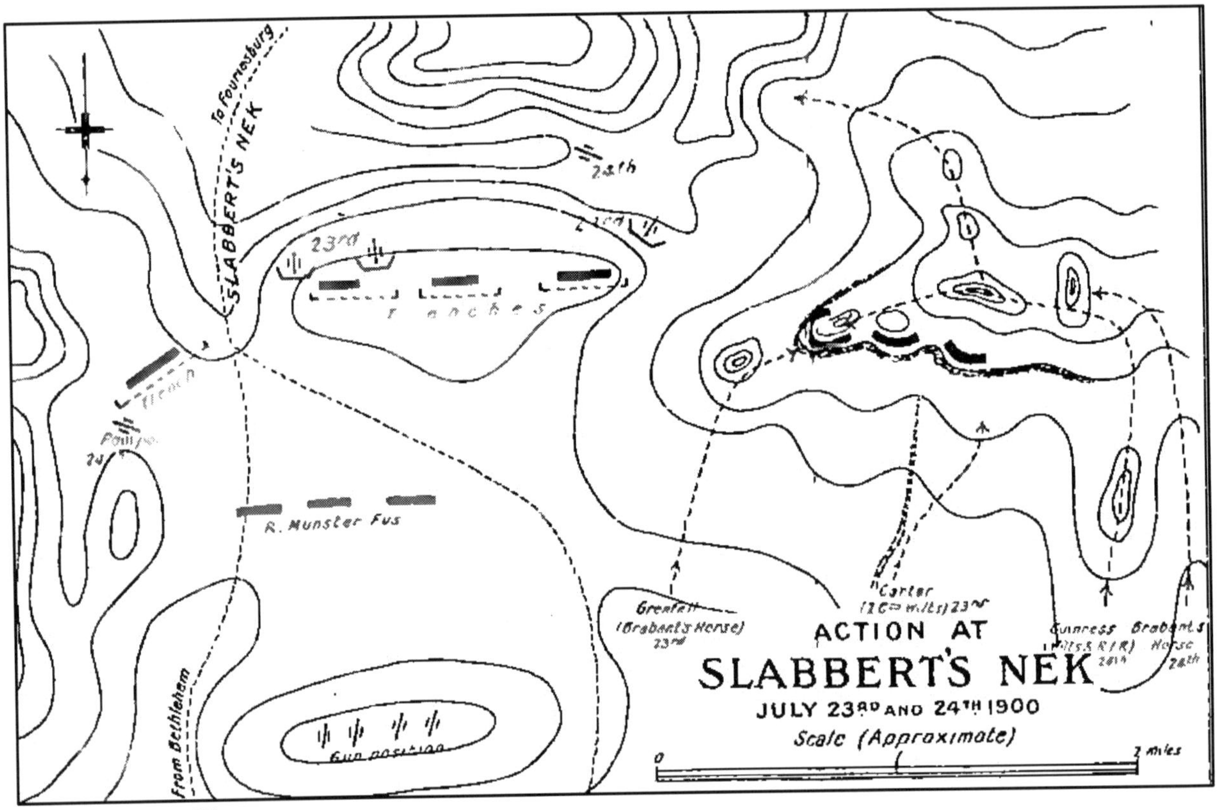

Slabbert's Nek.[193]

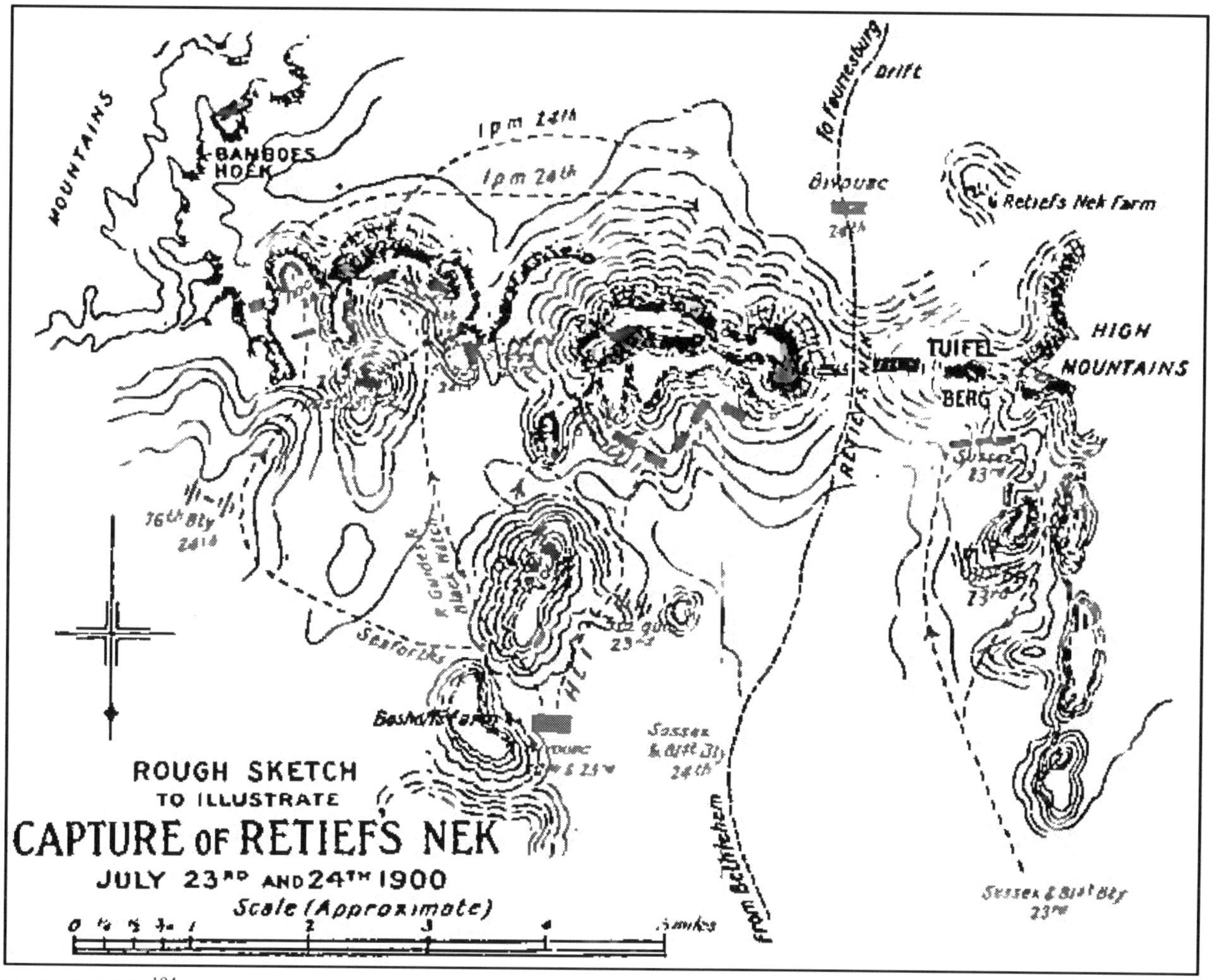

Retief's Nek.[194]

A more senseless course of action could hardly be imagined. The Boer Army, as anybody could see, was in a very tight place. Did its officers think that the English would be so foolish as to grant an armistice at such a time as this—when all that the burghers wanted was a few days in which to effect their escape? Either the officers were remarkably short-sighted, or ... something worse.

It was still possible for the commandos to retire in the direction of Oldenburg or of Witzeshoek. But instead of getting this done with all speed, Mr. Prinsloo began a correspondence with General Hunter about this ridiculous armistice, which the English general of course refused to grant.

It was on July the 29th, 1900, that Prinsloo, with all the burghers on the mountains, surrendered unconditionally to the enemy.[167]

The circumstances of this surrender were so suspicious, that it is hard to acquit the man who was responsible for it of a definite act of treachery; and the case against him is all the more grave from the fact that Vilonel, who was at that time serving a term of imprisonment for high treason, had a share in the transaction.

Prinsloo's surrender included General Crowther, Commandants Paul De Villiers, Ferreira, Joubert, Du Plooij, Potgieter, Crowther, Van der Merve, and Roux; and about three thousand men.[168]

[167] 7/30/1900 Surrender of Generals Marthinus Prinsloo, A. J. Villiers, and Crowther, and 4000 Boers to General Hunter, also, Commandant Frans Joubert, (brother of General Piet Joubert) who commanded at Bronkhorst Spruit. FYI Jacobis Prinsloo is the other commandant.

[168] De Wet indeed escaped escorting president Steyn, while the remainder of the Boer army instead failed to defend the pass Slabbert's Nek on 15 July 1900 and gave up Retief's Nek after a fight on 23–24 July 1900. https://en.wikipedia.org/wiki/Brandwater_Basin

Passing the Flag. Surrender of 5,000 Boers near Fouriesburg, to General Hunter. July 31st, 1900.[195] [169]

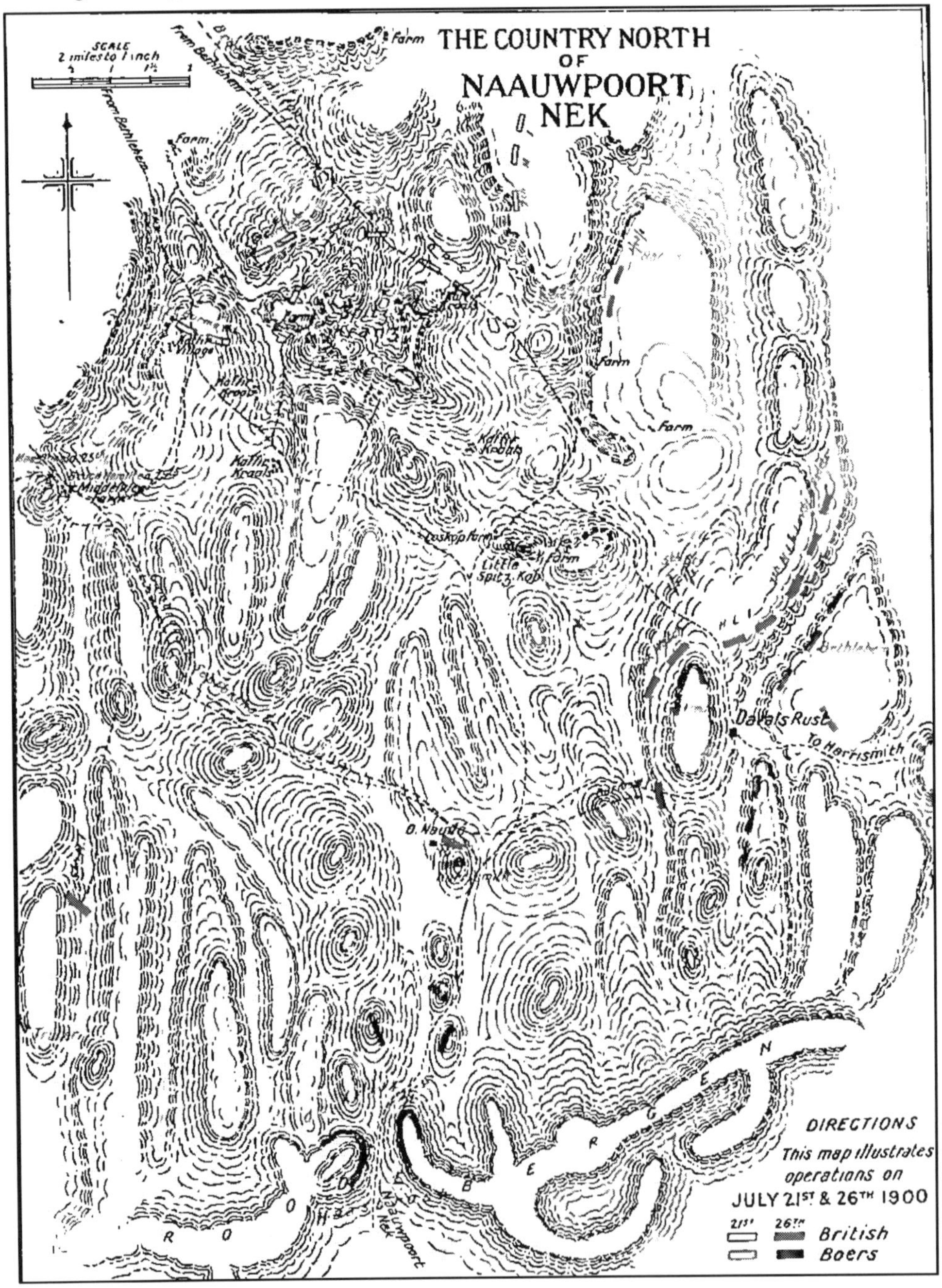

Naauwpoort Nek.[196]

[169] See: Brandwater Basin: https://en.wikipedia.org/wiki/Brandwater_Basin

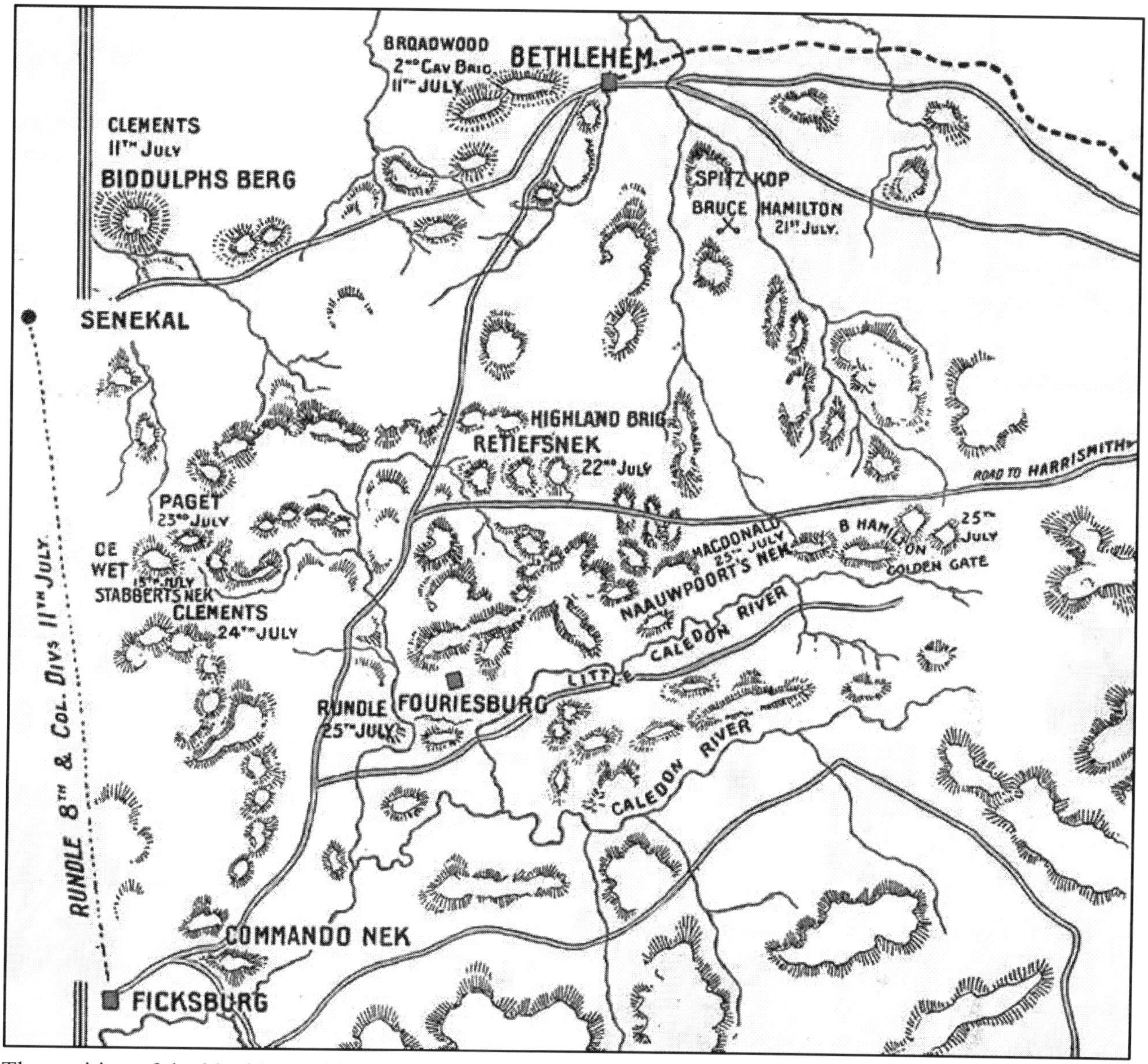

The position of the blocking British troops before the surrender of Prinsloo is shown. Note Nauwpoort's Nek.[197]

Sandkol, hot sandy hill near Surrender Hill, on the R711 road, Slaapkrans, Orange Free State. Photo 2015.[198]
Surrender Hill monument, beside the Clarens-Fouriesburg road in the eastern Free State, South Africa.[199]

A Scene in the Boer Lines before the Surrender. The Despirited Crowd Awaiting a Decision as to the Terms [200]

Surrender of Prinsloo's Boer forces in Brandwater basin, 1900.[201]

Boer Prisoners on the March to the Base.[202]

Group of Captured Boer Leaders L. to R. Joubert, Dupluoy, Potgieter, Crowther, and Joukal.

Jonathan Crowther.[203 170]

Starting on the Long and Weary Way. The Captured Boers Entraining under British Guard for the Journey South.[204] General Marthinus J. Prinsloo.[205 171 172]

The most melancholy circumstance about the whole affair was that, when the surrender was made, some of the burghers had reached the farm of Salamon Raath, and were thus as good as free, and yet had to ride back, and to go with the others to lay down their arms.

As to Roux, the deposed Commander-in-Chief, there is a word to be added. I had always heard that he was a very cautious man, and yet on this occasion he acted like a child, going in person to General Hunter's camp to protest against the surrender, on the ground that it was he (Roux), and not Prinsloo, that was Commander-in-

[170] Jonathan Crowther (also Jan Crowther, 20 October 1853–24 November 1926) was a Boer war general. https://en.wikipedia.org/wiki/Jonathan_Crowther_(general)

[171] Prinsloo Marthinus General Born in 1838, he was serving as the Commandant of the Winburg Commando in 1867. He was a man of wealth and served as an elder of the church. At the start of the Boer War, he commanded Free State commandos in Natal. In the attack on Caesar's Camp on 6 Jan 1900, he was accused of indecision. He commanded the defensive line on the Tugela after the battle of Spion Kop. When Ladysmith was relieved and Buller pushed north, Prinsloo commanded the Free State commandos at Van Reenen's Pass. Now aged 62, he was seen as too old for combat and retired to his farm. He re-engaged and operated in the Free State. On 30 July 1900 he was captured near Fouriesburg in the Brandwater Basin with over 4,000 men, 2,800 cattle, 4,000 sheep, 5,000 horses and 2 million rounds. He died in 1903. See Also: Marthinus Prinsloo - Wikipedia

[172] The British surrounded Prinsloo by also blocking the passes of Witnek, Kommandonek, Noupoortsnek (Nauwpoortsnek) and finally the Golden Gate pass to the east on the Little Caledon River, so that Prinsloo felt forced to surrender with his army to general Archibald Hunter on 30 July 1900. Some 4300 troops including Prinsloo, Roux and Crowther were taken prisoner of war near Fouriesburg, most of them at Surrender Hill. See: https://en.wikipedia.org/wiki/Brandwater_Basin See Also: https://en.wikipedia.org/wiki/Archibald_Hunter Boer general Piet Fourie or general Jan Hendrik Olivier had got away with 1500 men and several commanders in the east over the Golden Gate pass in time. See: https://en.wikipedia.org/wiki/Piet_Fourie And: https://en.wikipedia.org/wiki/Jan_Hendrik_Olivier And https://en.wikipedia.org/wiki/Brandwater_Basin

Chief. One can hardly believe that he really thought it possible thus to nullify Prinsloo's act. But he certainly behaved as if he did, and his ingenuous conduct must have afforded much amusement to the English general.

If any one is in doubt as to what was the result of General Roux's absurd escapade, I have only to say that the English had one prisoner the more!

Those who escaped were but few. Of all our large forces, there were only Generals Froneman, Fourie and De Villiers (of Harrismith); Commandants Hasebroek, Olivier, Visser, Kolbe, and a few others; a small number of burghers, and six or seven guns, that did not fall into the hands of the English.

What, then, is to be our judgment on this act of Prinsloo and of the other chief officers in command of our forces behind the Roodebergen?

That it was nothing short of an act of murder, committed on the Government, the country, and the nation, to surrender three thousand men in such a way. Even the burghers themselves cannot be held to have been altogether without guilt, though they can justly plead that they were only obeying orders.[173]

The sequel to Prinsloo's surrender was on a par with it. A large number of burghers from Harrismith and a small part of the Vrede commando, although they had already made good their escape, rode quietly from their farms into Harrismith, and there surrendered to General Sir Hector Macdonald.—One could gnash one's teeth to think that a nation should so readily rush to its own ruin!

Breaking up Mauser rifles at Harrismith.[206] [174]

Paul Hendrik Roux.[207] [175]

Boer Prisoners attending church service at Diyatalwa Camp, Ceylon.[208]

[173] After the departure of De Wet and his party the remaining Boers dithered and the British captured the mountain passes of Slabbert's Nek and Retief's Nek on 23 and 24 July 1900. They locked Prinsloo, Roux and Crowther in by also blocking the passes of Witnek, Kommandonek, Noupoortsnek (Nauwpoortsnek) and the Golden Gate. Prinsloo felt forced to surrender with all his troops, including Crowther's, to the British general Archibald Hunter.

[174] 8/4/1900 Surrender of Harrismith to General MacDonald.

18. I Am Driven into the Transvaal

As I have already stated, I led my commando, on the 15th of July, through Slabbertsnek, out of the mountain district. My force amounted to the total of two thousand six hundred burghers. The Government travelled with us, and also alas! four hundred waggons and carts. Whatever I did, it seemed as if I could not get rid of the waggons!

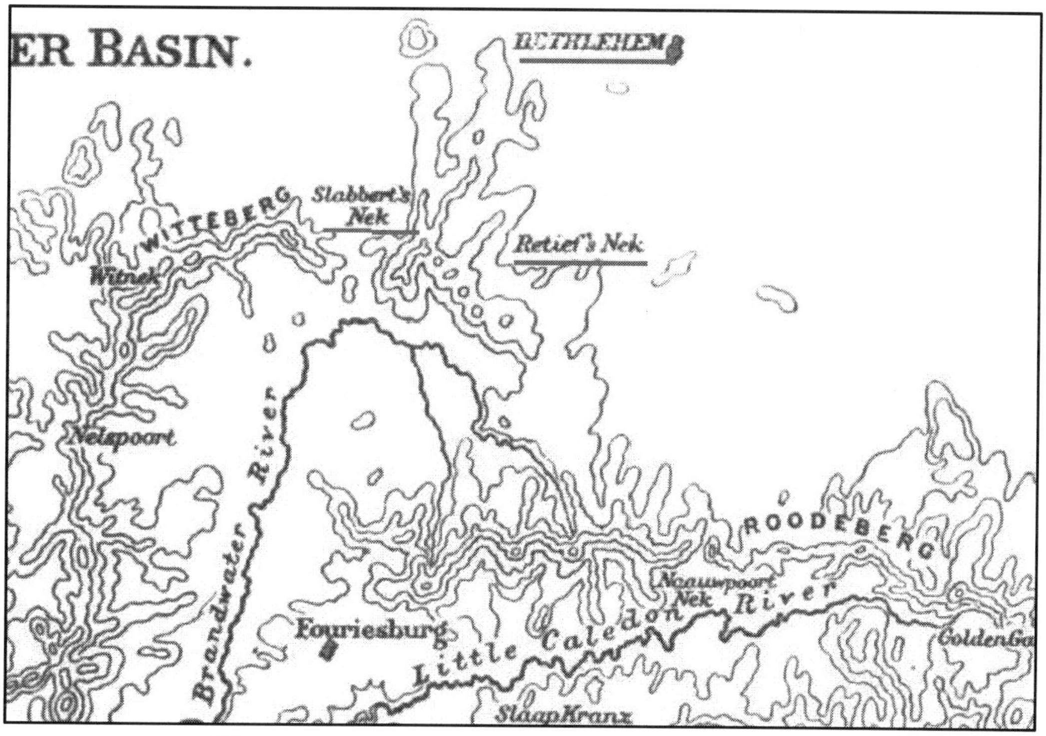

Brandwater Basin.[209] Note Slabbert's Nek (top center) and Retief's Nek south of Bethlehem.

That night we reached a farm six miles to the east of Kaffirs Kop; during our march we passed a column of the enemy that had left Bethlehem in the afternoon.

On the following day [16th] I came into contact with some English troops, who were marching in the direction of Witnek. They sent out a body of cavalry to ascertain what our plans might be. It was very annoying to me that they should thus discover our whereabouts, because it made it impossible to carry out my intention of attacking one or other of the English forces.

However, nothing was done that day, as neither we nor the enemy took up the offensive.

In the evening we pushed on to the east of Lindley, and the following day [17th] remained at the spot we had reached.

[175] English: Boer prisoners of war at the Sunday service in Diyatalawa camp, Ceylon, who were mostly taken captive at the Brandwater basin, Orange Free State, under general Prinsloo. Prinsloo's surrender was a major setback for the Boer cause during the war. Reverends Petrus Postma from Pretoria and Paul Hendrik Roux from Senekal stand side by side just right of center, and general J.H. Olivier is visible at middle, right. One caption to the photo was as follows:

The Boer Prisoners at Service in Ceylon.

The prisoners are guarded by the King's Royal Rifles, under Colonel Gore-Brown, Colonel Vincent being Commandant, and Colonel Jesse Coope in immediate charge of them. Temporary hospital huts have been erected and brightened with pictures and illustrated papers, and officials of the local branch of the Bible Society have distributed Bibles and portions of the Scriptures in Dutch. These were welcomed and specially acknowledged by a letter of thanks by a prisoner known as "the fighting parson [Petrus Postma]." Colonel Jesse Coope, who is very popular, fosters productive manufactures and artistic activity among the men, disposing of their work through an agent. Tanks for the storage of water being required, the prisoners were invited to volunteer for the work at a reasonable rate of pay, and many availed themselves of the offer. The population of Ceylon does not exceed 6,000 [Europeans?], and the settlement of the Boer prisoners has had a wholesome effect, not only on themselves, but on the Cingalese. The minister who is officiating (in the above photograph), is the "fighting parson" alluded to–the Rev. Mr. Postma–and General Roux stands beside him. Olivier can be identified nearer to the right margin of the picture and several rows further back.

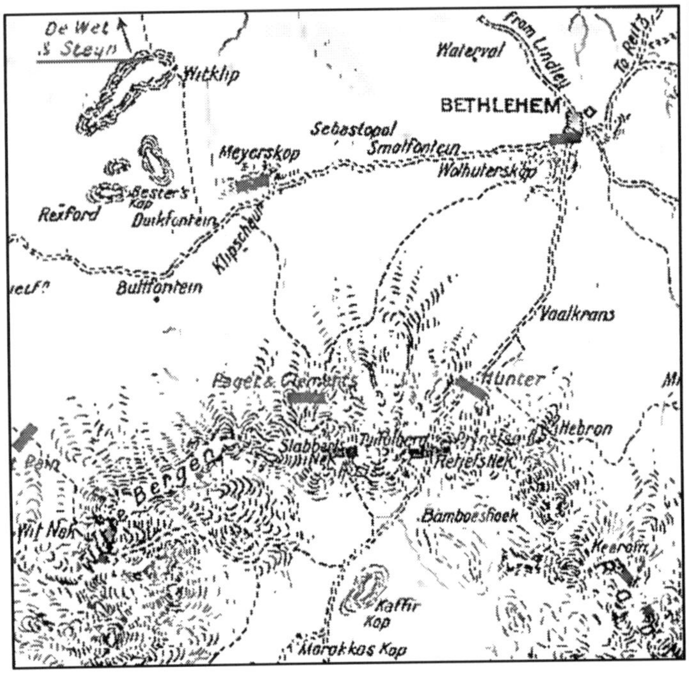

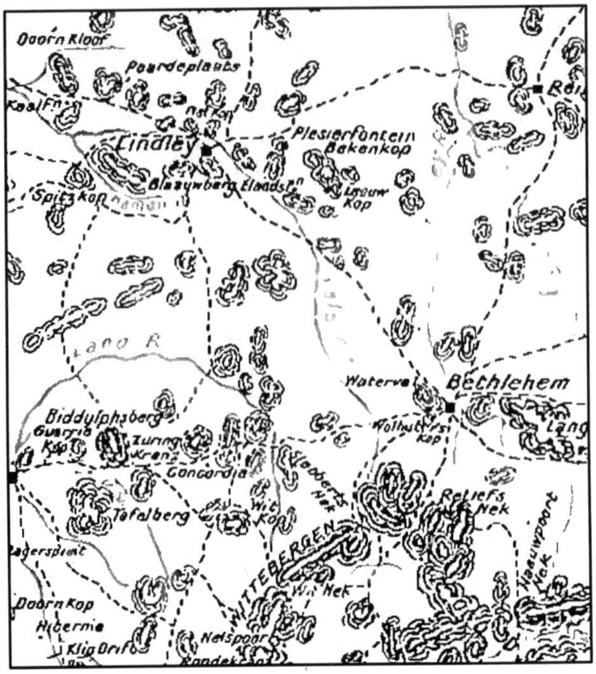

De Wet's route out of the Basin. Note 'De Wet, & Steyn top left. Another view showing their route.

The next evening we marched to the farm of Riversdale; and the night of the 18th found us on the farm of Mr. Thomas Naude, to the north-west of Lindley. We discovered that the English had all left this village and gone to Bethlehem. My scouts reported to me, the following day, that an English force, some four hundred men strong, was approaching Lindley. Need I say that these men had to be captured? With five hundred burghers and two guns I went out to do this. When I was only a short distance from my camp, I received a report that a large force of cavalry, numbering seven or eight thousand men, had arrived on the scene from Bethlehem. This compelled me to abandon the idea of capturing those four hundred men, and, instead, to try to escape in a westerly direction from this large body of mounted troops.[176]

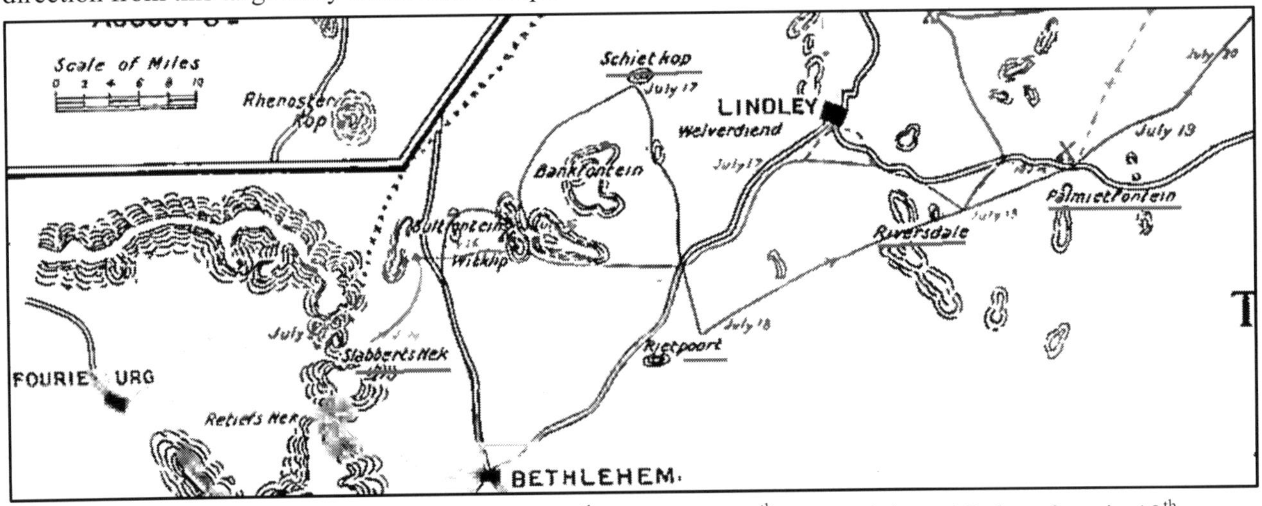

Starting at Slabbert's Nek to Schiet kop on July 17th, Rietpoort 18th, Riversdale and Palmetfontein 19th.

That evening we reached the farm of Mr. C. Wessels, at Rivierplaats. The next day [19th] we were forced to move on, for the mounted troops were coming nearer to us. They marched, however, somewhat more to the right in the direction of Roodewal; whereas I went towards Honingspruit, and halted for the night at the farm of Paardenkraal.

[176] 7/19/1900 General Little engaged De Wet near Lindley, and broke up his forces.

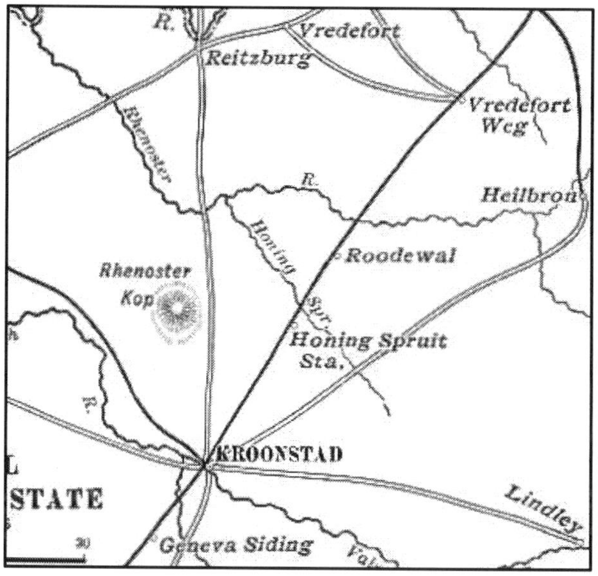

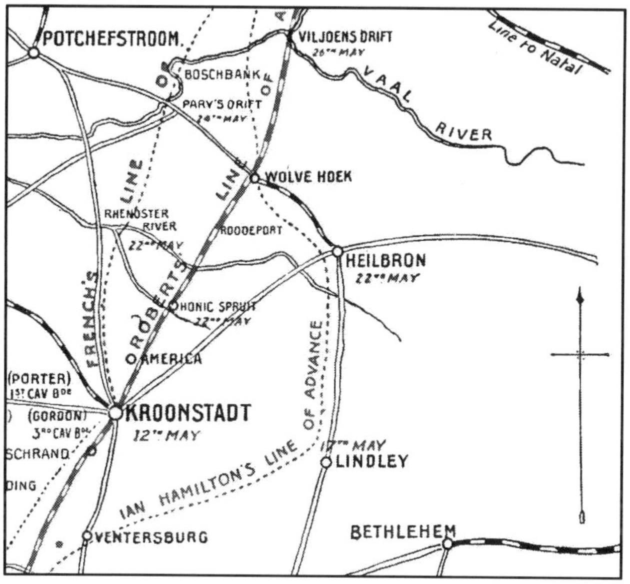

Honing Spruit Station and Roodewal; shown as Roodeport and Honic Spruit on second map.

On the following morning, the 20th of July, I let the commando go on, whilst I stayed behind to reconnoitre from a neighbouring kop. The President, and also some members of the Government, remained with me. We had the opportunity of accepting the invitation of Mr. C. Wessels to take breakfast at his house. It was there that General Piet de Wet came to me and asked if I still saw any chance of being able to continue the struggle?

The question made me very angry, and I did not try to hide the fact.

"Are you mad?"[177] I shouted, and with that I turned on my heel and entered the house, quite unaware that Piet de Wet had that very moment mounted his horse, and ridden away to follow his own course.[178]

After breakfast we climbed the kop; and when we had made our observation we followed after the laager. On reaching the commando, I gave orders to outspan at twelve o'clock.

While this was being done I heard from my sons that Piet de Wet had told them that we should all be captured that night near the railway line. He had not known that it was my intention to cross the railway that night, but he had guessed as much from the direction I let my commando take.

At two o'clock I received a report that two divisions of English troops were drawing near. One division was six miles to the left, and the other eight miles to the right of the road along which we had come.

I gave orders immediately that the laager should break up. What an indescribable burden this camp, with four hundred and sixty waggons and carts, was to me! What a demoralizing effect it had upon the burghers! My patience was sorely tried. Not only were we prevented from moving rapidly by these hampering waggons, but also, should we have to fight, a number of the burghers would be required to look after them, and so be unable to fire a shot.

A Boer Encampment "Waggons Massed on the Veldt." [210]

[177] I put down here the very words I used, for any other course would not be honest.

[178] 7/26/1900 Philip de Wet, younger brother of Christian de Wet, surrendered at Kroonstad. General Hunter occupied Fouriesburg. General MacDonald, after fighting a rearguard action, blocked Naauwpoort Nek.

We marched to the farm of Mr. Hendrik Serfontein, on Doornspruit, and whilst I was there, waiting for darkness, some burghers, who were not my scouts, brought a report that there were English camps both at Honingspruit and at Kaallaagte.

This alarmed the President [Steyn] and the members of the Government, because, should this report prove true, we should be unable to cross the railway line without hard fighting, and besides there would be a considerable risk of being taken prisoner.

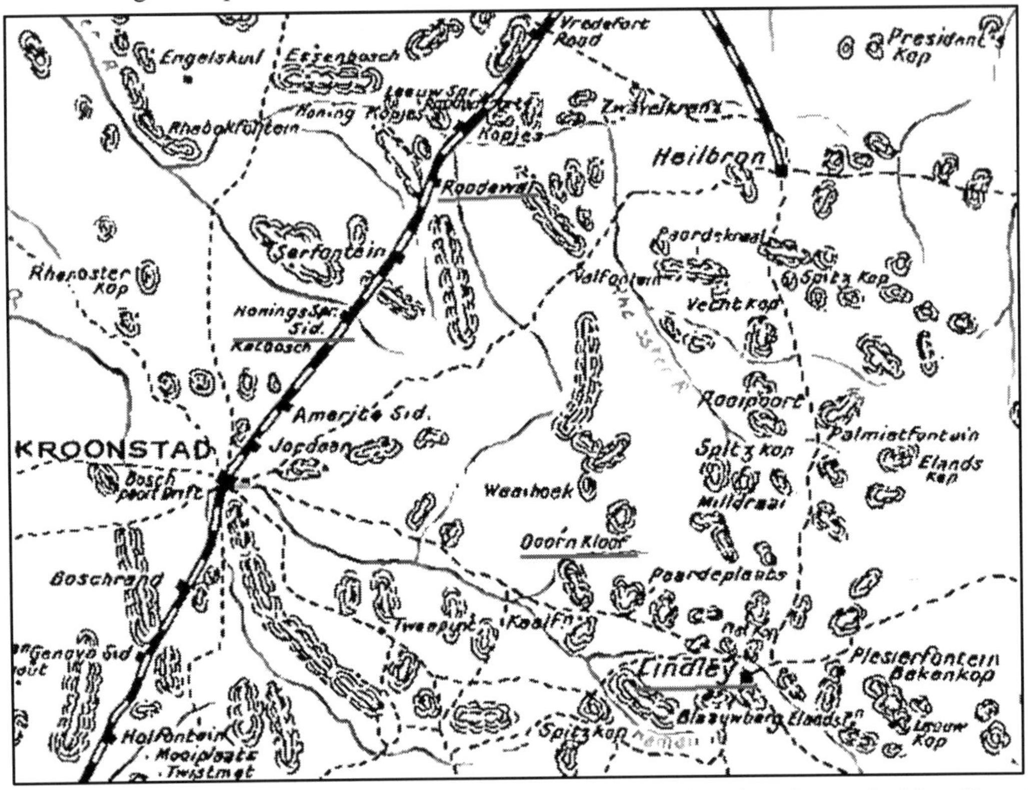

From Lindley and west to Hornings Spruit siding, and Roodewal to the north. Note Dorn Kloof and small 'spruit' probably Dorn Spruit.

For myself, I did not pay any attention to these burghers. I relied on my own scouts, and I waited for their reports. I knew that if there had been any truth in what we had been told, that I should have heard the news already from the men whom I had sent out in the morning in that direction. At last some of Captain Scheepers' men appeared—he was scouting in front, and Captain Danie Theron in the rear—and reported that the railway line was clear, with the exception that at Honingspruit there were half a dozen tents, and four in the Kaallaagte[179] to the north of Serfontein, and a few small outposts. This information came as a great relief to the President and the members of the Government.

If I was to escape from the large force which was dogging my footsteps, it was now necessary to cross the railway. I had made all preparation for this move. I had left behind me, that afternoon, on the banks of Doornspruit a commando of burghers, with orders to keep the enemy back until we should have crossed the line. And now I only waited until the darkness should come to my assistance.

As soon as the night came I ordered the waggons to proceed in four rows, with a force on each side, and with a rearguard and vanguard. Immediately behind the vanguard followed the President and myself. When we were about twenty minutes' march from the railway line I ordered the two wings of my force, which were about three miles apart, to occupy the line to the right and left of Siding.

Before we had quite reached the railway I ordered the vanguard to remain with the President, whilst I myself, with fifteen men, rode on to cut the telegraph wire. Whilst we were engaged in this task a train approached at full speed from the south. I had no dynamite with me, and I could neither blow it up nor derail it. I could only place stones on the line, but these were swept away by the cowcatcher, and so the train passed in safety.

[179] Kaallaagte—a barren hollow.

I had forbidden any shooting, for an engagement would have only produced the greatest confusion in my big laager.

Just as the last waggon was crossing the line, I received a report that Captain Theron had captured a train to the south of us. Having ordered the waggons to proceed, I rode over to see what had happened. When I arrived at the scene of action I found that the train had come to a standstill owing to the breaking down of the engine, and that on this the English troops had at once opened fire on my men, but that it had not been long before the enemy surrendered. Four of the English, but only one of our burghers, had been wounded.[211]

It was very annoying that the laager was so far off, but it was impossible to carry off the valuable ammunition which we found on the train.

I gave orders that the four wounded soldiers, who were under the care of the conductor of the train, should be taken from the hut in which I had found them, and placed in a van where they would be safe when I set fire to the train. After the burghers had helped themselves to sugar, coffee, and such things, I burned everything that was left. My ninety-eight prisoners I took with me.

We had not gone far when we heard the small arms ammunition explode; but I cannot say that the sound troubled me at all!

Thus we crossed the line in safety, and Piet de Wet's prediction did not come true. He knew that we had a large force behind us, and believing that the railway line in front of us would be occupied by troops, he had said: "This evening you will all be captured on the railway line." Yet instead of finding ourselves captured, we had taken ninety-eight prisoners, and destroyed a heavily-laden train! How frequently a Higher Power over-rules the future in a way we least expect!

That night we reached the farm of Mahemsspruit. From there we moved on to the Wonderheurel; and on the 22nd of July we arrived at the farm of Vlakkuil. I remained here for a day, for I wished to find out what the English troops (they had remained where we left them by the railway line) were intending to do.

Whilst I was waiting I despatched some corn on a few of my waggons to Mr. Mackenzie's mills near Vredefort, giving orders that it should be ground.

During the afternoon it was reported to me that a strong column of English were marching from Rhenosterriviersbrug to Vredefort, and that they had camped on the farm Klipstapel, some eight miles from my laager.

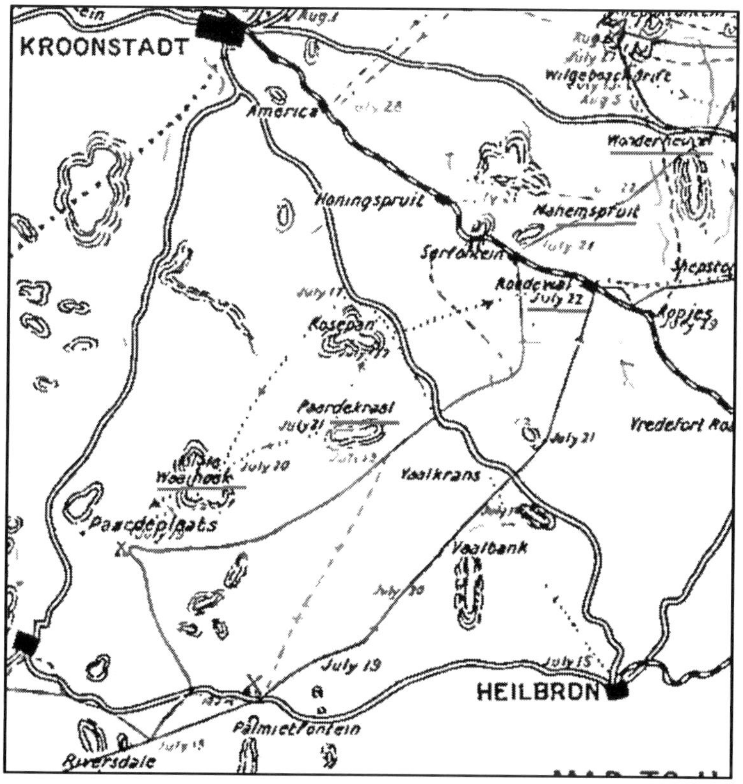

North is to the Right. Near Vaalbank 20th, Paardekraal on the 21st to Roodewal 22nd, then Mahemspruit and Wonderheure. North is to the right side.

De Wet's Drift on the Rhenoster River.

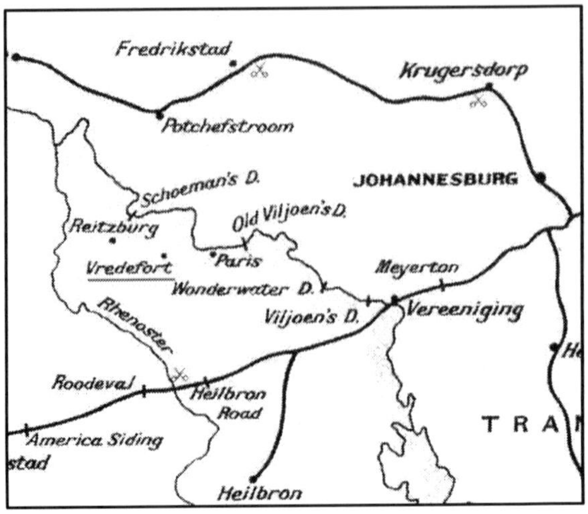

Vredefort.

Shortly after sunrise the following morning a second report was brought to me. It appeared that the enemy had sent out a force to capture our grain waggons, and had nearly overtaken them.

In an instant we were in our saddles, but we were too late to save our corn.

When the enemy saw us they halted at once; and meanwhile the waggons hurried on, at their utmost speed, to our camp.

The English numbered between five and six hundred men, whilst we were only four hundred. But although we were the smaller force, I had no intention of allowing our waggons to be captured without a shot, and I ordered my burghers to charge.

It was an open plain; there was no possible cover either for us or for the English. But we could not consider matters of that sort.

The burghers charged magnificently, and some even got to within two hundred paces of the enemy. They then dismounted, and, lying flat upon the ground, opened a fierce fire. One of the hottest fights one can imagine followed.

Fortunately a few paces behind the burghers there was a hollow, and here the horses were placed.

After an hour's fighting, I began to think that any moment the enemy might be put to rout. But then something happened which had happened very often before—a reinforcement appeared.

This reinforcement brought two guns with it; thus nothing was left to me but retreat. Our loss was five killed and twelve wounded. What the loss of the English was I do not know, but if the Kaffirs who lived near there are to be trusted, it must have been considerable.[180]

In the evening I moved my camp to Rhenosterpoort; whilst the English went back to Klipstapel.

And now the English concentrated their forces. Great Army Corps gathered round. From Bethlehem and Kroonstad new columns were constantly arriving, until my force seemed nothing in comparison with them.

I was stationed on the farm of Rhenosterpoort, which is situated on the Vaal River, twenty miles from Potchefstroom. At that town there was a strong force of the enemy, on which I had constantly to keep my eye.

But, notwithstanding their overpowering numbers, it seemed as if the English had no desire to follow me into the mountains of Rhenosterpoort. They had a different plan. They began to march around me, sending troops from Vredefort over Wonderheurel to Rhenoster River, and placing camps all along the river as far as Baltespoort, and from there again extending their cordon until Scandinavier drift was reached.

[180] 77/21/1900 Advance begun from Pretoria east, along Delagoa Bay Railway. A supply train, with 100 Welsh Fusiliers, captured near Honing Spruit.
7/23/1900 The Black Watch capture a hill at Retief's Nek. The Highland Light Infantry were compelled to retire from a steep hill above the Nek.
7/25/1900 Roberts's force reached Balmoral on the way to Middelburg. French's Cavalry and Hutton's Mounted Infantry put Boers to flight six miles south of Balmoral. General French crosses Oliphant's River.

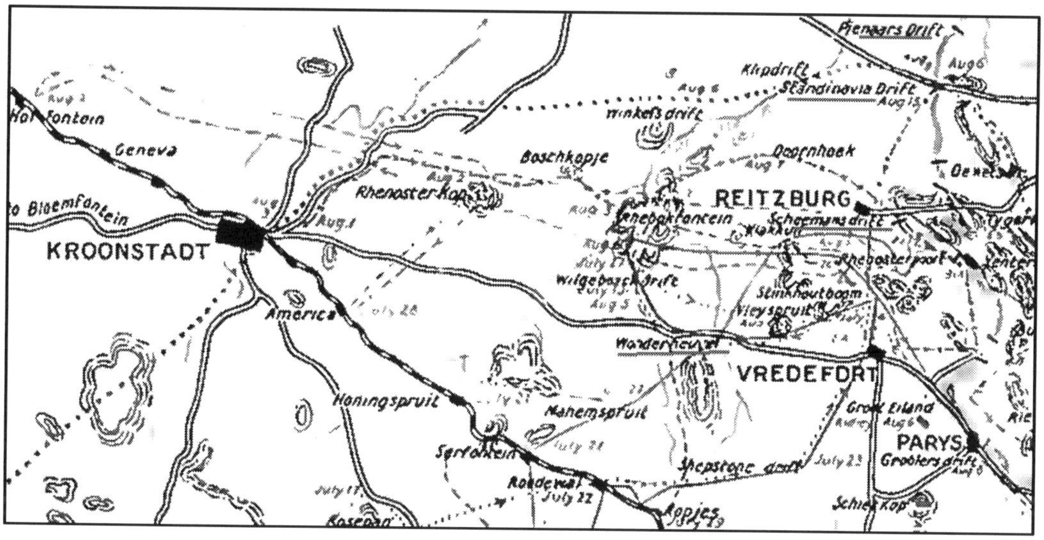

Wonderheurel, Schoeman's Drift located under Reitzburg. Standinavia drift is at the top near North is to the right near Pienaars Drift.

We were forced now either to break through this cordon, or to cross the Vaal River into the South African Republic. The Free-Stater preferred to remain in his own country, and he would have been able to do so had we not been hampered by a big "waggon-camp" and a large number of oxen. As these were with us, the Boers found it hard to make up their minds to break through the English lines as a horse-commando, as it necessitated leaving all these waggons and oxen in the hands of the enemy. But there we were between the cordon and the Vaal River.

Almost every day we came into contact with the enemy's outposts, and we had an engagement with them near Witkopjes Rheboksfontein. On another occasion we met them on different terms, in Mr. C.J. Bornman's house. Some of his "visitors" were, unfortunately for themselves, found to be English scouts—and became our prisoners.

We remained where we were until the 2nd of August. On that day we had to drink a cup of bitterness. It was on the 2nd of August that I received the news that Prinsloo had surrendered near Naauwpoort.[181]

A letter arrived from General Broadwood in which he told me that a report from General Marthinus Prinsloo addressed to me had arrived through his lines. The bearer of it was General Prinsloo's secretary, Mr. Kotze. And now the English General asked me if I would guarantee that the secretary should be allowed to return, after he had given me particulars of the report he had brought.

Mr. Prinsloo's secretary must certainly have thought that he was the chosen man to help us poor lost sheep, and to lead us safely into the hands of the English! But I cannot help thinking that he was rather too young for the task.

I had a strong suspicion that there must have been some very important screw loose in the forces which we had left stationed behind the Roodebergen, for on the previous day I had received a letter from General Knox, who was at Kroonstad, telling me that General Prinsloo and his commandos had surrendered.

In order to gain more information I gave General Broadwood my assurance that I would allow Mr. Prinsloo's secretary to return unhurt.

[181] 7/23/1900 The Black Watch capture a hill at Retief's Nek. The Highland Light Infantry were compelled to retire from a steep hill above the Nek.
7/26/1900 Piet de Wet, younger brother of Christian de Wet, surrendered at Kroonstad. General Hunter occupied Fouriesburg. General MacDonald, after fighting a rearguard action, blocked Naauwpoort Nek.

When I had done this the President and some members of the Government rode out with me to meet the bearer of this report. We did not wish to give him any opportunities to spy out our positions. Half way between the English lines and our own we met him. He presented us with this letter:—

Hunter's Camp, 30th July, 1900.

To the Commander-in-Chief, C.R. de Wet.

Sir,—

I have been obliged, owing to the overwhelming forces of the enemy, to surrender unconditionally with all the Orange Free State laagers here.

I have the honour to be, Sir,

Your obedient servant,

M. Prinsloo,

Commander-in-Chief.

I sent my reply in an unclosed envelope. It ran as follows:—

In the Veldt, 3rd August, 1900.

To Mr. M. Prinsloo.

Sir,—

I have the honour to acknowledge the receipt of your letter dated the 30th of last month. I am surprised to see that you call yourself Commander-in-Chief. By what right do you usurp that title? You have no right to act as Commander-in-Chief.

I have the honour to be,

C.R. de Wet,

Commander-in-Chief.

Hardly had I written this letter before two men on horseback appeared. They proved to be burghers sent by General Piet Fourie, who was with Prinsloo at the time of his surrender. These burghers brought from Generals Fourie, Froneman, and from Commandant Hasebroek and others, a fuller report of the surrender of Prinsloo. We learnt from the report that not all of the burghers had surrendered, but that, on the contrary, some two thousand had escaped. This news relieved our minds.

President Steyn and myself determined to despatch Judge Hertzog to the commandos which had escaped, giving him instructions to bring them back with him if possible. We had been told that these commandos were somewhere on the Wilgerivier, in the district of Harrismith.

My position had now become very difficult. It seemed, as far as I could discover, that there were five or six English generals and forty thousand troops, of which the greater part were mounted, all of them trying their best to capture the Government and me.

My force numbered two thousand five hundred men.

On the afternoon when I received the above-mentioned letter, there was still a way of escape open to me, through Parijs[182] to Potchefstroom. This road crossed the Vaal River at Schoemans drift, and then followed the course of the stream between Parijs and Vanvurens kloof. It was now, however, somewhat unsafe, for that same afternoon a large force of the enemy was marching along the Vaal River from Vredefort to Parijs. These troops would be able to reach Vanvurens kloof early the following morning; whilst the force at Potchefstroom, which I have already mentioned in this chapter, would also be able to arrive there at the same hour.

I led my burghers that evening across the Vaal River to Venterskroon, which lies six miles from Schoemans drift. The following morning my scouts reported that the English were rapidly approaching from Potchefstroom in two divisions; one was at Zandnek; the other had already reached Roodekraal on its way to Schoemans drift. One of these divisions, my scouts told me, might be turning aside to Vanvurenskloof.

[182] Parijs is situated on the Vaal River.

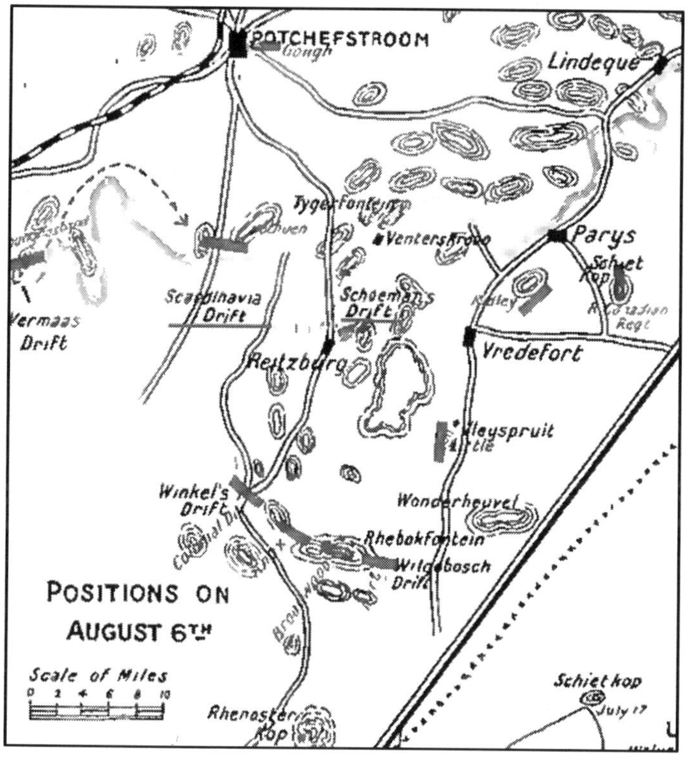

Scandihavia Drift below Potchefstroom and Schoemans Drift, lower center above Reitzburg. August 6th.[212]

Now the road from Venterskroon passed between two mountain chains to the north of Vanvurenskloof; and I feared that the English would block the way there. I had to avoid this at all costs, but I had hardly a man available for the purpose. The greater part of my burghers were still to the south-east and south-west of the Vaal River.

There was nothing left for me to do except to take the burghers who remained with me, and, whilst the laager followed us as quickly as possible, to advance and prevent the enemy from occupying the kloof. This I did, and took a part of my men to Vanvurenskloof, whilst I sent another body of burghers to Zandnek.

Everything went smoothly. The enemy did not appear and the laager escaped without let or hindrance—and so we camped at Vanvurenskloof.

I must have misled the English, for they certainly would have thought that I would come out by the road near Roodekraal. But I cannot understand why the force in our rear, which had arrived at Parijs the previous evening, remained there overnight, nor why, when they did move on the following morning, they marched to Lindequesdrift, eight miles up the Vaal River, and not, as might well have been expected, to Vanvurenskloof.

The burghers whom I sent in the direction of Roodekraal had a fight with the enemy at Tijgerfontein. A heavy bombardment took place; and my men told me afterwards that the baboons, of which there were a large number in these mountains, sprang from cliff to cliff screaming with fright—poor creatures—as the rocks were split on every side by the lyddite shells.

The burghers came to close quarters with the enemy, and a fierce engagement with small arms took place.

It appeared later that the enemy's casualties amounted to more than a hundred dead and wounded. Our loss was only two men.

As I have already stated, we camped at Vanvurenskloof. The next morning, while we were still there, we were surprised by the enemy—an unpleasant thing for men with empty stomachs.

I did not receive any report from my scouts[183] until the English were not more than three thousand paces from us, and had already opened fire on the laager, not only with their guns, but also with their rifles. We at once took

[183] The reason why Captain Scheepers was so late in sending his report was because he himself was engaging the enemy with six of his men near Zandnek. He had come across a convoy of fourteen waggons and thirty men, and had, after an hour's fight, nearly brought them to the point of surrendering, when reinforcements arrived. He was thus forced to retire, and then discovered that the enemy were approaching our laager; and he had a hair's breadth escape from capture in bringing me the report.

the best positions we could find; and meanwhile the waggons got away as quickly as possible. They succeeded in getting over the first ridge, and thus gained a certain amount of shelter, whilst we kept the English busy.

The enemy approached nearer and nearer to us with overpowering forces. Then they charged, and I saw man after man fall, struck down by our merciless fire. We were quite unable to hold the enemy back, and so we had to leave our positions, having lost one dead and one wounded.

That night we marched ten miles to the east of Gatsrand, on the road to Frederiksstad Station, and the following morning we arrived at the foot of the mountain. Here we outspanned for a short time, but we could not wait long, for our pursuers were following us at a great pace. It was not only the force from the other side of Vanvurenskloof with which we had to deal. The united forces of the English had now concentrated from different points with the purpose of working our ruin.

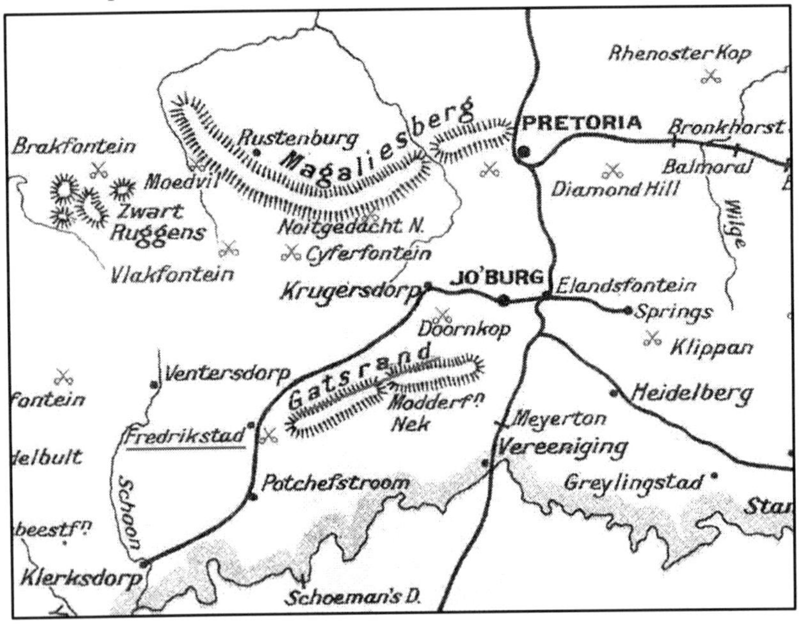

Gatsrand.

The English were exceedingly angry that we had escaped from them on the Vaal River, for they had thought that they had us safely in their hands. That we should have succeeded in eluding them was quite beyond their calculations; and in order to free themselves from any blame in the matter, they reported that we had crossed the river at a place where there was no ford, but this was not true; we had crossed by the waggon and post ford—the well-known Schoemans drift.

But whether the enemy were angry or not, there was no doubt that they were pursuing us in very large numbers, and that we had to escape from them. That evening, the 7th of August, we went to the north of Frederikstad Station, and blew up a bridge with two spans and wrecked the line with dynamite.

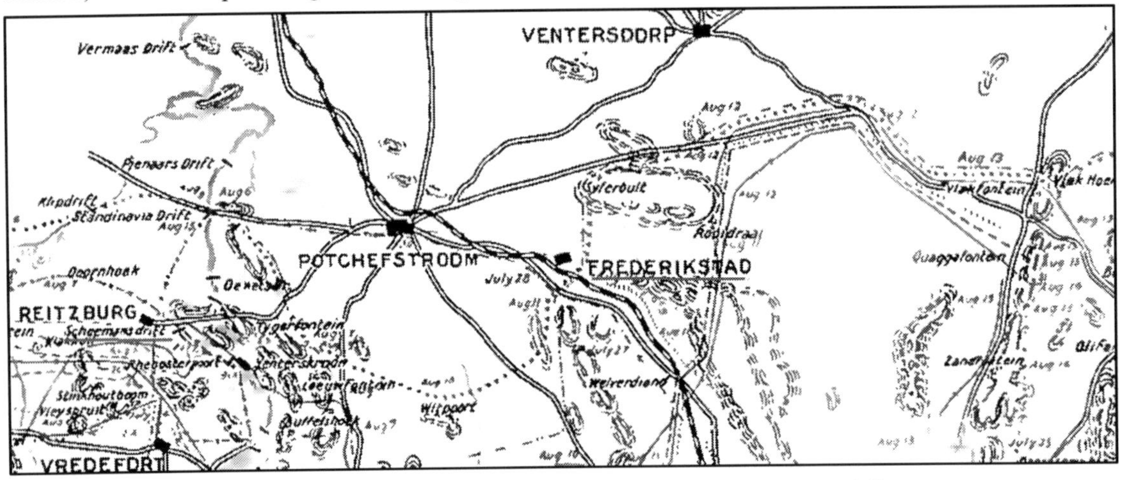

Schoeman's Drift is under Reitzburg. Frederikstad to the right near railroad line.

The following day we arrived at the Mooi River. This river is never dry winter or summer, but always flows with a stream as clear as crystal. It affords an inexhaustible supply of water to the rich land that lies along its bank. It is a fitting name for it—the name of Mooi.[184] [185]

At the other side of this river we found General Liebenberg's commando, which, like ourselves, was in the trap.

The General joined us on our march, and the following day we were nine miles from Ventersdorp.

Early that morning a report came that the English were approaching and were extended right across the country.

"Inspan!"

No man uttered a word of complaint; each man did his work so quickly that one could hardly believe that a laager could be put on the move in so short a time. And away the waggons and carts skurried, steering their course to Ventersdorp.

It was impossible to think of fighting—the enemy's numbers were far too great. Our only safety lay in flight.

We knew very well that an Englishman cannot keep up with a Boer on the march, and that if he tries to do so, he soon finds that his horses and oxen can go no further. Our intention was then to march at the very best pace we could, so that the enemy might be forced to stop from sheer exhaustion. And as the reader will soon see, our plan was successful.

Nevertheless we had to do some fighting, to protect our laager from a force of cavalry that was rapidly coming up with us.

They wanted to make an end of this small body of Boers, which was always retreating, but yet, now and again, offering some slight resistance—this tiny force that was always teaching them unpleasant lessons; first at Retiefsnek, then to the north of Lindley, then on the railway line, then near Vredefort, then at Rhenosterpoort, and then again at Tijgerfontein. Yes; this sort of thing must come to an end once for all!

We attacked the approaching troops, and succeeded in checking their advance. But our resistance could not last long, and soon we had to retreat and leave one of our Krupps behind us.

Had I not continued firing with my Krupp until it was impossible to save it, then, in all probability, the laager would have been taken. But with the loss of this Krupp we saved the laager.

I withdrew my burghers; I released the prisoners whom I had with me.

And now it was my task to make it as difficult as possible for my pursuers. The winter grass on the veldt was dry and very inflammable, and I decided to set fire to it, in order that the English might find it impossible to obtain pasture for their oxen and cattle. I accordingly set it alight, and very soon the country behind was black.

We hurried on until we reached Mr. Smit's farm, which is one hour on horseback from the southern slopes of the Witwatersrand—the great dividing chain of mountains that runs in the direction of Marico. Crossing this range, we continued on the march the whole night until, on the morning of the 11th of August, we arrived at the southern side of the Magaliesberg.

In the afternoon we went over the saddle of the mountain and across the Krokodil River.

My idea was to remain here and give our horses and oxen a rest, for the veldt was in good condition, and we could, if it were necessary, occupy the shoulder of the mountain behind us.[186]

General Liebenberg took possession of the position to the west, near Rustenburg; but hardly had he done so, before the English made their appearance, coming over another part of the mountain. He sent me a report to this effect, adding that he was unable to remain where he was stationed.

Thus again we had to retreat, and I was unable to give my animals the rest I had intended to give them.

We now took the road from Rustenburg to Pretoria, and arrived the following evening close to, which we soon found was held by an English force.

I left the laager behind and rode on in advance with a horse-commando. When I was a short distance from the enemy, I sent a letter to the officer in command, telling him that, if he did not surrender, I would attack him. I

[184] "Mooi" means beautiful in the Taal language.

[185] https://en.wikipedia.org/wiki/Mooi_River_(Vaal)

[186] 8/12/1900 De Wet escaped.

did this in order to discover the strength of the English force, and to find out if it were possible to attack the enemy at once, and forcing our way through the Nek, get to the east of the forces that were pursuing me.

My despatch rider succeeded in getting into the English camp before he could be blindfolded. He came back with the customary refusal, and reported that although the enemy's force was not very large, still the positions held were so strong that I could not hope to be able to capture them before the English behind me arrived.

I had therefore to give up the thought of breaking through these and flanking the English. Thus, instead of attacking the enemy, we went in the direction of Zoutpan, and arrived a few hours later at the Krokodil River.

I had now left the English a considerable distance behind me; and so at last—we were able to give ourselves a little rest.[187]

View of Magaliesbergen valley and reservoir of Hartbeespoort Dam.[213]

The Magaliesberg Range.[214] [188] Opposite Page: The Magaliesberg District.[215] Again Note North to the right side. Also Crocodile River north of Reitfontein. Note Olifants's Nek at top and Commando Nek.

[187] On July 16th, 1900 De Wet with about 1,600 men broke out of the Brandwater Basin, and Broadwood with the 2nd Cavalry Brigade went in pursuit. Other troops joined him, and De Wet then made for the Reitzburg Hills, south of the Vaal, where he remained for about three weeks. On the 27th July the Colonial Division was taken from Rundle's district and marched via Kroonstad to the Rhenoster River to watch the drifts there. On the night of the 6th August De Wet with his force crossed the Vaal, and the Colonial Division followed, and joining Lord Methuen, who had been posted on the north of the river, they accompanied that General in the chase to the Megaliesberg, where De Wet escaped through Olifant's Nek, from which, by some mistake, the blocking force had just been removed. The Colonial Division now accompanied Lord Methuen to Zeerust, in the north-west of the Transvaal [Lord Robert's despatch of 10th October 1900, para 28]. On the 25th August the division along with the 3rd Cavalry Brigade started on their return journey to the Orange River Colony, Colonel Dalgety of the CMR being in command of the force. They met with much opposition between Zeerust and Krugersdorp. The losses of the division on this march were 10 men killed and 5 officers and 20 men wounded. About the middle of September the division was concentrated at Rhenoster, in the Orange River Colony.
https://www.angloboerwar.com/?option=com_content&view=article&id=316

[188] The Magaliesberg of northern South Africa, is a modest but well-defined mountain range. It rises at a point south of the Pilanesberg National Park to form a curved prominence that intersects suburban Pretoria before it peters out some 50 km (31 mi) to the east, just south of Bronkhorstspruit. The highest point of the Magaliesberg is reached at Nooitgedacht, about 1,852 metres (6,076 ft) above sea level. Magaliesberg - Wikipedia

19. I Return to the Free State

Whilst we were encamped on the Krokodil River, President Steyn expressed a wish to pay a visit, with the Members of his Government, to the Government of the South African Republic, which was then at Machadodorp. This was no easy task to accomplish, for one would have to pass through a part of the Transvaal where there was a great scarcity of water—it was little better than a desert—and where in some places the Kaffirs were unfriendly. In other words, one would have to go through the Boschveldt.[189] There would also be some danger from the English, since the President would have to cross the Pietersburg Railway, which was in that direction.

However, this plan was approved.

I decided not to accompany the President, but to return at once with two hundred riders to the Orange Free State. I intended to make it known on the farms which I passed on the way that I was going back, hoping thus to draw the attention of the English from our laager.

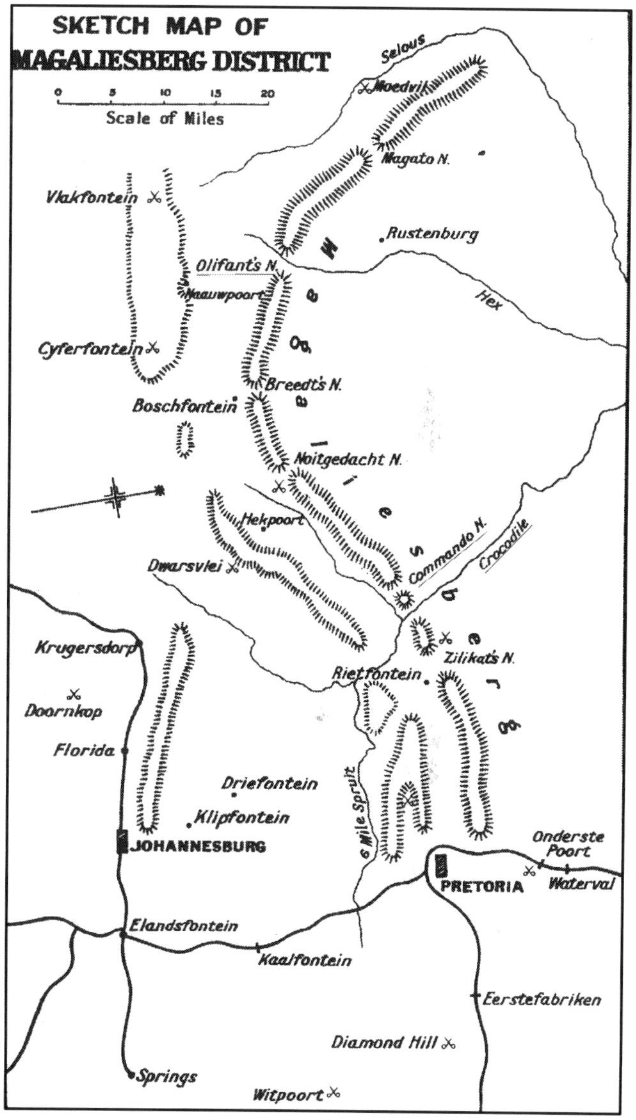

I called together the Commandants, and informed them of my intention. They agreed that the course I proposed was the right one. Commandant Steenekamp was then nominated to act as Assistant Commander-in-Chief, with the duty of conducting the laager through the Boschveldt.

On August the 14th President Steyn left the laager on his way [East past Middleburg to] Machadodorp; and I myself took my departure three days later. I took with me General Philip Botha and Commandant [Michel] Prinsloo, and 200 men, and also Captain Scheepers with his corps, which consisted of thirty men. With the addition of my staff we numbered altogether 246 men.

Thus our ways parted—the President going to the Government of the South African Republic, the laager to the north, and I back to the Free State. I had now to cross the Magalies Mountains. The nearest two passes were Olifantsnek and Commandonek. But the first named was too much to the west, and the second was probably occupied by the English. I therefore decided to take a footpath that crossed the mountains between the two saddles. I was forced to choose this middle road because I had no means of ascertaining whether Commandonek was, or was not, in the hands of the enemy.

On August 18th we arrived at a house where some Germans were living—the parents and sisters of Mr. Penzhorn, Secretary to General Piet Cronje. They were exceedingly friendly to us, and did all in their power to make us comfortable.

We did not stay here for long, but were on the march again the same day. Soon after we had mounted our horses we came in sight of a large English camp, which was stationed on the road from Rustenburg to Pretoria, between Commandonek and Krokodil River. This camp lay about six miles to the south-east of the point where we first saw it. Another great camp stood about seven miles to the north-west.

[189] Boschveld: Literally means 'bush field'. It usually refers to a specific part of SA which includes parts of Limpopo and North West Provinces (two of our most northern provinces) but also stretches into Botswana and Zimbabwe, so areas close to the Tropic of Capricorn, sub-tropical and mainly open grassland, with bush.

The enemy could see us clearly, as it was open veldt, with only a few bushes cropping up here and there. We now rode on in the direction of Wolhuterskop, which is close to the Magalies Mountains. I thought I should thus be able to reach the great road from Rustenburg to Pretoria, which was eight or nine miles from the footpath across the Magaliesberg. When we were about two miles east of Wolhuterskop we suddenly came upon two English scouts. One of them we captured; and he told us that there was a great force of the enemy in front of us and marching in our direction. What could we do now? It was impossible to proceed along the footpath because that road was closed by the enemy. North and west of us there were other bodies of troops, as I have already said; and there, directly in front of us, were the chains of the Magaliesbergen. Thus we found ourselves between four fires.

In addition to this, I was much troubled by the thought that our horses were now exhausted by all this endless marching. I knew this was also the case with the English horses, but for all I knew, they might have obtained fresh ones from Pretoria. They could at all events have picked the best horses from each camp, and thus send an overpowering force against me. This was one of those moments when a man has to keep his presence of mind, or else all is lost.

Whilst I was still thinking the matter over, troops began to come out of the camps, about two miles to the west of us on the road between Wolhuterskop and Magaliesberg. The scout who had escaped might now be with that force. I had therefore to act at once.

I decided on climbing the Magalies Mountains, without a path or road!

Near by there was a Kaffir hut, and I rode up to it. When the Kaffir came out to me, I pointed to the Magalies Mountains, and asked:—

"Right before us, can a man cross there?"

"No, baas,[190] you cannot!" the Kaffir answered.

"Has a man never ridden across here?"

"Yes, baas," replied the Kaffir, "long ago."

"Do baboons walk across?"

"Yes! baboons do, but not a man."

"Come on!" I said to my burghers. "This is our only way, and where a baboon can cross, we can cross."

With us was one Adriaan Matthijsen, a corporal who came from the district of Bethlehem, and was a sort of jocular character. He looked up at the mountains, 2,000 feet above him, and sighed:—

"O Red Sea!"

I replied, "The children of Israel had faith and went through, and all you need is faith. This is not the first Red Sea we have met with and will not be the last!"

What Corporal Matthijsen thought I do not know, for he kept silence. But he pulled a long face, as if saying to himself:—

"Neither you, nor anybody else with us, is a Moses!"

We climbed up unobserved to a bit of bush which, to continue the metaphor of the Red Sea, was a "Pillar of Cloud" to hide us from the English.

We then reached a kloof[191] running in a south-westerly direction, and ascended by it, still out of sight of the English, till we reached a point nearly half-way up the mountain. There we had to leave the kloof, and, turning to the south, continue our ascent in full view of the enemy.

[190] Master.
[191] Ravine.

Looking West along the Magaliesberg in the Skeerpoort Area.[216]

A ravine in the Magaliesberg.[217]

Ridgeline of the Magaliesberg.[218]

It was now so precipitous that there was no possibility of proceeding any further on horseback. The burghers had therefore to lead their horses, and had great difficulty even in keeping their own footing. It frequently happened that a burgher fell and slipped backwards under his horse. The climb became now more and more difficult; and when we had nearly reached the top of the mountain, there was a huge slab of granite as slippery as ice, and here man and horse stumbled still more, and were continually falling.

We were, as I have said, in view of the enemy, and although out of reach of the Lee-Metfords, were in range of their big guns!

I heard burghers muttering:—

"Suppose the enemy should aim those guns at us—what will become of us then? Nobody can get out of the road here!"

I told them that this could only be done if the English had a Howitzer. But I did not add that this was a sort of gun which the columns now pursuing me were likely enough to possess.

The 5-inch Howitzer or Siege Gun.[219] [192]

But nothing happened. The English neither shot at us, nor did they pursue us. Corporal Matthijsen would have said that they were more cautious than Pharaoh.

We now reached the top of the mountain—entirely exhausted. I have ascended many a mountain—the rough cliffs of Majuba, the steep sides of Nicholson's Nek—but never before had I been so tired as I was now; yet in the depths of my heart I was satisfied. All our toil was repaid by the glorious panorama that now stretched out before us to the south. We saw the undulating veldt between the Magaliesbergen where we stood, and the Witwatersrand. Through a ravine we had a view extending for many miles, but wherever we cast our eyes there was no sign of anything that resembled the enemy.

As it was now too late to off-saddle, we began, after having taken a little rest, to descend the mountain on the other side, my object being to reach a farm where I hoped to get some sheep or oxen for my men, who not only were tired out, but nearly famished.

We went down the mountain—well, somewhat quicker than we had climbed it; however, we could not go very fast, as the incline was steep. In an hour and a half we reached a Boer farm.

One can imagine how the burghers recovered their spirits as they ate their supper, and what it meant for them to give their tired limbs a rest.

The following morning we found good horse-provender, and plenty of it. It was not as yet the habit of the English to burn everything they came across—they had not yet begun to carry out that policy of destruction.

I now felt quite easy about the safety of our camp. The attention of the English would be turned in quite another direction.

I was quite right in this view of the matter. For I heard a few days later that the enemy had not been able to pursue the laager as their draft-cattle and horses were so completely exhausted, that they had fallen down dead in heaps. I heard also that they had soon been made acquainted with the fact that I was on my way back to the Free State, where I would soon begin again to wreck railway lines and telegraph wires. They had also discovered that President Steyn had left the laager and was on the road [East] to Machadodorp.

It was on the 18th of August, 1900, that we were able to eat our crust of bread in safety on the farm just mentioned, and to let our horses have as much food as they wanted. It seemed that for the time being a heavy burden had fallen from our shoulders. That afternoon we crossed the Krokodil River, and stopped at a "winkel"[193] under the Witwatersrand, which had been spared as yet, although it was nearly empty of stores. Fodder, however, was plentiful, and thus, again, we could give our horses a good feed.

I now received a report that a strong contingent of the enemy was on the march from Olifantsnek to Krugersdorp, and accordingly we rode off in the night. We found that this force was the very one that had flanked our laager the previous week, when we were passing Ventersdorp. The road which the enemy were taking was the same which Jameson had marched when he made his inroad into the South African Republic.

[192] The howitzer is an artillery weapon that falls between a cannon (or field gun) and a mortar. It is capable of both low angle fire like a field gun and high angle fire like a mortar.

[193] General Store.

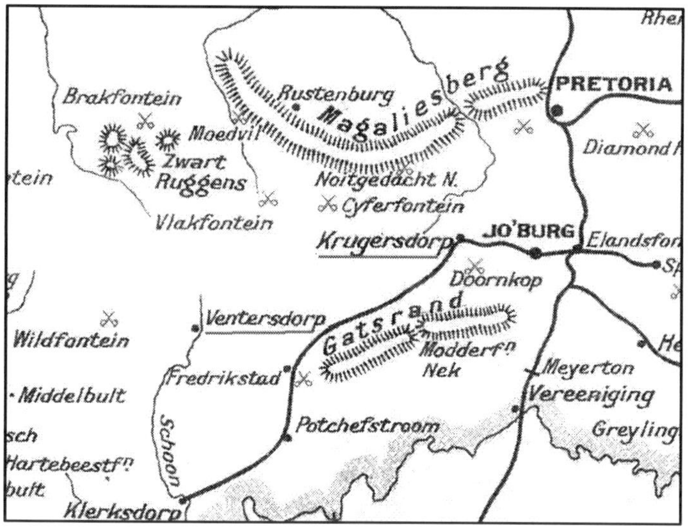

Krugersdorp and Ventersdorp.

My intention was to cross the enemy's path before daylight the following morning, which I succeeded in doing; and we heard no more of this force. I proceeded now in the direction of Gatsrand.

From there I still went on, and crossed the Krugersdorp-Potchefstroom Railway, about eight or ten miles to the north of Bank Station.

The line was then not guarded everywhere. There were small garrisons at the stations only, and so one could cross even in the day time. To my vexation, I had not a single cartridge of dynamite, or any implements at hand with which I could wreck the line. It was painful to see the railway line and not be able to do any damage to it! I had made it a rule never to be in the neighbourhood of a railway without interrupting the enemy's means of communication.

We arrived now at the farm of Messrs. Wolfaard, who had been captured with General Cronje; and here I met Commandant Danie Theron, with his eighty men.

He had come to this place to avoid the troops lying between Mooi River and Ventersdorp. His horses, although still weak, were yet somewhat rested, and I gave him orders to join me in a few days, in order to reinforce me until my commandos should come back. My intention was not to undertake any great operations, for my force was not strong enough for that. I intended my principal occupation to be to interrupt the communications of the enemy by wrecking the line and telegraph.

With regard to the main line in the Free State I must remark here that things there were in a different condition from what they were on the Krugersdorp line, which we had crossed. The Free State railway was Lord Roberts' principal line of communication, and he had provided guards for it everywhere.

During the night of August 21st, we arrived at Vanvurenskloof. How delightful it was when the sun rose to see once more the well-known mountains to the south of the Vaal River in our own Free State!

"There is the Free State," we called out to each other when day broke. Every one was jubilant at seeing again that country which of all the countries on the earth is the best. From here I despatched General Botha with the purpose of collecting the burghers of Vrede and Harrismith who had remained at home, and of bringing them back to join me.

We remained only as long as was necessary to rest the horses, and then at once went on. The same evening we arrived at the farm of Rhenosterpoort, where our laager had waited since we had crossed the Vaal River more than a week ago.

The proprietor of the farm of Rhenosterpoort was old Mr. Jan Botha. It could not be that he belonged to the family of Paul Botha, of Kroonstad, for Jan Botha and his household (amongst whom was his son Jan, an excellent veldtcornet) were true Afrikanders. And even if he did belong to the family of Paul Botha, then the difference in his feelings and actions from those of other members of his family was no greater than that, alas! which frequently occurred in many families during this war. One member put everything at the disposal of his country, whereas another of the same name did everything possible against his country and his people. But there was no such discord here. The two old[er] brothers of Mr. Botha, Philip and Hekky, were heart and soul with us.

Potchefstroom was not at that time in the hands of the English. I rode over to the town, and then it was that the well-known photo was taken of me that has been spread about everywhere, in which I am represented with a Mauser in my hand. I only mention this so as to draw attention to the history of the weapon which I held in my hand. It is as follows:—

When the enemy passed through Potchefstroom on their way to Pretoria, they left a garrison behind them, and many burghers went there to give up their arms, which forthwith were burnt in a heap. When the garrison left the dorp the burghers returned. Amongst them were some who set to work to make butts for the rifles that had been burnt.

"This rifle," I was told by the man who showed it to me, "is the two hundredth that has been taken out of the burnt heap and repaired."

This made such an impression on me that I took it in my hand, and had my photo taken with it. I am only sorry that I cannot mention the names of the burghers who did that work. Their names are worthy to be enrolled on the annals of our nation.

General Christiaan de Wet.[220] [194] Christiaan is posing with a restored rifle that was handed in to the British as part of the amnesty and laying down of arms.[221] Destroying Captured Boer Rifles.[222] [195] General Christiaan de Wet.[223] (A second photo taken in the same sitting. His hat is on the table.)

A third picture taken on the same day.[224]

[194] The photo was taken in Potchefstroom on the 27th of August 1900, six days after he and his commando of 246 men escaped from near encirclement north of the Magaliesberg File:Christiaan de Wet4.jpg - Wikimedia Commons

[195] PF: When I found this picture I never thought I'd use this in this sense. Taking a second look at the 'conflagration' it is not surprising that some useful parts survived. Also, a close look at the original photo shows the barrel of Christiaan's weapon to be pitted on the outside and a close look at the bolt shows the same. What verve to adopt this weapon by choice!

Uncropped Picture with the staff identified.[225] The Scout is Jonas, Van Grahn is far right.[226]

After having provided myself with dynamite, I left Potchefstroom and returned to my commando, then quietly withdrew in the night to Rhenosterkop. From there I sent Veldtcornet Nicolaas Serfontein, of the Bethlehem commando, in the direction of Reitz and Lindley, to bring the Kaffirs there to a sense of their duty, for I had heard that they were behaving very brutally to our women. The remainder of the Bethlehem burghers under Commandant Prinsloo and Veldtcornet Du Preez, remained with me to assist me in getting under my supervision the commandos which had escaped from behind the Roodebergen. These were under the command of General Fourie, and some were in the south of the State. I left Captain Scheepers behind me with orders to wreck the line every night.

That evening I went to Mr. Welman's farm, which was to the south-west of Kroonstad.

There I received a report that the commandos under General Fourie were in the neighbourhood of Ladybrand. I sent a despatch to him and Judge Hertzog asking them to come and see me, with a view to bringing the burghers under arms again, in the southern and south-western districts of the State.

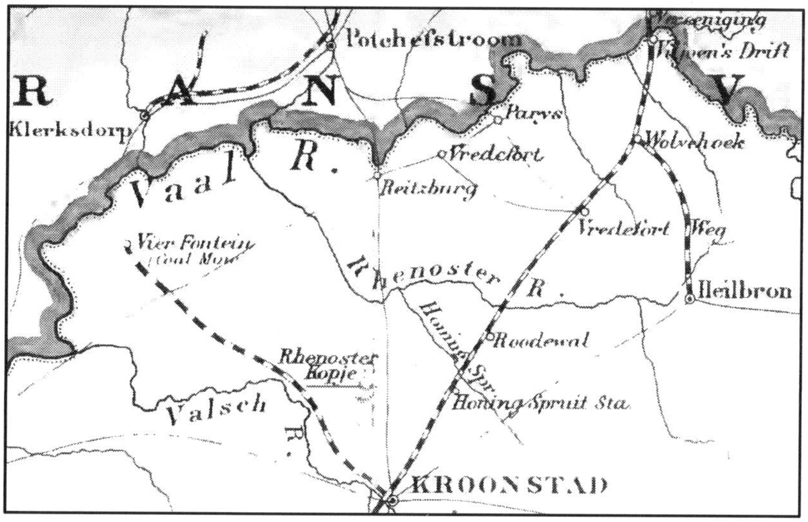

Rhenoster Kopje.

This letter was taken by Commandant Michal Prinsloo and some despatch riders to General Fourie. The night that he crossed the line a train was passing, and he wrecked the railway both in front of it and behind it. The train could thus neither advance nor retreat, and it fell into the hands of Commandant Prinsloo, who, after having taken what he wanted, burnt it.

With regard to myself, I remained in the neighbourhood of Commandant Nel's farm.

Here I had the most wonderful of all the escapes that God allowed me in the whole course of the war.

On the third evening at sunset, a Hottentot came to me. He said that his "baas," whose family lived about twelve miles from the farm of Commandant Nel, had laid down their arms, and that he could not remain in the service of the wife of such a bad "baas." He asked me if he could not become one of my "achterrijders."[196]

As he was still speaking to me, Landdrost Bosman from Bothaville, came to pay me a visit.

"Good," I said to the Hottentot, "I shall see you about this again." For I wished to cross-question him. I then went into the house with the Landdrost, and spent a good deal of time in writing with him. Late in the evening he went back to Bothaville and I to bed exactly at eleven o'clock.

I had scarcely laid down when the Hottentot came back to my thoughts, and I began to grow uneasy. I got up and went to the outhouse where my Kaffir slept. I woke him up and asked him where the Hottentot was. "Oh, he is gone," he replied, "to go and fetch his things to go with the baas."

I at once felt that there was something wrong, and went and called my men. I told them to saddle-up, and went off with my staff to the farm of Mr. Schoeman on the Valsch River, to the east of Bothaville.

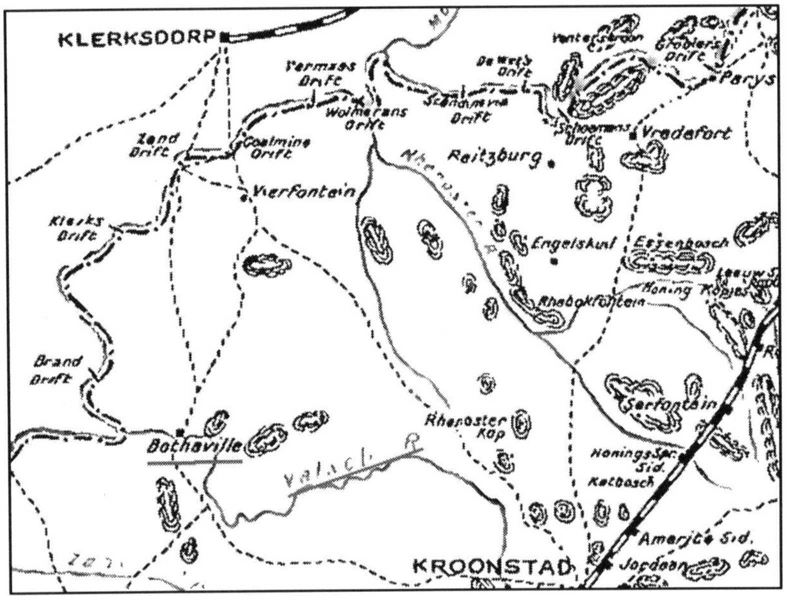

Bothaville and Valsch River.

On the following morning before daybreak, a force of two hundred English stormed the farm of Commandant Nel. They had come to take me prisoner.

From Schoeman's farm I went [East] to the Rhenoster River and found Captain Scheepers there. He reported that he had wrecked the line for four or five consecutive weeks, as I had told him.

I also received there the sad news of the death of the never-to-be-forgotten Danie Theron,[197] in a fight at Gatsrand. A more brave and faithful commander I have never seen.

So Danie Theron was no more. His place would not be easily filled. Men as lovable or as valiant there might be, but where should I find a man who combined so many virtues and good qualities in one person? Not only had he the heart of a lion but he also possessed consummate tact and the greatest energy. When he received an order, or if he wished to do anything, then it was bend or break with him. Danie Theron answered the highest demands that could be made on a warrior.

One of Commandant Theron's lieutenants, Jan Theron, was appointed in his place.

From there I went with Captain Scheepers to the railway line, where I burnt a railway bridge temporarily constructed with sleepers, and wrecked a great part of the rails with dynamite.

[196] Achterrijder, achterryder A mounted groom or other attendant; a military servant.

[197] September 5th, 1900.

Rhenoster Bridge with just cribbing [sleepers] supporting it.[227] The Railway Bridge over the Sunday River (Natal).[228]

I then proceeded to various farms in the neighbourhood, and after a few days, with Commandant Michal Prinsloo, who had joined me, I returned to the same part of the railway in order to carry out its destruction on a larger scale.

At twenty-five different places a charge of dynamite was placed with one man at the fuse, who had to set light to it as soon as he heard a whistle, that all charges could be ignited at the same time, and every one be out of the way when the pieces of iron were hurled in the air by the explosion.

When the signal was heard the lucifers were struck everywhere, and the fuses ignited.

The English, keeping watch on some other part of the line not far from us, on seeing the lights fired so fiercely on the burghers that they all took to their horses and galloped off.

Only five charges exploded.

I waited for a moment, but no sound broke the silence.

"Come on!" I said, "we must fire all the charges."

On reaching the line we had to search in the darkness for the spots where the dynamite had been placed. And now again the order was given that as soon as the whistle was blown every one had to ignite his fuse.

Again there was a blunder!

One of the burghers ignited his fuse before the signal had been given, and this caused such a panic that the others ran away. I and a few of my staff lay flat on the ground where we were until this charge had exploded, and then I went to fetch the burghers back.

This time everything went off well, and all the charges exploded.

Engineers repairing a Railway destroyed by the Boers near Kroonstadt.[229]

The bridge I had destroyed had been rebuilt, and so I was forced to burn it again. When this was done we departed and rode on to Rietspruit, where we up-saddled, and then pushed on to Rhenosterpoort.

Commandant-General Christian De Wet.[230]

Commandant -General Christian H. De Wet.[231]

20. The Oath of Neutrality

Arriving at Rhenosterpoort, I found there Commandant F. Van Aard, with his commando. He told me that after I had left the laager, the burghers had not been troubled again by the English. He had gone on to Waterberg, and after having stayed there for a short time, he had returned to the laager. He still had some of his waggons with him, but in many cases the oxen had been so exhausted that the waggons had to be left behind, the burghers returning on horseback, or even on foot. He also told me that Vice-Commander-in-Chief Steenekamp had, just before my arrival, crossed the line in the direction of Heilbron, in which district there were then no English.

Generals Fourie and Froneman, with Hertzog, were also at Rhenosterpoort, having left their commandos behind, in the district of Winburg.

They had much to tell me which I had heard already, but which I now obtained at first hand. It appeared that the burghers who had been taken prisoner with General Prinsloo had been sent to Ceylon, notwithstanding the promise that had been given them that their property would be safe, and that they would be allowed to return to their farms.

It was now that I conceived the great plan of bringing under arms all the burghers who had laid down their weapons, and taken the oath of neutrality, and of sending them to operate in every part of the State. To this end I went with these officers to the other side of the railway line, in order to meet General Philip Botha in the country to the south-east of Heilbron, and also, if possible, General Hattingh, who was in command of the Harrismith and Vrede burghers.

We succeeded in crossing the railway between Roodewal and Serfontein siding, but not without fighting. Before we came to the railway line the English opened a cross fire on us from the north-east, from the direction of Roodewal; and almost directly afterwards another party fired on us from the south. We succeeded, however, in getting through with the waggons which Commandant Van Aard had with him, but we lost one man killed, and three wounded.

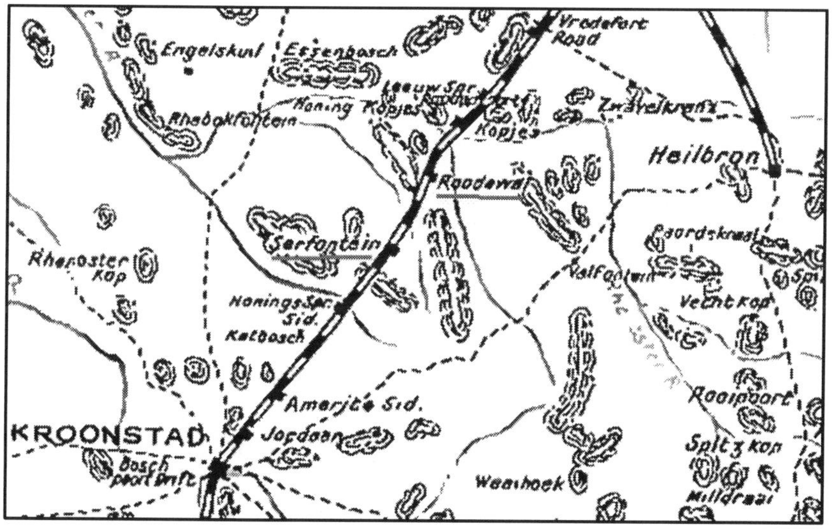

Serfontein is south of Roodenwal station.

On the following day I gave Commandant Van Aard the order to go to his district (Midden Valsch River) in order to give his burghers an opportunity of getting their clothes washed, and of obtaining fresh horses, if any were to be had. For although the enemy already had begun to burn down our houses, and to carry away our horses, things had not as yet reached such a pitch that the columns spared nothing that came in their way.

Commandant Van Aard started off on his errand, but alas! a few days afterwards I heard that he—one of the most popular of all our officers—had been killed in a fight near his own farm between Kroonstad and Lindley. He was buried there, where he had fallen, on his own land.

And now began the great work which I had proposed to accomplish.

I gave instructions to Vice-Commander-in-Chief Piet Fourie to take under his charge the districts of Bloemfontein, Bethulie, Smithfield, Rouxville, and Wepener, and to permit the burghers there, who had remained behind, to join us again. He was not, however, to compel anybody to do so, because I was of opinion that a coerced burgher would be of no real value to us, and would besides be untrustworthy. The following officers were to serve under Fourie: Andrias, Van Tonder and Kritzinger. The last-named had been appointed in the place of Commandant Olivier, who had been taken prisoner at Winburg.[198]

I had appointed Judge Hertzog as a second Vice-Commander-in-Chief, to carry out the same work in the districts of Fauresmith, Philippolis and Jacobsdal. He had under him Commandant Hendrik Pretorius (of Jacobsdal) and Commandant Visser. The latter was the man who, when the burghers from Fauresmith, even before the taking of Bloemfontein, had remained behind, broke through with seventy or eighty troops. He had always behaved faithfully and valiantly until, in an engagement at Jagersfontein, he gave up his life, a sacrifice for the rights of his nation. His name will ever be held in honour by his people.

These two Vice-Commanders-in-Chief had no easy task to perform. In fact, as every one will admit, it was a giant's burden that I had laid upon their shoulders. To lighten it a little I made the following arrangement: I sent Captain Pretorius, with a small detachment, in advance of General Fourie, to prepare the road for him, and Captain Scheepers to do the same for Judge Hertzog. The first had to say: "Hold yourselves in readiness! Oom Pieter![199] is coming." The other had to say: "Be prepared! The Rechter[200] is at hand!"

All went well. General Fourie set to his task at once and did excellent work. He had not been long in his division before he had collected seven hundred and fifty men, and had had several skirmishes with the enemy. It was on account of his acting so vigorously that the English again put garrisons into some of the south-eastern townships, such as Dewetsdorp, Wepener, and others.

With General Hertzog things went even better. He had soon twelve hundred men under arms. General Fourie had not succeeded in getting together an equally large force in his division, because many burghers from these districts had been taken prisoner at the time of the surrender of Prinsloo. General Hertzog also fought more than one battle at Jagersfontein and Fauresmith.

I ought to add that after I had crossed the Magaliesberg I had sent Veldtcornet C.C. Badenhorst, with twenty-seven men, on a similar errand to the districts of Boshof and Hoopstad. I promoted him to the rank of commandant, and he soon had a thousand troops under him, so that he was able to engage the enemy on several occasions. He had not been long occupied in this way, before I appointed him Vice-Commander-in-Chief. The reader who has followed me throughout this narrative, may very naturally ask here how it could be justifiable for nearly three thousand burghers thus to take up arms again, and break their oath of neutrality? I will answer this question by another—who first broke the terms of this oath?—the burghers or the English military authorities? The military authorities without any doubt; what other answer can one give?

Lord Roberts had issued a proclamation saying that, if the burghers took an oath of neutrality, and remained quietly on their farms, he would give them protection for their persons and property. But what happened? He himself ordered them to report to the British military authorities, should any Boer scout or commandos come to their farms, and threatened them with punishment if they did not do so.

Old people also who had never stirred one step from their farms were fined hundreds of pounds when the railway or telegraph lines in their neighbourhood were wrecked. Besides, instead of protection being given to the burghers, their cattle were taken from them by the military, at prices they would never have thought of accepting, and often by force. Yes; and from widows, who had not even sons on commando, everything was taken away. If then the English, on their part, had broken the contract, were not the burghers perfectly justified in considering themselves no longer bound by the conditions which the oath laid on them?

And then if one goes further into the matter, and remembers that the English had been employing such people as the National Scouts, and had thus been arming men who had taken the oath of neutrality, how can one think that the Boer was still under the obligation of keeping his oath?

There is also the obligation which every one is under to his own Government; for what Government could ever acknowledge an oath which their citizens had no right to take?

[198] Commandant Van Tender had been made prisoner at the same time, but he eluded the vigilance of his captors, and running for his life under a shower of their bullets, got away in safety.

[199] Uncle Peter.

[200] Judge.

No! taking everything into consideration, no right-minded burgher could have acted otherwise than to take his weapons up again, not only in order to be faithful to his duty as a citizen, but also in order not to be branded as a coward, as a man who in the future could never again look any one in the face.

I arranged various matters at Doornspruit, in the district of Kroonstad, on the 23rd of September, 1900, and then went from there in the direction of Rietfontein, in order to meet the commando which I had ordered to be at Heilbron on the 25th.

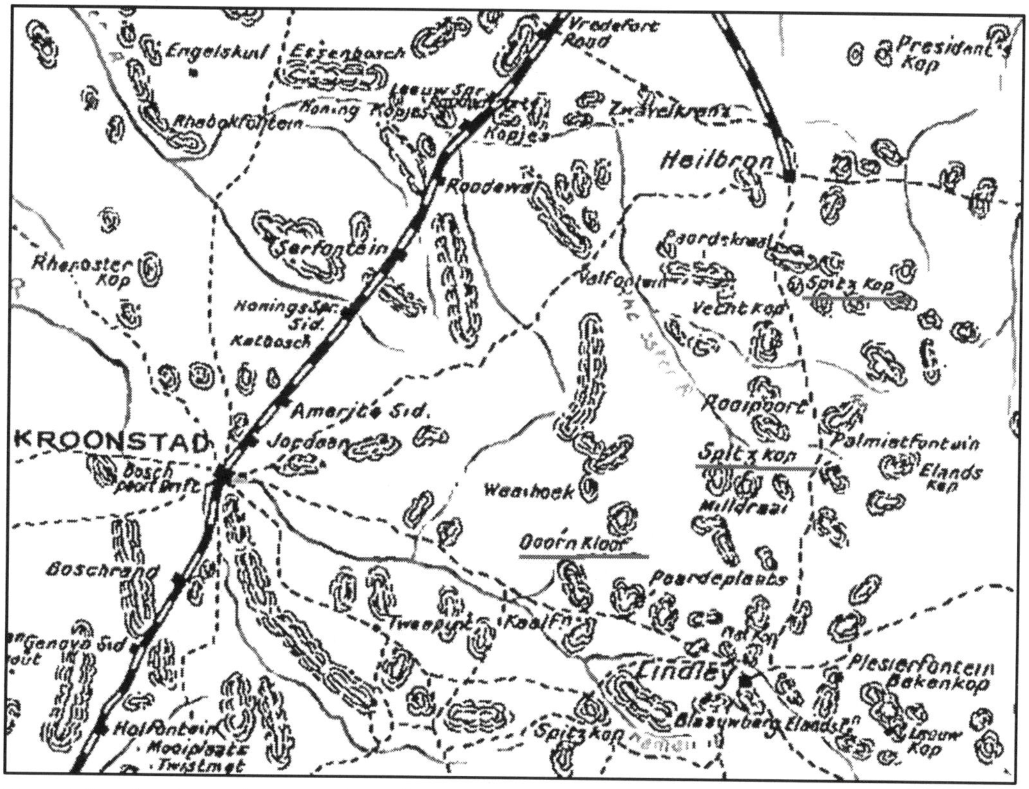

Dorn Kloof and small stream 'spruit,' probably Dorn Spruit. Also note Spitz Kop north of 'Findley' Lindley and "Spitskopje, seven miles to the south-east of Heilbron."

21. Frederiksstad and Bothaville

When I was on the road to Heilbron, I heard that the commandos under General Hattingh (those, namely, of Harrismith and Vrede) were near the Spitskopje, seven miles to the south-east of Heilbron. I therefore went out of my course and proceeded in the direction of these commandos. They were among those who had stood the crucial test, and had not surrendered with Prinsloo.

It was a real pleasure to me to meet the Harrismith burghers, and to talk with them over bygone days. This was our first meeting since December, 1899. The last time we had seen each other was when we were encamped round Ladysmith, where we were, so to speak, neighbours—our positions being contiguous.

But what a shock went through my heart when I saw the cumbersome waggon-camps which had come both from Vrede and Harrismith! For I remembered what trouble and anxiety the waggons and carts had already caused me, and how my commandos, in order to save them, had been forced to fly 280 miles—from Slabbertsnek to Waterberg. As Commander-in-Chief, I was now determined to carry out most strictly the Kroonstad regulation and have nothing more to do with the waggons.

I did not think that I should have any difficulty in convincing the commanders of Harrismith and Vrede that the best thing would be to do away with these unnecessary impediments, because, shortly before, the English themselves had given me a text to preach from, by taking away a great number of waggons from Commandant Hasebroek at Winburg and at Vet River. Nevertheless, my words fell on unwilling ears.

It was not long after I had arrived in the camp when I got the burghers together and spoke to them. After thanking the officers and men for not having surrendered with Prinsloo at Naauwpoort, I congratulated them on their success at Ladybrand, where they had driven the English out of the town and forced them to take refuge in the caverns of Leliehoek. I then went on to tackle the tender subject—as a Boer regards it—of sacrificing the waggons. No! I did not say so much as that—I only insisted on the waggons being sent home. Now this was very much the same as saying: "Give up your waggons and carts to the enemy"—an order which, expressed in that bald manner, would have given offence.

However, I was resolved to have my way, and at the end of my speech, I said, "I may not ask you, and I will not ask you what you will do with regard to the waggons. I only tell you that they must disappear."

On the following day I called the officers together, and gave them direct orders to that effect. I was very polite, but also very determined that the waggons should be sent off without a moment's delay. I also gave orders that the Harrismith and Kroonstad burghers under General Philip Botha should occupy themselves in cutting the English lines of communication between Kroonstad and Zand River. The Bothaville burghers were to carry out similar operations in their own district.

On that same afternoon I rode with my staff to the Heilbron burghers, who now had returned to their farms. (They had had permission to go home after they had got back from Waterberg.) They had assembled in very strong force.

The enemy also had arrived in this part of the country, and we were therefore obliged at once to get ourselves ready to fight in case it should be necessary, or to retreat if the enemy should be too strong for us.

With the Heilbron, Harrismith and Vrede commandos, I had now a very considerable force at my command.

When I met the burghers on the 25th of September I found that I must send a force in the direction of Kroonstad, in order to oppose outposts which the enemy had stationed some six miles from that town.

I at once sent orders to General Hattingh that he was to come over to me with his burghers. But what did I hear? The burghers had not been able to make up their minds to part with their waggons; most of the men from Vrede and Harrismith had gone home with these waggons, although there was a Kaffir driver and a leader for almost every one, and although I had given express orders that these Kaffirs were to be the ones to take back the waggons. How angry I was! At such moments as these one would be well nigh driven mad were there not a Higher Power to hold one back.

And, to make the situation still more serious, the English now came on from all sides, and I had no troops! The Kroonstad burghers were in their own district. I allowed those from Bethlehem to leave me in order to carry on operations in their part of the country; the same likewise with the Winburgers and the valiant Commandant Hasebroek, while the burghers of Vrede and Harrismith had gone home.

I had therefore with me only a small contingent from those districts, in addition to the burghers from Heilbron.

The reader will understand that, under these circumstances, the forces which now began to concentrate on us were too great for us to withstand; and that no other course lay open to me than to go through Schoemansdrift; and, in case I should be pursued, to Bothaville, in order to enter the zandveld (desert) through which it would be difficult for the enemy to advance.

We continued in the direction of Wolvehoek Station, and on the following night crossed the line between Vredefortweg and Wolvehoek, where I wrecked the railway at various points, and also took prisoner a small force of thirteen who had been lying asleep in their tents. This last incident happened early in the morning of September 30th.

We had crossed the line, and were about three miles on the further side of it, when a train came up and bombarded us with an Armstrong and a Maxim-Nordenfeldt, without however doing any damage.

The 6 -In Guns Taken to Warrenton, Mounted by Major Phillips, R.A.[232]

Naval 6-Inch Gun on it's specially-designed truck.[233]

Our guns were too far behind the vanguard, and the poor horses too tired to go back for them, or we should have answered their fire. However, we got an opportunity of using our big guns against 200 mounted men, who had pursued us, but who, when they saw we were ready to receive them, turned round and—took the shortest road to safety!

Free State Artillery & Heliograph.[234]

Walker's Maritzburg Artillery with Hotchkiss Gun.[235]

That evening we marched to a place a little to the south of Parijs, and the following day to the kopjes west of Vredefort. There we stayed a few days until the enemy again began to concentrate at Heilbron.

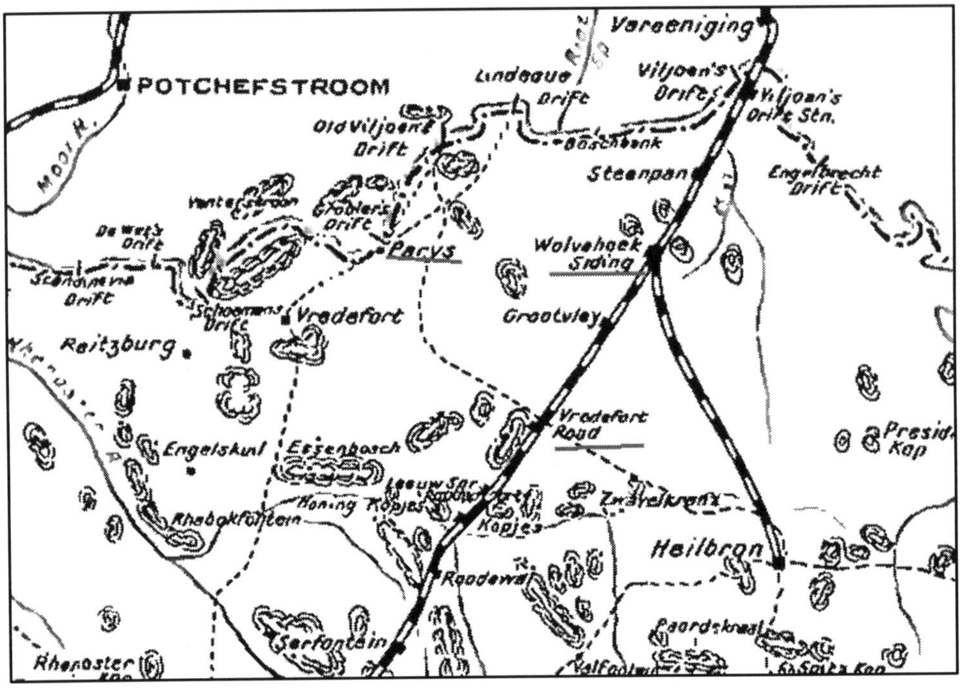

Wolvehoek siding and Vredefort Road south of it. Also Parys to the NW and the kopje's to the west of it.

I then divided my commando into two parts. One part I took with me, while I sent the Harrismith burghers (those at least who had not gone home with the waggons) under General Philip Botha, in the direction of Kroonstad, where he would meet the commando of that district, which had received orders to operate to the west of the railway line. General Philip Botha nominated Veldtcornet P. De Vos as Commandant of the Kroonstad contingent instead of Commandant Frans Van Aard. He made a good choice, for Commandant De Vos was not only a valiant officer, but also a strictly honourable man.

For some days the enemy remained encamped on the farm called Klipstapel, which lies to the south-east of Vredefort. Then they attacked us. We held our own for a day and a half, but at last had to retreat to the Vaal River, whither the English, doubtless thinking that we were again going to Waterberg, did not pursue us. This was on the 7th of October, 1900.

I now received a report from General Liebenberg that General Barton and his column were in the neighbourhood of Frederiksstad Station. He asked me (as he was too weak to venture anything alone) whether I would join him in an attack upon the English General. I decided to do so, and sent him a confidential letter saying that I would join him in a week's time.

In order to mislead the English, I retreated ostentatiously through Schoemansdrift to the farm of, which stands on the banks of the Rhenoster River, fifteen miles from the drift. The following night I returned by the way I had come, and crossed the river a little to the west of Schoemansdrift.

When on the following night we were again in the saddle I heard from many a mouth, "Whither now?"

Our destination was Frederiksstad Station, where we were to engage General Barton. Previous to an attack, thorough scouting should always take place. Accordingly I sent out my scouts, and discovered that General Liebenberg had entirely cut off the English from their communications, so that, except for heliographic messages, they were entirely out of touch with the rest of their forces. Now I do not know if they had "smelt a rat," but they were certainly well entrenched near the station on ridges to the south-east and to the north.

We had therefore to besiege General Barton in his entrenchments. For the first five days we held positions to the east, to the south, and to the north-west. On the fifth day I agreed with General Liebenberg that we should take up a new position on the embankment north-west of the strongest part of the English encampment. This position was to be held by two hundred men, of whom I gave eighty to General Froneman and one hundred and twenty to General Liebenberg. It was a position that we could not leave during the day without great danger, and it needed a large force to hold it, for its garrison had to be strong enough to defend itself if it should be attacked.

Transvaal side of the border. Note Frederikstad just north of Potchefstroom.

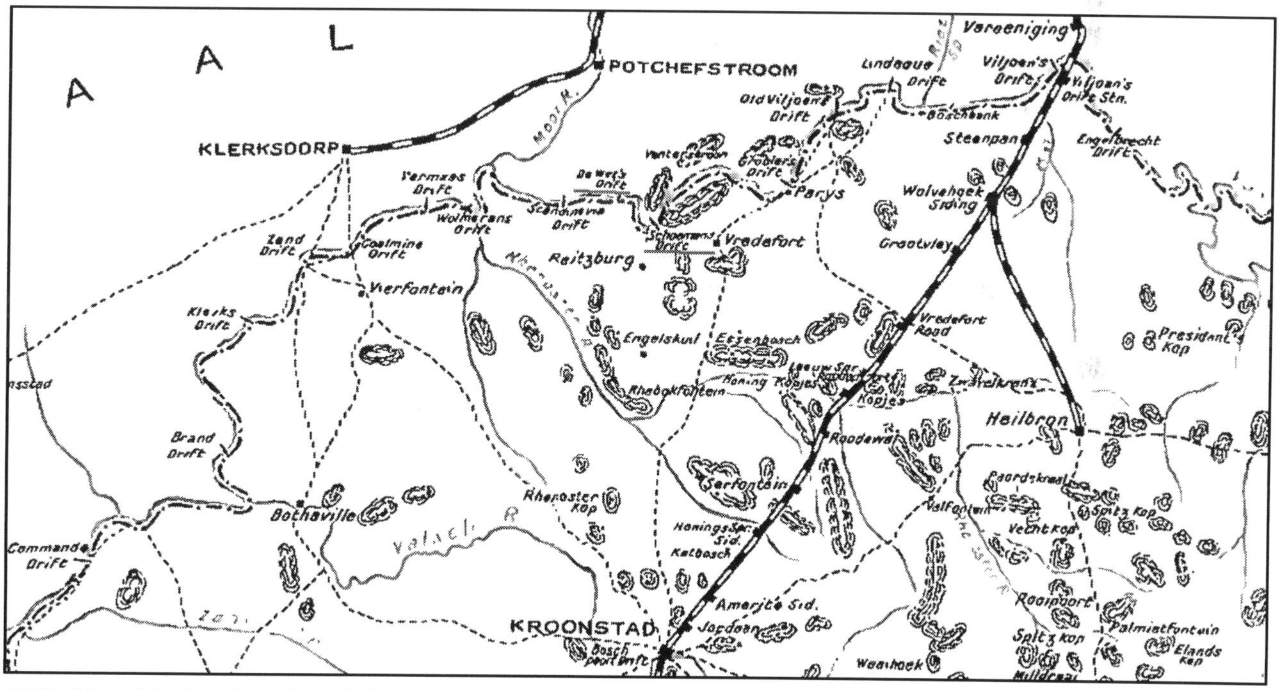

OFS side of the border. Note Schoemansdrift and also, barely legible is De Wet's Drift right next to it. Also note Bothaville.

If only my arrangements had been carried out all would have gone well.

But what happened?

I thought that two hundred men had gone in accordance with my orders to that position. Instead of this there were only eighty there when, on the following morning, a very strong reinforcement of English, ordered up by General Barton, appeared from the direction of Krugersdorp. I did not hear of this reinforcement till it was so close that there was no chance for me to keep it back. In fact, when I got the report the enemy were already storming the unfortunate handful of burghers and firing fiercely upon them. If these burghers had only had enough ammunition they would have been able to defend themselves, but as they were obliged to keep up a continuous fire on the storming party their cartridges were speedily exhausted. When this happened there was nothing for them to do but to fly. This they did under a fierce fire from three guns, which had been bombarding them continuously since the morning—doing but little damage however, as our burghers were behind the railway embankment. But now they had to fly over open ground, and on foot, as they had gone down without their horses because there was no safe place for the animals.

If two hundred burghers—the number I had arranged for—had been in the position, there would have been no chance of the enemy's reinforcement being able to drive them out: and in all probability General Barton would have been obliged to surrender. Instead of this we had a loss of thirty killed and wounded, and about the same number were taken prisoners. Among the dead was the renowned Sarel Cilliers, grandson of the worthy "voortrekker"[201] of the same name. Veldtcornet Jurie Wessels was the most distinguished of the prisoners.

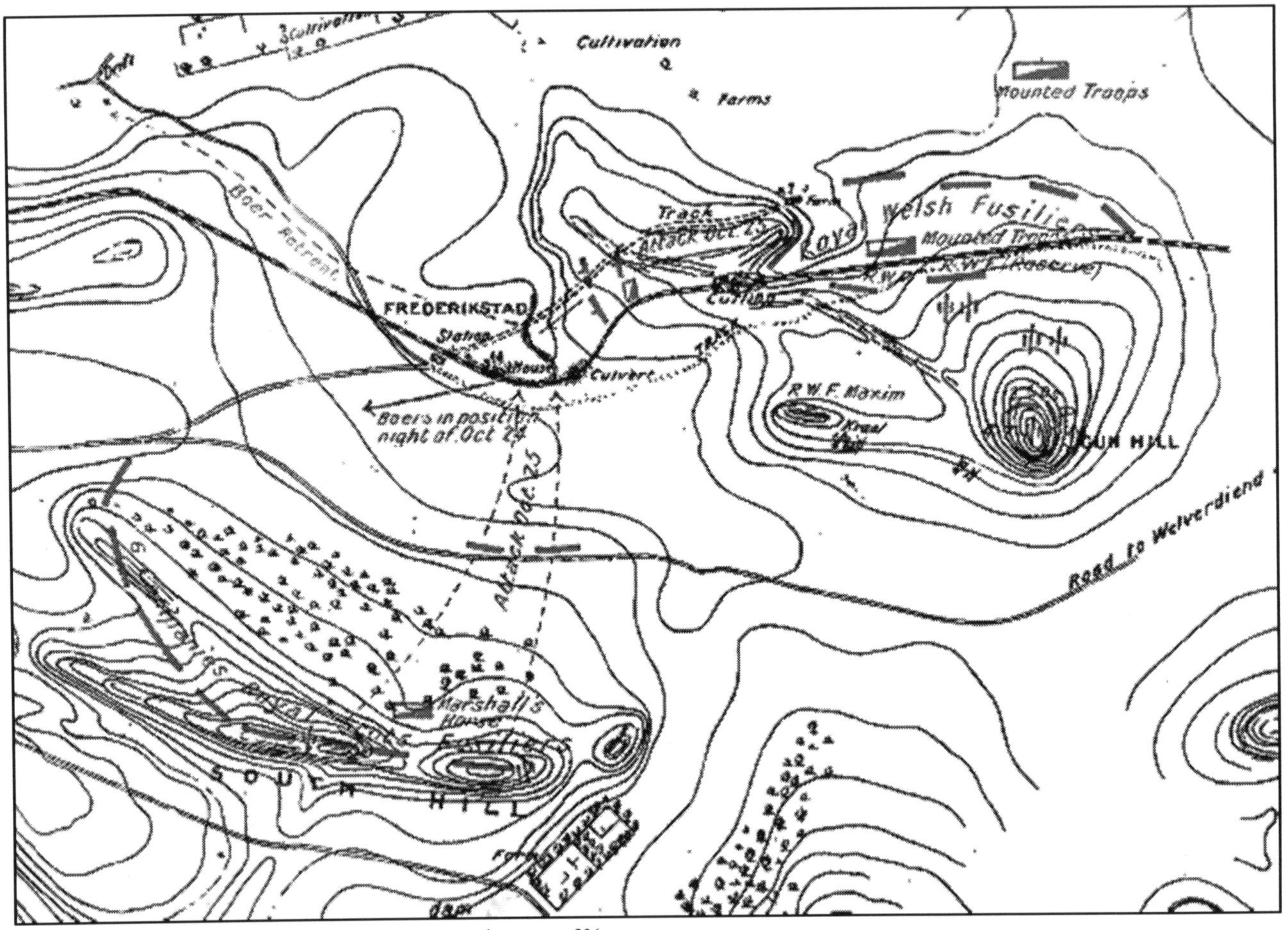

Frederikstad attack from October 21st -25th, 1900.[236]

It was a miserable affair altogether: General Froneman ought to have called his men back when he saw that General Liebenberg had not sent his contingent. I have heard however that Captain Cilliers refused to leave the position until it became no longer tenable. It was hard indeed for him to lose a battle thus, when it was nearly won, and to be compelled to retreat when victory was all but within his grasp.

We retired towards Vanvurenskloof, and on arriving there the following evening heard that a great English force had come from Schoemansdrift and captured Potchefstroom, that another force was at Tijgerfontein, and a third at Schoemansdrift.

Early next morning we crossed the Vaal River at Witbanksfontein. There we off-saddled.

Now I had sent out scouts—not, however, Commandant Jan Theron's men, but ordinary burghers whom the Commandants had sent out—and just as we had partaken about noon of a late breakfast, these burghers came hurriedly into the camp, shouting: "The enemy is close at hand!"

It was not long before every one had up-saddled, and we were off. The English had taken up positions on the kopjes due north of the Vaal River, whilst we had for our defence only kraals and boundary walls. As these

[201] Pioneer.

offered no shelter for our horses, we were forced to retreat. And a most unpleasant time of it we had until we got out of range of their guns and small arms. During this retreat we lost one of our guns. This happened while I was with the left wing. One of the wheels of the carriage fell off, and the gun had to be left behind. Another incident of our flight was more remarkable. A shell from one of the enemy's guns hit an ox waggon on which there were four cases of dynamite, and everything was blown up.

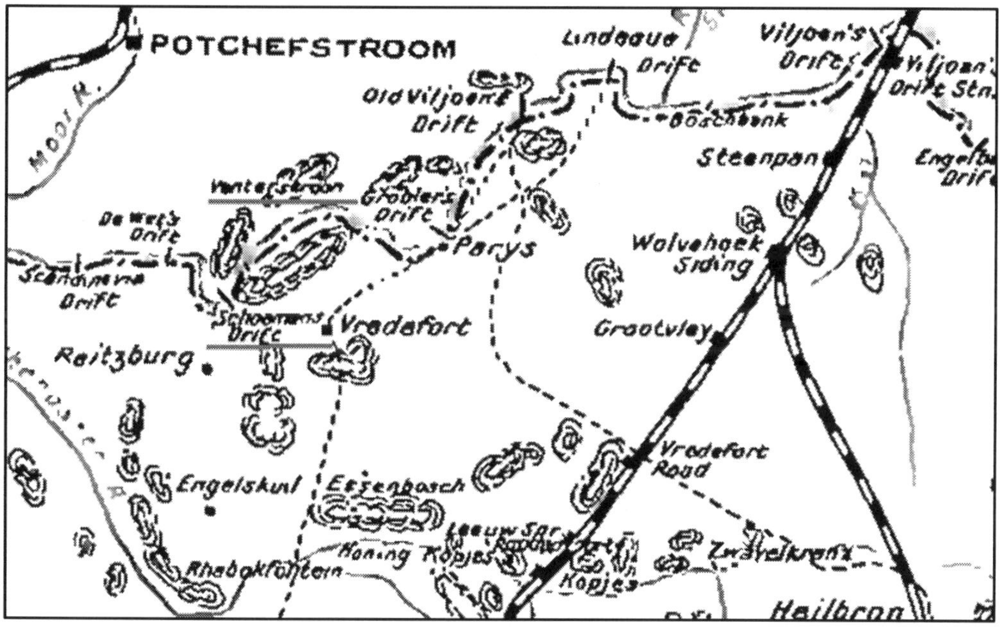

Barely legible, Venterstroom north of Schoeman's drift.[202]

The oxen had just been unyoked and had left the waggon, or else a terrible catastrophe would have occurred.

We lost also two burghers, who, thinking that it would be safe to go into a dwelling house, and hide themselves there, gave an opportunity to some English troops who were on the march from Schoemansdrift, to take them prisoner.

We retired for some distance in an easterly direction, and when it became dark, swerved suddenly to the west, as if aiming for a point somewhat to the south-west of Bothaville. The following evening we stayed at Bronkhaistfontein, near the Witkopjes. From there we went on next morning to the west of Rheboksfontein, remaining that night at Winkeldrift, on the Rhenoster River.

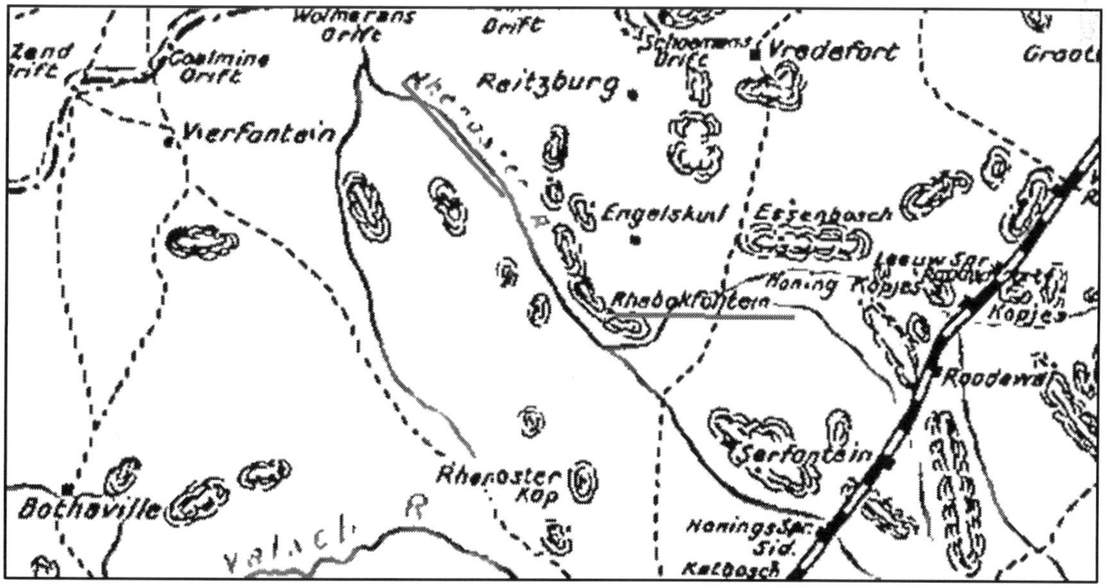

Rheboksfontein, center kopje's and Rhenoster River.

[202] 10/9/1900 De Wet driven north, across the Vaal, at Venterstroom.

There I received a report that President Steyn with his staff was coming from Machadodorp, where he had met the Transvaal Government. The President requested me to come and see him, and also to meet General De la Rey, who would be there.

I told the commandos to go on in the direction of Bothaville and went with my staff to the President. We met on the 31st of October near Ventersdorp. From him I heard that when he came to Machadodorp President Kruger was just ready to sail from Lourenco Marques, in the man-of-war Gelderland, which had been specially sent by Queen Wilhelmina to bring him over to the Netherlands. This was shortly before Portugal ceased to be neutral—the old President got away only just in time.

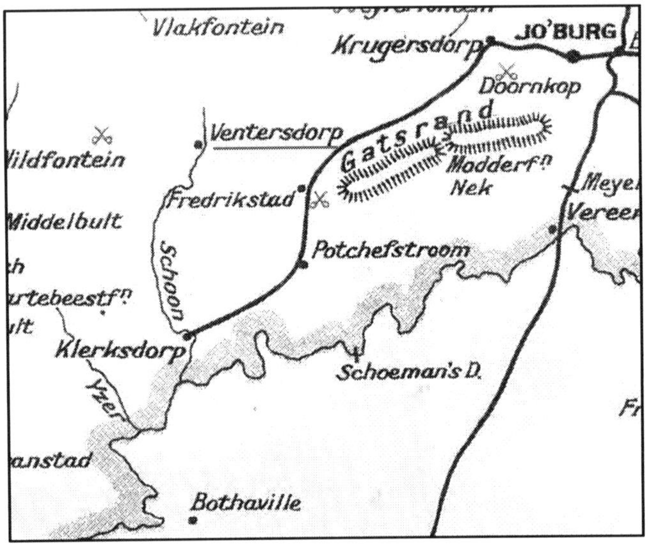

Ventersdorp.

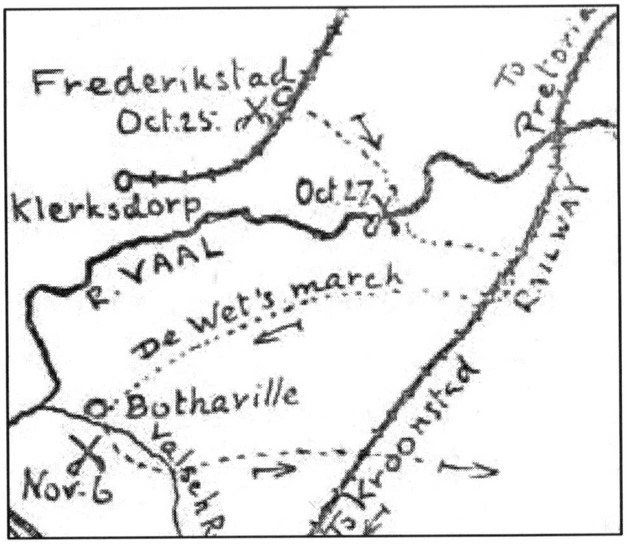

Excerpt from De Wet's map infra, showing his movements.

General De la Rey had been prevented from coming: and on the 2nd of November I went with the President towards Bothaville.

I had received reports from General Fourie, Judge Hertzog, and Captain Scheepers, that the burghers in their districts had rejoined; this made me think that the time had now come to make another dash into Cape Colony. President Steyn had expressed a wish to go with us.

We marched on with the intention of crossing the railway line somewhere near Winburg. On the morning of the 5th we arrived at Bothaville, where we found General Froneman, who had been marching with the commandos from Rhenoster River. Little did we know that a terrible misfortune was awaiting us.

That very afternoon a strong English force, which indeed had been in pursuit of us all the time, came up, and a skirmish took place, after which the English withdrew out of reach of our guns, while we took up a position under cover of the nearest hill. Without suspecting any harm we went into camp about seven miles from the English, keeping the Valsch River between us and them.

I placed an outpost that night close to the river and told them to stay there till the following day. The burghers of this watch returned in the morning and reported that they had seen nothing but wreaths of smoke ascending from the north bank of the river. They believed that these came from the English camp.

We were still safe then—so at least we all believed.

But the corporal who had brought this report had but just left me, and was scarcely one hundred paces off when I heard the report of rifles. I thought at first that it was only some cattle being shot for food, but all at once there were more shots, and what did we see? The English were within three hundred paces of us, on a little hill near Bothaville, and close to the spot from whence my outpost had just returned.

It was early morning. The sun had not risen more than twenty minutes and many of the burghers still lay asleep rolled up in their blankets.

The scene which ensued was unlike anything I had ever witnessed before. I heard a good deal about panics—I was now to see one with my own eyes. Whilst I was looking for my horse to get him up-saddled a few of the burghers were making some sort of a stand against the enemy. But all those who had already up-saddled were riding away at break-neck speed. Many even were leaving their saddles behind and galloping off bare-back. As I up-saddled my horse I called out to them:—

“Don’t run away! Come back and storm the enemy’s position!” But it was no use. A panic had seized them, and the victims of that panic were those brave men who had never thought of flight, but only of resisting the enemy!

The only thing I could do was to leap into the saddle and try to persuade the fugitives to return. But I did not succeed, for as I stopped them at one point others galloped past me, and I was thus kept dodging from point to point, until the whole commando was out of range of the firing.

The leader of the enemy’s storming party was Colonel Le Gallais, without doubt one of the bravest English officers I have ever met. On this occasion he did not encounter much resistance, for only a very few of the burghers attacked him, and that only at one point of his position. Among these burghers were Staats-Procureur Jacob De Villiers, and Veldtcornet Jan Viljoen. As for the rest of our men, it was useless to try to get them to come back to the fight. The gunners however did everything they could to save their guns, but had not enough time to get the oxen inspanned.

Our loss was, as far as I could make out, nine killed, between twenty and thirty wounded, and about one hundred prisoners. Among the dead were Veldtcornets Jan Viljoen, of Heilbron, and Van Zijl, of Cape Colony; and among the wounded, Staats-Procureur Jacob De Villiers and Jan Rechter, the latter of whom subsequently died. The wounded who managed to escape included General Froneman, who was slightly wounded in the chest; Mr. Thomas Brain, who had been hit in the thigh; and one of my staff who was severely wounded, his shoulder being pierced by a bullet.

According to English reports, Dr. De Landsheer, a Belgian, was killed in this engagement. The English newspapers asserted that the doctor was found dead with a bandolier round his body. I can vouch for the fact that the doctor possessed neither rifle nor bandolier, and I am unable to believe that he armed himself on the battlefield.

Six of our Krupp guns were captured in this battle, but as our ammunition for these pieces was nearly exhausted, the loss of them made little difference to us.

I feel compelled to add that, if the burghers had stood shoulder to shoulder we should certainly have driven back the enemy, and the mishap would never have occurred. We were eight hundred men strong, and the enemy numbered not more than one thousand to one thousand two hundred. But a surprise attack such as theirs had been usually produces disastrous consequences.[203]

[203] 11/6/1900. On 6 November, De Wet camped at Bothaville on the Valsch River with 800 Orange Free State commandos. His party included the president of the Free State, Marthinus Steyn. De Wet was aware that Maj. Gen. Charles Knox’s greatly superior British all-arms force was camped 7 miles (11 km) away and believed that his outposts would give him adequate warning of any enemy moves. What he did not know was that the men at his main outpost had fallen asleep.
Shortly after dawn, right after De Wet received a reassuring report from a scout, Knox’s 600-man advance guard, the 5th and 8th MI under Lieutenant-Colonel P. W. J. Le Gallais appeared only 300 yards (270 m) from the Boer camp. For once, De Wet was surprised and a panic ensued in the Boer camp. Because his adjutant had kept a horse saddled, President Steyn was able to escape. The bulk of De Wet’s commando fled for their lives on horseback, but a hard core of about 150 men stayed behind and fought it out with the MI. The two sides went to ground or occupied farm buildings and blazed away with rifles and even field guns at close range. “The heroism displayed on both sides made it one of the most ferocious and gruesome little actions of the war.”
The battle went on for four hours before Knox belatedly showed up with his infantry force. At that point, one surviving officer, Major William Hickie, led the MI in a bayonet charge. This quickly produced a Boer surrender. The MI wanted to shoot two Boers who were found with dum-dum bullets, but Knox humanely stopped this. To the irritation of the MI officers, Knox and his staff seemed preoccupied with dividing the spoils in the captured Boer camp. There was no pursuit.
The Boers lost 25 killed and about 130 captured, including 30 men wounded. De Wet also had to abandon four Krupp field guns, a pom pom, and two artillery pieces captured from the British at the battles of Colenso and Sanna’s Post.
British losses were also serious: about 38 men either wounded or killed in action. Le Gallais died that night of his wounds. Lieutenant-Colonel Wally Ross of the 8th MI was severely wounded in the face.
Because of Knox’s failure to pursue, most of De Wet’s force got away intact and soon began operating against British garrisons and supply convoys again. A furious Hickie wrote, “The General is an old woman... If Knox had the same dash as Le Gallais we should have taken the whole lot, bagged the whole crowd.” Serious losses of materiel had also been inflicted on the Boers, including six field-guns, a Maxim gun and a pom-pom, all de Wet’s wagons carrying gun and small-arms ammunition as well as clothing and other supplies. Despite this, de Wet remained in the field and, within a fortnight, struck back at the British at the head of 1,500 burghers. https://en.wikipedia.org/wiki/Battle_of_Bothaville

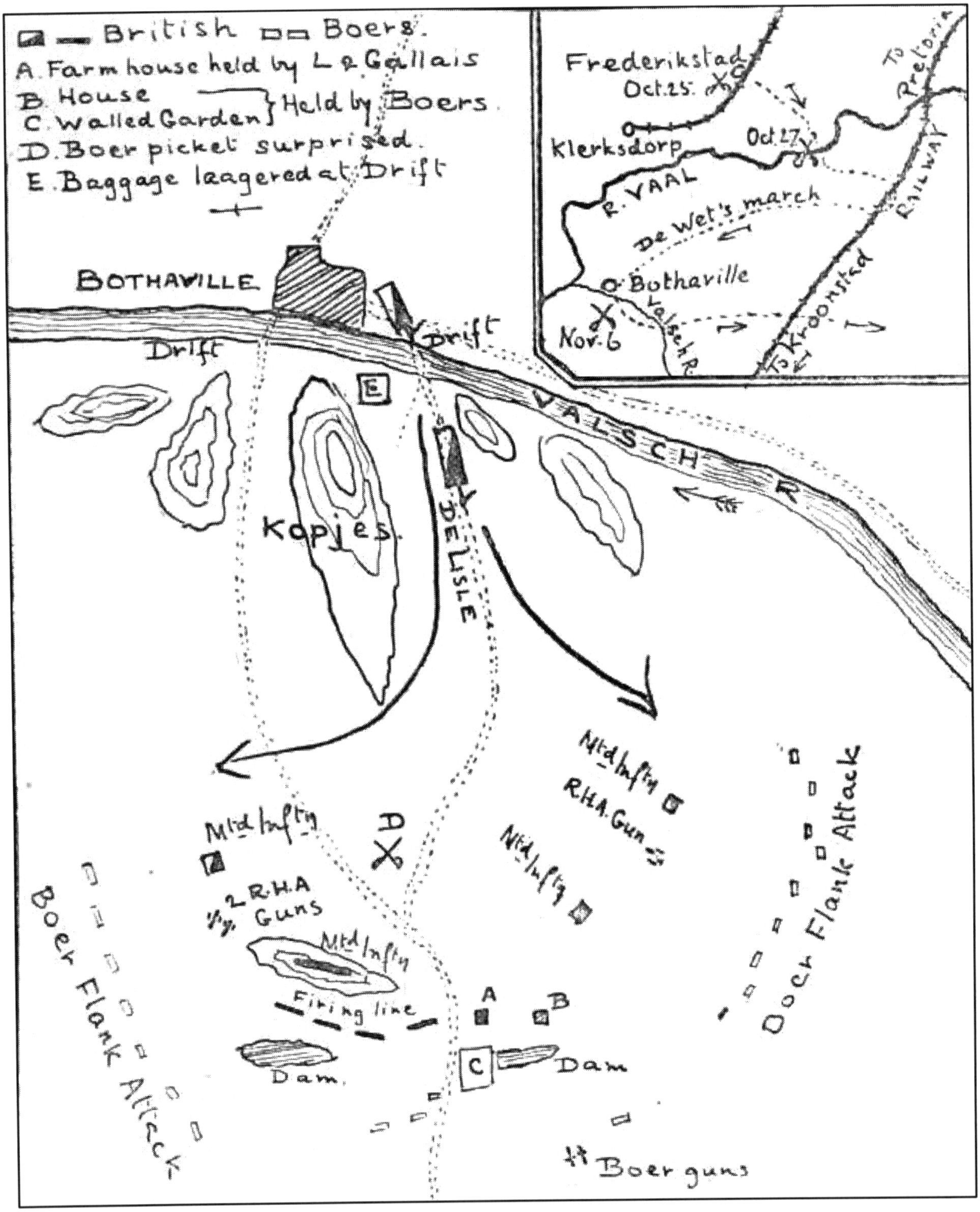

Battle of Bothaville.[237]

22. My March to the South

The horses of the burghers were in a very weak condition; and as the Boer is only half a man without his horse—for he relies on it to get him out of any and every difficulty—I had now to advance, and see if I could not find some means of providing my men with horses and saddles. I went on this errand in the direction of Zandriviersbrug to the farm of Mr. Jacobus Bornman.

Here, however, I divided the commandos. General Froneman, with the Vrede and Heilbron burghers, I sent back to cross the railway lines between the Doorn and Zand Rivers, with orders to operate in the northern districts of the State. I took with me Commandant Lategan of Colesberg, with about one hundred and twenty men, and Commandant Jan Theron, with eighty men, and proceeded on the 10th or 11th of November across the railway line between Doorn River and Theronskoppen, with the intention of executing my plan of making an inroad into the Cape Colony.

We wrecked the railway line and blew up a few small bridges, and then proceeded in the direction of Doornberg, where I met Commandant Hasebroek and his burghers. I sent orders to General Philip Botha to come with the Harrismith and Kroonstad burghers, which he had with him. They arrived about the 13th of November.

We then marched, with about fifteen hundred men, in the direction of Springhaansnek, to the east of Thaba'Nchu. At the northern point of Korannaberg, Commandant Hasebroek remained behind, waiting for some of his men to join him.

Korannaberg is the only landmark I can identify other than Tahba'Nchu and Springhaan's Nek at the bottom.

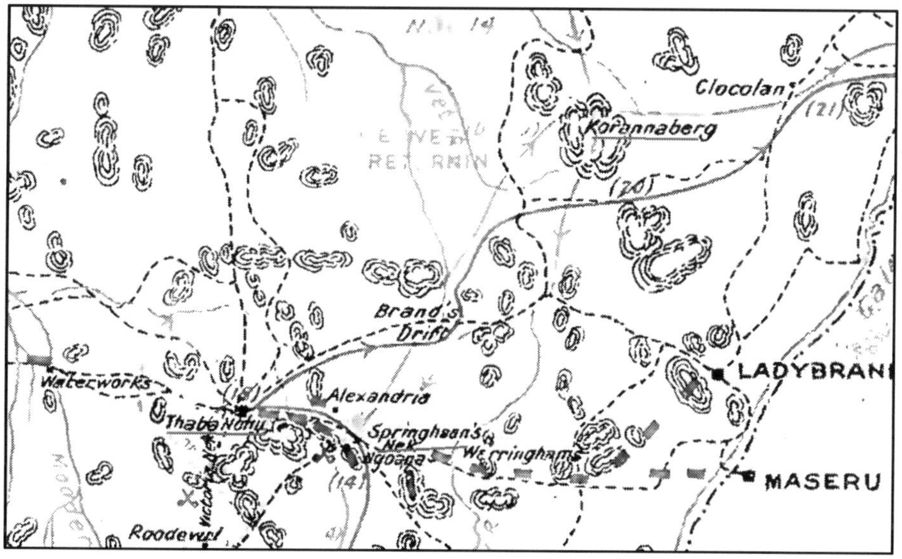

Second Pursuit map of the same area. Note: Springhaansnek.

We took with us one Krupp with sixteen rounds—that was our whole stock of gun ammunition!

By the afternoon of the 16th we had advanced as far as Springhaansnek. The English had built a line of forts from Bloemfontein to Thaba'Nchu and Ladybrand. And just at the point where we wanted to pass them, there were two forts, one to the south and the other to the north, about 2,000 paces from each other, on the shoulder of the mountain.

My first step was to order the Krupp to fire six shots on one of these forts; and, very much to the credit of my gunners, almost every one of these shots found its mark. Then I raced through.

All went well. The only man hit was Vice-Assistant-Commandant Jan Meijer, of Harrismith, who received a wound in the side. He was shot while sitting in a cart, where he had been placed owing to a wound which he had received a few days before, in the course of a hot engagement, which General Philip Botha had had at Ventersburg Station.

We now rode on through Rietpoort towards Dewetsdorp, staying, during the night of the 17th of November, at a place on the Modder River. The following day we only went a short distance, and halted at the farm of Erinspride.

On the 19th I made a point of advancing during the day, so as to be observed by the garrison at Dewetsdorp.

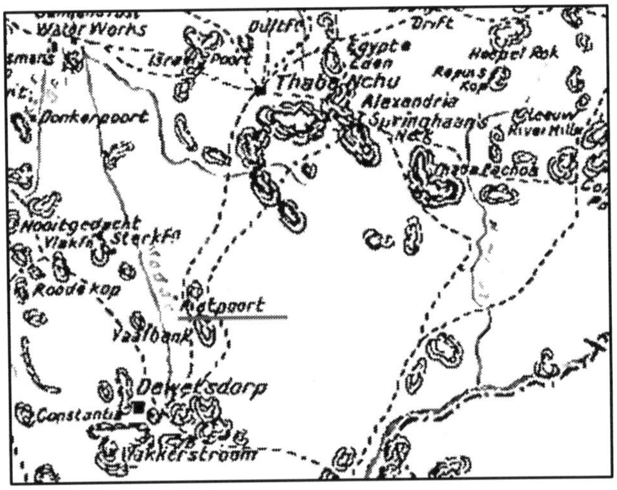

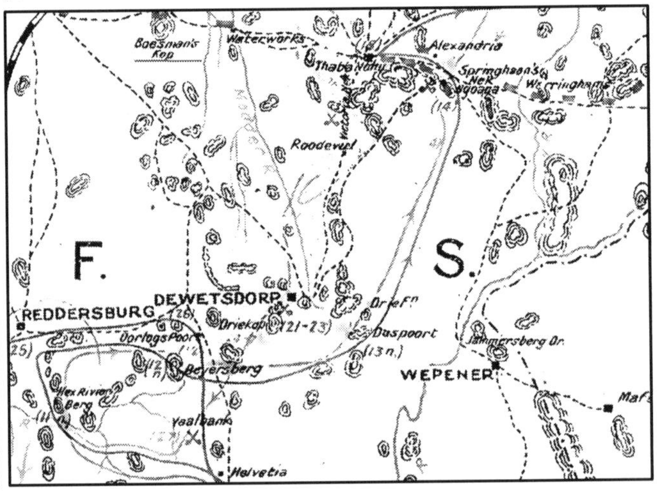

Reitpoort is north of Dewetsdorp. Second Pursuit view of the same general area. Note Boseman's Kop, top left.

My object was to lead the garrison to think that we did not want to attack them, but wished first to reconnoitre the positions. This would have been quite an unnecessary proceeding, as the town was well known to me, and I had already received information as to where the enemy was posted.

The garrison could only conclude that we were again flying, just as we were supposed to have done—by readers of English newspapers—at Springhaansnek. They would be sure to think that after reconnoitring their positions at Dewetsdorp we had gone on to Bloemfontein. Indeed, I heard afterwards that they had sent a patrol, to pursue us to the hills on the farm of Glengarry, and that this patrol had seen us march away in the direction of Bloemfontein. In fact the enemy seemed to have a fixed impression that I was going there. I was told that they had said: "De Wet was either too wise or too frightened to attack Dewetsdorp; and if he did, he would only be running his head against a wall." And again, when they had received the telegram which informed them that I had gone through Springhaansnek, they said: "If De Wet comes here to attack us, it will be the last attack he will ever make."

We came to the farm of Roodewal, and remained there, well out of sight, the whole of the 20th of November. Meanwhile our friends at Dewetsdorp were saying: "The Boers are ever so far away."

But on the evening of the same day I marched, very quietly, back to, and crept up as close as I dared to the positions held by the enemy's garrison. My early days had been spent in the vicinity of this town, which had been named after my father by the Volksraad; and later on I had bought from him the farm [204] where I lived as a boy.

[204] Nieuwjaarsfontein.

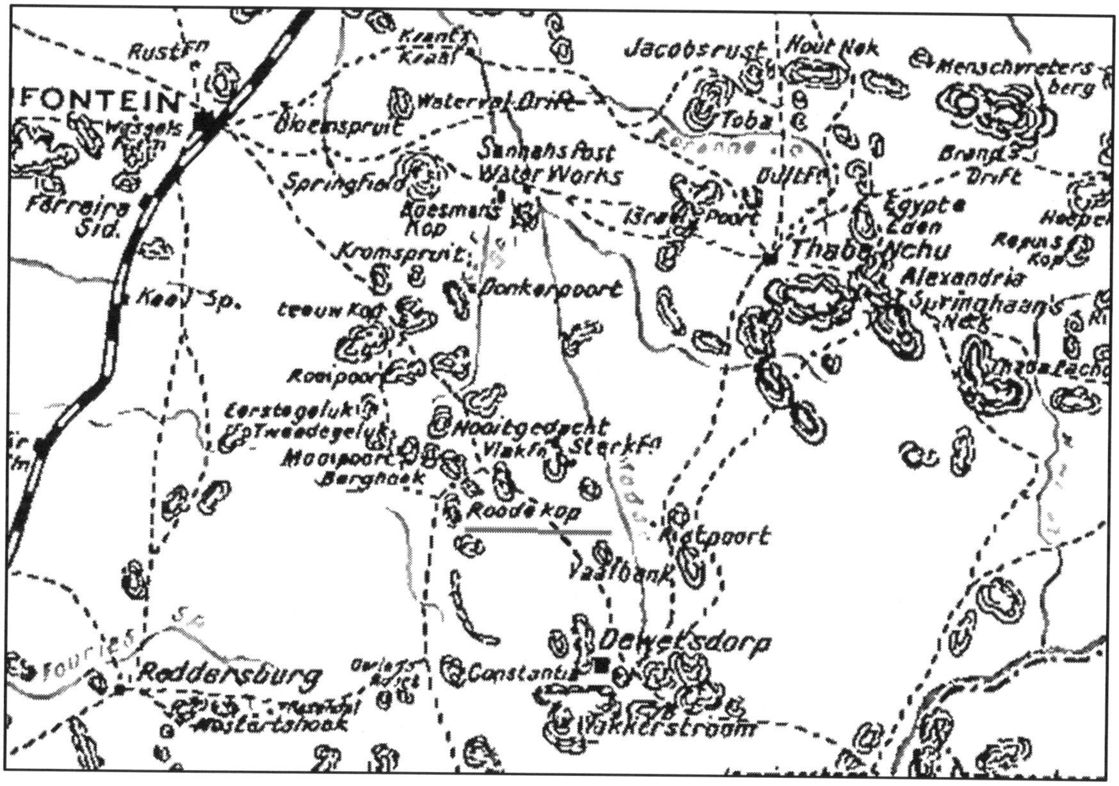

Roode Kop NW of Dewetsdorp and Vaal bank.

By day or by night, I had been accustomed to ride freely in and out of the old town; never before had I been forced to approach it, as I was now, like a thief! Was nothing on this earth then solid or lasting? To think that I must not enter Dewetsdorp unless I were prepared to surrender to the English!

I was not prepared to surrender to the English. Sooner than do that I would break my way in by force of arms.

At dawn, on the 21st of November, we took possession of three positions round the town.

General [Phillip] Botha, who had with him Jan and Arnoldus Du Plessis as guides, went from Boesmansbank to a tafelkop,[205] to the south-east of the town. On this mountain the English had thrown up splendid schanzes, and had also built gun forts there, which would have been very advantageous to us, if we had only had more ammunition. The English had undoubtedly built these forts with the intention of placing guns there, and thus protecting the town on every side should danger threaten. But they did not know how to guard their own forts, for when General Botha arrived there he found only three sentries—and they were fast sleep! Two of them escaped, leaving their clothes behind, but the third was killed.

Commandant De Vos and I occupied a position on the ridge which lies to the north of the town; from this point we could shoot into the town at a range of about 1,600 paces.

Commandant Lategan was stationed on the hill to the west of the town, close to the farm of Glengarry, whose owner, Mr. B.W. Richter—father of my valiant Adjutants, B.W. and Jan Richter—must have been much surprised that morning when he discovered that something very like an attack was being made on Dewetsdorp.

The enemy held strong positions on points of the ridge to the south-east (above the Kaffir location) to the south-west and to the north-west. Their schanzes were built of stones, and provided with trenches. On the top of the schanzes sandbags had been placed, with spaces left between them for the rifles.

[205] A table-shaped hill.

Royal Munster Fusiliers fighting from behind the redoubt at Honey Nest Kloof, South Africa.[238]

Of Major Massey, who was in command, and his force, consisting of parts of the Gloucestershire regiment, the Highland Light Infantry, and the Irish Rifles, five hundred all told, I have only to say that both commanding officer and men displayed the greatest valour.

Although Commandants Hasebroek and Prinsloo had not arrived, nevertheless I had as many as nine hundred men. But I was obliged to send a strong patrol to Roodekop, eighteen miles from us in the direction of Bloemfontein, in order to receive reports in time, should reinforcements be coming up to the help of the English. I had also to send men to keep watch out towards Thaba'Nchu, Wepener and Reddersburg; nor could I leave the President's little camp (which I had allowed to proceed to the farm called "Prospect") without some protection. Thus it was that of my nine hundred men, only four hundred and fifty were available for the attack.

It delighted me to see how courageous our burghers were at Dewetsdorp. As one watched them creeping from schanze to schanze, often without any cover whatever, and in danger at every moment of falling under the enemy's fire, one felt that there was still hope.

On the first day we advanced until we were close to the schanzes on the south-east and on the north; we remained there during the night in our positions, our food being brought to us.

The second day, November 22nd, firing began very early in the morning, and was kept up until the afternoon. Our most advanced burghers, those of Harrismith, had come to within about one hundred paces of the first schanze.

I saw one of our men creeping on till he was close under the enemy's fort. Directly afterwards I observed that rifles were being handed over the schanze to this man. Later on it appeared that the man who had done this valiant deed was none other than Veldtcornet Wessels, of Harrismith. He was subsequently promoted to the rank of Commandant, to take the place of Commandant Truter; later on again, he became Vice-Commander-in-Chief.

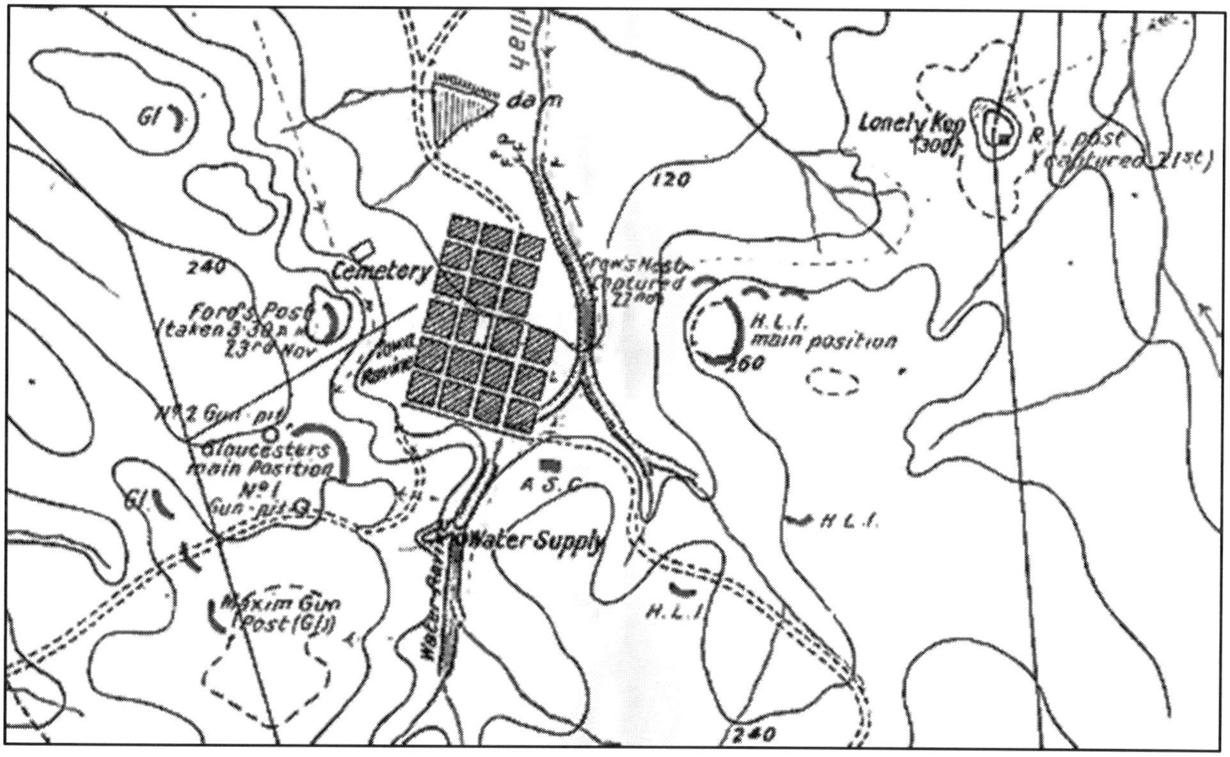

Dewetsdorp Captured November 23rd, 1900.[239]

Our burghers could now enter this fort without incurring much danger. But they had hardly done so, when the two English guns, which had been placed to the west of the town, opened fire on them. When this happened, I gave orders to my men that a great schanze of the English, about eighty paces from the one which we had just taken, should be stormed. This was successfully carried out by Veldtcornet Wessels, who had with him about twenty-five men.

The enemy meanwhile kept up a heavy fire on our storming party, from some schanzes which lay still further away; our men, therefore, had nothing left them but to take these also. Then while our men kept in cover behind the fort which they had just taken, the English left the schanzes upon which the storming party had been firing so fiercely; this, however, Veldtcornet Wessels and his burghers did not know, because, after having rested a little, and desiring to renew the attack, they only saw that everything was quiet there, and that they were now only under the fire of guns from the western forts, which lay right above the town. I also had not observed that the forts had been abandoned.

Just as the sun was setting, and when it was too late to do anything, General Philip Botha, with his two sons, Louis and Charlie, rushed up to Veldtcornet Wessels and told him what the real state of affairs was.

I now saw columns of black smoke rising from the mill of Mr. Wessels Badenhorst, to the south of the town. Everybody was saying: "The English are burning their commissariat; they are going to surrender!"

The English had a strong fort on the north, near the place where Commandant De Vos was stationed. In order to take this schanze one would have been obliged to cross 200 metres of open ground. Moreover, it was so placed that it was the only part of the English possession which De Vos's guns commanded. Accordingly, when the sun had gone down, I sent orders to him that he was to storm this schanze before daybreak on the following morning.

My orders were duly carried out.

Commandant De Vos crept stealthily up to the fort, and was not observed by the enemy until he was close to them. They then fired fiercely on him, killing two of his burghers, but our men would not be denied; they leapt over the schanze and compelled the enemy to surrender. The English losses on this occasion were six killed, a few wounded, and about thirty taken prisoner.

While this was going on, Veldtcornet Wessels, in accordance with orders which I had given him the previous evening, had taken possession of the river bank exactly opposite to the town, which he was now preparing to storm.

The English had only a few schanzes to the west of him, and these were not more than two hundred paces off.

I had been to the laager at "Prospect" the night before, with the intention of returning so as to be in time for the storming of the town. I had arranged to go there very early in the morning, because my journey could be accomplished with much less risk if carried out in the dark. Unfortunately, however, daylight overtook me when I had got no further than the Kaffir location, and I had to race from there, over country where I had no sort of cover, to the ravine near the town. From this ravine to where Veldtcornet Wessels was waiting for me on the river bank, I rode in comparative safety.

The reader can easily imagine how delighted I was to meet again the Dewetsdorp folk, to whom I was so well known. But I could not show myself too much. That would not have been safe. After I had visited three houses—those of the Schoolmaster, Mr. Otto, of Mr. Jacobus Roos, and of old Mr. H. Van der Schijf—and had partaken in each of a cup of coffee, I hurried off to my burghers.

The remaining English schanzes had been so well constructed that their occupants could still offer a very stubborn resistance, and they did so. It was not until about three o'clock on the afternoon of the 23rd of November that we saw the white flag go up, and knew that the victory was ours.

We took four hundred and eight prisoners, amongst whom were Major Massey and seven other officers. We also took fifty Kaffirs. Two Armstrong guns with more than three hundred rounds of ammunition, some waggons, horses and mules, and a great quantity of Lee-Metford cartridges also fell into our hands.[206]

We never knew the exact numbers of the English dead and wounded, but they must have lost something between seventy and one hundred men.

Our own loss was heavy. Seven of the burghers were killed and fourteen wounded; most of these, however, slightly.

The sun had already set before we had put everything in order, and it was late in the evening when we returned to our laager at "Prospect." There I received a report that a great column was marching from the direction of Reddersburg, in order to relieve Major Massey—but they were too late!

Very early the following morning we made preparations to intercept the advance of this column. We took up positions to the west of Dewetsdorp, and the day was spent in exchanging shots with the enemy's guns. During the night we remained in our positions, but when the sun rose I discovered that the column, which was already too strong for us, was expecting a reinforcement, and as no attack was attempted on their side, I decided to leave the position quietly, and to march on.[207] My inroad into Cape Colony must no longer be delayed.

Our positions at Dewetsdorp were so situated that I could leave them unnoticed. I thought it well, however, to leave behind a small number of burghers as a decoy, so that the English should not pursue us at once.

[206] 11/23/1900 Garrison at Dewetsdorp surrendered to De Wet.

[207] 11/27/1900 General Charles Knox in touch with De Wet at Beyersberg.

23. I Fail to Enter Cape Colony

The enemy gave us plenty of time in which to effect our escape, and by nightfall we had abandoned our positions at Platkop. Taking with us the prisoners of war (whom I intended to set free on the far side of the Orange River), we marched towards Vaalbank, arriving there on the following morning. That day the English attacked us unawares. While I was at Dewetsdorp, Captain Pretorius had come up to give me a report of his recent doings. I had sent him, two months previously, from the district of Heilbron to Fauresmith and Philippolis, in order to fetch two or three hundred horses from those districts; he had told me that he had brought the horses, and that they were with his 200 men at Droogfontein.

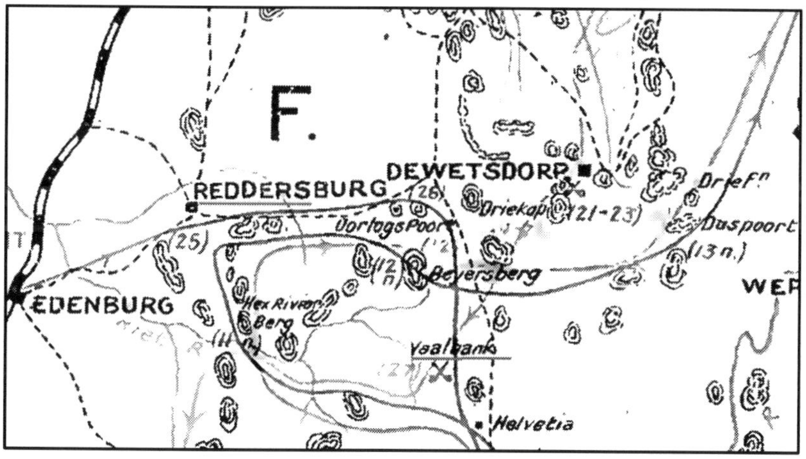

Vaalbank and Reddersburg.

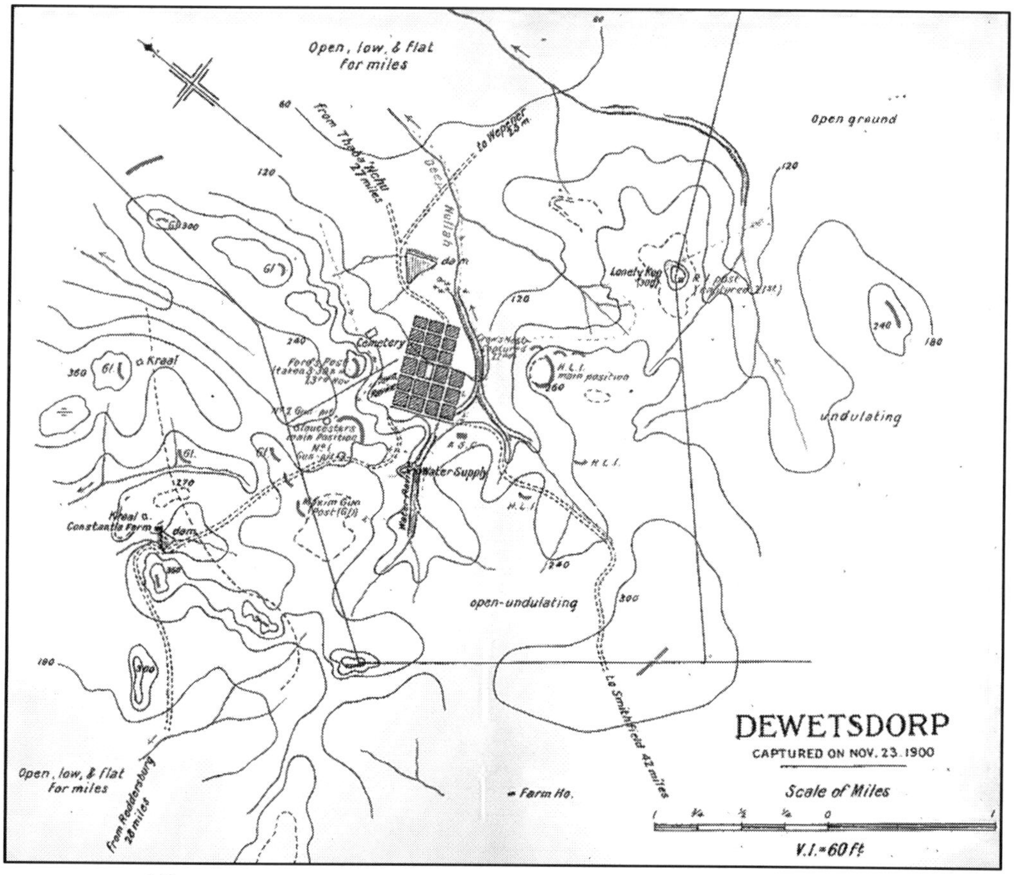

Dewetsdorp.[240]

It was about eight o'clock in the morning after our night march that our outpost at Vaalbank saw a mounted commando riding from in the direction of Reddersburg. I was at once informed of this, but as I was expecting Pretorius from that direction, I merely said: "It is sure to be Captain Pretorius."

"No; this is an English commando."

English or Australian—it made very little matter—they were enemies.

I had no need to give the order to off-saddle, the burghers did it at once of their own accord. But before we were ready for him, the enemy opened fire on us from the very ridge on which our outpost had been stationed.

Off went the burghers, and I made no effort to stop them, for the spot where we were did not command a good view of the surrounding country, and I already had my eye on some ridges, about half an hour's ride away. There we should be able to reconnoitre, especially towards Dewetsdorp, whence I expected the enemy at any moment. During the retreat Veldtcornet de Wet was severely wounded. Moreover, some of our horses had to be left behind, being too exhausted to go any further.

We marched on [South] towards Bethulie. When in the neighbourhood of this town, and of the farm of "Klein Bloemfontein," I fell in with General Piet Fourie and Captain Scheepers, and took them with me. While on this farm I set free the Kaffirs whom I had taken prisoner at Dewetsdorp; they pretended they had not been fighting, but were only waggon-drivers. I gave them a pass to go into Basutoland.

Oxen drawn wagon, stuck in drift.[241]

We then proceeded towards Karmel, and just as we were approaching the farm of "Good Hope," we caught sight of an English column which had come from Bethulie, and was making for Smithfield. I at once opened fire upon them from two sides, but they were in such good positions that we failed that day to drive them out. On the morrow, early in the morning, the fight began afresh.

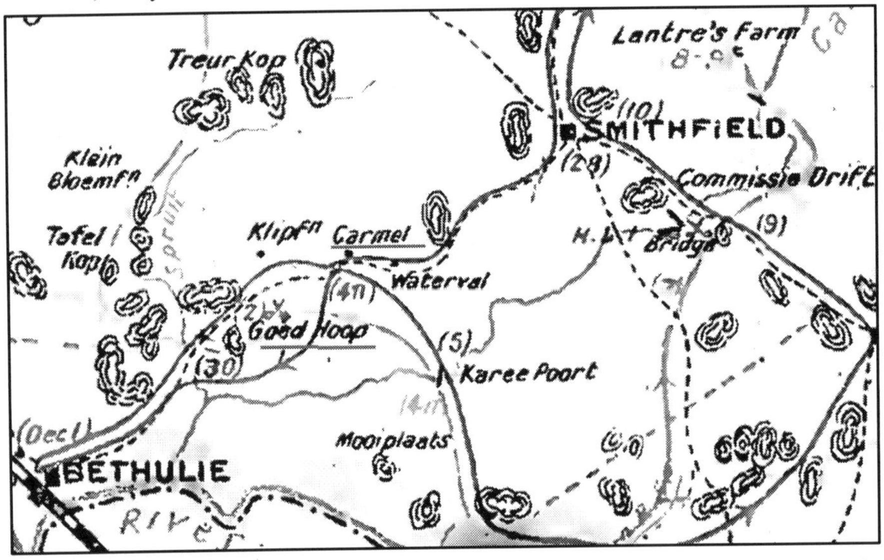

'Carmel' and 'Good Hoop.' The Caledon River flows through Karee Poort into the Orange River.

About four o'clock in the afternoon General Charles Knox, with a large reinforcement, arrived from Smithfield, and we had once more to retire. It was here that I sustained a loss upon my staff—my nephew, Johannes Jacobus de Wet. It was sad to think that I should never again see Johannes—so brave and cheerful as he had always been. His death was a great shock to me.

Our only other casualties were four burghers wounded, whereas the enemy, unless I am much mistaken, must have lost heavily.

Whilst this fight was in progress General Hertzog joined me. We arranged that he should with all speed make an inroad into Cape Colony, between the Norvals Pont and Hopetown railway bridges, and that I should do the same between the railway bridges at Bethulie and Aliwal North. He was to operate in the north-western part of the country, I in the eastern and midland parts.

That night we continued our march towards Karmel, under a heavy downpour of rain. Next morning it was still raining when we started to continue our march; later on in the day we off-saddled for a short time and then went on again, so as to be able to cross the Caledon River before it became impossible to do so. I can assure you that it rained so hard while we were fording the Caledon, that, as the Boers say, "It was enough to kill the big devils and cut off the legs of the little ones." We then marched on—still through heavy rain.[208]

Commandant Truter, who was in command of the rear-guard, had left a Krupp and an ammunition waggon behind. I was not at all pleased about this, but, as we had not a single round of Krupp ammunition left, the gun would only have hampered us.

That evening we reached the Orange River, at a point some three miles to the north of Odendaalsstroom, but, alas! what a sight met our eyes! The river was quite impassable owing to the floods, and, in addition, the ford was held by English troops stationed on the south bank.

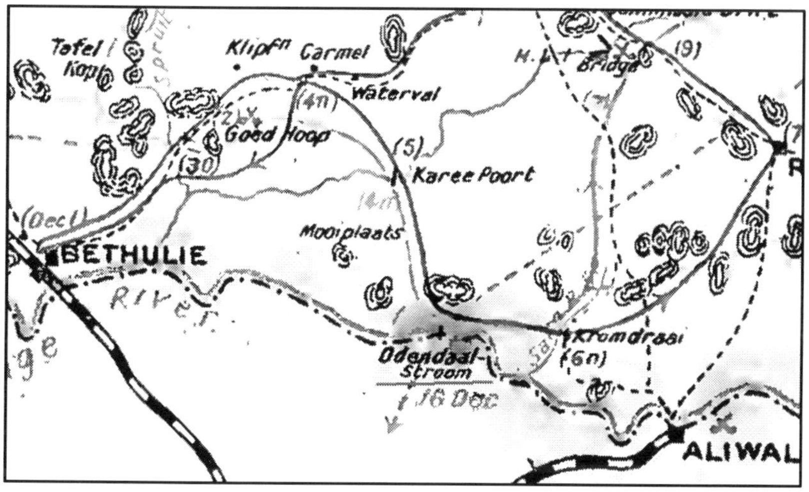

Odendaalsstroom.

Our position was beginning to be critical, for there was an English garrison at Aliwal North, so that I could not cross the Orange River by the bridge there. It was also highly probable that the Caledon would be in flood, and I knew that General Charles Knox had left a division of his troops at Smithfield—they would be sure to be holding the bridge over the Caledon at Commissiedrift. Moreover, Jammerbergsdrift, near Wepener, was doubtless well guarded, so that there, too, I would have no chance of crossing the river. There was still Basutoland, but we did not wish to cross its borders—we were on good terms with the Basutos and we could not afford to make enemies of them. Surely we had enough enemies already!

Bridge over the Orange River at Aliwal North.

[208] 12/5/1900 De Wet crossed the Caledon with a view to entering Cape Colony.

To make the best of a bad job I sent Commandant Kritzinger[209] and Captain Scheepers, with their three hundred men, to march in the direction of Rouxville with orders that as soon as the Orange River became fordable, they were to cross it into Cape Colony without delay. I entertained no doubt that they would succeed.[210]

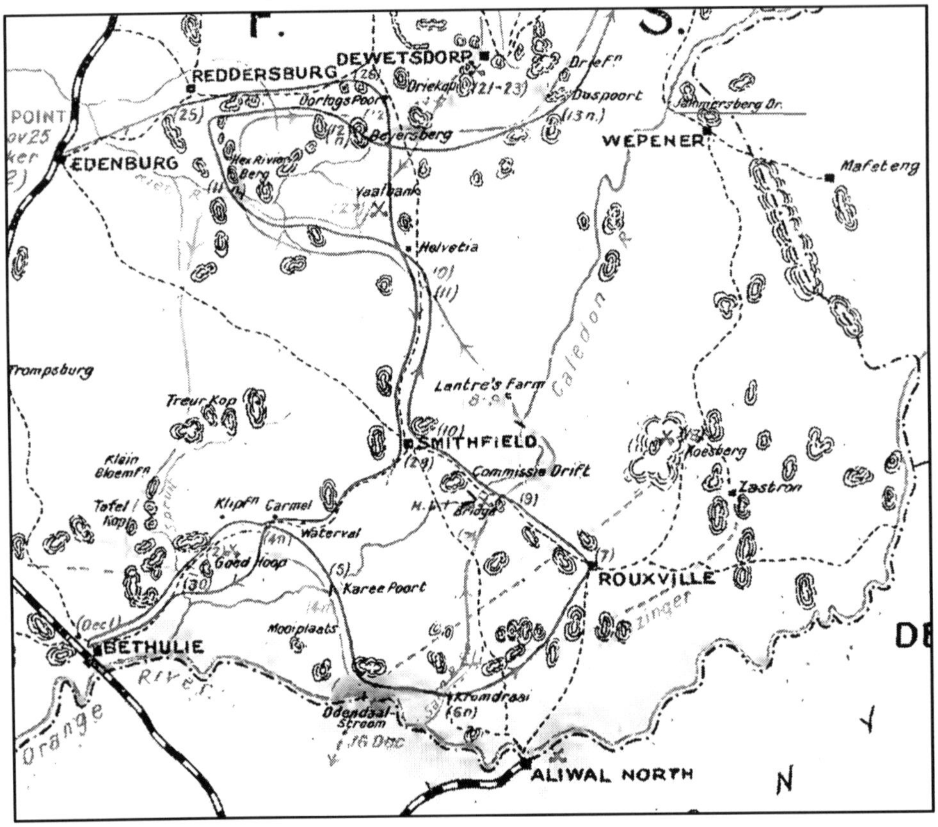

Wepener top right and Jammersberg Drift.

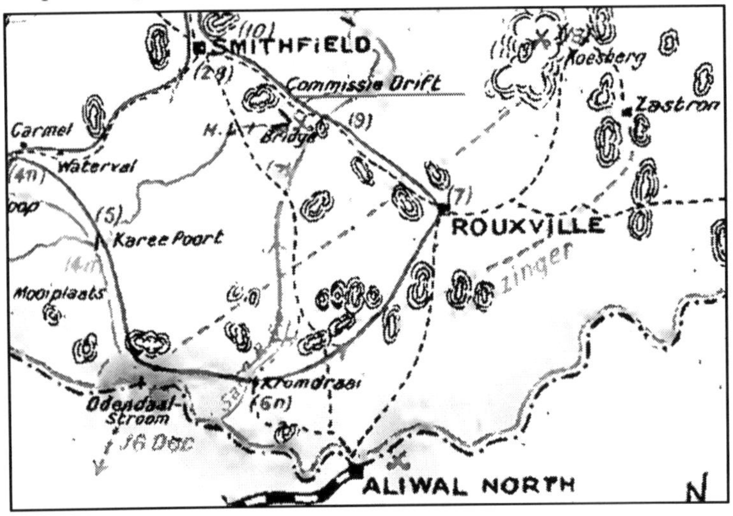

Commisser Drift SE of Smithfield and Zastron to the east.

Everything is as it must be, and unless one is a sluggard—who brings trouble upon himself by doing nothing to avoid it—one has no reason to complain.

Such were my thoughts as I contemplated our situation.

The Orange River was in flood—the Government and I, therefore, could not possibly remain where we were for long. The English were so fond of us that they would be sure to be paying us a visit! No, to wait there until the river was fordable was not to be thought of.

209 He was subsequently appointed Vice-Commander-in-Chief in Cape Colony.

210 12/5/1900 De Wet crossed the Caledon with a view to entering Cape Colony.

The reader will now perceive how it was that my projected inroad into Cape Colony did not become a fact. My dear old friend, General Charles Knox, was against it, and he had the best of the argument, for the river was unfordable. What then was I to do? Retreat I could not, for the Caledon also was now full. Again, as I have already explained, it would not do for me to take refuge in Basutoland. But even that would be better than to attempt to hold out where I was—in a narrow belt of country between two rivers in flood—against the overpowering force which was at General Knox's disposal, and which in ten or twelve days would increase tenfold, by reinforcements from all parts of the country.

I knew that the Orange and the Caledon Rivers sometimes remained unfordable for weeks together. How could I then escape?—Oh, the English had caught me at last! They hemmed me in on every side; I could not get away from them. In fact they had "cornered" me, to use one of their own favourite expressions. That they also thought so appears from what I read afterwards in the South African News, where I saw that Lord Kitchener had given orders to General Charles Knox "not to take any prisoners there!" For the truth of this I cannot positively vouch; but it was a very suspicious circumstance that Mr. Cartwright, the editor of the newspaper to which I have referred, was afterwards thrown into prison for having published this very anecdote about Lord Kitchener.

Our prospects were then by no means bright; I knew very well that those trusty counsellors of the English—the National Scouts—would have advised their masters to seize the bridges and thus make escape impossible for Steyn and De Wet.

Without delay I proceeded to the Commissiedrift bridge over the Caledon. As I feared, it was occupied by the enemy. Entrenchments had been dug, and schanzes thrown up at both ends.

Foiled here, I at once sent a man down to the river to see if it was still rising. It might be the case that there had not been so much rain higher up. The man whom I had sent soon returned, reporting that the river was falling, and would be fordable by the evening. This was good news indeed.

On the other hand, our horses were exhausted. They had now for three days been obliged to plough their way through the wet, muddy paths. We had no forage to give them, and the grass was so young as yet that it did not seem to strengthen them at all.

Nevertheless, we had to be off. And there was but one road open to us—we must somehow get across the Orange River and thus obtain elbow-room. Accordingly we returned to make for Zevenfontein, a ford ten or twelve miles further up the river. If it were not already in the enemy's hands, we would surely be able to get across there. Shortly before sunset, on the 8th of December, we arrived at Zevenfontein. To our immense joy, it was unoccupied and fordable.

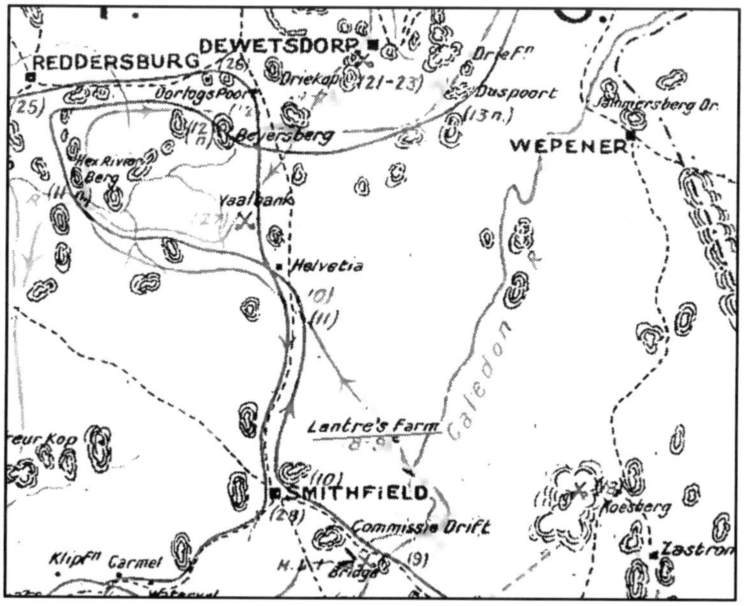

On this map De Wet crosses the Caledon at Lantre's Farm on the 8th.

I at once marched towards Dewetsdorp, intending, if only General Knox and his huge force would give me the chance, to rest my horses, and then make another attempt to enter Cape Colony.

But it was not to be.

The English were afraid that if President Steyn and I were in Cape Colony their troubles would be doubled. General Knox therefore concentrated all his available forces in order to drive us northwards. It was disappointing, but there was a bright side to it. If the English were pursuing me, they would have to leave Commandant Kritzinger and Captain Scheepers, who would thus be able to cross the Orange River.

These two officers, however, were not left entirely in peace. While they rested for a time near, Zastron in order to give their horses a chance of recovering their strength, there came a division of Brabant's Horse to pay them a visit. The result was that about sixty of the visitors were wounded or taken prisoner, while the rest found it as much as they could do to get back to Aliwal North, whence they had started. Commandant Kritzinger and Captain Scheepers had then another opportunity for rest until the day should come when they could make an inroad into Cape Colony according to my instructions.

Although, as I have already said, the English were passionately devoted to President Steyn and myself, I was deprived of their endearments for the space of two whole days, during which I was at Wilgeboomspruit. Here I was joined by Commandant Hasebroek with his commando, and all of us—horses as well as men—enjoyed a little rest. But very soon General Knox was again at our heels, and, to escape him, I marched west in the direction of Edenburg, hoping at last to be able to get into Cape Colony. Not only were the forces of General Knox behind us, but, when we arrived at the farm of "Hexrivier," and thus were within two hours' march of Edenburg, I heard from my scouts, whom I had sent on in advance, that there was a great English column in front of us at that town.

In the evening, therefore, I turned off towards the east, and marched in the direction of Wepener.[211]

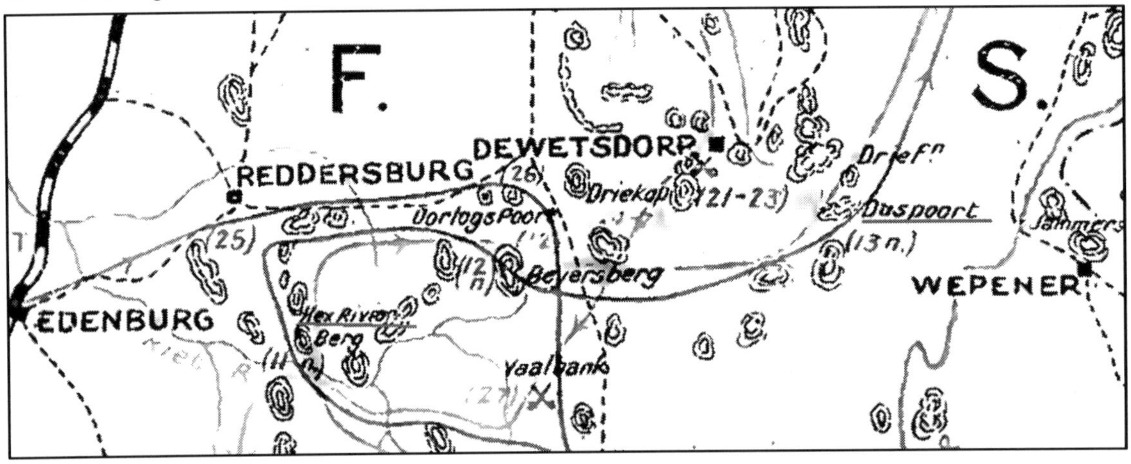

'Hex Riviers Berg' and Daspoort.

The following morning the enemy was again on our track; but, as we had covered twenty miles during the night, we were so far ahead that it was unnecessary for us to move very fast during that and the following day.

At mid-day, the 13th of December, we took up excellent positions—placed in a line of about eight miles from end to end—on the farm called "Rietfontein," which is in the district of Wepener, north-east of Daspoort. We were so strongly posted that the enemy had to halt and wait for the arrival of the rearguard. I had calculated on this, and knew that darkness would come to our aid before the English were ready to attack us. But in front of us there was a strong line of forts, extending from Bloemfontein through Thaba'Nchu and Springhaansnek, [Sprinkaannek] to Ladybrand. Through this line we should have to fight our way; this would be difficult enough, and it would never do to have General Knox at our heels, to increase the difficulty. Our only plan, then, was to make a long night march, and thus to get well out of the way.

Accordingly, I gave orders to the men to hold their positions until dark, and to let the enemy see that they were doing so. I had even had schanzes built, so as to impress them with the idea that I intended to attack them the following day if they advanced towards my positions. And just before the night came on, I ordered the burghers to show themselves from behind all our schanzes.

Then night fell, and I at once gave orders to march off.

[211] 12/11/1900. De Wet, after being turned northward by General Knox, moved towards Reddersburg.

The burghers could not understand this, and began to grumble about it—what could their General mean? Why this sudden change in his plans? I said nothing, but thought to myself, "You shall know why to-morrow."

We marched directly towards Springhaansnek. It was very slow work, for many of the burghers' horses were so weak that their owners had to go on foot. General Philip Botha and I were with the rearguard, and did not expect to reach the line of forts until ten o'clock on the following morning.

We had not advanced very far before we were joined by Commandant Michal Prinsloo, who had with him three hundred of the Bethlehem burghers. He had come down from Springhaansnek, and as his horses were in good condition I ordered him to go in advance of us, to pass through Springhaansnek, and then to occupy positions to the north of the lines of forts and east of Thaba'Nchu.[212]

My object in making this arrangement was that when on the following morning we were crossing the mountains, he might be able to hinder the enemy at Thaba'Nchu from either checking our advance, or sending reinforcements to the Springhaansnek forts.

And in point of fact, Prinsloo's commando proved to be our salvation; for the English, from their high position at Thaba'Nchu, spied us as soon as day broke, and indeed sent troops to reinforce the point for which we were making. But Prinsloo succeeded in holding them in check, so that when we arrived at Springhaansnek we had to fight against strong positions, but against nothing else—but I must not anticipate.

Before it began to be light on the morning of the 14th of December, Commandant Prinsloo passed through the enemy's lines between the forts. The English fired upon him, but he did not turn back. Then a small outpost of the enemy, which lay half-way between the forts, made an attempt to turn the oncoming burghers by shooting at them from the front. The Commandant only gave strict orders that the men must force their way through. The consequence was that two of the enemy, who did not get out of the way in time, were literally ridden over. The burghers thought that these two unfortunate men had been trodden to death by the horses, but it was not likely that any of them would dismount to see if this were actually the case.

As I have already said, General Botha and I were in the rearguard. We knew, however, that Vice-Commandant-in-Chief Piet Fourie—a man whom nothing on earth would stop, if he had once made up his mind—was leading the van, and that he was supported by Veldtcornet Johannes Hattingh, who was as resolute and undaunted as his chief.

Fourie did not wait for us to catch him up, but at once went down the mountain side. When we saw this, General Botha and I rode with all speed ahead, telling the burghers to come on more gently with their weary horses. I did not fear thus to leave them behind, because I knew that General Knox was still a long way in the rear.

Just as General Fourie, leading the first storming-party, had passed between the forts, we came up with him, our burghers still straggling on behind us. As soon as we had crossed over the first piece of rising ground, I halted my men, and ordered them to leave their horses out of sight of the enemy, and to return to the brow of the hill, so as to be able to fire into the forts on the right and left hand, which were from eight hundred to nine hundred paces from us. From this hill we kept up as fierce a fire as we could, and this to a great extent prevented the enemy in those forts from firing on our burghers who were still coming on in a long train.

It is necessary, in order that the reader may understand the task which we had set ourselves to accomplish, to say a few words about Springhaansnek. At either side of the way by which we must pass, there were two strong forts, at a distance of from a thousand to twelve hundred paces from each other. In the space between them there was absolutely no cover; and the distance from the point where the burghers were first visible to the men in these forts, to the point where they again disappeared from view, was at least three thousand paces.

Over these terrible three thousand paces our burghers raced, while a storm of bullets was poured in upon them from both sides. And of all that force—eight thousand strong—no single man was killed, and only one was wounded!

Our marvellous escape can only be described to the providence and irresistible protection of Almighty God, who kept His hand graciously over us.

What the enemy's loss was I never heard.

[212] See Also: 12/13/1900 Reverse to General Clements De la Rey, J.C. Smuts and C.F. Beyers overran a British camp under the control of General Clements near the Magaliesberg. Brabant's Horse mishap near Zastron. Battle of Nooitgedacht https://en.wikipedia.org/wiki/Battle_of_Nooitgedacht

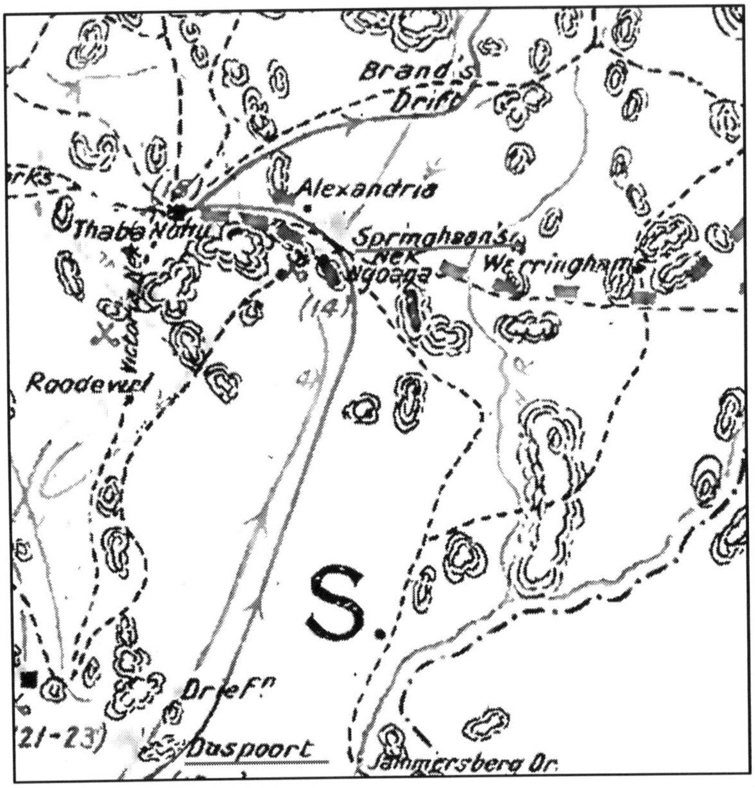

From Daspoort north through the line of forts at Springhaan's Nek.

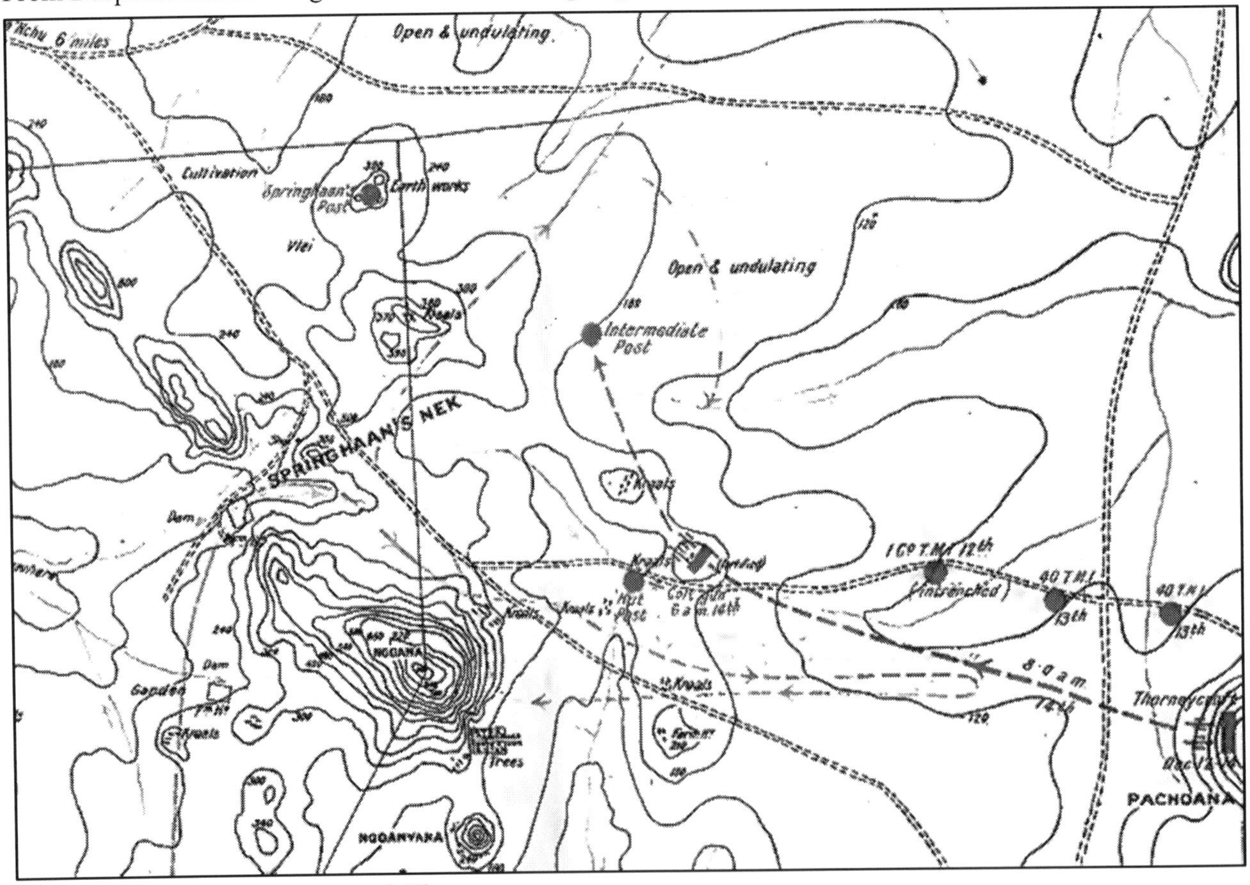

Springhaan's Nek. December 14th.[242]

In addition to the burghers, a few carts and waggons, as well as one of the two guns which had been taken at Dewetsdorp, got safely through the English lines. The other gun was left behind by the sergeant of the artillery, before he reached the fighting line. He sent the horses of the gun-carriage with the gunners, back to Commandant Hasel, who subsequently followed us to Ijzernek, to the west of Thaba'Nchu.

My ambulance with Dr. Fourie and Dr. Poutsma, were stopped by the English. Dr. Fourie had, as was quite proper, remained outside the fighting line, with the intention of coming through afterwards. This he was permitted to do on the following day. He brought me a message from General Knox to the effect that Commandant Hasebroek had lost heavily in an engagement with Colonel White, who had marched out from Thaba'Nchu. But I had already received information that the Commandant had got through the enemy's lines unhurt, and that on the contrary it was he who had killed some of Colonel White's men, while they were attacking him.

We decided to retreat still further, in order to reach a place of safety where we might rest our horses, in preparation for that long dash into Cape Colony, which I still intended to carry out on the first opportunity. I felt sure, however, that my commandos would be allowed no rest by the enemy as long as the President and I were with them. Accordingly I planned that as soon as we got to the north of Winburg he and I should absent ourselves from the commandos for some time, while I proceeded to arrange certain matters (to be set down in a later chapter) by which I hoped to effectually "settle"[213] the English.

On our arrival at a certain farm to the south of Senekal we discovered that General Knox was once more at our heels. We had several small engagements with him, in one of which a son of Commandant Truter, of Harrismith, was killed.

On the afternoon of Christmas Day, 1900, we left the farm, and rode on to the Tafelkop, nine miles to the west of Senekal.

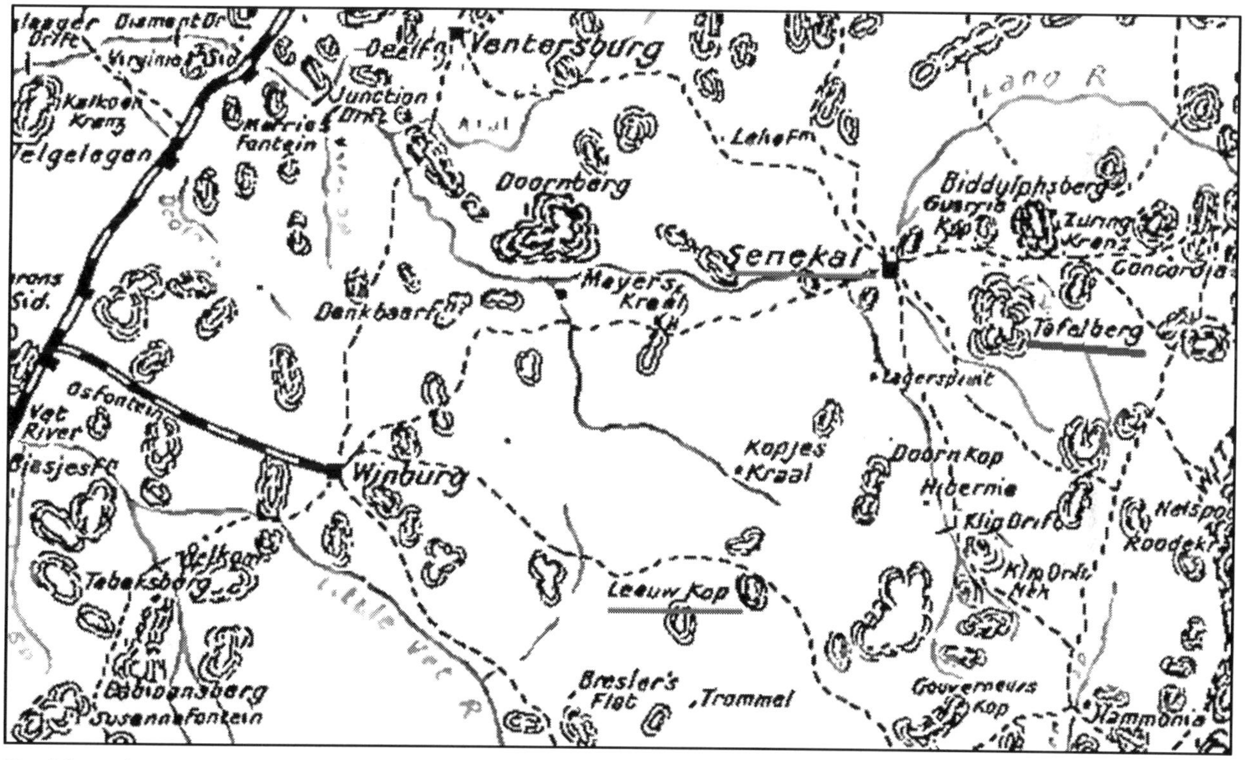

Unable to locate Tafelkop. Tafelberg upper right is East. Leeuw Kop, lower center.[214]

[213] In the original a Kaffir word is used here. The literal meaning of the phrase is "to throw the knuckle bones"—the Kaffir equivalent for dice.

[214] 12/22/1900 Boer movement in Cape Colony checked.

12/26/1900 General Charles Knox engaged with De Wet near Leeuw Kop.

12/28/1900 De Wet, frustrated in his attempt to break through to the south, withdrew to Senekal. Cape raiders driven northward.

12/29/1900 Boer Forces led by Generals Ben Viljoen and Chris Muller won a victory over the British Forces, led by Major Cotton. British garrison at Helvetia was captured. Battle of Helvetia: https://af.wikipedia.org/wiki/Slag_van_Helvetia

24. Wherein Something Is Found about War against Women

It was decided here, on the 26th December, to divide the large commando into two. The one part was to be under the command of Assistant-Chief-Commander P.H. Botha, and the other Assistant-Chief-Commander Pete Fourie.

I entrusted to President Steyn a bodyguard under Commander Davel, who went with the Government in the direction of Reitz.

As regards myself, I went to Assistant-Chief-Commander C.C. Froneman, who was with the Heilbron Commander, L. Steenekamp, in the neighbourhood of Heilbron. It was my intention to take with me from there a strong escort, and to dig up the ammunition at Roodewal taken on the 7th of June, as both our Mauser and our Lee-Metford ammunition were nearly exhausted, although we still had a fairly large supply of Martini-Henry Giddy cartridges.

I then started from Tafelkop, on the 27th of December, and arrived two days later at General Froneman's commando, close to Heilbron.[215]

I had to wait there till the evening of the 31st December, until the necessary carriages and oxen had been got together for carrying the ammunition with us. Carriages were now no longer to be got easily, because the British had not only taken them away from the farms, but had also burnt many of them. Where formerly in each farm there were at least one carriage and a team of oxen, and in some two, three or even more, there were now frequently not a single one. Even where there were carriages the women had always to keep them in readiness to fly on them before the columns of the enemy, who had now already commenced to carry the women away from their dwellings to the concentration camps within their own lines, in nearly all villages where the English had established strong garrisons.

Proclamations had been issued by Lord Roberts, prescribing that any building within ten miles from the railway, where the Boers had blown up or broken up the railway line, should be burnt down. This was also carried out, but not only within the specified radius, but also everywhere throughout the State. Everywhere houses were burnt down or destroyed with dynamite. And, worse still, the furniture itself and the grain were burnt, and the sheep, cattle and horses were carried off.

Wagon Loading (PF: Confiscation?).[243] A Domiciliary Visit Firing the outhouse [outbuilding] of a Boer who has been shooting natives.[244]

[215] See Also: The Battle of Helvetia took place on 29 December 1900 during the Second War of Independence at Helvetia, 10km north of Machadodorp. Here the Boer forces led by Generals Ben Viljoen and Chris Muller won a victory over the British Forces, led by Major Cotton. The Boer force was 350 men and the British force 350 men strong and the British had one 120mm cannon at their disposal. https://af.wikipedia.org/wiki/Slag_van_Helvetia

Burning of a farmhouse near Dewetsdorp in May, 1900.[245]

Farms set alight.[246]

Bales of Wool. Worth Carrying Away. Confiscated wool which afterwards fetched £120 for Government.[247] A useless haul. A captured wagon for which no oxen are available is burnt.[248]

A Boer family looks on at their house that was set alight by the British forces.[249216]

[216] Original Footnote: The Times History of the War says:
The policy of burning down farmhouses and destroying crops as a measure of intimidation had nothing to recommend it and no other measure aroused such deep and lasting resentment. The Dutch race is not one that can be easily beguiled by threats, and farm-burning as a policy of intimidation totally failed, as anyone acquainted with the Dutch race and the Dutch history could have fore seen. Applying this system against a white race defending their homes with a bravery and resource which has rightly won the admiration of the world was the least happy of Lord Roberts' inspirations and must plainly be set down as a serious error of judgment....'

"Rough Justice in Wartime."[250] "The Penalty of Foul Play. South African Light Horse Burning a Farm which has fired on Our Troops."

Ruined town-appears to be Frankfort, northern Orange Free State.[251]

Inevitable Reprisals in South Africa.[252] "A Bonfire of Forage. Burning hay [and barn?] found on a surrendered Boers Farm." [But not the dwelling.]

Nor was it long before horses were shot down in heaps, and the sheep killed by thousands by the Kaffirs and the National Scouts, or run through by the troops with their bayonets. The devastation became worse and worse from day to day. And the Boer women—did they lose courage with this before their eyes? By no means, as when the capturing of women, or rather the war against them and against the possessions of the Boer commenced, they took to bitter flight to remain at least out of the hands of the enemy.

In order to keep something for themselves and their children, they loaded the carriages with grain and the most indispensable furniture. When then a column approached a farm, even at night, in all sorts of weather, many a young daughter had to take hold of the leading rope of the team of oxen, and the mother the whip, or vice versa. Many a smart, well-bred daughter rode on horseback and urged the cattle on, in order to keep out of the hands of the pursuers as long as at all possible, and not to be carried away to the concentration camps, which the British called Refugee Camps (Camps of Refuge).

Family of Boer refugees passing by a flock of sheep.[253]

Loaded Ox Wagon.[254]

How incorrect, indeed! Could any one ever have thought before the war that the twentieth century could show such barbarities? No. Any one knows that in war, cruelties more horrible than murder can take place, but that such direct and indirect murder should have been committed against defenceless women and children is a thing which I should have staked my head could never have happened in a war waged by the civilized English nation. And yet it happened. Laagers containing no one but women and children and decrepit old men, were fired upon with cannon and rifles in order to compel them to stop.

I could append here hundreds of declarations in proof of what I say. I do not do so, as my object is not to write on this matter. I only touch upon it in passing. There are sufficiently many righteous pens in South Africa and England to pillory these deeds and bring them to the knowledge of the world, to remain on record for the future. For what nation exists, or has existed, which has not a historical record whether to its advantage or to its disadvantage? I cannot do it here as it should be done. And too much cannot be said about this shameful history.

I had to unburden my heart. Now let me proceed.

On the evening of the 1st of January, 1901, I pushed on towards Roodewal Station, for I had obtained all the waggons I needed for my purpose.[217] Perhaps that night the outposts were asleep; but however that may be, we reached the railway without the enemy being aware of our movements. The hour was growing late, and so we had no choice but to remain where we were, nine miles from the spot at which we aimed. But the following evening we were again on the march, and reached the place where the ammunition had been buried. We found it untouched, and just where we had left it, a few miles from the railway, and quite close to the English camp, at Rhenosterriviersbrug.

[217] 1/1/1901 General Charles Knox and others continued the pursuit of De Wet.

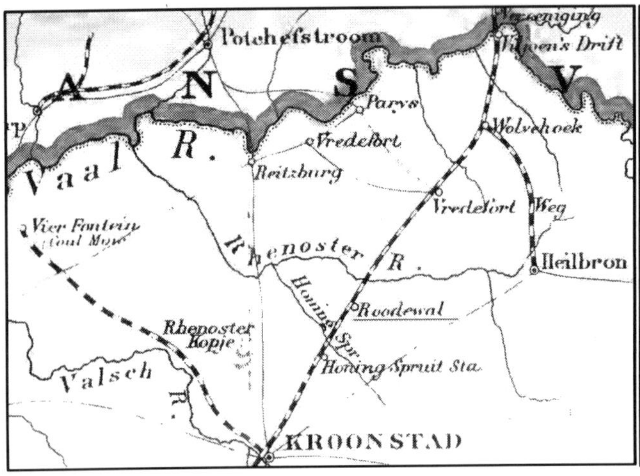

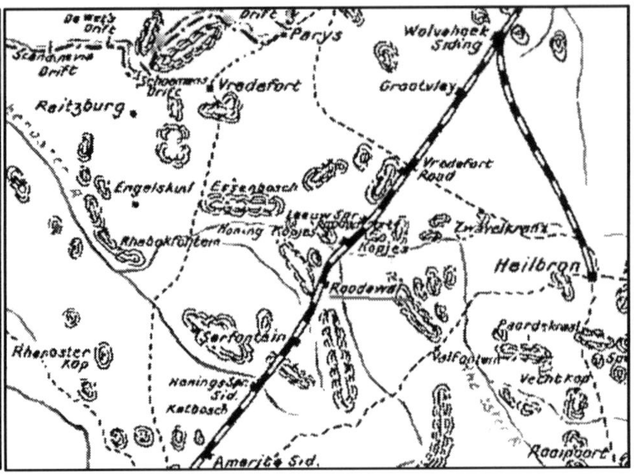

Roodenwal and General Area.

We were very careful to recover every cartridge, since it was clear that the war must still continue for a long space of time. We could have no thought of giving up the struggle, whilst the pride of England would not allow her to turn back.

We loaded our waggons with the ammunition, and I gave to General Froneman the task of conducting it across the railway line. I myself proceeded to the Vredefort commandos, which were stationed some fifteen miles away, for the state of affairs amongst these commandos called for my presence. On the 4th of January, when night had fallen, I crossed the railway near Vredefortweg, unnoticed by the enemy.

Two days later I was back again with General Froneman's commando, where I found that the ammunition had arrived in safety. I was informed that General Knox had divided his forces into three parts, one of which had engaged General Fourie and Commandant Prinsloo, near Bethlehem. We had given the enemy a good beating, but had lost two men in the affair. I regret to say that one of them was that clever officer, Vice-Commandant Ignatius du Preeij. He was a man whom every burgher loved, for he was goodness personified. The second of General Knox's division had set out in the direction of Heilbron, whilst the third had pursued General Philip Botha along the Liebenbergsvlei.[218]

This division had attempted to mislead General Botha by all sorts of tricks, but on January the 3rd he had put up notices outside different farmhouses, stating that he did not like such familiarity.

On one occasion the General, with only fifty burghers, had charged one hundred and fifty of the bodyguard, and had taken one hundred and seventeen prisoners, leaving the whole of the remainder either killed or wounded.

A panic now occurred among General Knox's forces. The division that was marching to Heilbron suddenly turned aside towards Kroonstad, only to meet with General Botha, who left them in anything but an undamaged condition.

The division which had been despatched to deal with General Fourie and Commandant Prinsloo entered Senekal.

When I arrived at General Botha's camp, which was situated six miles to the east of Lindley, I found that General Knox had already taken Kroonstad.

After this we allowed ourselves a rest.

On the 8th of January I received reports from Commandant Kritzinger and Captain Scheepers dealing with the state of affairs in Cape Colony. They informed me that they had safely crossed the Orange River by a foot-path. There was another footpath, more to the south, which an English outpost of eight men was guarding. These soldiers occupied a house near by, and the first warning they had that we had crossed the river was when the door of their abode opened, and they heard the order to "hands up."[219]

[218] Vlei—a valley with stagnant water in it.

[219] See Also: 1/7/1901 Boers attacked Belfast, Wonderfontein, Nooitgedacht, Widfontein, and Pan, and after sharp fighting were dispersed.
1/10/1901 Machadodorp attacked by night. Post gallantly defended.
1/12/1901 Boers driven eastward from Witwatersberg by General French. Activities in Cape Colony to frustrate Hertzog's advance.

Commandant Kritzinger and Captain Scheepers also assured me that the sympathies of the Colonial burghers were strongly with us. Like every other right-minded man, I had expected this to be the case, for "blood is thicker than water."[220]

Gideon James Scheepers.[255 221]

Although the Colonials were well aware what a dangerous course they would be pursuing if they joined us, and how, later, they would be sure to be treated as rebels, they nevertheless threw in their lot with ours.

From Judge Hertzog I received a very encouraging report as to the burghers in the north-western parts of Cape Colony. This news decided me on leaving behind, in their own districts, parts of the commandos from all the various divisions, and on taking others to join with me in a second expedition into Cape Colony.

The following were the officers I took with me, ordering them to assemble at Doornberg, in the district of Winburg, on the 25th of January, 1901: Generals Piet Fourie, Philip Botha and Froneman; Commandants Prinsloo (Bethlehem), Steyn (Ficksburg), Hasebroek (Winburg), De Vos (Kroonstad), Merve (Parijs), Ross (Frankfort), Wessel Wessels [222] (Harrismith), Kolbe (Bloemfontein), and Jan Theron, with the renowned Theron Scouts.

From the 8th to the 25th of January we were in the north-western districts of the Free State. We were waiting for a suitable opportunity to make a dash into Cape Colony.

[220] The Boer proverb is:—"Blood creeps where it cannot walk."

[221] Scheepers Gideon J, Commandant Scheepers was born in 1878 near Middleburg. In 1894 or 1898, he joined the Staatsartillerie as a heliograph operator. He was seconded to the Free State Artillery to help develop their system of communications. He led General de Wet's scouts during the Boer War. His area of operation was the Brandwater Basin. In 1900 he was in the Transvaal and undertook attacks against the British lines of communications. He re-entered the Cape Colony in 1901 and operated around Aliwal North and Graaff-Reinet. His hope of starting a revolt in the Cape came to nothing. His commando reached Mossel Bay. After destroying local property, he withdrew to the Ladismith district. He was captured in October 1901. He was charged with offences including murder, flogging a British subject, placing prisoners in the firing line, etc. A guilty verdict resulted in his execution at Graaf-Reinet on 18 January 1902. He was ill on the day of his execution and had to be tied to a chair. https://www.angloboerwar.com/name-search

[222] I had appointed him in place of Commandant Truter, who had resigned.

25. I Again Attempt to Enter Cape Colony

I was now about to make a second attempt to march into Cape Colony. I had great fears that my plans would leak out, since I was obliged to mention them to the commandants. But I was not able to confine all knowledge of my future movements entirely to the commandants. For I had sent many a burgher home to fetch a second horse; and the burghers began to make all sorts of guesses as to why they had to fetch the horses; and one could hear them mutter: "We are going to the Colony."

But nevertheless they were all in good spirits, with the exception of some, who had for commander a most contradictory and obstinate officer.

By January the 25th nearly the whole of my commandos had assembled; only General Philip Botha, with the burghers from Vrede under Commandant Hermanus Botha, had yet to arrive in order to complete our numbers; and he had been prevented coming.

President Steyn and the Government decided to go with me and my two thousand burghers.

At Doornberg the council of war was called together by the Government. President Steyn then communicated to the meeting that his term of office would soon expire. He pointed out that the provisions of the law designed to meet this contingency could not be carried out, because a legally constituted Volksraad could not be summoned at the present moment.

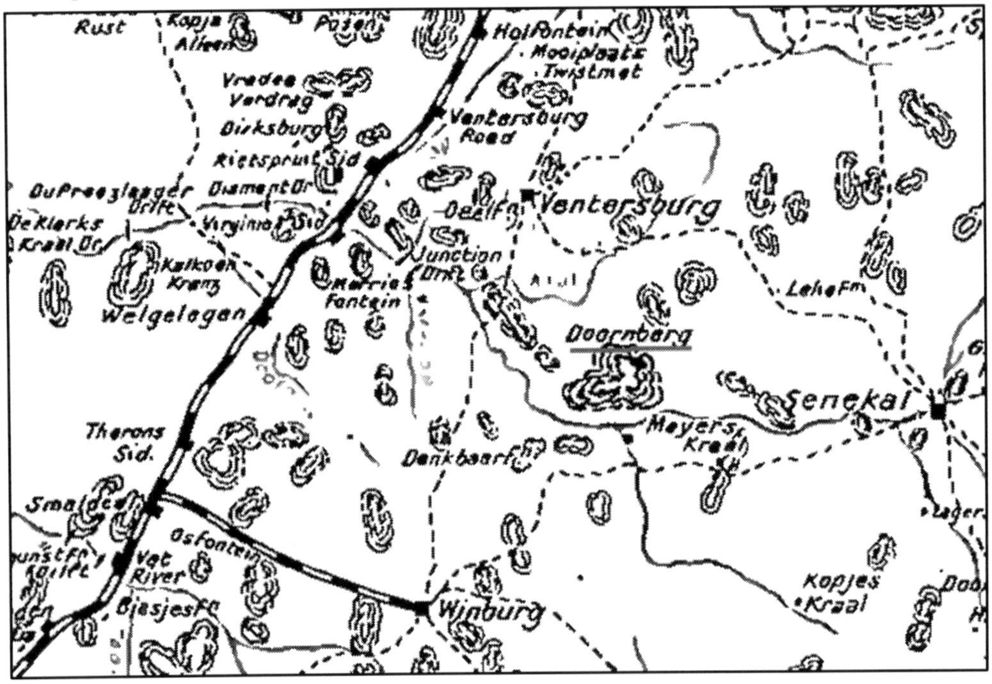

Doornberg Center.

The council of war decided to propose a candidate to the burghers without any delay, at the same time giving them the option of nominating candidates of their own. Further, it was decided that the candidate who should be elected should be sworn in as Vice-States President, and retain that title until the time arrived when the condition of the country should make it possible to hold an election in conformity with the law.

After the voting had taken place, it was found that the former President, Marthinus Theunis Steyn, had been unanimously re-elected.

At the burghers' meeting the voting resulted in the same way, except at a meeting at which Mr. Cecil Rhodes was proposed as a candidate. This proposal was not seconded!

President Steyn was declared elected. And he was then sworn in.

The executive Raad now consisted of the President, as chairman, with T. Brain, Secretary of State, W.J.C. Brebner, Secretary of State, A.P. Cronje, Jan Meijer and myself as members. Mr. Rocco De Villiers was Secretary of the War Council, and Mr. Gordon Fraser, Private Secretary to the States President.

No States-Procureur had been appointed since Mr. Jacob De Villiers had been taken prisoner at Bothaville; but the Council appointed Mr. Hendrik Potgieter, Landdrost of Kroonstad, as Public Prosecutor.

Various causes had made it impossible for a legally constituted Volksraad to sit. Some members had, as we called it, "hands-upped"; others had thought that they had done quite enough when they had voted for the war. I would be the last to assert that they had done wrong in voting thus. The whole world is convinced that, whatever the Boers might have done, England was determined to colour the map of South Africa red! And England succeeded beyond her expectations! For South Africa was stained with the blood of burghers and defenceless women and children, and with the blood of English soldiers who had died in a quarrel for which they were not responsible, and which could have been avoided!

There were other members—and I had no patience with them—who had said: "We will give our last drop of blood for our country," and then had taken good care that no one should have a chance of getting even the first drop! They preferred to remain quietly at home, and wait for the English to come and make them prisoners of war!

Only a minority of the members had remained faithful to our cause, and these did not constitute a quorum; and so no sitting could take place. This small party, as far as I can recollect, consisted of the following ten members: C.H. Wessels Bishop, Chairman; Wessel Wessels (Vrede); J.B. Wessels (Winburg); A.P. Cronje (Winburg); Jan Steijl (Bloemfontein); Jan Meijer (Harrismith); J.J. Van Niekerk (Fauresmith); Daniel Steyn (Heilbron); Hendrik Ecksteen (Vrede); and Hendrik Serfontein (Kroonstad).

We marched from Doornberg on the 26th of January to Commandant Sarel Hasebroek's farm, which is eight miles to the north of Winburg.

There was a strong English force seven or eight miles to the east of Winburg, and another body of the enemy eleven or twelve miles still further to the east. In addition, a column was marching northwards from Ventersburg, west of our position.

It was perfectly plain that the enemy were aware of our intentions; but this, as I have already said, could not be helped. Our army was so constituted that no secret could be kept; and I decided for the future to tell no one of any further plans I might form.

On the 27th of January I reconnoitred to the east of Winburg, and took care to let myself be seen, for I wished to make it appear that it was my intention to proceed in that direction in the evening. Meanwhile I secretly sent my scouts to the west.

That night I marched to the west of Winburg, crossing the branch railway without meeting with any opposition, and arrived on the following morning at the Vet River—to the south of the town. We did not advance very fast,[223] as we expected that we should soon once more have to face the difficulty of marching with exhausted horses.

In the afternoon we continued our way till we had passed Tabaksberg. The following morning, January 28th, I received a report that the English were advancing in two divisions. I ordered my burghers to up-saddle and to occupy positions to the east of Takasberg.

The enemy's right wing was to the east, and we stationed ourselves on some ridges that lay in front of them, but were unable to deliver an attack. We charged their left wing, however, and captured a Maxim-Nordenfeldt, which was in perfect order, at the cost of one killed and three wounded. Our other losses amounted to a very small number.

As to the enemy's losses, they took some of their dead and wounded away, but they left behind them several of their dead at the spot where we had captured the gun.

To remain there and continue the fighting the next day could not even be thought of; for if we had waited the English would have had time to bring up reinforcements, and my plan of entering Cape Colony would have been rendered impossible.

[223] Our forethought proved later on to have been of little avail. For notwithstanding the bountiful rains which had fallen at the end of November and the beginning of January in the southern and western parts of the State we found, when we arrived there, that the grass had been entirely destroyed by the locusts. Neither could we obtain any fodder; and so the difficulty of providing for our horses was as great as ever.

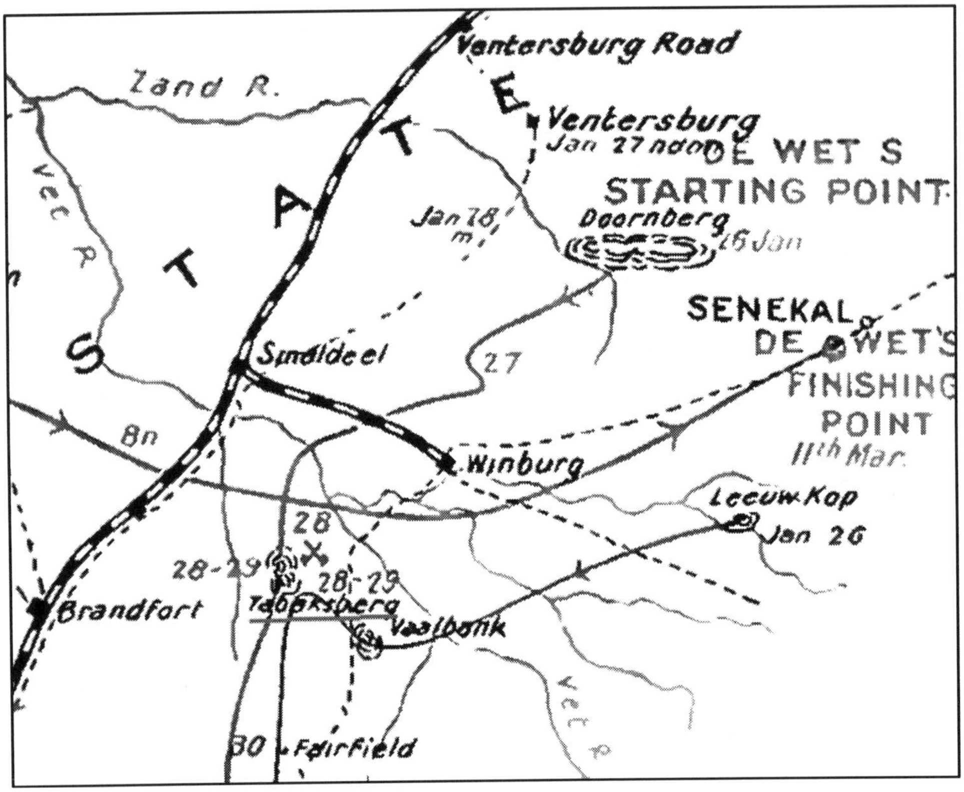

Starting point and action at Tabaksberg.

Our position was difficult enough. The enemy were at our heels, and we had to get away as best we could. In front of us there was the line of fortifications from Bloemfontein to Ladybrand, which had been greatly strengthened since we had forced our way through it at Springhaansnek. It was impossible to get through at Springhaansnek now.

I decided to march towards Thaba'Nchu. But in order to deceive the English I sent a strong patrol on the following day in the direction of Springhaansnek, ordering them to make no attempt to conceal their movements.

I could advance for eight miles without attracting the enemy's notice; but if I had gone further I should have been seen from the forts. I need scarcely say that it was greatly to my advantage not to give the English a chance of seeing me. And so when we had covered eight miles we off-saddled. If I had allowed the English to discover what I was doing they would have brought up troops from Thaba'Nchu, Sanna's Post and Bloemfontein; and these troops in combination with the force behind me might have put me into a very awkward position.

My old friend, General Knox, whose duty it had been to prevent me entering Cape Colony on a previous occasion, was again entrusted with the same task. Any person who has had dealings with this General will acknowledge that he is apt to be rather a troublesome friend; for not only does he understand the art of marching by night, but he is also rather inclined to be overbearing when he measures his strength with that of his opponents.

And now, as we were in camp, congratulating ourselves that we were safe for the time being, my scouts reported that this same General Knox was approaching. I at once ordered the burghers to up-saddle, and to inspan the ten waggons we had with us laden with ammunition and flour.

I left behind me a portion of my commando under General Fourie, whose duty it was to check General Knox, whilst I myself was going forward to clear a road through the enemy's forts.

It was lucky for us that General Knox had been deceived by the strong patrol I had sent in the direction of Springhaansnek, and that he had come to the conclusion that my commando was marching to the same place. He therefore started off in that direction and continued until he discovered his mistake. Then he turned aside and came in contact with General Fourie. Our men held him back for a few hours, and lost two men, very badly wounded in the engagement.

Whilst this was occurring I had reached the forts between Thaba'Nchu and Sanna's Post. When I was there a reinforcement of cavalry approached from the direction of Bloemfontein.

I immediately opened fire (with a gun and a Maxim-Nordenfeldt at a range of 4,000 paces) on the fort, which obstructed my road. After we had fired a few shots the English abandoned that fort and fled to the nearest fort to the east. Shortly afterwards this fort was also abandoned.

The fort to the west was captured by Commandant Steenekamp and the Heilbron burghers. They succeeded in taking a few prisoners; but most of the enemy fled to Sanna's Post. Only one of the Heilbron burghers was wounded—Piet Steenekamp, the son of the Commandant.

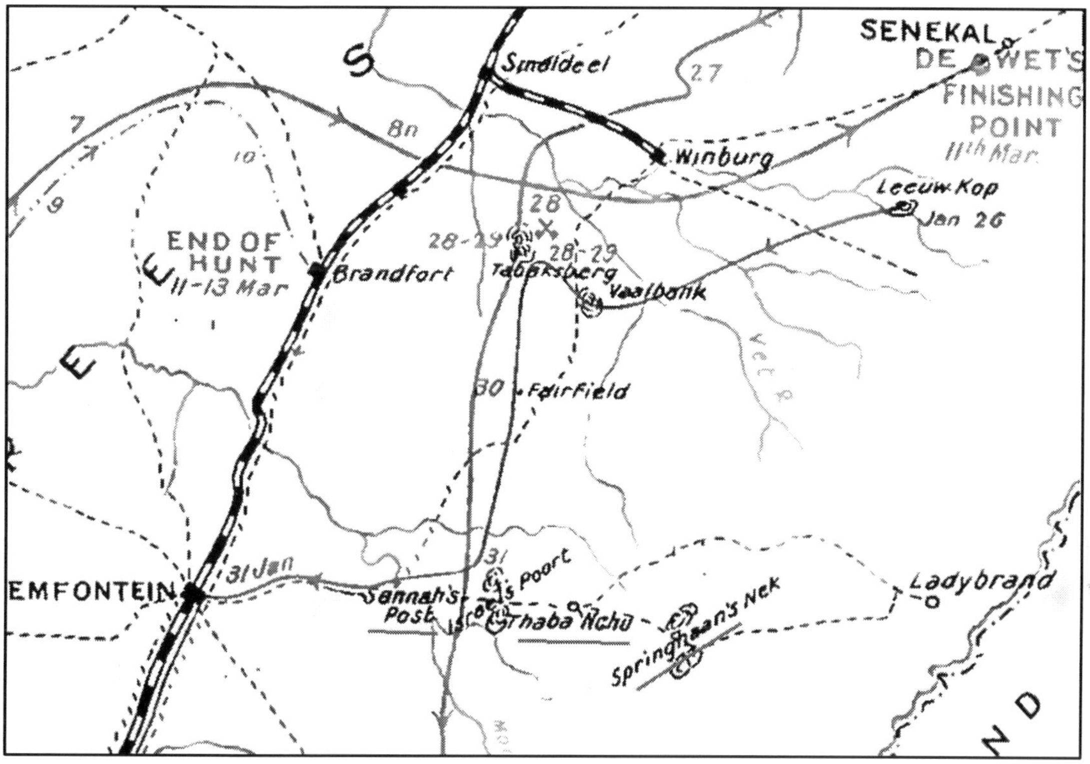

Springhaansnek.

And now our road was clear; and we passed through! General Fourie joined us two hours after sunset. Then we marched on [South] to Dewetsdorp[224] where we arrived on January 31st.[225]

General Knox, I heard, proceeded to Bloemfontein; thence he sent his troops to the railway bridge across the Orange River, near Bethulie. He was now aware that we were determined to enter the Colony at all costs, and so he stationed troops everywhere to turn us back. He placed forces not only at Bethulie railway bridge, but also at Springfontein, and Norvalspont. Thus he could easily prevent us crossing at the fords.

I had now to find some trump card which would spoil the game he was playing!

I ordered General Froneman to proceed from the source of the Kaffir River in the direction of Jagersfontein Road Station, to the west of Dewetsdorp.

General Fourie I despatched in the direction of Odendaalsstroom, on the Orange River, to the farm of Klein Kinderfontein, to the west of Smithfield.

I then sent scouts to the neighbourhood of Odendaalsdrift. They told me that there was an English patrol at the drift, and that they had heard that the enemy expected that we should try and cross into Cape Colony at that spot.

[224] At this date the English had not re-garrisoned the town.

[225] 1/30/1901 De Wet breaks through the Bloemfontein-Ladybrand line going south.

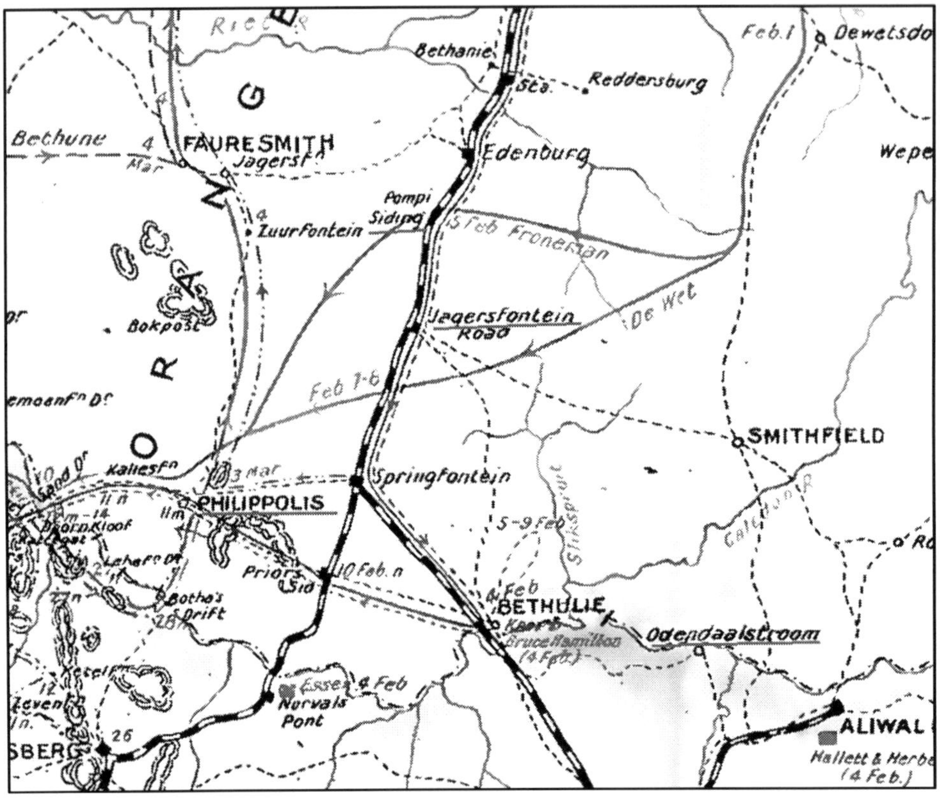

Froneman at Pompi Siding, Jagersfontein, Odendaalsstroom and Philippolis.

The following day I ordered a patrol to ride up and down the river; and I caused a report to be spread to the effect that I considered it too dangerous to cross the Orange River below its junction with the Caledon, owing to the river being already very full and quite unfordable if there was any rain at all; and that I had for this reason decided to recall General Froneman, and to take Odendaalsstroom by force, or else to attack the enemy at the Aliwal-north Bridge.

I felt quite sure that this rumour would reach General Knox that very day, for he had plenty of friends in the neighbourhood of the Caledon and the Orange River.

General Froneman had orders to march in the direction of Zanddrift, which is about half-way between Norvals Pont railway bridge and that of Hopetown. He succeeded in capturing a train close to Jagersfontein Road Station, by the simple device of blowing up the line both in front of it and behind it. In this train the burghers found a great quantity of things they greatly needed.

It should not be forgotten that there were scarcely any factories in South Africa, and this was more especially the case in the two Republics. And, as all imports had been stopped for some considerable time, it was natural that any booty which consisted of such things as saddles, blankets and ammunition was very acceptable.

When the burghers had helped themselves to what they wanted, the train was burnt.

For the space of a day I remained quiet, so that I might be quite sure that the English had received the report I had spread.

I soon discovered that my plan had been quite successful. The English marched off in the direction I wished, believing, no doubt, that the rumours they had heard were true; whilst I, on the evening of the 5th of February, 1901, took some of the burghers, with the guns and waggons, to a spot between the stations of Springfontein and Jagersfontein, and the following day remained in hiding.

I left General Fourie behind me with a horse-commando, with orders to remain there for two days, and to carry on manoeuvres in the direction of Odendaalsstroom.

I crossed the railway line that evening without any mishap to my force, but to my great sorrow the valiant Lieutenant Banie [226] Enslin, one of the best of my scouts, was severely wounded the same night, and fell into the hands of the English. He had ridden in advance with one of Theron's Scouting Corps, with the object of finding a favourable spot where he could lead us across the railway. The night was very dark, and he had lost his way. We crossed, as I have already said, without hindrance; but he and his companions rode into an outpost of the enemy a few miles to the north. The English opened fire on them, with the unhappy result that the estimable Banie was so seriously wounded that he had to be left behind. His comrades joined us the following morning, bringing the sad news with them.

We now continued our march at as rapid a pace as was possible; but the road was so soaked by rain that it was difficult for the oxen and the mules to draw the waggons and the guns.

On the 8th of February we overtook General Froneman at Lubbesdrift, six miles to the north of . Philippolis.

We pushed on that evening towards Zanddrift, which we reached on the 10th of February. Then we crossed over into Cape Colony.[227]

'De Wet crossing the Orange River.'[228]

When we had crossed the [Orange] river, I received a report from my scouts that there were about twenty of the enemy in a strong schanze on a kopje, which was about half an hour's march further up stream. I gave orders that a veldtcornet and twenty-five men, among whom was one of my staff, Willem Pretorius, should go and capture the schanze.

The veldtcornet preferred not to approach beyond a certain distance, and consequently Willem Pretorius and four other men were left to do the work.

Willem climbed the hill from one side, and the others, dividing into two, climbed it from the other side at two different points. They were met by a severe fire from the fort, but when they got to close quarters up went the white flag, and the English shouted "We surrender!"

Thus Willem Pretorius and four burghers captured twenty prisoners and a like number of horses, saddles, bridles, rifles and bandoliers, not to mention some three thousand cartridges.

When the veldtcornet at last arrived with his twenty men, he certainly proved himself very useful in carrying away the booty!

[226] Barend.

[227] 2/10/1901 "Call to arms" at Cape Town. De Wet, after many contests with the British forces in Orange River Colony, succeeded in crossing the river at Sand Drift.

[228] After Pretoria: The Guerilla War Vol 3 Feb, 1901.

This veldtcornet was shortly afterwards "Stellenbosched."[229] I then nominated in his place Willem Pretorius[230] as veldtcornet.

We left the river that afternoon behind us, and marched south to Mr. Bezuidenhout's farm. The following day we waited there for General Fourie to join us. He arrived the next day—and now we were ready to begin the game once more!

Our position was embarrassing, for not only was there a large English force at General Fourie's heels, but also there were two strong columns on the north from Colesberg, which were making for Hamelfontein. And these two columns were some twelve miles from us.

I at once set out in the direction of Hamelfontein, and the following day I discovered that the enemy's columns had divided into two parties; one of them had gone in a westerly direction, whilst the other was marching straight towards us. Meanwhile the force which had pursued General Fourie had crossed the river at Zanddrift.

My intention had been to divide my force into three divisions directly I arrived in the Colony. But I had been obliged to wait till General Fourie could join me; and when he had come, there was such large numbers of the enemy on every side that they gave me no opportunity of carrying out my original intention.

I may mention here that Lieutenant Malan, who became afterwards Commandant, and ultimately Vechtgeneraal, had penetrated into the Colony with fifty or sixty men, and had advanced considerably farther than I had done.

That afternoon I ordered the small waggon to proceed to a point between Philipstown and Petrusville.

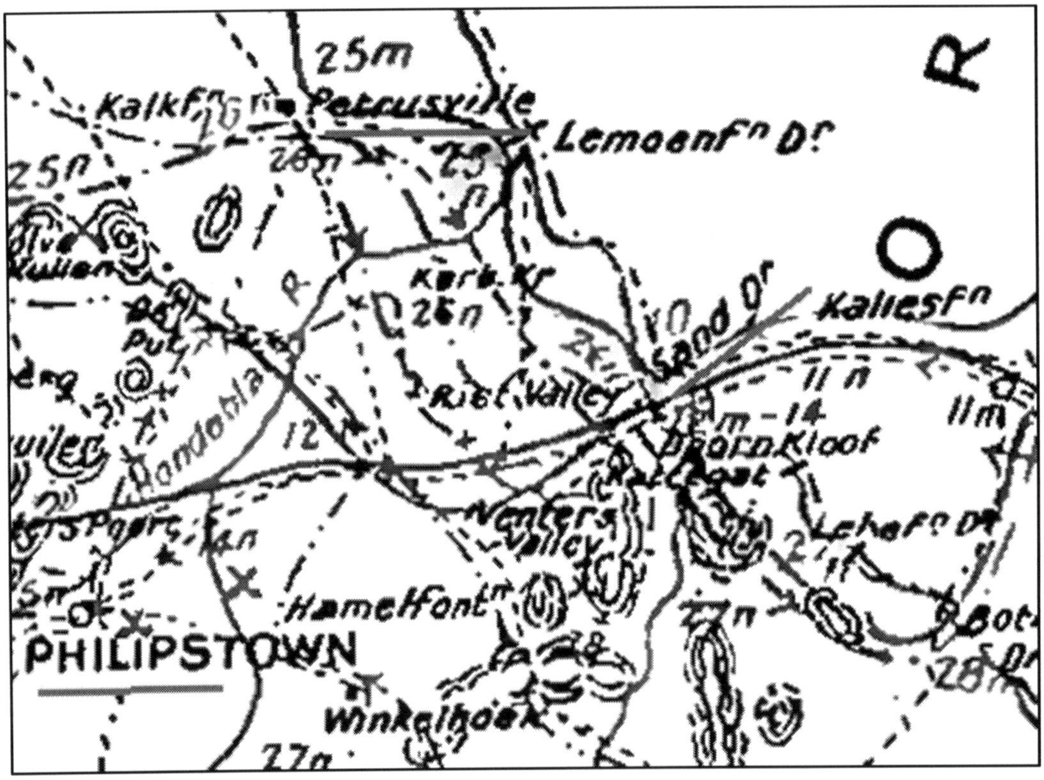

Philipstown and Petrusville top. Also 'Sand Dr.'

[229] Stellenbosched: this was the word the English applied to officers, who, on account of inefficiency, or for other reasons, had to be dismissed. Stellenbosch was a place where only very unimportant work was performed. [Stellenbosch, also known as the Oak City, is the second oldest European settlement in South Africa after Cape Town, originally established as an agricultural colony. It is located in the Western Cape province.] https://af.wikipedia.org/wiki/Stellenbosch

[230] I must give a short account of Willem Pretorius, for he was a dear friend of mine. He had only reached the age of twenty when I made him a Veldtcornet. His courage certainly could not be surpassed, yet he never let it go beyond his reason. About twenty days before the conclusion of Peace, he was killed by a bullet at a range of 1,100 paces. Throughout the whole previous course of the war fortune had favoured him almost miraculously: six horses had been killed and many more wounded under him; yet he had never received more than a scratch. But in the end he, like so many other brave men, was destined to die for the country that he loved so dearly. Poor Willem! You and the other heroes in our struggle will live for ever in our memories.

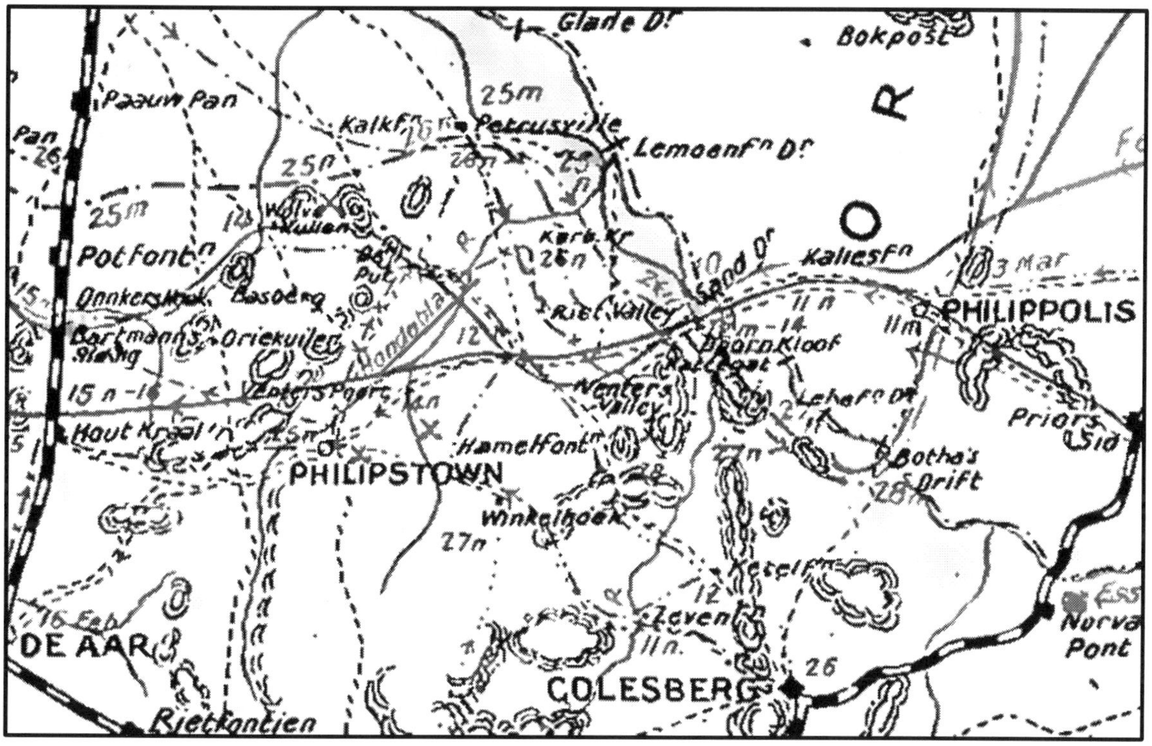

Larger area of operations. Note Hondoblaf River

We had several slight skirmishes with the English; and at sunset we nearly fell into their hands, but fortunately we were successful in holding the enemy in check until our small laager had passed.[231]

During that night we marched to Hondeblaf River. The following morning we found that there was no grass for the horses, for the locusts had eaten it all. The horses, poor creatures, were very hungry, and also much exhausted by all those forced marches. When we had been at Winburg, the pasture had been very poor although it had rained every day. This, of course, was very good for the veldt; but unfortunately it did not rain grass—the veldt required time to produce it.

All this was most unlucky. Already some of my men had to go on foot, and there were no horses to be obtained in that district.

The number of my burghers had now been diminished by nearly six hundred men. Commandant Prinsloo had remained behind with three hundred men, Vice-Commandant Van Tonder with one hundred, and lastly, Commandant De Vos at the Orange River with two hundred.

There was now only one course open to us—and that was to cross as quickly as possible the railway line near Hopetown, for if an English force was brought down by rail, it would mean our utter destruction.

We accordingly moved away at once from Hondeblaf River. The following day the English were again hot on our track. I ordered General Fourie and General Froneman to oppose the enemy, for it was necessary that something should be done to save our rearguard from being cut off. These Generals had several sharp engagements with the English, resulting in the capture of a number of prisoners, and a considerable loss in dead and wounded to the English.

After we had been on the march for a short time, a "Broodspioen"[232] came rushing up to me. (Had not my scouts been riding in a different direction they would have given me notice of his proximity.) He told me that he and a friend of his of the same calling had gone to a farm near by to buy bread, but when they had approached the house, a number of English soldiers appeared at the door and called out "hands up!" His friend had been captured, but he having been some fifteen paces from the house, had managed to escape under a hail of bullets. He had had to gallop one thousand paces before he could get out of range behind a ridge that stretched between

[231] 2/14/1901 Animated chases after De Wet.

[232] Broodspioen: literally a bread spy. This was the name applied to a burgher, who, with or without an order from his officer, rode in advance of his commando to obtain bread for himself and his comrades. He was frequently a man who placed the interests of his stomach before the safety of his commando.

us and the farm. I ordered the burghers to halt behind the ridge, and sent a small body of men ahead to determine the strength of the enemy. We could now see that the English had hidden their horses behind some fruit-trees. When they caught sight of our men on the top of the ridge, they took up positions behind kraals and a dam-wall not far from the house, knowing well that escape was impossible.

I thought it best to send a note to this handful of men, advising them to surrender, for I did not wish that any of my burghers' lives should be sacrificed in an unnecessary attack. Whilst I was writing the letter they punctuated it by an incessant fire, to which the burghers replied by a few shots, although none of the enemy were visible. As soon, however, as my despatch rider appeared with a white flag, their firing ceased. The answer they returned left something to be desired—"We shall not surrender!"

I immediately ordered fifty of my men to attack them. Hardly had I given the order, when a number of young burghers sprung on their horses and galloped at break-neck pace towards the kraals.

And now there was an end to all boasting, for without firing a single shot the enemy surrendered.

We took twenty prisoners there, and an equal number of rifles and bandoliers. The horses we captured—again twenty in number—were in excellent condition, and all up-saddled. We now had made ninety men our prisoners since we crossed the Orange River.

The joy of the [Other, captured] Broodspioen, who had been for fifty minutes in the hands of the English, was very great; and I believe he never returned again to his very doubtful profession.

The following day we came to a farm about six miles to the east of Houtkraal Station, which we christened Moddervlei,[233] on account of the experience we had on the night following our arrival.

The great English force was close behind us, and when night fell the enemy were not more than five miles from us.

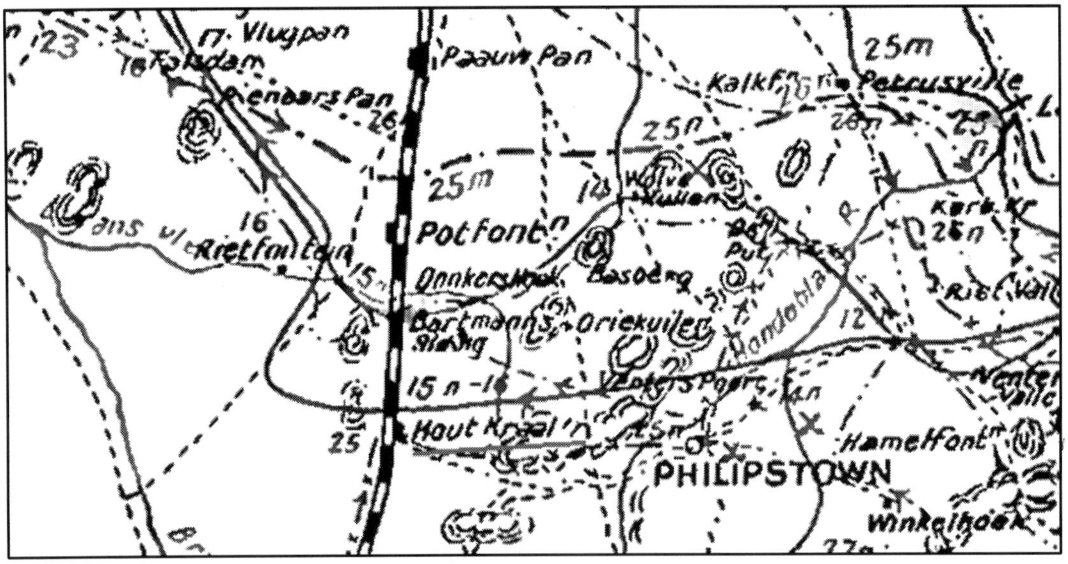

Hout Kraal [station] bottom of map on rail line.

It was at the hour of sunset, shortly before we came to the swamp, which I shall presently describe, that my scouts came across fifteen of the enemy. When the English saw our men they turned round at once. But they did not get away before one was shot from his horse, and another seriously wounded, and several of them taken prisoner.

I now sent two patrols to blow up the railway, seven miles at each side of the point where I intended to cross. I had no wish that an armoured train should appear and prevent my crossing.

[233] A swamp.

Wounded soldiers awaiting for medical help in the cold and damp.[256]

Pattern 1 Gun Truck and Train.[257] Pattern 2 train.[258]

But, before we could reach the railway line a swamp lay in our way. This swamp was about one thousand paces broad, and was covered knee deep with water, and in some places even deeper; for heavy rain had fallen during the afternoon. The water, however, would have been a matter of very little consequence, had it not been that the bottom of the swamp was of such a nature that the horses sank in it up to their knees, and even sometimes up to their girths. But we fourteen hundred riders had to get over it somehow or other!

Let the reader try to picture to himself the condition of the swamp when the last burgher had crossed!

Many of the men lost their balance as their horses struggled in the mud, and several of the burghers had to dismount and lead their poor tired-out animals.

The guns and the waggons caused us a great deal of trouble. We inspanned thirty oxen to each gun; but if it got stuck fast in the mud, fifty oxen were sometimes not sufficient to move it.

Drakensberg. Ox wagon 16 oxen crossing a drift near Mont-Aux-Sources.[259]

At last we got the guns through, and succeeded in getting a trolley, and the little waggon which carried my documents and papers, safely to the other side. But the ammunition and flour-waggons were impossible to move when they had once entered the swamp.

It was a night which I shall never forget!

We had now to determine what we should do with the waggons. The day would soon break and we could only cross the railway line when darkness covered our movements. It would be disastrous to us if, while we were still between the swamp and the railway, troops should be brought up by rail from De Aar and Hopetown.

It was perfectly clear that those who had crossed the swamp must go on. And so I advanced, at the same time giving General Fourie orders to remain behind with a hundred of the men whose horses were less exhausted than those of the other burghers, and to try to get the waggons through. In the event of the enemy arriving before his task was completed, I told him to leave the waggons and make his escape to the south.

Having given these orders, I proceeded with my commando to the railway line. Only the weakest of the horses were with us, so that many of my burghers had to go on foot.

The ninety prisoners we had taken were with me. I could not release them, because I did not want them to tell the enemy how exhausted our horses were. Should the English know this they would know exactly where our weak point lay.

I pitied the poor "Tommies," but what else could I do but order them to march with me? I treated them as well as I could, and made no difference between them and the burghers. And after all, many of our own men had to go on foot.

Any delay was dangerous, and so we hurried on as fast as possible. When we reached the railway line, day had already begun to break. Fortunately, we met with no opposition; the patrols had followed my orders and broken the line.

When the sun rose one could see what a terrible condition the burghers were in. On every man's face utter exhaustion could be read. But how could it have been otherwise? The men had had fighting to do the previous day, and had only once been able to off-saddle, and that not long enough to cook a piece of meat. Rain had also been falling in torrents, and most of the men were wet to the skin, for very few of them had waterproofs. And to make matters still worse, the burghers were covered with the mud from the swamp that still clung to them.

Twenty-four hours had passed without the men being able to lie down and rest; and sleep, of course, had been entirely impossible.

Three miles beyond the railway line I gave orders to off-saddle, although there was no grass for the horses. Hardly had we dismounted when I was told that we should find grass about one hour's ride further on. And so we mounted again, fatigued though we were, and found pasture at last for the poor animals. I thought it better that the masters should endure more hardships than that the horses should go without grass. We were rewarded for our short ride by the knowledge that our horses had something to eat, and we could sleep in peace without having to think that our animals were starving.

But before we could sleep hunger compelled us to kill a sheep which we had bought from a farmer living near. In that part of Cape Colony sheep-farming is almost the only occupation, and so well adapted is this district for rearing sheep that it is quite an exception to see a lean one. It may interest some of my readers to know that the African sheep has a very remarkable peculiarity; it possesses a huge tail, which sometimes weighs as much as ten pounds.

We were unable to obtain bread, and our flour had remained behind in the waggons. The sound of an explosion had told us that General Fourie had not been able to save them, and that by now they must have been burnt.

I heard later on that General Fourie had been attacked by the English and had not been able to set fire to the waggons himself. But the English, so my scouts informed me, had done the work for him, and so thoroughly that they had also burnt some of their own waggons which had got into the swamp.

After we had helped ourselves to a good "African boutspan," and had slept with our saddles as pillows, we were all in good spirits again, although we could not forget our experiences in the swamp.

The burghers whom I had with me were of the right stamp, and were prepared to sacrifice everything for the freedom of the people. If any one had asked them whether they were ready to undergo any further hardships, they would have replied that a hundred swamps would not discourage them. They knew that freedom was a pearl of such value that no man since the world began had been able to set a price upon it.

When General Fourie had abandoned the waggons, he retreated to the south, crossing the railway at De Aar. He joined me again near Petrusville when I was returning to the Free State.

As the English had to march round the swamp, leaving their waggons behind, we were not pressed for time, or obliged to march very far. We took advantage of this respite to give our horses a little rest.

I now proceeded to the west of Hopetown, in the direction of Strijdenburg. The following day the English were again on our heels in greater numbers than ever, and advancing more speedily than before. I was obliged to engage their vanguard for nearly the whole of that day.

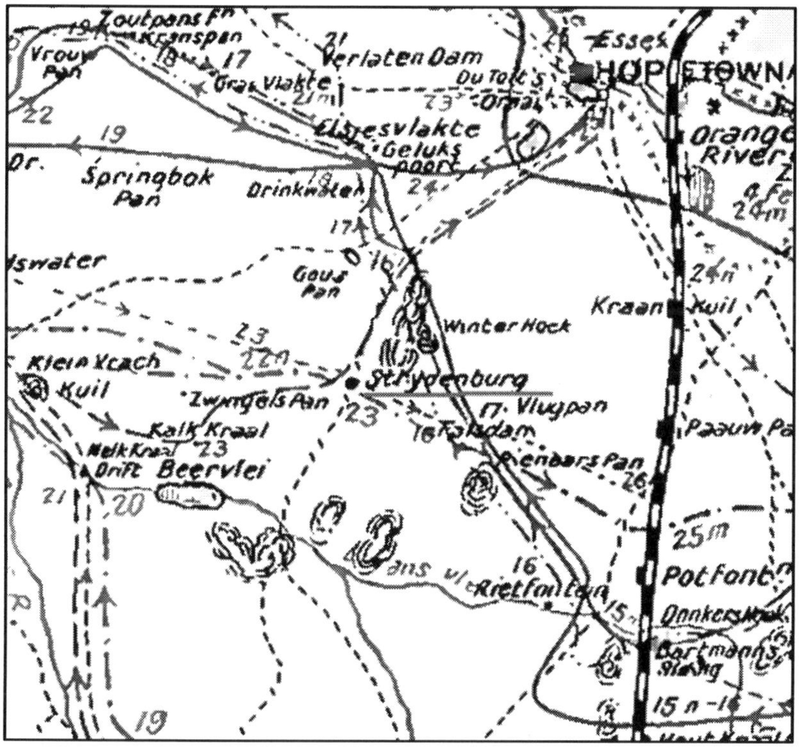

Stryndenburg in center of map.

That evening we arrived at a spot about ten or twelve miles to the north-west of Strijdenburg. Here I left Commandant Hasebroek behind with three hundred men, till the following morning, with orders to watch the enemy and hold them back if necessary. This would give my burghers who were on foot, or whose horses were exhausted, a chance of getting away.

I might here explain to the uninitiated our methods of checking the advance of the enemy.

The burghers who had the best horses would remain behind any rise or kopje they could find in the neighbourhood. When the enemy approached and saw ahead of them two or three hundred burghers they would halt and bring their guns (which were usually placed in the middle of the column) to the front. When they had got the guns in position, they would bombard the ridge behind which the burghers were stationed. But as our men had no wish to remain under fire, they would then quietly withdraw out of sight. But the English would continue bombarding the hill, and would send flanking parties to the right and left. Sometimes it would take the English several hours before they could make sure that there were no Boers behind the rise.

It was tactics such as the above that gave my burghers who were handicapped by the condition of their horses, time to retreat.

It sometimes happened, in these rearguard actions, when the position was favourable, that the enemy were led into an ambush, and then they were either captured or sent racing back under our fire to bring up their guns and main force. Had we not acted in some such way as this, all my men would have been taken prisoner in this and in many other marches.

The large forces which the English on all occasions concentrated round me deprived me of any chance of fighting a great battle; and I could only act in the way I did.

If the reader is eager to know how it was that I kept out of the enemy's hands until the end of the war, I can only answer, although I may not be understood, that I ascribed it to nothing else than this:—It was not God's will that I should fall into their hands.

Let those who rejoice at my miraculous escapes give all the praise to God.

26. Darkness Proves My Salvation

Commandant Hasebroek held the enemy in check whilst we continued our march to a place called Vrouwpan. On the following day we struck the Brak River at a point ten miles south-east of its confluence with the Orange River, to the east of Prieska. It was not fordable, and we had to off-saddle.

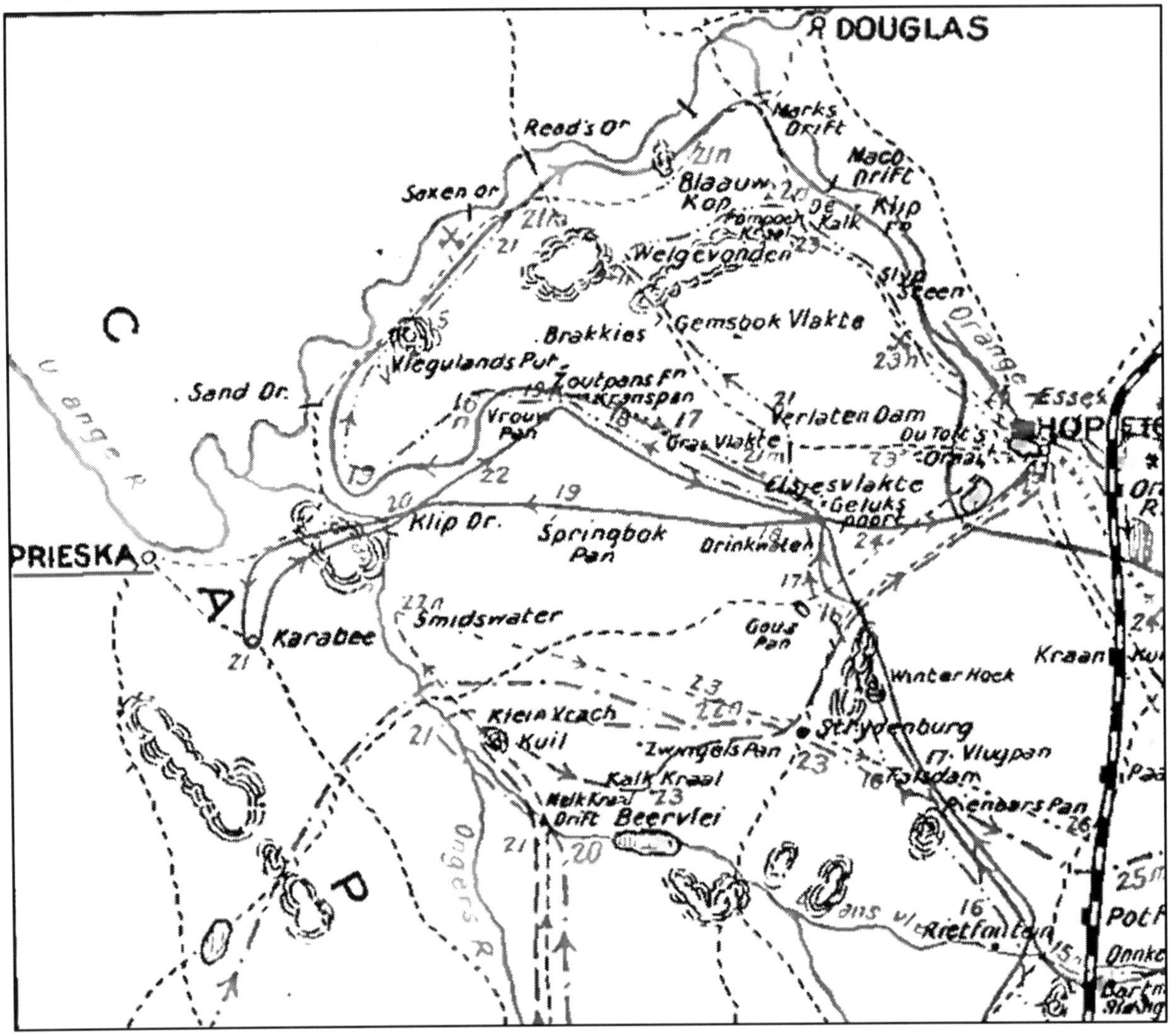

Prieska and Orange River.

There was absolutely no chance of getting across—the best of swimmers would have been helpless in that swollen torrent, which rushed down to the Orange River, its great waves roaring like a tempestuous sea.

About two hours before sunset Commandant Hasebroek reported that the English were rapidly approaching. The question was, "Which way shall we go?" It was impossible to escape either to the south of the river or in the direction of the enemy, for the veldt was too flat to afford us any cover. If we were not to be cornered against an impassable torrent, we must make our way down stream to the north-west; and even then we should be in danger of being driven on to the Orange River, which was only ten miles distant. By taking this road the English would not see us, on account of a ridge which lay between us and them.

My plan was to get behind this ridge and to march under its shelter until darkness came on; then, proceeding up the Orange River, to attack the enemy in the rear. They were, however, only nine miles from us, and should their advance be rapid, they would reach the friendly ridge before night came on; and the danger would then be that before I had fulfilled my purpose, we should be hemmed in between two swollen rivers with the most fatal consequences. The risk was great, but no other course was open to us. There was no time to seek advice from any one; I had but a moment to spare in which to acquaint President Steyn with my scheme. He said at once: "General, do as you think best."

My mind had been already made up; but my respect for the President was so great, and we had always worked in such harmony, that I did not like to do anything without his knowledge; besides which, his advice was often of great value. Joshua of old prayed that the day might be lengthened: but here the case was different; we had reason to be thankful that the day was passed and night had begun to fall before the vanguard of the enemy had reached the ridge, from the summit of which they might have observed us.

That night was the darkest I had ever known. And this was in our favour. Very quietly we retreated in a line parallel with the English column until, on the following morning, we were not only out of sight but a good nine or ten miles behind the enemy, who were marching on, fully expecting to corner us between the two rivers.

The English army had been enormously reinforced, and it was clear that now more than ever they were putting forth all their powers to silence President Steyn and myself effectually.

From their point of view they were right; for had things turned out in such a way that we could have remained in Cape Colony, then I am convinced we should have made matters very awkward for them.

But what were we to do now? With so many burghers on foot or provided only with worn-out horses, it was useless to think of circumventing the enemy, and thus getting once more to the south of them; whereas to go up stream along the banks of the Orange River until we could discover a ford, and then to return across it into the Free State, would mean the upsetting of my plan of campaign.

I was obliged to make the best of a bad bargain; and I decided to find a way across the Orange River before the enemy had discovered my whereabouts.

That day, the 20th of February, we set out along the river, looking for a ford. The river was falling, but as there was no feasible crossing we had no choice but to go on, trusting that we should find one near the confluence of the two rivers. Here again we were disappointed; the punts[234] which should have been there had been destroyed some time before by the English, but we heard of a boat six miles higher up, so on we marched. When found, it was only a small boat, capable of holding, at most, twelve men, but we got to work at once, and by the evening of the 22nd there were two hundred dismounted burghers on the other bank of the river. Some crossed by swimming, in attempting which a man of the name of Van de Nerwe was drowned.

A few of those who crossed in the boat succeeded in pulling their horses after them.

On the morning of the 23rd I received a report that the English forces were close on our heels. We did not expect them so soon, but they had made a long night's march. Without delay we off-saddled, and proceeded along the river, while the rearguard covered our retreat. The force of the enemy was, however, too great, and the rearguard had, after a short engagement, to give way.

Fortunately the veldt was broken, and we could (as we had done a few days previously) march ahead out of sight of the enemy. Towards two o'clock in the afternoon we were obliged to off-saddle, but could only do so for one hour, for the English were upon us again. Our gun and Maxim-Nordenfeldt we had to leave behind for the enemy; the draught cattle had become exhausted, and we had no dynamite with which to blow up the guns.

But what did it matter? England had already so many big guns that two more could not make much difference, if added to the four hundred which that country—one of the oldest and strongest of Empires—had brought against a small nation, fighting only to defend its sacred rights.

Nevertheless, it cut me to the heart to give up my guns[235] on that day—the 23rd of February—the commemoration day of the independence of the Orange Free State. In happier times we had celebrated this day amongst our friends, to the accompaniment of salvoes of rifles. Now we were obliged to celebrate it by giving up the only two guns with which we could still shoot, and which we were now to see turned upon ourselves.

[234] Flat Bottom Boats.

[235] There were still two Krupps left, but we had no ammunition for them.

Rustle for Themselves.

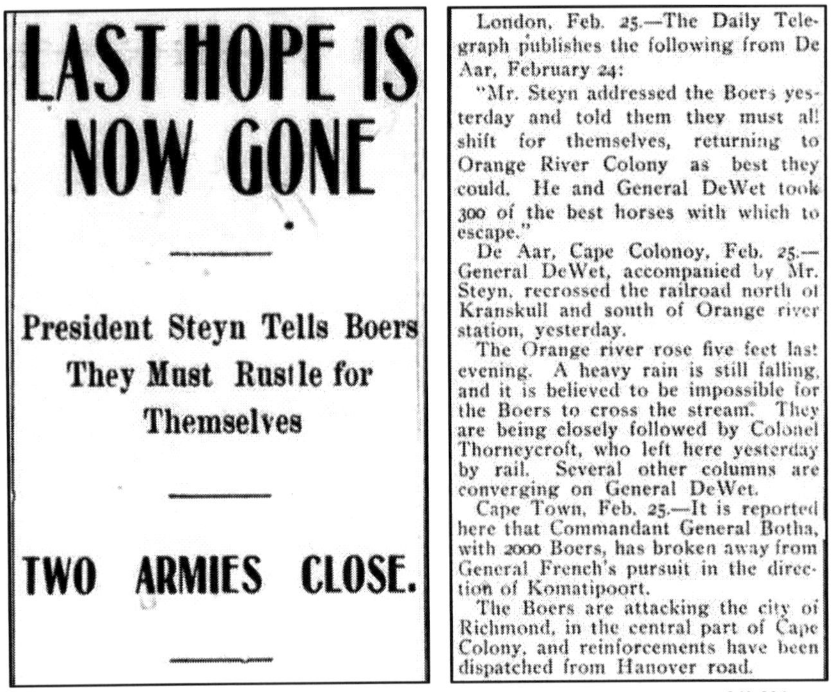

LAST HOPE IS NOW GONE

President Steyn Tells Boers They Must Rustle for Themselves

TWO ARMIES CLOSE.

London, Feb. 25.—The Daily Telegraph publishes the following from De Aar, February 24:

"Mr. Steyn addressed the Boers yesterday and told them they must all shift for themselves, returning to Orange River Colony as best they could. He and General DeWet took 300 of the best horses with which to escape."

De Aar, Cape Colonoy, Feb. 25.—General DeWet, accompanied by Mr. Steyn, recrossed the railroad north of Kranskull and south of Orange river station, yesterday.

The Orange river rose five feet last evening. A heavy rain is still falling, and it is believed to be impossible for the Boers to cross the stream. They are being closely followed by Colonel Thorneycroft, who left here yesterday by rail. Several other columns are converging on General DeWet.

Cape Town, Feb. 25.—It is reported here that Commandant General Botha, with 2000 Boers, has broken away from General French's pursuit in the direction of Komatipoort.

The Boers are attacking the city of Richmond, in the central part of Cape Colony, and reinforcements have been dispatched from Hanover road.

President Steyn Tells Boers They Must Rustle for Themselves.[260] [236]

President Steyn Tells Boers They Must Rustle for Themselves

Two Armies Close.

Brief News of the Outside World. London, Feb. 25. The Daily Telegraph publishes the following from De Aar, February 24: "Mr. Steyn addressed the Boers yesterday and told them they must all shift for themselves, returning to Orange River Colony as best they could. He and General DeWet took 300 of the best horses with which to escape."

De Aar, Cape Colony, Feb. 25. General DeWet, accompanied by Mr. Steyn. recrossed the railroad north of Kranskull and south of Orange river station, yesterday. The Orange river rose five feet last evening. A heavy rain is still falling, and it is believed to be impossible for the Boers to cross the stream. They are being closely followed by Colonel Thorncycroft, who left here yesterday by rail. Several other columns are converging on General DeWet.

Cape Town, Feb. 25. It is reported here that Commandant General Botha, with 2000 Boers, has broken away from General French's pursuit in the direction of Komatipoort. The Boers are attacking the city of Richmond, in the central part of Cape Colony, and reinforcements have been dispatched from Hanover road.

[236] Times History misleading Jingoistic Reference: 2/22/1901 Extraordinary proclamation signed by Steyn and De Wet published. 2/23/1901 De Wet succeeded in recrossing the river after losing 200 prisoners, all his guns, ammunition, and waggons.

My feelings on that day I can never forget! Those Englishmen who go by the name of "Pro-Boers" are the best fitted to describe the anguish which then overpowered me, for they stood up for justice even against their own people. And this not because they were hostile to their Government, or to the greatness of England's power, but only because they were not without moral sense, because they could not stifle conscience at the expense of justice, nor identify themselves with iniquitous actions.

But the day will come—of this I am convinced—when not Pro-Boers only, but all England will acknowledge our rights—the rights which we shall then have earned by our quiet faithfulness and obedience. I cannot believe that any father will look without pity on a child who comes to him as a child should—obedient and submissive.

The 23rd of February, 1901, the forty-seventh anniversary of the Orange Free States, had been a disastrous day for us indeed, but it was to end in another miraculous escape, for in the darkness of that evening it again happened that we were delivered from an apparently unavoidable misfortune. As I have said already, the English were firing on my rear-guard; at the same time my scouts came in to tell me that, just in front of us, at a distance of not quite four miles, there was another great army of the enemy.

I had intended to march that night to the west of Hopetown. But now if I went in that direction I should only run straight on to this army. If we went to the left we could only advance 2,000 paces before being visible to the English on the kop close to Hopetown, from where they could make known our movements by heliograph. At our front, at our back, on our left, the outlook was hopeless; and to the right lay the cruel river. Stand still we could not—the enemy were upon us—it was impossible that anything could save us—no, not impossible—a rescue was at hand.

The sun was just going down, and by the time we could be seen from Hopetown, night would have covered us with its sheltering wings.

We should then be able to execute a flank movement, and make a detour round the enemy who were before us.

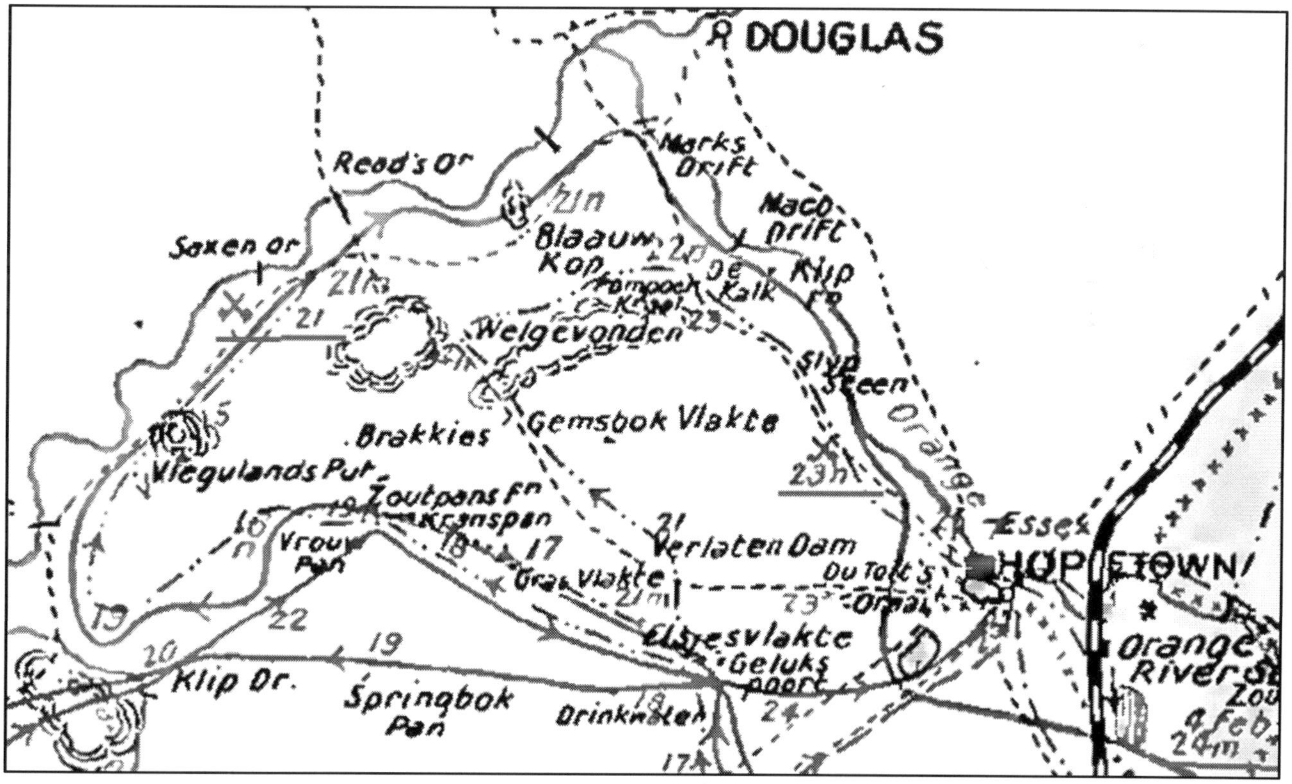

Area of Operations showing engagements on the 21st and 23rd then crossing the rail line on the 24th.

But now I knew that we must be prepared to march nearly the whole night through, in order that we might be able, early on the following morning, to cross the railway lines. If we did not do this, then we should have the enemy close in our rear, and perhaps an armour train threatening us in front. But ... there were the burghers on foot and those who had weak horses; and I had not the heart to make them march on foot for so long a time, yet the thought of allowing such trustworthy patriotic burghers to fall into the hands of the enemy was unbearable. I therefore decided on letting them take a cross road to the north, to the banks of the Orange River about five miles from our position. There, on the banks of the river, were many bushes amongst which they could hide themselves until the enemy had passed by.

They could then proceed along the banks of the river and cross it by means of the boat. I cautioned them not to march in one troop, or in one trail, but to spread out, so that the English could not easily follow their tracks. In this the poor burghers succeeded; they already, on that memorable and sad day, had marched eighteen miles; but they had yet to cover another five miles to the river before they could take their night's rest. They accomplished this feat (on the second day) under the valiant and true Commandant Hasebroek, whose horse, although tired, was still able to proceed.

As for me, I marched away in the evening, and after we had rested that night for a few hours, we arrived at a place a short distance to the south of Hopetown. About eight o'clock we crossed the line, which was fortunately at that point not as yet guarded by forts, and off-saddled about six miles beyond. We had eaten nothing since the previous day, and it will easily be understood that we were so hungry that we, as the Boer proverb says,—"could have eaten off a nail's head." There we got some sheep, and it was not long before they were killed, broiled, and eaten; what a meal we made!

Towards mid-day [24th] we headed once more for the Orange River. We thought that by the time we arrived it would be fordable, for we had seen on the previous morning that it was falling rapidly, but what was our disappointment! there must have been rain higher up the stream, as the river had become fuller, and there was still no chance of crossing.

The English were approaching. We had, however, to use our field glasses to enable us to see them, as we were fifteen or sixteen miles in front of them. Once more there were burghers whose horses were tired and who had to march on foot. We thought now that there would be a better chance at Limoensdrift; and every one who knew this ford said that it was a shallow one. The following day [25th] saw us there, and—the river was quite full! We then tried higher up, [South] still with the same result—every drift was unfordable.

At last we reached the Zanddrift, where we had crossed seventeen days before. We knew that this was a shallow drift, and on arriving there I got two young burghers,—of whom the one, David Heenop, was an excellent swimmer,—to make a trial. The water had not appeared to be so deep as we found it to be, when the two burghers plunged into it. They could not remain on their horses' backs, but had to swim alongside of them to the other side of the river.

All thought of their return was out of the question; they had risked their lives in crossing, and I gave them orders from my side of the river not to attempt the passage back. But they had not a stitch of clothing on them, for they had stripped themselves before entering the water! In this state, then, they were obliged to mount their horses and proceed, and this under a burning sun, which scorched them with its rays. About three-quarters of an hour's ride from there was a Boer farm; their only course, they thought, was to ask for gowns from the ladies there, in which to dress themselves. When they arrived at a short distance from the house (such was the account they gave on joining me later on) they halted and shouted to the house for clothing. A Boer vrouw[237] named Boshof, sent to each one through her son—not a gown, but a pair of trousers and a shirt of her husband's, which she had been able to hide from the English, who had passed there, and who generally took away, or burnt, all male attire.

The enemy had, in the meantime, approached quite close to us, and we were again obliged to look for a drift up stream. We had hopes that if the river did not all of a sudden rise, we should find one. We came so close to the English that we had to open fire on their advance guard before we could proceed.

Here General Judge Hartzog met us with his commandos from the south-west of Cape Colony, and with him, General Fourie.

[237] Farmer's wife.

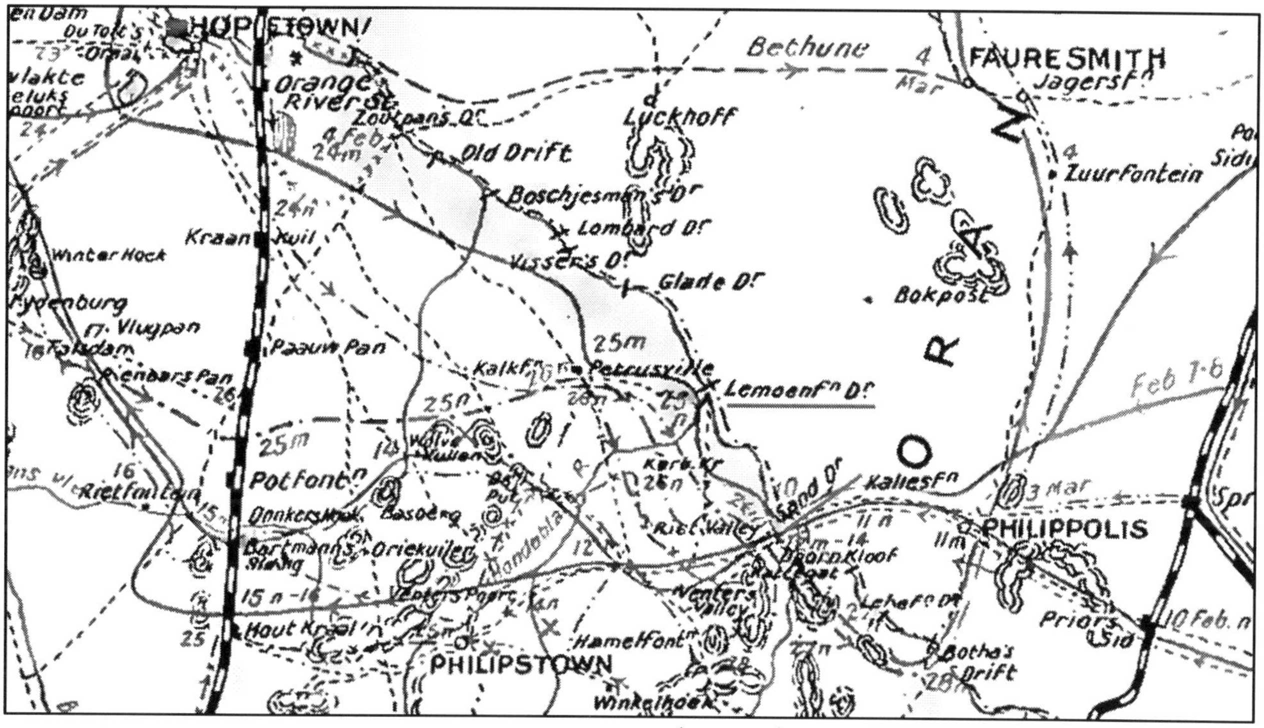

Lemoenf Drift [Limoensdrift] Center and Sand Drift on 25th and 26th.

That night we marched about fourteen miles.

In the night, after crossing the Zeekoe River, we arrived at a Boer farm, to which (we are told) twenty English scouts had paid a visit shortly after sunset, and, having asked for information concerning us, had gone away by the same road we were following. About four or five miles from there we had to cross a ridge. It was dark, and I had forgotten those twenty English. I had sent out no scouts before me, but rode, as was my habit, with my staff, in front of the commandos. As we approached the summit of the mountain I saw a group of horses fastened together, and some men lying in front of them. The horses and men were not twenty paces to the left of the path, among the bushes. I thought at first that they were some of my burghers who had ridden on in advance, and were now lying there asleep; I myself had rested for a while at the foot of the mountains, to give the burghers, who were on foot, a chance of coming up with me. The thought angered me, for it would have been against all orders that any burghers, without special permission, should go in advance. I proceeded to wake them up.

"What do you mean by riding ahead like this?" I called out to them. Nearly all with one accord sprang up and asked, "Who are you?" "Hands up!" I called out; as one man their hands went up. They explained that they were seven of the twenty scouts before mentioned,—but here the remainder opened fire upon us from about two hundred paces to the front. I called out to the burghers, "Charge!"

The burghers did so, but as they came to the little hill where we had seen the sparks from the guns they found nobody. The English had fled, and, as the moon had just gone down, it was too dark to pursue them. Taking with us the seven prisoners, we continued on our way until the following morning. We allowed them to retain their clothes. It was still before the "uitschuddings"[238] period.

[238] Stripping.

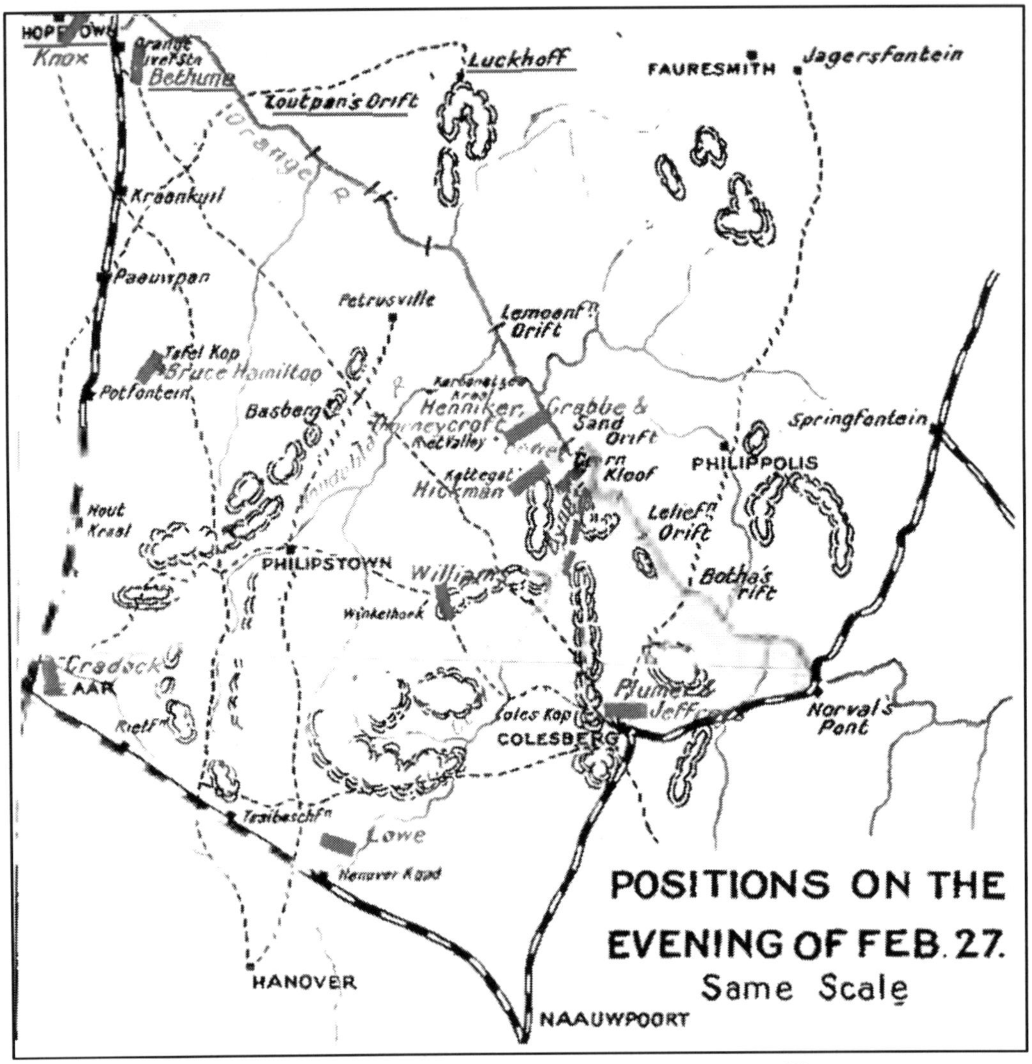

Hopetown, Bethune and Zoutpan's Drift heading towards Luckhoff.[239]

The day broke, and after having been turned back on the banks of the Brak River, we marched to the fifteenth ford. "If we could only get across here," we said. We knew that once across we should have a respite from the enemy, and could with thankful hearts take breath even if it were only for three or four days.

When we came to the river I at once ordered a few burghers to undress and go in. Alas! when the horses entered the ford, the water came over their backs, and they had almost to swim. "Now they will have to swim!" we cried, but presently we saw that the farther they went the shallower it became, and that they walked where we expected them to swim, until at last the water reached only to the horses' knees.

What a scramble there was now among the burghers in order to cross! Soon the river was one mass of men from bank to bank.

I can hardly describe the different exclamations of joy, the Psalms and the songs that now rose up from the burghers splashing through the water. "Never will we return," "No more of the Colony for me," "The Free State," "On to the Free State!" "The Free State for ever!" Then again, "Praise the Lord with cheerful song," "Hurrah!" These were among the expressions which met my ears.

Although this was only an old waggon-ford, which had not been used for the last few years, my little waggon and a few carts got across. One of the carts was drawn by two small donkeys. Somebody told me that the little donkeys had to swim a short distance where it was deep, and at one time disappeared beneath the water; but that the driver was so full of joy—or of fear—that he went on whipping the water!

[239] 3/2/1901 De Wet was forced over the Orange River with the loss of his guns and convoy.

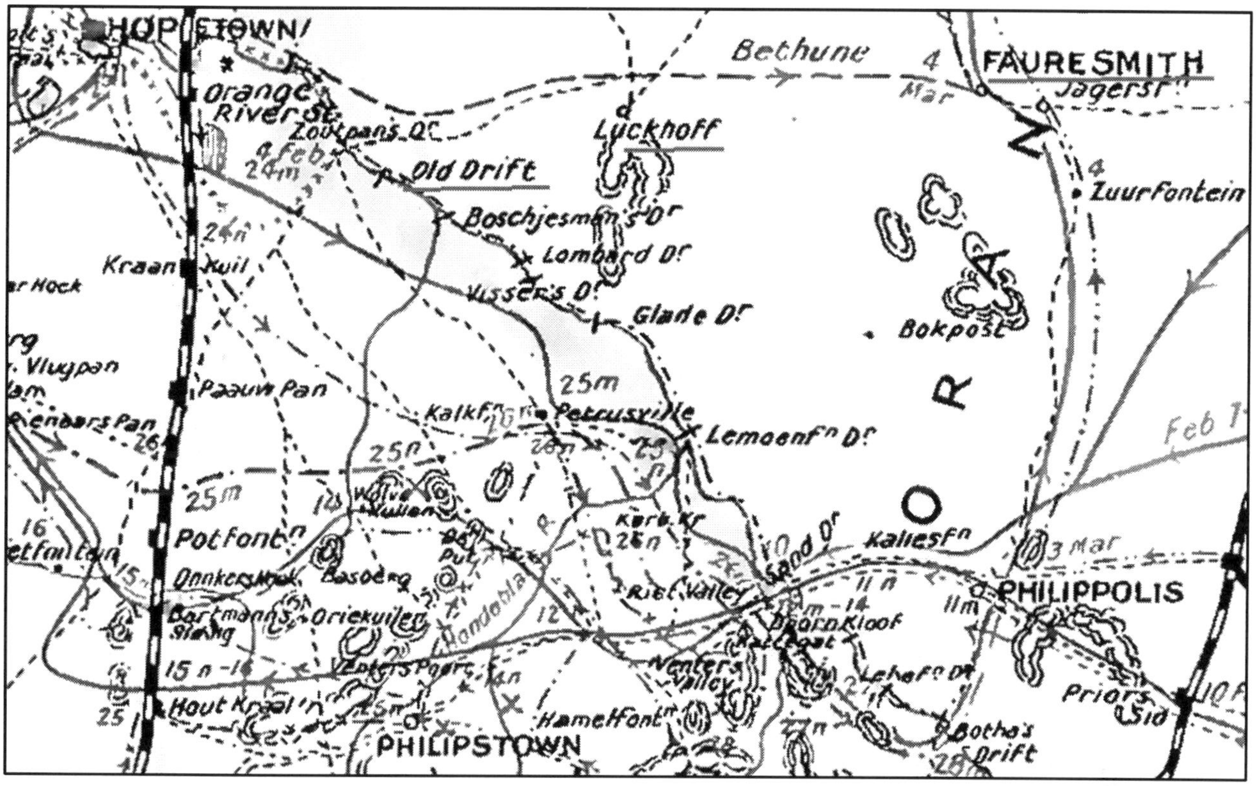

'Old Drift' in addition to other landmarks, reaching Fauresmith March 4th.

A fearful experience we had had! We asked each other in wonder, "Is it possible? How could we have endured it?" But as I have only been hinting at things, the reader will perhaps say, "O come! it hasn't been as bad as all that!"

Give me leave then, dear reader, to place before you the whole of the circumstances. England's great power pitted against two Republics, which, in comparison with European countries, were nearly uninhabited! This mighty Empire employed against us, besides their own English, Scotch and Irish soldiers, volunteers from the Australian, New Zealand, Canadian and South African Colonies; hired against us both black and white nations, and, what is the worst of all, the national scouts from our own nation sent out against us. Think, further, that all harbours were closed to us, and that there were therefore no imports. Can you not see that the whole course of events was a miracle from beginning to end? A miracle of God in the eyes of every one who looks at it with an unbiased mind, but even more apparent to those who had personal experience of it. Yet, however that may be, I had to declare again that if there had been no national scouts and no Kaffirs, in all human probability matters would have taken another turn. But as things have turned out, all that can now be said is, that we have done our best, and that to ask any one to do more is unreasonable. May it be the cry of every one, "God willed it so—His name be praised!"

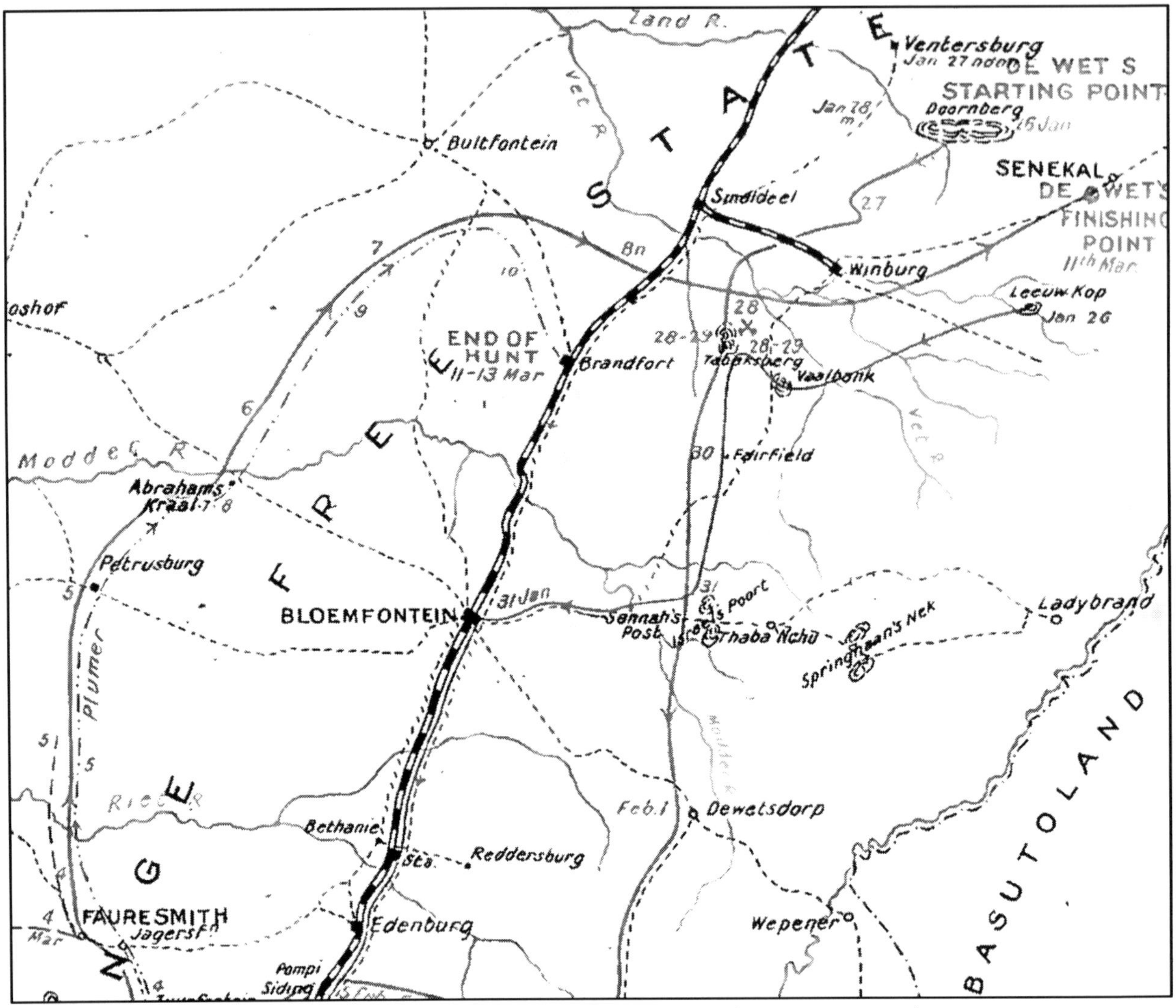

Showing De Wet's exit route north from Fauresmith from March 4th until March 11th at Senekal.

THE INVASION OF CAPE COLONY AND THE GREAT DE WET HUNT

(December 1900–March 1901)

Times' History Coverage of this operation. See Times History Vol 5, Chapter 5.

27. Was Ours a Guerrilla War?

Something almost miraculous now happened! Hardly had we been three hours across the river when it became completely unfordable!

We knew that we should have now a few days at least in which to rest ourselves, and we marched slowly to the farm of Lubbeshoop. From there I sent General Fourie to operate in the south-eastern districts, where he had been before, and despatched Judge Hertzog to the south-western districts.

We were of the opinion that we should be able to do better work if we divided the commandos up into small parties. We could not risk any great battles, and, if we divided our forces, the English would have to divide their forces too.

The commandos were now divided as follows:

1. The district of Kroonstad: the men under Commandants Philip De Vos, Jan Cilliers and Maree.

Sub-district of Heilbron: the men under Commandants F.E. Mentz, Lucas Steenekamp and J. Van de Merwe.

All of these were under Vice-Commander-in-Chief Johannes Hattingh.

2. The district of Vrede: the men under Commandants Ross and Manie Botha.

Sub-district of Harrismith: the men under Commandants Jan Meijer, Jan Jacobsz,[240] and (at a later period) Brukes.

All of these were under Vice-Commander-in-Chief Wessel Wessels.

3. The district of Winburg: the men under Commandant Hasebroek.

The sub-district of Ladybrand: the men under Commandant Koen.

The sub-district of Ficksburg: the men under Commandant Steyn.[241]

The sub-district of Bethlehem: the men under Commandant Michal Prinsloo.

All of these men were under Vice-Commander-in-Chief C.C. Froneman.

4. The district of Boshof: the men under Commandant J.N. Jacobsz, P. Erasmus and H. Theunissen.[242]

Sub-district of Hoopstad: the men under Commandants Jacobus Theron (of Winburg) and A.J. Bester (of Brandfort).

All of these were under Vice-Commander-in-Chief C.C.J. Badenhorst.

5. The district of Philippolis: the men under Commandants Munnik and Hertzog.

Sub-district of Fauresmith: the men under Commandant Charles Nieuwouwdt.

Sub-district of Jacobsdal: the men under Commandant Hendrik Pretorius.

Sub-district of Petrusburg: the men under Commandant Van du Berg.

All of these were under Vice-Commander-in-Chief Judge J.B.M. Hertzog, who also was in command of the western part of Bloemfontein.

6. The district of the southern part of Bloemfontein: the men under Commandants Ackerman and Willem Kolbe.

Sub-district of Thaba'Nchu: the men under Commandant J.P. Strijl (a member of the Volksraad).

Sub-districts of Bethulie and Smithfield: the men under Commandant Gideon Joubert.

Sub-district of Rouxville: the men under Commandant Frederik Rheeders.

Sub-district of Wepener: the men under Commandant R. Coetzee.

All of these were under Vice-Commander-in-Chief Piet Fourie, and later on under George Brand.

Not long after this arrangement had been made the district under General Froneman was divided into two divisions, and Commandant Michal Prinsloo was promoted to be Vice-Commander-in-Chief of Bethlehem and Ficksburg as separate sub-districts. Bethlehem was then given three Commandants, namely, Commandants Olivier, Rautenbach and Bruwer.

[240] Veldtcornet Franz Jacobsz was afterwards appointed in the place of this Commandant, who resigned.

[241] When this Commandant resigned, Veldtcornet J.J. Van Niekerk was appointed in his place.

[242] When, at a later period, Commandant Theunissen was put in command of the burghers of Fauresmith, Commandant Mijburg was appointed in his place. This latter Commandant was afterwards killed.

All this new arrangement of our forces made it impossible for great battles to be fought; it offered us the opportunity of frequently engaging the enemy in skirmishes, and inflicting heavier losses upon them than would otherwise have been the case. For the same reason our losses grew larger from month to month, but they did not increase in the same proportion as those of the enemy. Again, we captured more prisoners than formerly. It is much to be regretted that we were unable to keep them, for had we been in a position to do so, the world would have been astonished at their number. But unfortunately we were now unable to retain any of our prisoners. We had no St. Helena, Ceylon or Bermuda, whither we could send them. Thus, whilst every prisoner which the English captured meant one less man for us, the thousands of prisoners we took from the English were no loss to them at all, for in most cases it was only a few hours before they could fight again. All that was required was that a rifle should be ready in the camp on a prisoner's return, and he was prepared for service once more.

The fact that we fought throughout the Free State in small detachments, put the English to some trouble, for they felt themselves obliged to discover a vocabulary of names to apply to us!

Thus when Lord Roberts on the 24th of May, 1900, proclaimed the Orange Free State (and afterwards the Transvaal) as annexed by the British Crown, he described those who continued to fight as rebels. Then again we were called "Sniping Bands" and "Brigands." But the list of epithets was not exhausted yet, for it appeared that we were "Guerillas," and our leaders "Guerrilla Chiefs!"

Orange Free State Annexation May 24th, 1900.[261] [243]

I was always at a loss to understand by what right the English designated us "Guerillas." They had, however, to withdraw the soubriquet at the Peace Negotiations, when they acknowledged that our leaders formed a legal government.

Let me say a few words more about this term "Guerillas." We will suppose that England has captured New York, St. Petersburg, Berlin, Paris, Amsterdam, or any other capital of a free and independent State, Kingdom or Empire, and that the Government of such State, Kingdom or Empire still continues to defend itself. Would England then be entitled to call their antagonists "Guerillas"? Or, we will suppose that England's capital has been taken by another nation, but the English Government still remains in existence. Could England then be

[243] 5/20/1900 Colonel Bethune's Mounted Infantry ambushed near Vryheid. In Natal.

5/22/1900 General Ian Hamilton occupied Heilbron after a series of engagements. The main army, under Roberts, pitched its tents at Honing Spruit, and General French crossed the Rhenoster to the north-west of the latter place.

5/23/1900 Rhenoster position turned.

5/24/1900 British Army entered the Transvaal, crossing the Vaal near Parys, unopposed.

5/27/1900 The passage of the Vaal was completed.

5/28/1900 The Battle of Biddulph's Berg. Orange Free State formally annexed under the title of Orange River Colony.

considered to be annexed by the other nation, and could the enemy term the English "Guerillas"? Surely it would be impossible!

The only case in which one can use this word, is when one civilized nation has so completely vanquished another, that not only is the capital taken, but also the country from border to border is so completely conquered that any resistance is out of the question.

But that nothing like this had happened in South Africa is clear to every one who recalls the names of Lindley,[244] Roodewal, Dewetsdorp, Vlakfontein, Tafel kop[245] and Tweefontein, not to speak of many other glorious battle-fieldls on which we fought after the so-called annexation.

Nor must we forget to mention the defeat that Lord Methuen received at the hands of General De la Rey immediately before the conclusion of peace; a defeat which put the crown on all our victories.[246]

But, as I have already said, it very soon appeared that when England stamped us as "Guerillas," they really did not mean to use the word at all.

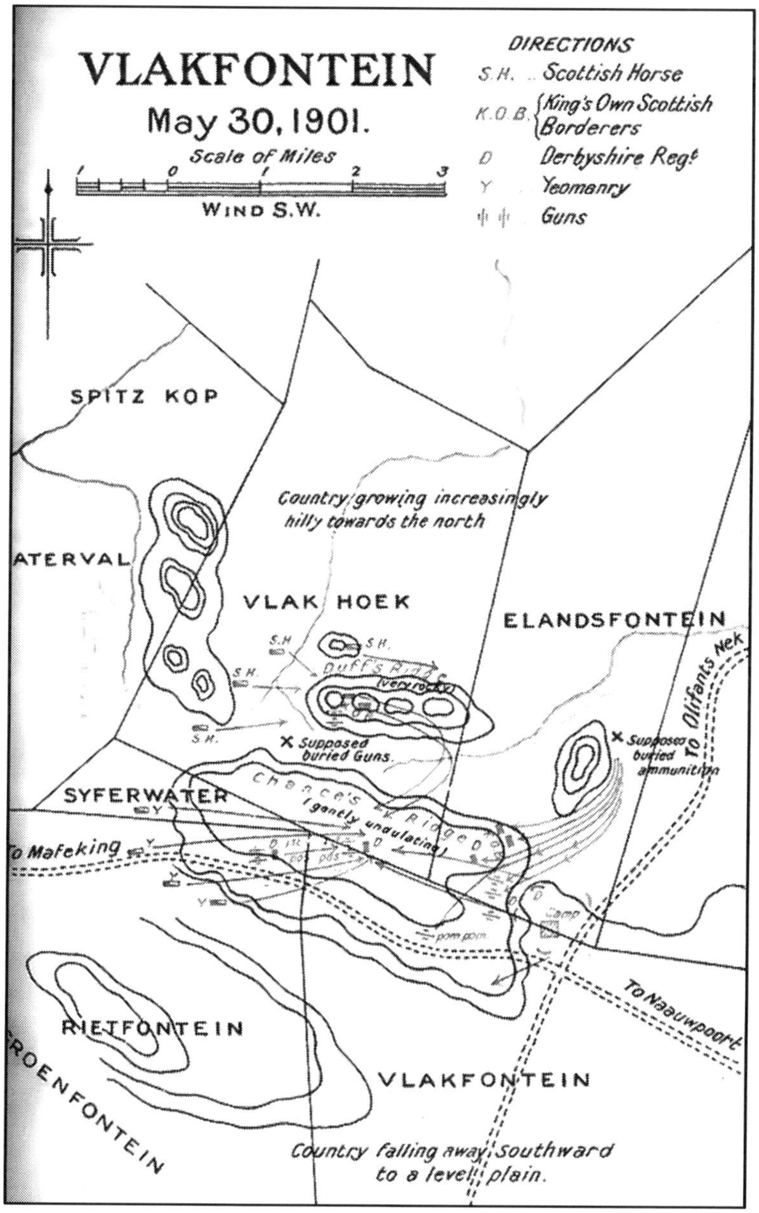

Vlakfontein.[247]

[244] Where the yeomanry were captured.

[245] (District Vrede)—encounter with Brabant's Horse.

[246] Successful attack by De la Rey on Methuen's force at Tweebosch. Methuen wounded and taken prisoner, released on parole. https://en.wikipedia.org/wiki/Battle_of_Tweebosch

28. Negotiations with the Enemy

It was the intention of President Steyn to remain for some time in the division of Vice-Commander-in-Chief Judge Hertzog. Meanwhile, I went to the northern commandos, in order to keep in touch with Generals Louis Botha and De la Rey and our Government. When I was about twelve miles to the south of Petrusburg, I received a letter from General Botha, informing me that Lord Kitchener desired to have a conference held, at Middelburg, in the middle of February, as the English Government wanted to make a Peace Proposal. General Botha asked the President and myself to come yet nearer, so that, in case we might be wanted, we should be within reach.

I sent on his letter to President Steyn, giving him my opinion of it, and asking if he would come. The President, who was always ready to do anything for his country or people, did not lose one moment, but came at once. Meanwhile, I went on ahead with my staff, taking with me also Captain Louis Wessels, and five of his men.

About the 15th of March I crossed the railway line, ten miles to the north of Brandfort, during the night. There we placed some charges of dynamite under the rails, but before we had completed our work, a train came up so quietly that one might call it a "scouting train." It was a dark night, and there was no lantern at the head of the engine, so that we did not see it until it was close upon us. We had, therefore, no chance to ignite the fuse. We retired to a distance of about one hundred paces from the line, when a fierce fire was opened upon us from the train. We replied to this as the train went past, to be succeeded immediately afterwards by a second one. As soon as this also had passed us, we fired the fuses and blew up the railway line at different places close to each other.

No. 3 Artillery Train at the railway station at Bronkhorstspruit, Gauteng Province.[262]

Immediately after this two trains came up, stopping close to the place where the explosions had occurred, and fired on us for about ten minutes without intermission. We paid them back in their own coin, and then each train went its way, leaving the repairing of the line to the following day.

From there we marched on, without accident, except that a German received a slight wound, and one horse was killed. We soon reached Senekal (which had been abandoned by the English), where for the first time I met Dr. Reich and his wife. The doctor received us very heartily; although he did not belong to our Field Ambulance, he did everything that he could for our wounded, as he had done for those of the enemy.

From Senekal I went on to pay a visit to the Heilbron commando, after which I proceeded to Vrede, arriving there on the 24th of February, [sic, March].

It was at Vrede that I had asked Louis Botha to meet me, if he could manage it, and the day after my arrival this meeting took place. The General told me that the negotiations between him and Lord Kitchener had resulted in nothing.[248]

[247] The Battle of Vlakfontein took place on 29 May 1901 during the Second War of Independence. A Boer force led by General Jan Kemp inflicted considerable casualties on a British force led by General Dixon. Vlakfontein is the farm where the Rustenburg/Ventersdorp and Koster/Hekpoort wagon roads intersect. General Kemp had a force of 400-500 citizens. https://af.wikipedia.org/wiki/Slag_van_Vlakfontein

[248] See Vol. 4 'Commando' for the article on the conference. Botha had reported back to Kitchener and declined the terms offered on March 23rd.

Although this was not very satisfactory, still it was just as well that I should meet the Commandant-General of the Transvaal. We had much to discuss and, after a long talk, we parted with the firm determination that, whatever happened, we would continue the war.

On the 27th General [Louis] Botha returned to the Transvaal, and I to the Heilbron commando. After a few days President Steyn came from the south of the Free State, in order to meet the Transvaal Government at Vrede. After this meeting had taken place he went off to a camp of his own, for it was thought better that he should not remain with the commandos any longer. I gave him fifty burghers, under the command of Commandant Davel, to serve as a bodyguard.

I had but just returned from my meeting with General Botha when a serious matter arose at Petrusburg, demanding my immediate presence there. It was three hundred and sixty miles there and back, and the journey promised to be anything but a pleasure trip—far less a safe excursion—for me; but the country's interest requiring it, I started on the 8th of April, although much fatigued by my inroad into Cape Colony.

My staff succeeded in capturing an outpost of sixteen men on the railway line near Vredefort, the English losing one killed and two wounded.

I visited the commando at Vredefort, arranged everything at Petrusburg, and started on my return journey on the 17th. I crossed the railway line between Smaldeel and Ventersburg Road Station, and after paying Commandant Hasebroek a short visit, I came back to the Heilbron commando.

Our tactics of dividing our commandos, and thus keeping the English busy in every part of the Free State, or, where they were too numerous for us, of refusing to allow them to give us battle, so enraged them that they no longer spared the farmhouses in the north and north-western districts. Even in the south and south-west many of the houses were wrecked, but the work of destruction was not carried out with the same completeness as in the afore-mentioned districts. The enemy, moreover, did not spare our cattle, but either drove them off or killed them for food. As for our women-folk—any of them who fell into the hands of the enemy were sent off to the concentration camps.

I have no space here, however, to write about the treatment of the women; it is such a serious matter that it would require whole chapters to deal with it adequately. Abler pens than mine will deal with it in full detail. I will only remark here that the Boer women were shamefully treated, and that if England wishes to efface the impression which these cruelties have left upon the hearts of our people, she will have to act as every great conquering race must act, if it is ever to be reconciled with the nations it has vanquished.

Concentration Camps

Please see the series Appendix for a full discussion regarding the British policy regarding internment of civilians during the conflict.

Over 100,000 women and children were held captive in 45 camps. Disease was rampant and about 22,057 children under the age of 16 years died. A total of 1,421 aged, (No sex, or age value provided in article) .4,200 women older than 16 years, as well as 210 boys older than 16 years died.[263]

Concentration Camp.[264]

Children in Camp.[265]

Transvaal	Orange Free St.	Natal	Cape
Barberton	Bloemfontein	Ladysmith	Aliwal-Noord
Balmoral	Brandfort	Colenso	Kimberley
Belfast	Bethulie	Howick	Norvalspont
Heidelberg	Heilbron	Pietermaritzburg	Doornbult
Irene	Harrismith	Wentworth	Kromelmboog
Johannesburg	Kroonstad	Merebank	Port Elizabeth
Klerksdorp	Ladybrand	Eshowe	Oos-Londen
Krugersdorp	Springfontein	Jacobs	Vryburg
Meintjieskop	Vredefortweg	Mooirivier	Uitenhage
Middelburg	Winburg		
Mafeking			
Nylstroom			
Pietersburg			
Potchefstroom			
Standerton			
Vereeniging			
Volksrust			

Camps in South Africa.[266]

Inside one of the British concentration camps.[267]

Boer women and children in a British concentration camp during the Boer war, circa 1901.[268]

Our winter season had now begun. We had no provisions except meat, bread and maize. Even these were rather scarce, but we could not yet say that we were altogether destitute. Coffee and sugar—except when we had an opportunity of helping ourselves from the enemy's stores—were unknown to us. With regard to the first-named commodity, however, the reader must know that in the district of Boshof there grows a wild tree, whose roots make an excellent substitute for coffee. Broken up into small pieces and roasted, they supplied us with a delicious beverage. The only pity was that the tree was so scarce that the demand for this concoction very greatly exceeded the supply. We therefore invented another drink—which we also called coffee—and which was composed of corn, barley, maize, dried peaches, sweet potatoes, and miscellaneous ingredients. My own favourite beverage was abundant—especially after heavy rain!

One of the Boer concentration camps in Winter, probably Bloemfontein judging from the image source.[269]

The question of clothing was now beginning to be a very serious one. We were reduced to mending our trousers, and even our jackets with leather. For the tanning of this leather the old and feeble were employed, who, as soon as the enemy approached, fled, and as soon as they had passed, returned to their tanning. At a later period the English had a trick of taking the hides out of the tanning tubs and cutting them to pieces, in the hope, I suppose, that we should then be compelled to go barefoot and unclothed.

It was to obviate such a catastrophe as this that the custom of Uitschudden[249] now came into force. The burghers, although against orders, stripped every prisoner. The English had begun by taking away, or burning, the clothes which the burghers had left in their houses—that was bad enough. But that they should cut up the hides, which they found in the tanning tubs, was still worse; and—the burghers paid them back in the same coin by stripping the troops.

Towards the end of May I crossed the railway line to Parijs and Vredefort, intending to go on from there to see General De la Rey, and discuss our affairs with him. I had come to the conclusion that it would be good policy to send small commandos into Cape Colony; for small bodies of men can move rapidly, and are thus able to get out of the way if they are threatened by overpowering numbers. Moreover, such small detachments would compel the English to divide their forces.

When I reached Vredefort I received a despatch from President Steyn, summoning me to him. I had thus to abandon my idea of visiting General De la Rey; instead of this, I wrote him a letter requesting him to come to the President. I also sent for Judge Hertzog.

De la Rey was the first to arrive, and, without waiting for Judge Hertzog, we at once proceeded to take into consideration the following letter from the Government of the South African Republic.

[249] Stripping.

1901 SAR Letter

Government Offices, In the Field, District Ermelo, South African Republic,

To the Government Secretary, O.F.S.

May 10th, 1901

Sir,—

I have the honour to report to you that to-day the following officers met the Government, namely, the Commandant-General, General B. Viljoen, General J.C. Smuts (Staats-Procureur), the last-named representing the western districts. Our situation was seriously discussed, and, among others, the following facts were pointed out:—

1. That small parties of burghers are still continually laying down their arms, and that the danger arising from this is becoming every day more threatening, namely, that we are exposed to the risk of our campaign ending in disgrace, as the consequence of these surrenders may be that the Government and the officers will be left in the field without any burghers, and that, therefore, heavy responsibility rests upon the Government and War Officers, as they represent the nation and not themselves only.

2. That our ammunition is so exhausted that no battle of any importance can be fought, and that this lack of ammunition will soon bring us to the necessity of flying helplessly before the enemy. And that through this same lack it has become impossible for us to afford adequate protection to our people and their cattle, with the result that the general population is being reduced to poverty and despair, and that even the troops will soon be unable to be supplied with provisions.

3. That through the above-mentioned conditions the authority of the Government is becoming more and more weakened, and that thus the danger arises of the people losing all respect and reverence for lawful authority, and falling into a condition of lawlessness. And that to prolong the war can only lead to hastening the ruin of the people, and making it clear to them that the only authority in the country is that of the enemy.

4. That not only is our nation becoming disorganized in the manner above referred to, but that it will also most certainly happen that the leaders of the nation, whose personal influence has hitherto kept it together, will fall into utter contempt, and lose that influence which is our only hope for reviving the national spirit in the future.

5. That the people are constantly demanding to be told what hope still exists of successfully prosecuting the war, and that they have the right to expect to be informed in an honest and straightforward manner that their cause is hopeless, whenever this has become evident to the Government and the Leaders.

Up to the present time the Government and the nation have been expecting that, with the co-operation of their Deputation and by the aid of European complications, there would be some hope for the success of their cause, and the Government feels strongly that before taking any decisive step, an attempt should again be made to arrive with certainty at the results of the Deputation and the political situation in Europe.

Having taken all the facts into consideration, the Government, acting in conjunction with the above-mentioned officers, have arrived at the following decision:

Firstly, that a request should be addressed this very day to Lord Kitchener, asking that through the intervention of ambassadors sent by us to Europe, the condition of our country may be allowed to be placed before President Kruger, which ambassadors are to return with all possible speed.

Secondly, that should this request be refused, or lead to no results, an armistice should be asked for, by which the opportunity should be given us of finally deciding in consultation with your Government, and the people of the two States, what we must do.

This second proposal is, however, subject to any solution which your Government, taking into consideration the above-mentioned grievances, may be able to suggest.

The Government feels very keenly that it would no longer be right to allow things to go on as they have been going on, and that the time has arrived for taking some definite steps; it will, therefore, be glad to receive an answer from your Government as soon as possible.

I have the honour to be,

Yours, etc., F.W. Reitz,

Secretary of State. [of the Transvaal]

The answer which the President [Steyn] sent to this letter was formerly in my possession, but has been lost with many of my documents. I am able, however, to give an extract, which I received from the Rev. J.D. Kestell. It was to the following effect:—

The President was much disappointed with the letter of the Transvaal Government; he said that although there had been in the past some surrenders in the Free State, this difficulty had now been overcome. Moreover, although the ammunition had for a long time been scarce, nevertheless, after every fight, there had been enough to begin the next with.[250]

To the question, What probability was there of their being able to continue the struggle? he would reply by asking another question—What hope had the two little Republics, at the beginning of the war, of winning the fight against the might of England? If they had trusted in God at the beginning, why did they not continue to trust in Him?

He also pointed out that if the Boer cause was really quite hopeless, the Deputation would have been sure to send word to that effect. Further, he assured the Transvaal Government that if an armistice were to be obtained, and if during it the people of the Free State were to be asked for their opinion, the decision of the burghers who were still in the field would be to continue the war.

He could not approve of the decision of the Transvaal Government to ask Lord Kitchener to allow ambassadors to be sent to Europe, for, by so doing, the Government would be showing its hand to the enemy; he added that he was very sorry that such a decision had been taken without first consulting the Free State.[251]

As to the fear expressed by the Transvaal Government, that the Authorities and the Officers in the field would be left without burghers, the President said, that even if the Government and the Officers of the Free State were to surrender, the nation would not do so.

It would be a great misfortune, he added, if the Orange Free State, which had not only lost its property and the lives of many of its burghers but also even its very independence, in the defence of the sister Republic, should now be abandoned by that Republic; that then all confidence in one another and all co-operation between Afrikanders would come to an end for ever: and that, under such circumstances, it would be too much to expect that the African nation should ever be able to rise again. If then the Boers wished to remain a nation, it was absolutely necessary to continue the war.

After having quoted various appropriate passages from the newspapers, the President went on as follows:—"All these considerations combine to make me believe that we should be committing a National murder if we were to give in now. Brethren! Hold out a little longer. Let not our sufferings and our struggles be in vain; let not our faith in the God of our fathers become a byword. Do all that you can to encourage one another."

The President concluded this very remarkable and powerful letter with the question:—

"Are we again to leave the Colonial burghers in the lurch? God forbid."

We decided to set out for the Transvaal in order to discuss the matter with the Government; and on the evening of the 5th of June we marched four or five miles from Liebenbergsvlei, to a place opposite Verkijkersdorp. We were, all told, between sixty and seventy men, including the staff and part of the bodyguard of President Steyn, the staff of General De la Rey, and eight of my staff officers.

[250] 3/26/1901 Victory by General Babington over Delarey at Ventersdorp. Nine Boer guns captured.

4/6/1901 General French, in his sweeping operations in the Eastern Transvaal, captured all the enemy's guns in that district.

4/8/1901 Colonel Plumer captured Pietersburg, the terminus of the railway running due north from Pretoria.

4/10/1901 Civil administration resumed in the Transvaal.

4/15/1901 Smuts' commando defeated near Klerksdorp. Two guns captured.

4/19/1901 Generals Plumer and Walter Kitchener co-operated with General French in clearing the Eastern Transvaal and Lydenburg district.

4/30/1901 General Blood discovered documents and banknotes of Transvaal (Government at Roosenekal, from which place Mr. Schalk Burger fled.

[251] The reply to the enquiry, facilitated by Kitchener, had not yet been received. 7/5/1901 In reply to Botha's inquires about ending the war, Kruger telegraphed to Botha to continue fighting.

Marthinus Theunis Steyn.[270]

President Steyn [Seated] with his Burghers.[271]

The following morning, an hour and a half after sunrise, a burgher came galloping up to tell us that the enemy had just captured a laager of women.[252]

It seemed impossible to ride over to the rescue of these women, for our horses had still to make the long journey into the Transvaal. I asked our guest, General De la Rey, what he thought about the matter. He at once replied that we must go and liberate the women. As we were already up-saddled in readiness for our march, I had nothing to do but to give the order to start. The President, with his staff and some of the bodyguard, remained behind; while General De la Rey, Commandant Davel and I, with fifty-five men, hurried off. The retired General, Piet Fourie, was also with us.

The enemy had marched with the laager on to a hill near the Kaffir kraal, consisting of four or five huts and a building made of sods.

Showing a circle of Zulu homes placed around a Kraal or cattle pen. The kraal is made of the smaller branches of trees trimmed and placed in the ground, then bound together near the top, forming a strong hedge.[272]
Building a Zulu Homestead. It is covered with clay and grass mixed together and then thatched over with long grass.[273]

We first caught sight of the English when we were at a distance of four miles from them; they were then busy drawing up the waggons of the women in rows of ten or twelve. The oxen belonging to the first row stood close against the kraal, as we saw later on; those of the second row being behind them, and so on.

The women told us afterwards that they had asked to be allowed to retire to a place where they would not run the risk of being shot by us (for the English had taken cover barely one hundred paces behind the waggons and

[252] The previous evening we had received a report of two English camps on the Wilge River: One at Duminy Drift, the other at Steildrift—under General Elliott. They were led by Piet de Wet and other National Scouts.

were preparing to fight us from there), but that they were ordered to remain behind the soldiers. They were thus exposed to the danger of being hit by us, if we shot a little too high. It was, they said, the most terrible day they had ever spent.

When we came within range of the English, they opened a hot fire upon us. We had to gallop over ground as smooth as a table with no cover until we were close up to them, and protected by a small hill. We left our horses here, and ran as fast as we could up the incline. At the top we were within forty paces of the place where the English were lying in wait for us. As soon as our heads appeared over the brow of the hill they fired on us; but there was only one round fired, for our reply was so sharp and severe that many of them were at once mowed down. The rest jumped up and retreated behind the last row of waggons, several of them, however, being killed during their flight.

Our men dashed through between the waggons, but the English were the first to reach the kraal. They had made loopholes in its walls, through which they now fired on us. The only shelter we had was a Kaffir hut, which as is well known, always has a round wall. There was no chance for us to make loopholes—the wall was too solid—so that if a burgher wanted to shoot he had to expose his whole body, while the English lay ready behind their loopholes to fire on us. So it happened that eleven burghers were killed and seven wounded. Among the dead was Captain Thijnsma, and among the wounded, Lieutenant H. Howell. In the meantime we had got the waggons away, except the row which was nearest to the kraal, and which were too close to the enemy for us to be able to approach them safely.

No sooner had the English taken refuge in the kraal than the women fled with the waggons; and it is astonishing to relate that only one little boy of thirteen years was killed, and a woman and a girl slightly wounded. One of the burghers whom the English had taken prisoner was also killed.

I have no exact figure as to the losses of the English, but judging from the number of dead and wounded lying on the battlefield, I should say that their casualties must have been about eighty.

The fight lasted from eleven till three o'clock, and then a reinforcement of cavalry, from eight hundred to one thousand men strong, appeared with some guns. The force with which we had been engaged, numbering about two hundred men, belonged to the column which was now coming up. As we could not drive the English from the kraal before the arrival of the reinforcements, we had to give way.

Although I had given orders that all the waggons which had managed to escape should be sent on to Reitz, in the actual event only a few carts went there. The women had left the waggons behind, close to the hill at the foot of the English position, where I could not see them, in order to await the result. They had forgotten what I had told them, namely, that they were to get away as quickly as possible. This order I had given in the expectation that a reinforcement might arrive at any moment.

After I had ordered a few men to bring the wounded into a safe place, I retired with the remainder, some forty-five in number. Among these was Veldtcornet Serfontein and his burghers.

The English now directed their fire upon the women's laager, to compel it to come to a standstill. Whether any of the women and children were killed or wounded I was unable to ascertain, but it was horrible to see the bombs bursting over their heads. Thus the women again fell into the hands of the enemy.

With four of my adjutants and Piet Fourie, I succeeded in driving away quite one thousand five hundred head of cattle. The bombs fell heavily on them also, but I got them safely away. Late that evening we arrived at the spot where we had left President Steyn, only to find that he had gone away. He had been obliged to retreat before the force which the previous evening had been at Duminy Drift, and which had passed near him during the day. The President had accordingly gone some twelve miles in the direction of Lindley.

It was one of the coldest nights we had that winter, and our pack-horses which were carrying the blankets were with the President. It was impossible for us to sleep without any covering on such a night as that, and so we were obliged to march on. We had moreover to look for something to eat, for we had had nothing since breakfast. Our horses had never had their saddles off from the time we went out to fight until we arrived about midnight at the President's camp.[253]

[253] 6/6/1901 De Wet severely defeated near Reitz. By General Elliot, who made large captures.

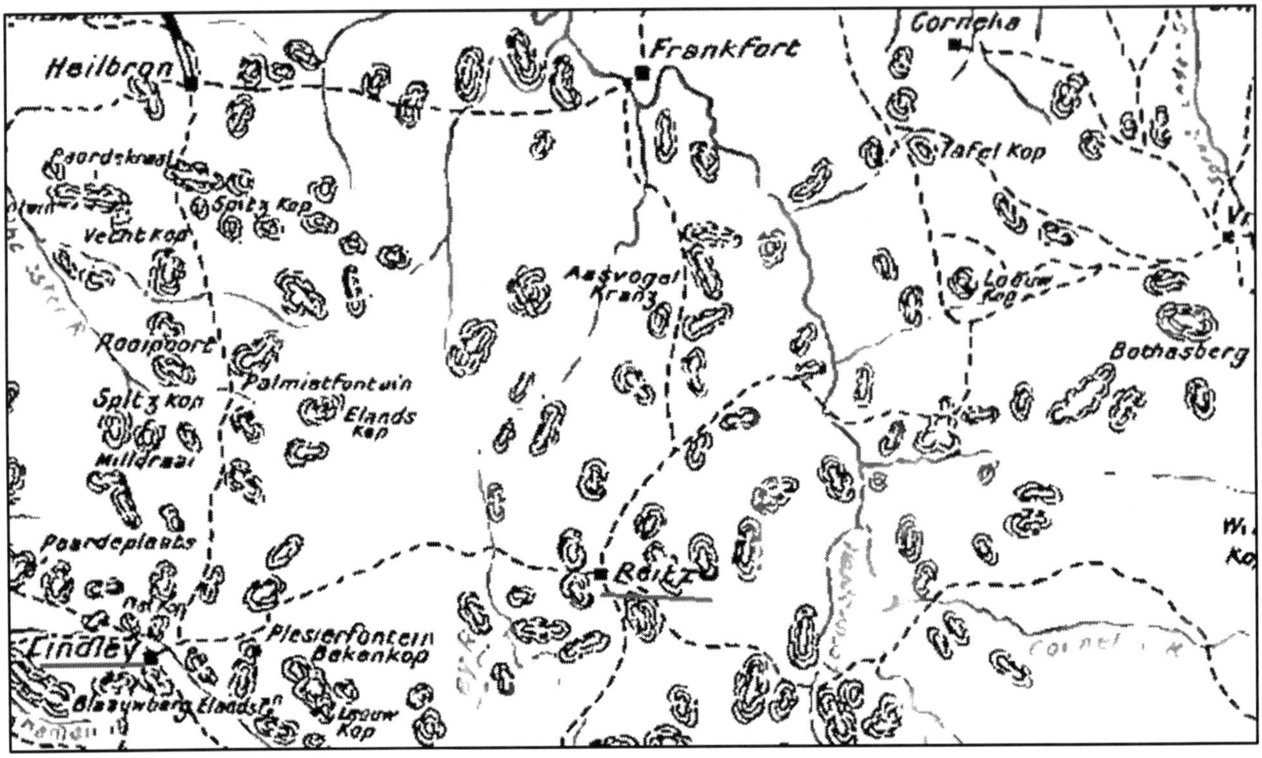

The action took place in this general area. Lindley lower left, Retz directly east.

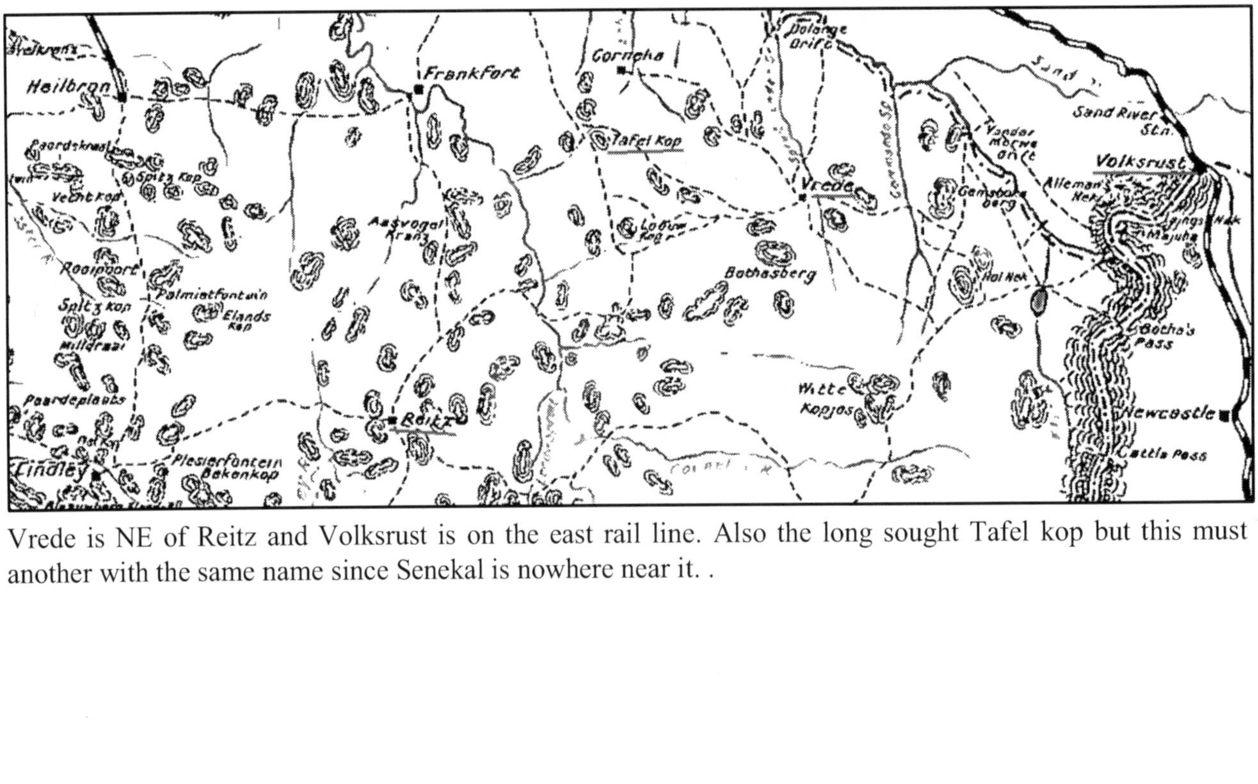

Vrede is NE of Reitz and Volksrust is on the east rail line. Also the long sought Tafel kop but this must be another with the same name since Senekal is nowhere near it. .

29. President Steyn's Narrow Escape

The following morning [June 7th] we had to continue our journey to the Transvaal. It being necessary to keep out of sight of the enemy, we marched first a short distance to the south, and then went south-east. After a few days we reached Vrede. There Commandant Manie Botha spared us a few burghers who knew this part of the country well to serve as guides across the railway line. We headed to the north of Volksrust, and on the second evening after we had left Vrede, we struck the railway line at a spot which was guarded by an outpost. They opened fire on us at once. General De la Rey and I then came to the decision that after the burghers had exchanged a few shots, we would quietly retreat a short distance, and then, with a sweep, try and cross the line at another spot. This ruse was successful and we crossed unobserved. But the first of our men had hardly got seventy paces from the railway line, when a fearful explosion of dynamite took place, not thirty paces from the spot where we had crossed. Whether this was managed by electricity or whether the hindmost horses had struck on the connecting wire of some trap set by the enemy, I cannot say; at all events, we escaped with only a fright.

On the fourth day after this [~June 11th] we met the Transvaal Government and held a conference at once, in accordance with the letter mentioned in my last chapter. It grieved us much that things should have taken this turn, for it nearly always happened that somehow matters of this sort came to the ears of the English.

But the Transvaal Government had again taken courage, as they had received an answer to the cable which they had sent to the Deputation, which answer instructed them to hold out; and also because two successful battles had taken place shortly before—one fought by General Kemp,[254] and the other by Commandant Muller.[255]

We remained there for two days, and after it had been settled by the two Governments that the war should be continued with all our might, and also that days of thanksgiving and humiliation should be appointed, we went away accompanied by the genial and friendly Commandant Alberts, of Standerton, who brought us across the Natal-Transvaal railway. Captain Alberts was renowned as a valiant soldier; we now also found him to be a most sociable man. He beguiled the time with agreeable narratives of events in which he had taken part, and almost before we realized it we had reached the railway line. We crossed in safety and took a hearty farewell of our friendly Commandant and his burghers.

On our march to Zilverbank—a farm on the Waterval River—I did not require any guide, for I knew the surroundings, having lived there for two years. After breakfast on the following morning we went on to within four or five miles south of Hexrivier farm, about three miles to the north of the Vaal River. There we off-saddled; and shortly after General De la Rey took leave of us.

He wanted to cross the railway at a place between Vereeniging and Meyerton Station. This would lead him by a shorter road to his commandos than if he went through the Free State. Our farewell was affectionate—all the more so because we did not know whether we should see each other again on this earth. Then we continued on our way with light hearts; having been inspirited, not only by the pleasant company of the last few days, but also by the decision taken by the two Governments, that, come what might, our independence should not be sacrificed by us.

I crossed the Vaal River at Villiersdorp and remained there that evening and through the following day. Then President Steyn and I parted. He went to Bezuidenhoutsdrift, and I, by way of, to the Heilbron commando. I remained at Frankfort for one night, with Commandant Ross and his men, and had a very enjoyable time.

With the Heilbron people I stayed a few days only, because I had important work to accomplish in the Winburg district; to this district therefore I went.

As the commandos were now so scattered there was enough work for each of us in his own district, and I had much more riding to do than formerly. I found Commandant Hasebroek and his men at Doornberg a few days later. Whilst there I received from President Steyn a report of his narrow escape at Reitz, on the 11th of July, 1901, when he and some of his bodyguard escaped, whilst, unfortunately, Commandant Davel and all the members of the Government, except Mr. W.C.J. Brebner, who was absent, were taken prisoners.

[254] 5/29/1901 Jan Kemp inflicted significant casualties on a British force led by General Dixon. Vlakfontein is the farm where the Rustenburg/Ventersdorp and Koster/Hekpoort wagon roads cross. General Kemp possessed a force of 400–500 citizens. Battle of Vlakfontein: https://af.wikipedia.org/wiki/Slag_van_Vlakfontein

6/1/1901 Severe engagement between General Dixon and Delarey at Vlakfontein, in the Magaliesberg. Enemy repulsed with heavy loss. Our casualties also heavy.

[255] 6/12/1901 General Beatson surprised near Middelburg (Transvaal). Loss of two pom poms.

From Winburg I paid a visit to Vice-Commandant-in-Chief J. Hattingh, of the Kroonstad commando, and then went to President Steyn. My joy in finding that the President was safe, was only equalled by my grief at the loss of such old friends as General Cronje, Member of the Executive Council; General J.B. Wessels; T. Brain, Secretary to the Government; Commandant Davel; Rocco De Villiers, Secretary to the Executive Council; Gordon Fraser, Private Secretary to the President; MacHardy, Assistant Secretary; Pieter Steyn, brother of the President and Veldtcornet of the staff; and my other friends in the bodyguard.

It was sad to think that such men were prisoners, and were lost to us so long as the war continued. We had become rather accustomed to such experiences, but what made this so hard to bear was that treachery had a hand in it—when the English took the Government and President Steyn's bodyguard prisoners, they had had a Free State burgher as their guide.

The vacant posts in the Government had now to be filled up, and the President appointed the following persons:—In the place of A.P. Cronje, General C.H. Olivier, as Member of the Executive Council; and in place of Mr. T. Brain, Mr. W.C.J. Brebner, as Government Secretary. Mr. Johannes Theron he appointed Secretary to the Executive Council, instead of Mr. Rocco De Villiers; and Mr. B.J. Du Plessis Private Secretary to himself in place of Mr. Gordon Fraser.

The President also decided to have, in future, only thirty burghers as his bodyguard, and appointed Captain Niekerk as their Commandant.[256]

Times History Account

Quote.[257]

On July 6 De Lisle, with the left flank column, passed through Reitz, finding in the deserted town a document proving that Boers were in the habit of leaving the place when a British column approached and of returning to it when the danger was past. Elliot turned this hint to excellent account.

On the 10th, when his entire line was well to the north of Reitz, he ordered Broadwood, who led the right flank column to double back by a forced night march and take the town by surprise. Broadwood, with 400 men and a pom-pom, carried out this order with skill. But for the loss of touch by a connecting file he would have actually surrounded Reitz by daylight on the 11th; as it was he had to gallop hard for the last two miles and had no time to block the roads leading out of the town.

It happened that Steyn himself, his bodyguard under Davel, and all the Free State Government officials, together with Generals A. P. Cronje and J. B. Wessels, and Field-cornet Steyn (the President's brother), were sleeping in Reitz that night, having left it at De Lisle's approach and reoccupied it when his back was turned. De Wet, unluckily for the Government party, was away at his favourite headquarters, the farm Blijdschap.

The slumbers of the President and his friends were sound and careless. When Broadwood's troopers galloped into the town there was a sauve qui peut.[258] It was too late. Twenty-nine prisoners, including the whole Government staff, together with Davel, Wessels, Cronje and the President's brother, were captured.

Only eight persons escaped—seven of the bodyguard and President Steyn himself, whose faithful Kaffir cook warned him just in time. Coatless, nightcapped, with a halter for a bridle, the President galloped down the street. He was pursued and shot at by a sergeant of the 7th Dragoon Guards; but the Boer pony was fresh and the British charger tired, so that the fugitive made good his escape.

He had lost his suite, his papers, his treasury—everything but his intense hatred of the British invader and his unquenched faith in the eventual triumph of the cause.

[256] 7/5/1901 In reply to Botha's inquires about ending the war, Kruger telegraphed to Botha to continue fighting.

7/6/1901 A train wrecked on the Pretoria- Pietersburg line.

7/15/1901 Capture of the so-called "Orange Free State Government " at Reitz. Important Boer papers seized. Steyn alone of the members of his "Government" escaped.

7/16/1901 Important success by General French in Cape Colony.

7/19/1901 Publication of Kitchener's despatch embodying contents of important documents seized at Reitz. Death of Mrs. Kruger.

[257] Times History, Vol. 5 pages 300–301.

[258] Sauve-qui-peut is a French expression that means stampede or mad rush. It is often used as an exclamation to warn people to run for their life. See: the Collins French-English dictionary.

30. The Last Proclamation

I now impressed upon my officers as forcibly as I could the importance of intercepting the communications of the enemy by blowing up their trains. A mechanical device had been thought of, by which this could be done. The barrel and lock of a gun, in connexion with a dynamite cartridge, were placed under a sleeper, so that when a passing engine pressed the rail on to this machine, it exploded, and the train was blown up. It was terrible to take human lives in such a manner; still, however fearful, it was not contrary to the rules of civilized warfare, and we were entirely within our rights in obstructing the enemy's lines of communication in this manner.

Owing to this, the English were obliged to place many more thousands of soldiers along the railway line, in order to keep the track clear. Even then, the trains, for a considerable time, could not run by night. The English soon discovered how we arranged these explosions, and the guards carefully inspected the lines each day to find out if one of these machines had been placed beneath the rails. We knew that one had been found and removed, whenever we saw a train pass over the spot without being blown up. This, however, only made us more careful. We went to the spot which we had fixed upon for the explosion, hollowed out the gravel, placed the machine under the sleeper, and covered it up again, throwing the gravel that was left to a good distance from the line. After this, the guards could not discover where the machine was placed. They trebled the troops on the line in consequence.

The month of July had passed, and we wondered what August held in store for us. The customary fights of the different commandos still went on; here five, here ten, here thirty of the English were killed, wounded or made prisoners. If these numbers had been put down they would have mounted up to a considerable total; but the war was not of such a nature that an office could be opened to record them. Reports of battles were sent to me, and after I had allowed them to accumulate for three or four weeks, they were sent to the different Vice-Commandants-in-Chief for their general information, and then torn up.

Many reports and much correspondence concerning the beginning of the war have been preserved. I gave them to a trustworthy friend with instructions to bury them, but do not know where he placed them, as he was taken prisoner later on, and I have never been able to find out where he was sent to. These documents are of great value, and ought to be published.

I was on the farm of Blijdschap, between Harrismith and Bethlehem—my English friends, Generals Knox, Elliott and Paget, with their Colonels Rimington, Byng, Baker, etc., etc., will not have forgotten where Blijdschap is—when I received a letter from Lord Kitchener, enclosing his Proclamation of the 7th of August, 1901.

Kitchener's August 7th, 1901 Proclamation

This proclamation was as follows:

"By his Excellency Baron Kitchener of Khartoum, G.C.B., K.C.M.G., General Commander-in-Chief of His Majesty's forces in South Africa; High Commissioner of South Africa, and Administrator of the Transvaal, etc.

"Whereas the former Orange Free State and South African Republic are annexed to His Majesty's possessions;

"And whereas His Majesty's forces have now been for some considerable time in full possession of the Government seats of both the above-mentioned territories, with all their public offices and means of administration, as well as of the principal towns and the whole railway;

"And whereas the great majority of burghers of the two late Republics (which number thirty-five thousand over and above those who have been killed in the war) are now prisoners of war, or have subjected themselves to His Majesty's Government, and are now living in safety, in villages or camps under the protection of His Majesty's forces;

"And whereas the burghers of the late Republics, now under arms against His Majesty's forces, are not only few in number, but have also lost nearly all their guns, and war requisites, and are without proper military organization, and are therefore not in a position to carry on a regular war, or to make any organized resistance against His Majesty's forces in any part of the country;

"And whereas the burghers who are now still under arms, although not in a position to carry on a regular war, continue to make attacks on small posts and divisions of His Majesty's forces, to plunder and to destroy farms, and to cut the railway and telegraph lines, both in the Orange River Colony and in the Transvaal and other parts of His Majesty's South African possessions;

"And whereas the country is thus kept in a state of unrest, and the carrying on of agriculture and industries is hindered;

"And whereas His Majesty's Government has decided to make an end of a situation which involves unnecessary bloodshed and devastation, and which is ruining the great majority of the inhabitants, who are willing to live in peace, and are desirous of earning a livelihood for themselves and their families;

"And whereas it is only just that steps should be taken against those who still resist, and principally against those persons who are in authority, and who are responsible for the continuance of the present state of disorganization in the country, and who instigate their fellow citizens to persist in their hopeless resistance against His Majesty's Government;

"I, Horatio Herbert Baron Kitchener, of Khartoum, G.C.B., K.C.M.G., General Commander-in-Chief of His Majesty's forces in South Africa; High Commissioner in South Africa, on behalf of His Majesty's Government, proclaim and make known as follows:

"All Commandants, Veldtcornets and leaders of armed bands—being burghers of the late Republics—still resisting His Majesty's forces in the Orange River Colony and the Transvaal, or in any part of His Majesty's South African possessions, and all members of the Government of the late Orange Free State and of the late South African Republic, shall, unless they surrender before the 15th September of this year, be banished for ever from South Africa; and the cost of maintaining the families of such burghers shall be recoverable from, and become a charge on, their properties, whether landed or movable, in both Colonies.

"GOD SAVE THE KING.

"Given under my hand at Pretoria, the seventh day of August, 1901.[259]

"Kitchener, General,

High Commissioner of South Africa."

[259] See Also Kitchener' Proclamation article in Viljon's Appendix and in the Century of Wrong Volume in this series.

Kitchener's Proclamation

This appeared in the Commando volume but is reproduced here was well..

LORD KITCHENER'S PROCLAMATION.

SURRENDER OR BANISHMENT.

GOVERNMENT DETERMINED TO END THE WAR.

A Parliamentary paper, issued to-day, contains the proclamation issued by Lord Kitchener on August 6 with the concurrence of the British Government. It orders surrender before September 15, or permanent banishment for all commandants, field cornets, leaders of armed bands, and all members of the Boer Governments. The full text of the proclamation is as follows :—

Whereas the late Orange Free State and the late South African Republic have been annexed to his Majesty's dominions, and whereas his Majesty's forces are and have been for some considerable time in complete possession of the seats of Government of both of the aforesaid territories with their public offices and the whole machinery of administration as well as of all the principal towns and the whole of the railway lines, and whereas the great majority of the burghers of the two late Republics to the number of 35,000 exclusive of those who have fallen in the war are now either prisoners or have submitted to his Majesty's Government and are living peaceably in towns or camps under the control of his Majesty's forces, and whereas the burghers of the late Republics still in arms against his Majesty are not only few in numbers but have also lost almost all their guns and munitions of war and are devoid of regular military organization and are therefore unable to carry on regular warfare or to offer any organized resistance to his Majesty's forces in any part of the country, and whereas those burghers who are still in arms though

UNABLE TO CARRY ON REGULAR WARFARE

continue to make isolated attacks upon small posts and detachments of his Majesty's forces to plunder or destroy property and to damage the railway and telegraph lines both in the Orange River Colony and Transvaal and in other portions of his Majesty's South African dominions, and whereas the country is thus kept in a state of disturbance checking the resumption of agricultural and industrial pursuits, and whereas his Majesty's Government is determined to put an end to a state of things which is aimlessly prolonging bloodshed and destruction and inflicting ruin upon the great majority of the inhabitants who are anxious to live in peace and to earn livelihood for themselves and their families, and whereas it is just to proceed against those still resisting and especially against those persons who being in a position of authority are responsible for the continuance of the present state of lawlessness and are instigating their fellow burghers to continue their hopeless resistance to his Majesty's Government: Now therefore I, Lord Kitchener, &c., under instructions from his Majesty's Government proclaim and make known as follows :—

"All Commandants, Field-Cornets, and leaders of armed bands, being burghers of the late Republics, still engaged in resisting His Majesty's forces, whether in the Orange River Colony and the Transvaal or in any other portion of His Majesty's South African Dominions, and all members of the Governments of the late Orange Free State and the late South African Republic, shall, unless they surrender before September 15 next, be permanently banished from South Africa; the cost of the maintenance of the families of all burghers in the field who shall not have surrendered by September 15 shall be recoverable from such burghers and shall be a charge upon their property moveable and immovable in the two Colonies."

LORD KITCHENER'S REPLY

Lord Kitchener replied on August 2nd:
"Ministers, Cape Town, entirely concur in proclamation. Have not yet heard from Natal." Natal."

On the following day Lord Kitchener telegraphed :—"The Government of Natal express approval of the proclamation."

The correspondence further shows that on July 24th, Sir H. E. McCallum, Governor of Natal, telegraphed to Mr. Chamberlain as follows :—

"I sent you by last mail long minute submitted by Ministers, containing the following suggestions :—(Begins), Protracted continuance of hostilities, vitally affecting interests of Natal, is viewed with grave concern. Raids into the Colony frequently render it impossible for loyalists to return to farms and avocations. Feeling of unrest among natives created by raids. Revenue suffering, trade paralysed, railways monopolised by military, towns overcrowded with refugees, and persons awaiting return to Transvaal; stock being infected with disease due to introduction of captured stock from the new Colonies; famine prices prevailing; Colony still subject to censorship and martial law. Under these circumstances Ministers advocate sterner measures to crush present guerilla warfare. They point out that Boers still fighting have little to lose, that their women and children are protected and well treated, and that their farms are safe from confiscation. Therefore, Boers free from anxiety and encouraged to continue in the field; growing accustomed to life of pillaging and looting, and communicate frequently with refugee camps, which thus are source of danger. Those who are not rebels know that if captured they will be treated as prisoners of war and released at conclusion of hostilities. Ministers believe that excellent effect would be produced if it were made generally known that if burghers now in field do not surrender before given date say within one month, cost of maintenance of all women and children will be chargeable against immovable property of burghers in field; also that Boer generals and leaders in field should be informed that unless they and their commandos surrender by date specified, they will be banished from South Africa for life when captured.

"In making these observations and recommendations Ministers disclaim intention of appearing to reflect on military operations, of which they realise the immense difficulties. They will continue to render Imperial Government every assistance to secure settlement and pacification." (Minute ends.)

Proclamation Text.[260] Kitchener's Reply to a request to obtain consent from the Colonial Governors.[261]

Lord Kitchener's Proclamation.
Surrender Or Banishment.
Government Determined To End The War.[262][263]

A Parliamentary paper, issued to-day, contains the proclamation issued by Lord Kitchener on August 6 with the concurrence of the British Government. It orders surrender before September 15, or permanent banishment for all commandants, field cornets, leaders of armed bands, and all members the Boer Governments. The full text of the proclamation is follows:

Whereas the late Orange Free State and the late South African Republic have been annexed to his Majesty's dominions, and

Whereas his Majesty's forces are and have been tor some considerable time in complete possession of the seats of Government of both of the aforesaid territories with their public offices and the whole machinery of administration as well as of all the principal towns and the whole of the railway lines, and

[260] Kitchener's Proclamation: Pall Mall Gazette August 9, 1901 p. 7 col. 1 BNA File BL_0000098_19010809_042_0007.pdf.
[261] Kitchener's Proclamation: The Northern Guardian, (Hartlepool) Friday August 9th, 1901 p. 3 col 3. BNA File: BL_0003039_19010809_040_0003.pdf.
[262] Kitchener's Proclamation: Pall Mall Gazette August 9, 1901 p. 7 col. 1 BNA File BL_0000098_19010809_042_0007.pdf.
[263] Fife Free Press - Saturday 10 August 1901 https://www.britishnewspaperarchive.co.uk/viewer/bl/0001062/19010810/032/0002.

Whereas the great majority of the burghers of the two late Republics to the number of 35,000 exclusive of those who have fallen in the war are now either prisoners or have submitted to his Majesty's Government and are living peaceably in towns or camps under the control of his Majesty s forces, and

Whereas the burghers of the late Republics still in arms against his Majesty are not only few in numbers but have alto lost almost all their guns and munitions of war and devoid of regular military organization and are therefore unable to carry on regular warfare or to offer any organized resistance to his Majesty's forces in any part of the country, and

Whereas those burghers who are still in arms though unable to carry or regular warfare continue to make isolated attacks upon small posts and detachments of his Majesty's forces to plunder or destroy property and to damage the railway and telegraph lines both in the Orange River Colony and Transvaal and in other portions of his Majesty's South African dominions, and

Whereas the country is thus kept in a state of disturbance checking the resumption of agricultural and industrial pursuits, and

Whereas his Majesty's Government is determined to put an end to a state of things which is aimlessly prolonging bloodshed and destruction and inflicting ruin upon the great majority of the inhabitants who are anxious to live in peace and to earn livelihood (or themselves and their families), and

Whereas it is just to proceed against those still resisting and especially against those persons who being in position of authority are responsible for the continuance of the present state of lawlessness and are instigating their fellow burghers to continue their hopeless resistance to his Majesty's Government:

Now therefore I, Lord Kitchener, &c., under instructions from his Majesty's Government proclaim and make known as follows "All Commandants, Field-Cornets, and leaders of armed bands, being burghers of the late Republics, still engaged in resisting His Majesty's forces, whether the Orange River Colony and the Transvaal or in any other portion of His Majesty's South African Dominions, and all members of the Governments of the late Orange Free State and the late South African Republic, shall, unless they surrender before September 15 next, be permanently banished from South Africa; the cost of the maintenance of the families of all burghers in the field who shall not have surrendered September 15 shall be recoverable from such burghers and shall Ire charge upon their property moveable and immovable in the two Colonies."

Times History.[264]

> Mr. Chamberlain's Message Mr. Chamberlain, in telegraphing the text of the proclamation to Lord Kitchener, on July 30th, said His Majesty's Government proposed to instruct Lord Kitchener to issue the proclamation with the least possible delay. "But," added the Colonial Secretary, "they desire you before doing so to communicate its terms to the Governors of Cape Colony and Natal, and to ascertain whether their Governments agree to them. You should state that in the opinion of His Majesty's

[264] Times History Summary of Proclamation of August 7th, 1901:[In mid 1901].............Kitchener, at the instance originally, of the Natal Government and with the approval of Milner and the Home Government, made his first and last attempt to end the war by a minatory proclamation.

Dated August 7, and issued while the great Free State drive was still in progress, it began by giving, in the form of a solemn legal preamble, an account of the military situation which must have seemed strangely unconvincing to the Boers, and which, it must be confessed, was replete with unconscious humour.

The fourth paragraph in particular, where the Boers were informed that they were not only few in number but had lost nearly all their guns and munitions of war and their proper organisation, and were therefore incapable of carrying on regular warfare, was a strange tissue of half-truths and perverted logic.

It might have been answered that the reduction of numbers had gone hand in hand with aggressive vigour; that the loss of guns, so far from being a weakness, had inspired more effective tactics; that .... the Boer organisation, if not "proper," was eminently suited to its purpose, and that whether the hostilities were called regular or irregular was not of much account.

The operative part of the proclamation demanded the surrender of all Boers before the date of September 15th under pain of certain stringent penalties.

These penalties were of two sorts. Government officials, commandants. Field cornets, and "leaders" were to be banished for ever from South Africa; burghers were to be rendered chargeable for the cost of their families' support in the British camps. The charge to be levied on their property and land. The immediate results of the proclamation were not encouraging.

Botha, Steyn and de Wet all sent defiant replies. [Received by Kitchener on August 27th, 1901]

Meanwhile the columns, worked by Kitchener at the highest possible pressure, endeavoured to supply the missing arguments for submission. Times History Vol 5 page 321–322.

Government its terms seem to be fully warranted by the existing situation, and calculated to have a good effect in bringing a more rapid termination of hostilities."

Lord Kitchener's Reply

Lord Kitchener replied on August 2nd: "Ministers, Cape Town, entirely concur in proclamation. Have not yet heard from Natal." On the following day Lord Kitchener telegraphed:— "The Government of Natal express approval of the proclamation." The correspondence further shows that on July 24th, Sir H. E. McCallum, Governor of Natal, telegraphed to Mr. Chamberlain as follows: "I sent you by last mail long minute submitted by Ministers, containing the following, suggestions:—(Begins),

Protracted continuance of hostilities, vitally affecting interests of Natal, is viewed with grave concern. Raids into the Colony frequently render it impossible for loyalists to return to farms and avocations.

Feeling of unrest. among natives created by raids.

Revenue suffering, trade paralyzed, railways monopolised by military, towns overcrowded with refugees, and persons awaiting return to Transvaal; stock being infected with disease due to introduction of captured stock from the new Colonies; famine prices prevailing; Colony still subject to censorship and martial law. Under these circumstances Ministers advocate sterner measures to crush present guerilla warfare. They point out that Boers still fighting lave little to lose, that their women and children are protected and well treated, and that their farms are safe from confiscation.

Therefore, Boers free from anxiety and encouraged to continue in the field; growing accustomed to life of pillaging and looting, and communicate frequently. with refugee camps, which thus are source of danger.

Those who are not rebels know that if captured they will be treated as prisoners of war and released at conclusion of hostilities.

Ministers believe that excellent effect would be produced if it were made generally known that if burghers now in field do not surrender before given date say within one month, cost of maintenance of all women will be chargeable against immovable property of burghers in field; also that Boer generals and leaders in field should be informed that unless they and their commandos surrender by date specified, they will be banished from South Africa for life when captured.

In Making these observations and recommendations Ministers disclaim intention of appearing to reflect on military operations, of which they realise the immense difficulties. They will continue to render Imperial Government every assistance to secure settlement and pacification"

Kitchener's Despatch of October 8th 1901

From Lord Kitchener to the Secretary of State for War, London, S.W. [From] Pretoria 8th October, 1901.[265]

Sir, In continuation of my despatch, dated 8th September, 1901, I have the honour to report that during the past month there has been a general recrudescence[266] of activity on the part of the enemy, which has made itself felt more particularly in the extreme south-east portions of the Transvaal.

The immediate cause of this was, undoubtedly, the necessity felt by the Boer leaders to tide over the 15th September, the date fixed by the Proclamation of 7th August [1901] as the limit of time within which by voluntarily surrendering, the leaders might avoid certain penalties threatened by that Proclamation. In this they have succeeded. There has been no general surrender, but the device to which the Commandant-General resorted for turning the thoughts of his burghers in another direction has probably cost him and his cause more heavily than a simple pursuance of the usual evasive tactics would have entailed.

[265] Kitchener Despatch October 8th 1901 p. 587 https://babel.hathitrust.org/cgi/pt?id=hvd.32044089026009&seq=587&view=1up

[266] From Late Latin recrudescentia, ("to become raw again");("to grow harsh or violent; to become worse") (from crūdus ("bleeding, bloody, raw") The word is cognate with French recrudescence. https://en.wiktionary.org/wiki/recrudescence

De Wet's Reply

I answered Lord Kitchener very carefully in the following words:—

"Excellency,—

"I acknowledge the receipt of your Excellency's missive in which was enclosed your Proclamation, dated the 7th August, 1901. I and my officers assure your Excellency that we fight with one aim only—our independence, which we never can or will sacrifice!"

It would have been childish to fear that letter and that Proclamation. From the short answer which I sent to Lord Kitchener, the reader will clearly see the opinion that I and my officers held concerning it: "Bangmaak is nog niet doodmaak,"[267] as our proverb says.

It was curious to see how this Proclamation was taken by the burghers. It had no effect whatsoever. I heard many burghers say that it would now be seen whether the officers had the cause of their country really at heart or not, and whether they were themselves to surrender and lay down their arms before the 15th of September. I must here declare that I know of no single case where an officer in consequence of this proclamation surrendered; on the contrary, when the day fixed by Lord Kitchener for the surrender had passed, the burghers had more reason to trust in their officers than before; and I can assure my readers that if at the beginning of the war we had had officers of the same kind as we had towards the end of the strife, it would have been easier to have maintained discipline.

September the 15th was thus fixed upon by Lord Kitchener as the last day on which we should have an opportunity of surrendering. The President and Commander-in-Chief of the Transvaal and the Orange Free State returned answer that they would still continue the war, and subsequent events put a seal to their answer.

Three battles were fought—one by General Brand at Blakfontein, another by General De la Rey in the west of the Transvaal, and yet another by General Botha at Itala, all in the month of September.[268]

Steyn's Reply

President Steyn sent Lord Kitchener a long letter, in which he showed most clearly what the causes of the war had been, and what was the condition of matters at that time. The letter was as follows:—

In the Veldt, August 15th, 1901.

To His Excellency, Lord Kitchener, etc.

Excellency,—

I have the honour to acknowledge the receipt of your Excellency's letter, dated Aug. 7th, 1901, enclosing your Excellency's Proclamation of the same date.

The conciliatory tone of your Excellency's letter encourages me to speak freely, and to answer it at some length. I have noticed that not only your Excellency in your letter asserts, but that also responsible statesmen in your country assert, that the declaration of war from the South African Republic, and the inroad on the British territory, had been the cause of the war.

I hardly believe it necessary to remind your Excellency that, in 1895, when the South African Republic was unarmed and peaceful, and had no thought but that their neighbours were civilized nations, an unexpected attack was made on them from the British territory. I do not consider it necessary to point out to your Excellency that the mad enterprise—for surely the instigators of it could not have been sane—miscarried, and the whole body of invaders fell into the hands of the South African Republic.

[267] Nobody dies of fright.

[268] 8/2/1901 More murders by Boers officially announced. One of the murdered men was an Imperial Yeoman.

8/8/1901 Commandant de Villiers and two Field Cornets surrendered at Warmbaths.

8/10/1901 Kitchener by proclamation called upon the Boer leaders to surrender on or before the 15th of September.

8/13/1901 Kitchener reported the largest return of Boer losses yet sustained in a week. More than 800 prisoners, 700 waggons, and 33,000 cattle.

The South African Government, trusting in the integrity of the English nation, handed over to His Majesty's Government all the persons whom they had taken prisoner, notwithstanding that, in conformity with international law, these persons had merited death.

I also do not consider it necessary to remind your Excellency that after an honest judge had condemned the leaders of this expedition to imprisonment, the most prominent of them were not compelled to serve the whole of their time, but, previous to its termination, were liberated for various most insufficient reasons.

Neither need I remind your Excellency that when a Parliamentary Commission was nominated, to investigate the causes and reasons of the said expedition, this Commission, instead of investigating the matter, would not allow the proofs to come to light, and that, when the Commission, notwithstanding the high influence at work during its sitting, had found the chief conspirator, Mr. Rhodes, guilty, and had reported him as such to Parliament, Mr. Chamberlain, who was one of the members of the Commission, contradicted his own report[269] by defending Mr. Rhodes.

Your Excellency will have to acknowledge that the South African Republic as well as the civilized world was perfectly justified in coming to the conclusion that the Jameson expedition, which we first believed to have been undertaken by irresponsible persons, and without the cognizance of His Majesty's Government, was well known, if not to all, yet still to some members of His Majesty's Government.

I need not remind your Excellency that since that time, not only has no reasonable indemnity been paid to the South African Republic, as was at that time promised, but also that the Republic has been harassed with despatches and threats concerning its internal Government.

I also need not tell your Excellency that outside influence was used in order that memorials to His Majesty's Government might be drawn up concerning alleged grievances, so that His Majesty's Government might have the desired opportunity of interfering with the inner policy of the South African Republic.

As I have said, I do not think it necessary to remind your Excellency of the above-mentioned facts, because I am of opinion that they are well known to you. I, however, should like your Excellency to be good enough to pay attention to the following facts:—

When, at the time of the circulation of the last-mentioned Memorial, I could see that a certain party was working hard to involve the British Government in a war with the South African Republic,[270] I stepped into the breach, and endeavoured, by bringing the par ties together, and by using my influence with the South African Republic, to induce the latter to give in to the demands of His Majesty's Government in order to maintain the peace.[271]

I succeeded in getting the Transvaal to yield, not because I was of the opinion that the English Government had any right to make such demands, but only in order to prevent bloodshed. When the British Government was still not satisfied, then the South African Government made concession after concession to the ever-increasing demands made upon them, until at last there came a request that the law on franchise should be laid before a Commission.

On the behest of the British Agent in Pretoria,[272] the South African Republic made a proposal granting far more than was demanded by the High Commissioner. As this proposal was not accepted by His Majesty's Government, who made yet further demands, the South African Republic withdrew their proposal, and declared themselves willing to accept England's proposal to lay the law before the Commission.

The British Government then closed all correspondence, and wrote to the South African Republic saying that they would make their demands later on.

[269] The report of the Commission of which he was a member.

[270] PF: Alfred Milner.

[271] See: Bloemfontein Conference. The Bloemfontein Conference was a meeting that took place at the railway station of Bloemfontein, capital of the Orange Free State from 31 May until 5 June 1899. The main issue dealt with the status of British migrant workers called "Uitlanders," who mined the gold fields in Transvaal.
The conference was initiated by Orange Free State president Martinus Theunis Steyn, in order to settle differences between Transvaal President Paul Kruger and British High Commissioner Alfred Milner. It was considered a last effort at reconciliation to prevent war between the two factions. https://en.wikipedia.org/wiki/Bloemfontein_Conference

[272] Connenghiam Greene.

In other words, the British Government then gave to the South African Republic an ultimatum, and it was clear that they were only prevented from commencing the war at once by the fact that they had not then landed sufficient troops in the country.

The Orange Free State Government then again came to the rescue, in order to attempt at the last moment to avoid the war, and cabled through the High Commissioner direct to the British Government, asking for information as to the nature of the demands which were to be made upon the South African Republic; which cable, to my sorrow, was never sent in its entirety. [See Appendices]

The only answer to my cable was the continual arrival of transports of troops from all quarters of the globe, which were massed, not only on the frontier of the South African Republic, but also on the frontiers of the still friendly Orange Free State. Then, when the South African Republic saw that England had no intention of repairing the alleged grievances, but had only brought them up as an excuse for depriving the Republic of its independence, they requested that the troops might be taken from their frontiers, and that all disputes might be settled by arbitration.[273]

This happened about three weeks after the British Government had issued their ultimatum, and about one month after the Orange Free State Government had received a wire asking them to remain neutral, thus clearly giving them to understand that the British Government intended to make war on the South African Government. This telegram was sent to the Orange Free State because they knew that the latter had made a defensive alliance with the South African Republic since the year 1899.

Then the South African Republic decided that they must defend their frontiers against the enemy who threatened their borders, and I was obliged to take a most painful step, namely, that of severing the bonds of friendship that existed between us and the British Government, and, true to our alliance with the Transvaal, to help the sister Republic. That we were perfectly correct in our surmise that the British Government had firmly decided to wipe out the two Republics has been clearly proved since the breaking out of the war. It was not only made evident from the documents that fell into our hands, although there it was easy to gather that since 1896, that is from Jameson's raid, the British Government was firmly determined to make an inroad into the two Republics: only lately it has been acknowledged by Lord Lansdowne[274] that he in June, 1899, had already discussed with Lord Wolseley (then Commander-in-Chief of His Majesty's troops)[275], the best time at which to make an attack on the two Republics.

Your Excellency will thus see that it was not we who drew the sword, but that we only put it away from our throats. We have only acted in self-defence—one of the holiest rights of man—in order to assert our right to exist. And therefore I think, with all respect, that we have a right to trust in a just God.

I again observe that your Excellency reverts to the impossibility of intervention by any foreign power, and that your Excellency interprets our resistance as only based on the hope of such intervention.

With your Excellency's permission, I should like to clear up our position with regard to intervention. It is this: We hope, and still are hoping, that the moral feeling of the civilized world would protest against the crime which England is now permitting in South Africa, namely, that of endeavouring to exterminate a young nation, but we were still firmly determined that, should our hopes not be realized, we would exert our utmost strength to defend ourselves, and this decision, based on a firm trust in a merciful God, is still unshaken in us.

I further notice that your Excellency thinks that our fight is hopeless. I do not know on what grounds this assumption is based. Let us for a moment compare our mutual situations of to-day with those of a year ago, after the surrender of General Prinsloo. Then, the Cape Colony was altogether quiet, and free from our commandos.

[273] See the simultaneous, 1899, British Guiana [Guyana] border dispute with Venezuela. "The United Kingdom and Venezuela went into arbitration with mediation from the United States, which resulted in the Paris Arbitral Award in 1899 and ruled largely in favour of Britain." https://en.wikipedia.org/wiki/Guyana%E2%80%93Venezuela_territorial_dispute And, https://en.wikipedia.org/wiki/British_Guiana

[274] Henry Charles Keith Petty-Fitzmaurice, 5th Marquess of Lansdowne, (14 January 1845–3 June 1927), was a British statesman who served successively as Governor General of Canada, Viceroy of India, Secretary of State for War and Secretary of State for Foreign Affairs. https://en.wikipedia.org/wiki/Henry_Petty-Fitzmaurice,_5th_Marquess_of_Lansdowne

[275] Field Marshal Garnet Joseph Wolseley, 1st Viscount Wolseley (4 June 1833–25 March 1913) was an Anglo-Irish officer in the British Army. He became one of the most influential British generals after a series of victories in Canada, West Africa and Egypt, followed by a central role in modernizing the British Army in promoting efficiency. https://en.wikipedia.org/wiki/Garnet_Wolseley,_1st_Viscount_Wolseley

The Orange Free State was almost entirely in your hands, not only as regards the principal townships, railway lines and villages, but also the whole country, except where Commandant Hasebroek was, with his commando. And in the South African Republic the situation was very similar. That country was also mainly held by you, except in the parts which General De la Rey and General Botha occupied with their commandos, far up in the Boschveldt.

How do matters stand now?

The Cape Colony is, so to speak, overrun by our commandos, and they are really in temporary possession of the greater part of Cape Colony. They go about there as they choose, and many of our nationality and others also are continuing to join us there, and uniting forces with us against the cruel injustice that is being done to the Republics.

In the Orange Free State I willingly acknowledge that your Excellency is in possession of the Capital, the railways, and some other towns not on the railways, but that is all that your Excellency has got. The whole of the Orange Free State, except the parts which I have just mentioned, is in our possession. In most of the principal towns there are landdrosts[276] appointed by us; thus in this State the keeping of order and the administration of justice are managed by us, and not by your Excellency. In the Transvaal it is just the same. There also justice and order are managed by magistrates appointed by our Government.

May I be permitted to say that your Excellency's jurisdiction is limited by the range of your Excellency's guns. If your Excellency will look on the matter from a military point of view then it must be acknowledged that notwithstanding the enormous forces that are brought against us in the field, our cause, in the past year, has made wonderful progress. Therefore we need be in no way discouraged, and, if your Proclamation is based on the assumption that we are so, then it has now even less justification than it had a year ago. I am sorry that anything I say should appear boastful, but the assertions in your Excellency's Proclamation compel me to speak in this manner.

With regard to the 35,000 men which your Excellency says are in your hands, I cannot speak as to the numbers, but this much I will say, I am not referring to those men who were led astray by the Proclamation of your Excellency's predecessor, and so failed in their duty to their Government; nor to those—thank God they are but few—who from treachery or other cause have gone over to the enemy; but of the remainder who have been taken, not too honestly, as prisoners of war, and are still kept as such.

Of these I will say that they are either old men and feeble, or young boys not yet of age, who were carried off by force from their farms by your Excellency's troops, and shut up against their will in your Excellency's camps. To say of these therefore, that they are "dwelling peacefully with you," is an assertion which can hardly be taken seriously. I am able to say with perfect truth, that except the prisoners, and the few who have gone over to the enemy, the overpowering majority of the fighting burghers are still under arms. As regards those who have gone over from us to the enemy—a rare occurrence now—I can only say that our experience is not unique, for history shows that in all wars for freedom, as in America and elsewhere, there were such: and we shall try to get on without them.

As regards the 74,000 women and children who, as your Excellency alleges, are maintained in the camps, it appears to me that your Excellency must be unaware of the cruel manner in which these defenceless ones were dragged away from their dwellings by your Excellency's troops, who first destroyed all the goods and property of their wretched captives. Yes, to such a pass had it come, that whenever your men were seen approaching, the poor sacrifices of the war, in all weathers, by day and by night, would flee from their dwellings in order that they might not be taken.

Does your Excellency realize that your troops have not been ashamed to fire (in the full knowledge of what they were doing) with guns and small arms on our helpless ones when they, to avoid capture, had taken flight, either alone or with their waggons, and thus many women and children have been killed and wounded. I will give you an instance. Not long ago, on the 6th of June, at Graspan, near Reitz, a camp of women, falsely reported as a convoy to your Excellency, was taken by your troops. This was rescued again by us, whilst your troops took shelter behind our women, and when your reinforcement came up, they opened fire with guns and small arms on that camp, notwithstanding the fact that they knew it contained women only.

[276] Resident Magistrates.

I can quote hundreds of cases of this kind, but I do not think it necessary, because if your Excellency will take the trouble to ask any soldier who respects the truth, he will be compelled to confirm my assertion. To say that the women are in your camps of their own free will is not in accordance with the facts, and for any one to assert that they are brought to the camps because the Boers are unwilling to provide for the maintenance of their families as it is said that His Excellency the Minister for War has asserted in Parliament, is to make himself guilty of calumny, that will do more harm to the calumniator than to us, and is a statement which I am sure can never meet with your Excellency's approval.

Now, as regards the Proclamation itself, I can give your Excellency the assurance as far as I am myself concerned, that it will make no difference to my fulfilling my duty faithfully to the end, for I shall be guided by my conscience and not by the enemy. Our country is ruined; our hearths and homes are wrecked; our cattle are looted, or killed by the thousand; our women and children are made prisoners, insulted, and carried away by the troops and armed Kaffirs; and many hundreds have already given their lives for the freedom of their fatherland. Can we now—when it is merely a question of banishment—shrink from our duty? Can we become faithless to the hundreds of killed and prisoners, who, trusting in our firmness, offered their lives and freedom for the fatherland? Or can we lose faith in a just God, who has so wonderfully upheld us till now? I am convinced that should we do so, we should be despised not only by your Excellency and all honest men, but also by ourselves.

I will close by giving your Excellency the assurance that no one is more anxious than I to see peace restored, and I am therefore ready to meet your Excellency at any time in order to discuss the terms on which this peace can be arranged; but in order that I may not mislead your Excellency, I have to say that no peace will be accepted by us which imperils the independence of the two Republics, or which does not take into consideration the interests of our Colonial brethren who have joined us.

If it is a crime to fight in one's self-defence, and if such a crime is to be punished, then I am of opinion that His Majesty's Government should be satisfied with the annihilation of the country, the misery of women and children and the general desolation which this war has already caused.

It is in your Excellency's power more than in that of any one else, to put a stop to this, and by doing so, to restore this unfortunate part of the world to its former happiness. We ask no magnanimity, we only demand justice. I enclose a translation of my letter in order to avoid any misinterpretation of it by your Excellency, as this happened not long ago when a letter which I had written to the Government of the South African Republic, and which at Reitz fell into your hands, was published in such a way that it was nearly unrecognizable, as not only was it wrongly interpreted in some places, but sentences were inserted which had never been written, and other parts were left out altogether, so that an entirely wrong meaning was given to the letter.

I have the honour, etc.,

M.T. Steyn,

State-President of the Orange Free State.[277]

Quote from Deneys Reitz in Commando and Times History note.

> There was another surprising feature in the [captured] mail-bag in the shape of a Proclamation by Lord Kitchener, wherein every burgher under arms after the 15th of September [1901] was sentenced to perpetual banishment from South Africa. This was news to us, and seeing that it was the 13th of that month by now, we were left with a bare two days in which to comply. This announcement was received with equal derision, when made known to the commando, and from what I have heard since, it had equally little effect up north in the Republics where the 'paper bomb', as it was called, got treated with the scorn deserved.[278]

[277] 8/27/1901 Kitchener received letters from Steyn and De Wet protesting against his proclamation.

[278] Original Footnote: The Times History says of this Proclamation, Vol. V, 321:

Kitchener made his first and last attempt to end the war by a minatory proclamation.

It began by giving in a solemn legal preamble an account of the military situation, which must have seemed strangely unconvincing to the Boers, and which, it must be confessed, was replete with unconscious humour. The fourth paragraph in particular where the Boers were informed that they were incapable of carrying on regular warfare was a strange tissue of perverted logic.... The operative art of the Proclamation demanded the surrender of all Boers before September 15th under pain of stringent penalties. The results were not encouraging: Botha, Steyn and de Wet sent defiant replies, and among the burghers at large there was a sullen silence.

31. Blockhouses and Night Attacks

While the great events recorded at the end of my last chapter were in progress, I paid a visit to the Harrismith burghers, who were under the command of Commandant **Jan Jacobsz**, and also to some of the Bethlehem men. On my return I learnt that the enemy were occupied in building a line of blockhouses from Heilbron to Frankfort.

It has always seemed to me a most unaccountable circumstance that England—the all-powerful—could not catch the Boers without the aid of these blockhouses. There were so many other ways in which the thing might have been done, and better done; and the following incident, which occurred during the war, serves to show that this policy of the blockhouse might equally well have been called the policy of the blockhead.

On the 27th of February, 1902, the English made one of their biggest "catches" in the Free State. They had made a great "kraal"—what they themselves call a "drive"—and stood, "hand in hand," one might almost say, in a ring around us, coming from Heilbron, Frankfort, Bethlehem, and Harrismith, and stretching, on the Transvaal side, from Vrede to the Drakensberg.

Narrower and narrower did the circle become, hemming us in more closely at every moment. The result was that they "bagged" an enormous number of men and cattle, without a solitary burgher (or, for the matter of that, a solitary ox) having been captured by means of their famous blockhouse system.

The English have been constantly boasting in the newspapers about the advantages of their blockhouses, but they have never been able to give an instance of a capture effected by them. On the contrary, when during the last stages of the war it happened, as it often did, that they drove some of our men against one or other of the great blockhouse lines which then intersected the country, and it became necessary for us to fight our way through, we generally succeeded in doing so. And that, with fewer casualties than when, as in the instance I have just given, they concentrated their forces, and formed a circle around us.[279]

The English then were busy when I returned from the south in building a blockhouse line from Heilbron to Frankfort. They accomplished this speedily, and then proceeded to the construction of other similar lines, not being contented until they had "pegged out" the country as follows:—

On the Natal frontier there was a line from Vrede to Bothaspas, continued westward by a series of forts to Harrismith, whence the line went on, still westward, to Bethlehem, and thence down to the Basutoland border at Fouriesburg.

Kroonstad was made, so to speak, the "axle," whence a series of "spokes" proceeded; one to the north-east, to Vrede; a second to the north-west, through Driekopjes Diamond Mine, to Winkledrift, and thence down the Rhenoster River to its confluence with the Vaal; a third, to the south-east, to Lindley; and a fourth, to the south-west, along the railway line, to the frontier of Cape Colony.

In the western districts there was a line along the left bank of the Valsch River to the point where it joins the Vaal, and another (also terminating at the Vaal River) starting from Zand River railway bridge, and running parallel to the Zand River. There was also a line from Boshof, across the Cape Colony frontier, to Kimberley.

Last, but not least, came the "White Elephant" with which the reader is already acquainted—the line from Bloemfontein to Ladybrand, through Thaba'Nchu.

All these lines were in the Free State. I make no mention here of the thousands of miles of similar blockhouse lines, which made a sort of spider's web of the South African Republic.

The blockhouses themselves were sometimes round, sometimes angular, erections. The roofs were always of iron. The walls were pierced with loop-holes four feet from the ground, and from four to six feet from one another. Sometimes stone was used in the construction of these walls, at other times iron. In the latter case the wall is double, the space of from six to nine inches between the inner and the outer wall being filled with earth.

These buildings stood at a distance of from a hundred to a thousand paces from one another; everything depended upon the lie of the ground, and the means at the enemy's disposal; a greater distance than a thousand paces was exceptional. They were always so placed that each of them could be seen by its neighbours on both sides, the line which they followed being a zigzag.

[279] 2/27/1902 Operations for driving Boers against Harrismith-Van Reenan's blockhouse line. Manie Botha killed ; 600 Boers killed, wounded, or prisoners. Defence by New Zealanders under Major Bauchop and New South Wales Mounted Infantry under Colonel Cox.

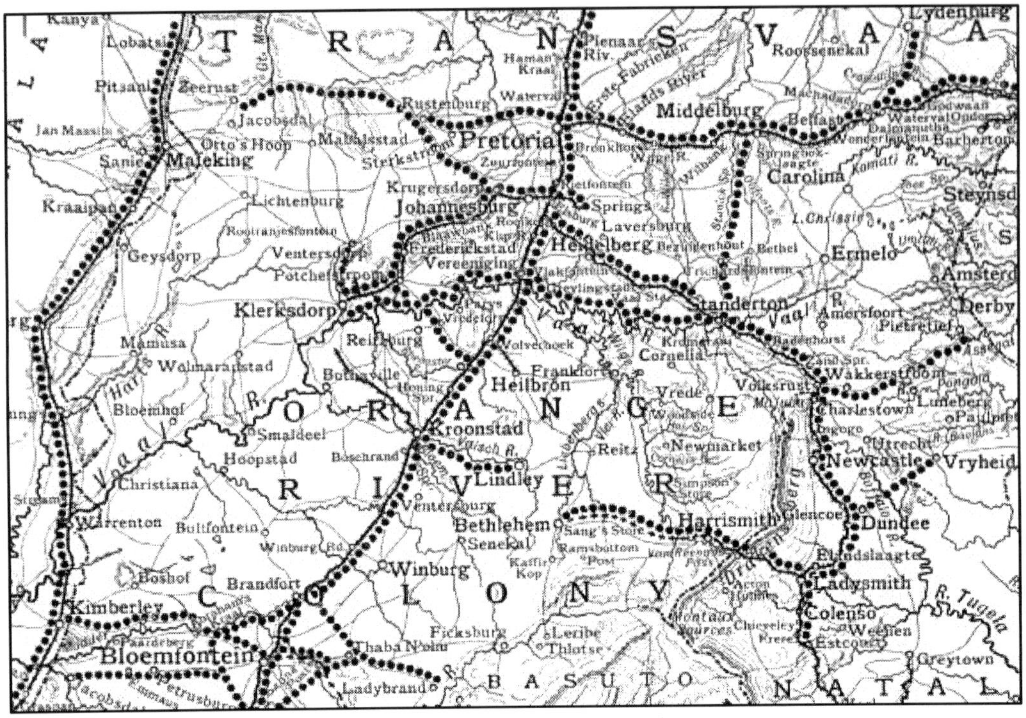

The portion of the Blockhouse System being discussed.[280]

Between the blockhouses were fences, made with five strands of barbed wire. Parallel with these was a trench, three feet deep and four to five feet across at the top, but narrower at the bottom. Where the material could be procured, there was also a stone wall, to serve as an additional obstacle. Sometimes there were two lines of fences, the upper one—erected on the top of the earth thrown up from the trench—consisting of three or four strands only.

There was thus a regular network of wires in the vicinity of the blockhouses—the English seemed to think that a Boer might be netted like a fish. If a wild horse had been trapped there, I should like to have been there to see, but I should not have liked to have been the wild horse.

The building of these blockhouses cost many thousands of pounds, and still greater were the expenses incurred in providing the soldiers in them with food, which had to be fetched up by special convoys. And it was all money thrown away! and worse than thrown away! for when I come to describe how I broke through these blockhouse lines (see next page), the reader will see that this wonderful scheme of the English prolonged the war for at least three months.

Blockhouse showing wire fence.[274]

Rock Blockhouse.[275]

[280]The distance between blockhouses was calculated to be able to extent the reach as far as possibly but at the same time to prevent the Boers from passing unnoticed between them. The average distance apart was about 1,000 yards (910 m). Line of sight was vital and this required a greater density of blockhouses in some areas. Anglo Boer War - Blockhouses

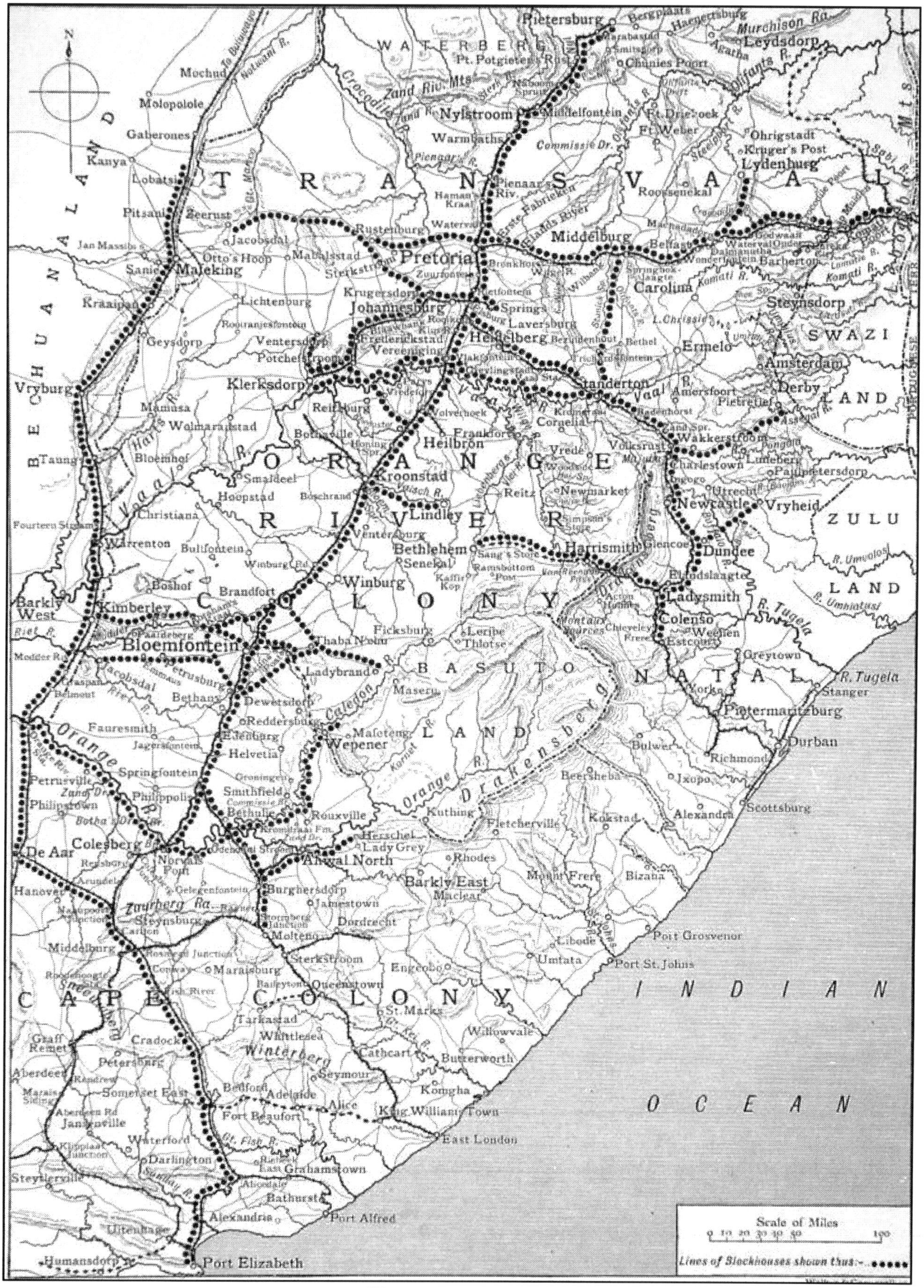

The Nearly Completed Blockhouse System.[276]

Type of Blockhouse.[277] Men guarding a [British] outpost "These men appear to be in charge of one of the ~8,000 blockhouses that were established to the countryside and rail lines. Note the wire fencing behind them and the sandbags or some form of protection from projectiles piled around the building. I believe they are armed with Martini-Henry Mark III single shot lever action rifles. One can identify them by the small circular appendage behind the trigger guard."[278]

Let us turn now to another, and a more successful device of the enemy.

From the first weeks of the winter, 1901—the reader must remember that our winter commences in May—the English began to make night attacks upon us; at last they had found out a way of inflicting severe losses upon us, and these night attacks grew more and more frequent during the last period of the war. But they would never have thought of them at all, if they had not been instructed in them by the National Scouts—our own flesh and blood!

These tactics were not always successful. It sometimes happened that the English got "cornered"; sometimes they had to "right about turn" and run for their lives. The latter was the case at Witkopjes, five miles to the south of Heilbron, and again, near Makenwaansstad. But on only too many occasions they managed to surprise troops of burghers on their camping places, and, having captured those who could not run away, they left the dead and wounded on the ground.

We soon discovered that these night attacks were the most difficult of the enemy's tactics with which we had to deal.

Sometimes the burghers, surprised by a sudden visit from the English at such an unconventional hour, found it necessary to run away at once as fast as their legs would carry them, so that they often arrived at the nearest camp without their hats. Indeed a series of these attacks produced such a panic among our men that I have known a Boer lose not only his hat, but also his head.

I come now, in the regular course of my narrative, to an engagement between my burghers and an English force which had marched from Bethlehem to Reitz, a distance of thirty miles. This force was guided by a son of one of the Free State Members of Parliament, and, marching all night, reached Reitz just as the day began to dawn. This was a smart piece of business; and though the guide to whom its success was due was my enemy, I fully appreciated the skill which he then displayed.

The English captured ten or twelve burghers at Reitz, whither they had perhaps gone in search of the President.

I was ten miles to the west, on the farm of Blijdschap, and did not receive reports of what had happened until towards noon.

What was I to do? I could not call up men from Heilbron, Bethlehem, Vrede, or Harrismith: it would have been at least twenty-four hours before they could have arrived. All I could do was to summon Veldtcornet Vlok with some of the Parijs commandos and Veldtcornet Louwrens, and Matthijs De Beer, and the men. With these and my staff we would not number more than sixty or seventy all told.

I at once gave orders to these veldtcornets to meet me at a certain place, and they were there by the appointed hour.

My intention was to deliver a flank attack upon the English while they retreated during the night; for, as they only numbered five hundred men, I felt sure that they would not care to remain thirty miles away from their column, but would fall back upon Bethlehem.

In the afternoon I marched to within a short distance of Reitz, in order to discover the enemy's plans; then, immediately after sunset, I sent a few burghers quite close to the town, with orders to meet me again at a certain point about two thousand paces to the south, and to inform me whither the enemy were going to march. The scouts returned at ten o'clock that night, and reported that the enemy was on the march towards **Harrismith.** In order to reach this town they would have to start by the Bethlehem road, from which the Harrismith road forks, at about eight thousand paces from the town.

Our horses stood ready up-saddled; I had only to give the order to mount.

I meant to cross the Bethlehem road and go to a deep hollow which I knew of near the Harrismith road; then, when the English appeared against the horizon, we would fire at them.

But my scouts had blundered. The English were not going to Harrismith after all. For as we came to the Bethlehem road, we nearly stumbled over them. They were riding quietly along only a short distance from us. As we were galloping they knew of our proximity before we were aware of theirs, and when we were less than two hundred paces from them they opened fire.

"Charge, burghers!"

They all heard me, but they did not all obey. About fifty of the most valiant of them galloped straight at the enemy. The rest fled.

After a short but fierce engagement we were forced to retire, as six of our men had been hit. Fortunately, their wounds were but slight, the most severe being that of my son Isaac, who had been shot through the leg below the knee.

We rode away a short distance, and saw looming through the darkness a company of horsemen approaching us from Reitz. I thought at first that they were some of my own burghers—the ones who had taken to their heels—but it turned out to be General Wessel Wessels,[281] who was nearer than I knew with his staff, in all some twenty men. I, however, could muster seventy, and we decided to cut off the retreat of the enemy.

But they had, in the meantime, been riding on so fast that we did not reach them until it had grown quite light. An engagement, short and fierce as the last, ensued, but as the enemy was from six to seven times as strong as we were, and had a gun and a Maxim-Nordenfeldt with them, we could not stand against them, and had to let them go on their road.

We were fortunate in suffering no loss there, and while the English marched on to Bethlehem we rode off in the opposite direction.

We had now a short period of repose. The English were so busy building blockhouses that they had no time to fight us. Our poor horses were in a miserable condition, for so little rain had fallen that the grass was very dry and sapless. But at least we could now give them the rest which they sorely needed.

[281] PF: Unable to identify this individual.

32. My Commando of Seven Hundred Men

Towards the end of September [1901] Commandant F.E. Mentz had an engagement with Colonel Byng's column near Heilbron. A portion of this officer's force had held a ridge where there were some Kaffir kraals for cover; and Commandant Mentz had with fifty burghers stormed this ridge, shooting down from thirty to forty of the enemy, and taking twenty-five prisoners. We lost two killed and three wounded. The Frankfort burghers under Commandant Ross had also not been idle, for they had attacked a division of Colonel Rimington's troops with the result that sixteen killed and wounded fell into their hands—among these were seven of the National Scouts.

Thus fighting was taking place all over the country. I do not give any report of the various engagements, as I was not present at them, and, as I have already said, I only wish to record my own experiences. But it will be easily seen, even from the scanty information I can give of these skirmishes, that our small commandos had a splendid record of success.

It is my intention to ask all my Vice-Commanders-in-Chief to narrate their experiences. And when the whole story is told I am convinced that the world will be astonished at what we were able to accomplish.[282]

[282] 9/2/1901 Another case of train-wrecking on the Pretoria-Pietersburg railway.
9/4/1901 Smuts invades the Cape.
9/7/1901 J. C. Lotter and his entire commando captured in Cape Colony. Battle of Groenkloof https://en.wikipedia.org/wiki/Battle_of_Groenkloof
9/17/1901 Battle of Elands River. (Modderfontein) https://en.wikipedia.org/wiki/Battle_of_Elands_River_(1901)
9/17/1901 General Louis Botha overran the Mounted Infantry under the command of Lieutenant-Colonel H de la P Gough. Battle of Blood River Poort: https://en.wikipedia.org/wiki/Battle_of_Blood_River_Poort And: https://af.wikipedia.org/wiki/Slag_van_Bloedrivierpoort
9/21/1901 Reverse at Vlakfontein, near Sanna's Post. Two guns lost. (Afterwards recovered.)
9/23/1901 The camp of Lovat's Scouts rushed by Kruitzinger near Herschell. Koch's commando captured near Edenburg. The Carolina commando captured by Colonel Benson.
9/26/1901 Attacks on Fort Itala and Fort Prospect. Boers repulsed with very heavy losses at both places. The attempt of Botha and De Wet to invade Natal foiled.
9/29/1901 Proclamation issued in Pretoria providing for the sale of the properties of Boers still in the field, in accordance with Kitchener's proclamation.
9/30/1901 Great attack by Delarey and Kemp on Colonel Kekewich's camp near Magato Nek, in the Magaliesberg. Boers repulsed. Severe losses on both sides. The Scottish Horse distinguished themselves and sustained severe loss.
10/6/1901 General Walter Kitchener and General Bruce-Hamilton engaged Botha's forces in the south-east of the Transvaal. Botha escaped to the north.
10/9/1901 Martial law extended to the whole of Cape Colony.
10/11/1901 Commandant Lotter sentenced to death. Death sentence on five members of his commando was commuted to penal servitude for life.
10/13/1901 Lieut.-Colonel J. Byng attacked laager at Jackfontein and captured eighteen prisoners.
10/15/1901 Major Damant took prisoner Adjutant Theron. Colonel de Lisle surprised laager at Wilge River and captured fifteen prisoners.
10/16/1901 Colonel Rawlinson returned to Standerton with twenty prisoners.
10/21/1901 Colonel Lukin surprised Vander Venter's laager near New Bethesda.
10/22/1901 Colonel Benson captured laager at Klippoortje.
10/23/1901 Gallant attack on laager in Pongola Bosch.
10/24/1901 Colonel von Donop's brilliant defeat of 1000 Boers at Kleenfontein.
10/25/1901 Botha's farm surrounded at Schimmelhoek. His papers captured.
10/26/1901 Colonel Benson repulsed attack on his rearguard on the Steenkool Spruit.
10/27/1901 Colonel Williams' force occupied the Witnek Pass and routed a strong body of Boers from the position.
10/30/1901 Attack on Colonel Benson's force at Bakenlaagte. Colonel Benson and Colonel Guinness killed. Colonel Kekewich captured a laager at Beestekraal. Battle of Bakenlaagte: https://en.wikipedia.org/wiki/Battle_of_Bakenlaagte
11/2/1901 Patrol under Captain Walker captured twenty-one prisoners near Wolvekop.
11/7/1901 Attack on Piquetberg repulsed by garrison under Major Wilson and Town Guard. General B. Hamilton commenced operations against Botha in the Eastern Transvaal.
11/8/1901 Major Wiggin (26th Mounted Infantry) surrounded laager near Mahamba. Fourteen prisoners secured.
11/9/1901 Line blown up at Myburg Siding by Fouche.
11/11/1901 Major Pack Beresford and detachment of South African Constabulary captured laager at Doornhock.
11/13/1901 Squadron Imperial Yeomanry detached from Hickie's force surprised and surrounded. Rescued by reinforcements.
11/14/1901 Rearguard of Colonel Byng's column attacked near Heilbron by 400 of the enemy under De Wet. Boers repulsed. British loss considerable.
11/16/1901 Further captures by Major Wiggin within Swaziland border.
11/20/1901 Engagement with Buys near Villiersdorp. Major Fisher killed. Buys captured by Colonel Rimington. Captain Elliot successfully engaged Boers in Grinqualand. Captain Elliot killed. Three officers wounded.

But however well these small commandos had fought, I myself believed that the time had now come to make a great stroke. With this object in view I gave orders that a number of the burghers should come to Blijdschap, in the district of Bethlehem, under the command of the following officers:—General Michal Prinsloo with Commandants Olivier, and Rautenbach of the Bethlehem Commando; Commandant David Van Coller, who was in command of the Heilbron burghers in the place of Commandant Steenekamp, who had resigned; Commandant Hermanus Botha of Vrede; Commandant Roen of Ladybrand; and Commandant Jan Cilliers of Kroonstad.

By the beginning of November I had a force of seven hundred burghers under me at Blijdschap.[283]

Although the spring was now far advanced, the veldt was in a very backward condition. I therefore ordered the various subdivisions of my commando to go and camp on the different farms in the neighbourhood. I spread the horses over a large area, as they would thus find better pasture and so the sooner recover their strength.

When November was drawing to a close I had an engagement with the English to the south of Lindley. I had with me at that time General Hattingh, General Wessel Wessels, and General Michal Prinsloo.

An English force had encamped two days previously on the farm of Jagersrust, which lies some ten miles to the south-east of Heilbron, and about the same distance from Blijdschap. I had wished to make an attack on them the night they arrived, but they were too near to Heilbron for me to venture on it.

The previous week three columns which came from Winburg and Kroonstad had been operating near the Liebenbergsvlei, and driving a large laager of women before them towards the north-east of the Liebenbergsvlei. But they had now left the laager alone and returned to Kroonstad. The women had arrived at Blijdschap at noon on November 28th on their way back to Lindley.

The morning following, two hours after sunrise, I received a report from General Hattingh, who with Commandant Cilliers and a hundred men was stationed close to Blijdschap. The General reported that the English from Jagersrust were hotly pursuing the women's laager. And it soon appeared that the women were being driven to the west of Blijdschap.

When General Hattingh heard that the English were hard by, he was some twenty minutes' ride from Blijdschap, but he mounted his horse at once and rode there as quickly as he could. On his arrival he immediately gave orders to up-saddle, and, having sent me a second report, he started in pursuit of the enemy.

As soon as I had received General Hattingh's reports, I followed him with General Wessels and a force of only a hundred men. I was at least five miles from General Hattingh, and the English were twelve miles ahead. General Michal Prinsloo was unfortunately a considerable distance away; and thus it was that I could not at once get together my whole force of six hundred burghers.

But General Michal Prinsloo had spent the time in attacking the English force on their left front. Shortly after he had engaged the enemy I came up behind them and delivered an attack on their right. But the veldt was very uneven and high hills and intervening hollows made any co-operation between us impossible, for one force could not tell where the other force was.

Meanwhile General Hattingh had attacked the enemy in the rear and thus compelled them to withdraw their vanguard, which was then not far from the women's laager and had nearly succeeded in capturing it. But now that the whole force of the enemy was opposed to General Hattingh, he was forced to give way and leave his positions. We lost two killed and three wounded. Among the dead was the valiant F.C. Klopper of Kroonstad.

When I, with General Wessels and Commandant Hermanus Botha hurried up, Commandant Hattingh was just on the point of retreating.

The English I saw numbered about a thousand mounted men and they had three guns with them. I determined to make a flank attack, and accordingly marched round to their right, at the same time sending orders to General

11/24/1901 General Dartnell, with Highland Light Infantry, engaged Boers near Harrismith. Captured twelve and killed two. Offer of Canadian Government to raise 600 more troops for service in South Africa accepted.

11/25/1901 General Dartnell's force surprised Boers near Bethlehem and took twelve prisoners.

11/26/1901 Basing engaged Joubert in Orange River Colony. Joubert wounded and captured. Major Pack Beresford attacked convoy near Paardeberg.

[283] A court-martial was held at this place, and several persons appeared before it. A certain De Lange was condemned to death for high treason.

Prinsloo to get in the rear, or if he preferred in front of the enemy, so that we might make a united attack upon them as they marched in the direction of Lindley.

It now began to rain and a little later a very heavy thunderstorm burst on our heads. This forced the English to halt on the farm of Victoriespruit.

The rain continued to fall in torrents and hindered General Prinsloo carrying out my orders.

And now the sun went down.

As our horses were quite exhausted by the hot pursuit after the English, and the burghers wet through to the skin, I decided to postpone the attack to the following day. I was also influenced in my decision by the consideration that as the English were so far from any point from which reinforcements could come, it was quite safe to let them alone until the morning. Nobody could have foreseen that they would escape that night.

We slept about five miles from them to the north-east, whilst General Prinsloo and his men were not very far away to the south-east.

That night we placed the ordinary outposts, but no "brandwachten."

When on the next morning I sent my scouts out to discover the movements of the enemy, what was my surprise when they reported that they had fled. They had gone, my scouts informed me, towards Heilbron, which was about eighteen miles off, and they had left behind them five laden waggons and one cart; and where they had crossed Karoospruit they had, very naturally, lightened their waggons, and flour, seed, oats, tarpaulins, and tents marked the point where they had crossed the spruit. The enemy were already so far ahead when I received this report that it was quite out of the question to catch them before they reached Heilbron; so all idea of pursuing them had to be abandoned.

So far as I was able to find out, this column was under the command of Colonel Rimington.[284]

As I was unable now to get in touch with the enemy, I set off with my commando to what was once the town of Lindley. Alas! it could not anymore be called a town. Every house was burnt down; not even the church and parsonage were spared.

We found the veldt in very good condition; the early spring rains and the downpours of the previous day had quite revived the grass. And so I decided to remain at Lindley as long as possible, to give our horses a chance of recovering their condition. It was impossible to provide them with forage, for the amount the English had left behind was entirely insufficient as a supply for the large number of horses we had with us.

For ten or twelve days we remained at Lindley,[285] and so the horses had a short breathing time, but not long enough to give the poor animals time fully to regain their strength. In addition to being overworked, some of our horses were suffering from a skin disease which we were quite unable to cure. This disease had never before been known in the Republics.

When I was at Lindley I sent Commandant Johannes Meijer, one of my staff, with forty men, to Cape Colony. With him went that brave soldier, Captain Willem Pretorius, of whom I have made mention previously. If Commandant Meijer had had sufficient time to collect a commando in the Colony, I am sure that he would have proved that the younger generation of Free-Staters, to whom he and Willem Pretorius belonged, possess qualities which were entirely unsuspected before the war began.

On the 8th of December three columns of the enemy appeared from Kroonstad.

[284] 11/27/1901 Attack on Colonel Rimington's rearguard by De Wet repulsed. Many prisoners taken. Separately, Imperial Light Horse under Colonel Mackenzie took twenty four prisoners, &c.

11/28/1901 Van Rensburg and thirteen burghers captured by Colonel Lowry Cole in Wepener district.

[285] 12/1/1901 General Elliot reached Kroonstad with 15 prisoners, 114 waggons, 89 carts, 2470 cattle, and 1280 horses.

12/3/1901 Colonel Colenbrander broke up Badenhorst's commando, and took fifteen prisoners and all the waggons.

12/4/1901 Laager surprised at Oshoek (twenty miles from Ermelo) by Spens' and Rawlinson's columns. Ninety three prisoners taken.

12/7/1901 Colonel C. Mackenzie, in night march towards Watervaal (Eastern Transvaal), took sixteen prisoners. Colonel Holland surprised Brand's laager and took six Boers.

12/11/1901 Badenhorst and twenty-two burghers secured by Colonels Colenbrander and Dawkins, near Zandriverspoort.

12/13/1901 Brilliant surprise of Boers by General B. Hamilton at Witkraus, Laager broken up. One of Benson's guns recovered.

12/16/1901 Haasbrock killed in encounter with Colonel Barker's men in the Doornberg. Capture of Kruitzinger by Colonel Dorans' and Charles Benlinck's colunms.

It had been my plan to remain at Lindley and wait my chance of dealing with **Colonel Baker**, for he had under him a certain National Scout, who constantly made raids from Winburg with a band of four or five hundred Kaffirs. A few months previously a division of Commandant Hasebroek's commando had been attacked at Doornberg by this man's Kaffirs, and four burghers had been murdered in a horrible manner.

More cases of this nature had taken place, and I only mention this one in passing. I am not in a position to give all the instances, but many of them were sworn to in affidavits, of which copies were sent to Lord Kitchener. The original affidavits fell into the hands of the English; but fresh ones shall be drawn up on my return to South Africa, so that I may be able to prove the statements I have made. The narration of these brutalities I prefer to leave to persons more conversant with the facts than myself. I have only alluded to the subject so as to make it clear why I like to keep my eye on Colonel Baker's column.

I must now continue my story where I left it.

I took up my position to the north-west of Lindley, in front of the columns which approached from Kroonstad. But after a few skirmishes with them, I returned to the east till darkness came on. When night had fallen I went round to the south, behind Kaffirskop, expecting to receive the news that Colonel Baker was coming up from Winburg, for he generally carried on his operations in conjunction with the forces at Kroonstad.

On the following day the enemy marched to Liebenbergsvlei, between Bethlehem and Reitz. Thence they took the road between Lindley and Reitz to Kroonstad.

Piet de Wet, of the National Scouts, was with these columns.

After we had remained two days at Kaffirskop, we crossed the Valsch River. The news then came that a column with a convoy was on the march from Harrismith to Bethlehem.

I felt that it was my duty to attack this column, but, although I advanced with all haste, I was not in time to catch the enemy before they reached Bethlehem. When I saw this, I decided to wait, at a distance of some fifteen miles to the north-east of Bethlehem, for I expected that the column would return to Harrismith.

The troops remained in Bethlehem till the morning of the 18th of December; they then marched out towards Harrismith.

I at once divided my commando into two parts, each consisting of two hundred and fifty men. One of these divisions I posted behind the eastern end of the Langberg, about forty miles from Bethlehem; the other on the banks of the Tijgerkloofspruit, at the point where the road to Harrismith crosses the stream.

I gave strict orders to both divisions that as soon as I opened fire on the English with the Maxim-Nordenfeldt, they were to charge down on them from both sides at the same time.

The enemy, I may mention, were about six or seven hundred men strong, and had two guns.

I myself, with the Maxim-Nordenfeldt, was now on a high round hill, on the eastern side of Tijgerkloof. I was very careful to be out of sight of the English, so that they might get quite close to the burghers before the gun disclosed my presence.

I succeeded in hiding my burghers so successfully that the English did not observe them until they were within about twelve hundred paces of my men in Tijgerkloof.

Some of the enemy's scouts rode on ahead, and when I judged that they must almost immediately see the burghers, I ordered Captain Muller, who was standing behind a rise, to come out of cover and open fire; then I jumped on my horse, and down the hill I went, at full gallop, to my burghers.

I had scarcely covered half the distance, when Captain Muller opened fire on the enemy.

As the sound fell on my ears, it seemed to me that nothing now could save them!

What was now my bitter disappointment when I saw that only one-third of my burghers were charging. The others were keeping under cover, and do what I would I could not drive them out.

Everything went wrong.

When the burghers who were charging the English discovered that the greater part of their comrades had remained, they turned round and retreated. But before this had happened they had attacked the English at four different points.

It had been a short but a very hot engagement.

There was no possibility of inducing my men to charge, and so I thought it wisest to retreat, swallowing my disappointment as best I could.

The burghers re-assembled to the south of the Langberg; and we found that our loss was two killed and nine wounded, of whom two subsequently died.

We could not ascertain the English losses, but we saw their ambulances very busy. We heard afterwards that they had suffered much more severely than we had done.[286]

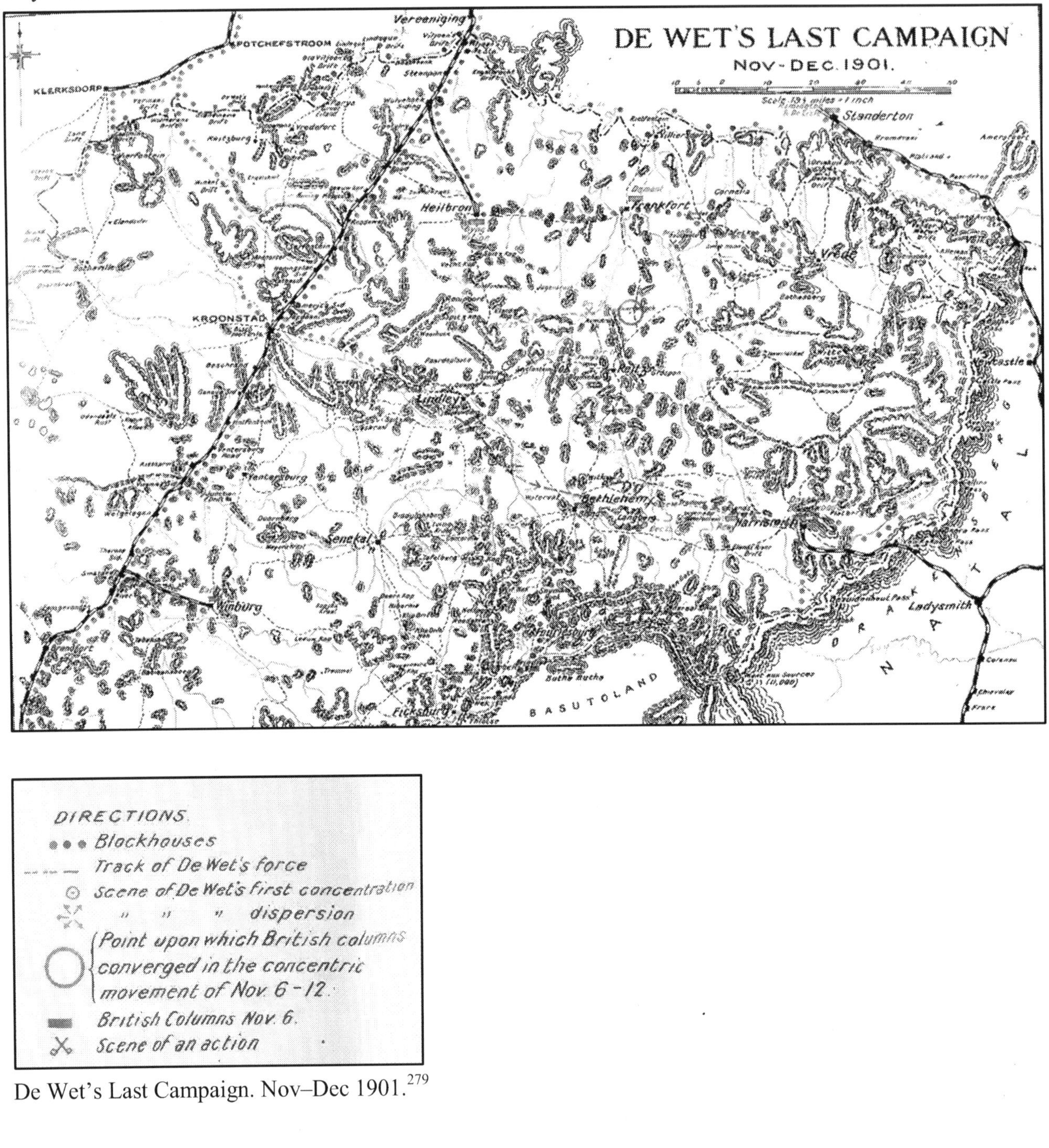

De Wet's Last Campaign. Nov–Dec 1901.[279]

[286] 12/18/1901 Colonel Steele, with South African Constabulary, captured thirty-six Boers in the region of the Magaliesberg. Four hours' fighting between De Wet and General Dartnell. Boers driven off. Methuen reported capture of thirty-two Boers.

12/19/1901 Colonel Allenby captured thirty-two of the enemy near Heidelberg.

12/20/1901 Colonel Damant attacked by 800 Boers. Two officers killed, three wounded. Boers repulsed.

12/21/1901 Capture of Smuts' convoy, near Bothwell, by Colonel Mackenzie.

12/22/1901 Seven hundred Cape raiders attacked columns of Colonels Wyndham and Crabbe. Were driven off with loss of five killed and twenty wounded.

12/23/1901 Successful attack on Grobelaar's laager by General B. Hamilton.

12/24/1901 Colonel Du Moulin surprised laager near Jagersfontein. Captured two Field-Cornets and twenty other Boers.

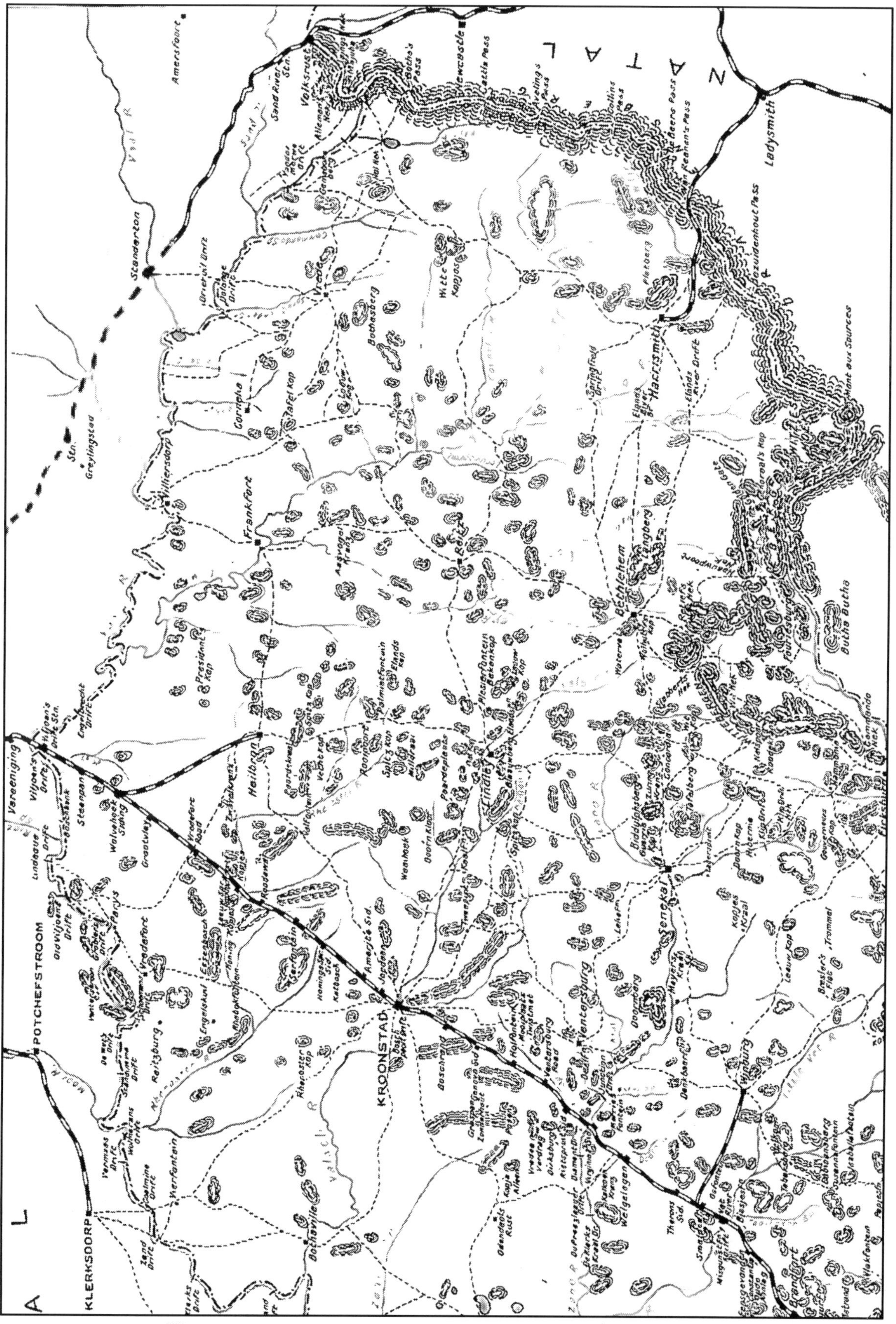

Orange Free State.[280] A Large cropped portion for reference in this section of the book.

33. A Success at Tweefontein

The column had marched to Harrismith.

It was time that I accomplished something further, and I determined that the next blow I struck should be a heavy one. I therefore retired to the north-east of Bethlehem, and concealed my men in the veldt round Tijgerkloof (which was suited to the purpose) whilst I made my plans.

Colonel Firman's brigade was camped between Bethlehem and Harrismith, at Elands River bridge, where he was building the line of blockhouses between the two towns. This camp was so well entrenched that there was no possibility of storming it, and I knew that so long as Colonel Firman thought I was still in the neighbourhood he would not dare to come out and give me an opportunity of attacking him.

I saw that a ruse was necessary to entice him out of his fortress. With this object in view I sent for Commandant Jan Jacobsz, with his fifty men from Witzeshoek. When he joined me I confided my secret to him, and ordered him to go back with his fifty men, and to let Colonel Firman see him doing so. He also had instructions to let some of his veldtcornets ride to the Kaffir kraals, which were close to the English camp, in order to tell these Kaffirs that he had had orders to come to me with fifty men, but that when he arrived I had commanded him to return to his district, because I was going to march with my commando to Winburg.

The following day Colonel Firman's scouts were, as might have been expected, informed by the Kaffirs of what they had heard from the burghers under Commandant Jacobsz; and the day after—that is, the 22nd of December—Colonel Firman's column, about six to seven hundred men strong, marched from Elands River to Tweefontein, half-way between Elands River and Tijgerkloof. On the farm of Tweefontein there was a mountain called Groenkop—which has since, for a reason which will soon be apparent to the reader, received the name of "Christmas Kop."

I gave Commandant Jacobsz orders to come to me with his fifty men on Christmas Eve, but this time with the strict injunction that he must conceal his march from the enemy. I also called up Veldtcornet Beukes, with his fifty men, from Wilge River, in the district of Harrismith. Veldtcornet Beukes was a brave man and trustworthy; he was shortly afterwards promoted to the command of a division of the Harrismith burghers.

My intention was to attack Colonel Firman early on Christmas morning.

Two days previously I had, with General Prinsloo and the Commandant, reconnoitred the neighbourhood of Groenkop, on which Colonel Firman was encamped. I approached as near as possible to the mountain, but could only inspect it from the west, north, and east, but on the following day I reconnoitred it also from the south.

My plan of making the attack early the next morning was somewhat spoilt by the fact that the English had already, on the 21st of December, quitted their camp on the mountain. Thus they had had four days in which to entrench themselves.

Whilst we were reconnoitring the mountain from the south, we saw three horsemen coming cautiously out of the camp, riding in a north-easterly direction, and thus giving us no chance to intercept them. Commandant Olivier and Captain Potgieter now made a detour, so that they could cut off the unsuspecting scouts from their camp, and could also get nearer to the mountain themselves. I knew that by doing so they would draw the fire of the two guns, which would tell me precisely where Colonel Firman's battery stood.

Before these officers could accomplish their purpose they were observed, and seeing that they could not cut off the three men, they turned their horses and galloped back. But when they saw that the three scouts had the temerity to pursue them, they faced round at the first rise and suddenly confronted them. The three (who were Kaffirs), seeing that the tables were turned, hastily wheeled round towards their own camp, but before they could reach it one of their number was caught and shot down. One gun and the Maxim-Nordenfeldt now fired upon our two officers as long as they were in sight, and thus we learnt that the guns were placed on the high western point of the mountain, from which they could shoot in all directions.

Let me describe Groenkop. On its western side was a precipice, on the north and south a steep descent, and on the east a gentle slope which ran down to the plain.

From which side should the attack take place?

Some of the officers were of the opinion that this should take place on the east, where it was the least steep, but I differed from them, for through our field-glasses we could see that the walls of the fort were so built that it was quite clear the enemy had thought that, should they be attacked, it would be from the east. The forts were built in a semicircle towards that side, and although this would be of little importance once the fight had begun (because the defenders had only to jump over the wall to find themselves still entrenched), still it was to the advantage of the attacking party to come from a side where they would not be expected.

These reasons brought me to the conclusion that the English would not be on the look-out for us from the west, and I therefore decided to make the attack from this side, the steep side of the mountain. But I did not then know how steep it really was.

On the western point there were four forts close to each other. Each was sufficient to give shelter to about twenty five men. To the south there were four forts, and to the east three.

The top of the mountain was not more than three to four hundred paces in diameter. To the east in a hollow the convoy was placed, and from every schanze we could rake it with our fire.

I remained on the spot from which I was reconnoitring, and sent word to the commando, in the afternoon of the 24th of December, to come to a certain place at Tijgerkloof, which they could do without being observed. I ordered them to remain there until nightfall, and then to advance within four miles of Groenkop, to the north, where I would meet them.

This was done. I found the commando at the appointed place, and also General Brand and Commandant Karel Coetzee, who had come on a visit that day to my commando. They also took part in the attack. My men consisted of burghers from General Michal Prinsloo, Commandants Hermanus Botha, Van Coller, Olivier, Rautenbach, Koen, Jan Jacobsz and Mears, in all six hundred men. Of these I left one hundred in charge of the Maxim-Nordenfeldt and the pack-horses.

We had not a single waggon with us; every man put what he had with him on his pack-horse, for long we had made it a rule not to be hampered with waggons. Yet whenever we picked up reports of engagements in the camping places of the English we repeatedly saw that they had taken a Boer camp—and their greatest delight was to say that it was one of De Wet's convoys.

They could not have been convoys of mine, because for the last fifteen months I had had no waggon-camp with me. If a waggon-camp was taken, it could only have been one consisting of women, who were flying in order to escape capture by the English, and to avoid being sent to the concentration camps. Everywhere in the State the women were taking to flight, and their terror was increased tenfold when the news came that many a woman and child had found an untimely grave in these camps.

The troops which had not remained with the pack-horses now advanced towards the mountain. Each commando was ordered to ride by itself, and to leave in single file. My orders were that they were to march quietly to the western foot of the mountain; here the horses were to be left behind, and the climb made on foot, the burghers keeping the same order as that in which they had been riding. Should the English, however, discover us before we reached the mountain, we must then storm it altogether, and leave the horses wherever we had dismounted.

We succeeded in coming to the mountain unobserved, and at once began the climb. It was exactly two o'clock in the morning of December 25th, 1901.

When we had gone up about half-way we heard the challenge of a sentry:—

"Halt; who goes there?"

Then followed a few shots.

My command rang out through the night—

"Burghers, Storm!"

The word was taken up by the burghers themselves, and on all sides one heard "Storm! Storm!"

It was a never-to-be-forgotten moment. Amidst the bullets, which we could hear whistling above and around us, the burghers advanced to the top, calling out, "Storm! Storm!"

The mountain, however, was so steep that it can scarcely be said that we stormed it; it was much more of a climb. Often our feet slipped from under us, and we fell to the ground; but in an instant we were up again and climbed on, and on, to gain the summit.

I think that after the sentry heard us, three or four minutes must have elapsed before the troops, who were lying asleep in their tents or on the veldt, were awakened and could come out, because their camp was about a hundred paces distant from our point of attack.

Directly we reached the top the deafening roar of a heavy fight began, and lasted from fifteen to twenty minutes. Shortly before this the Armstrong gun and the Maxim-Nordenfeldt had each fired two shots, but they fired no more; as we reached the top the gunners were shot down at their guns.

After a short but desperate struggle the English gave way, or surrendered, and we took possession of the Armstrong and Maxim-Nordenfeldt.

We continued to fire on the troops, who had retreated to a short distance. Again they gave way, and took up another position a little further on, and so it went on for about two thousand paces, and then the English took to flight.

As we had no horses with us and it was dark, we did not pursue the fleeing enemy, but returned to the camp. The whole engagement lasted, so far as I could judge, for about an hour. I cannot say for certain, because I made no note of the time.

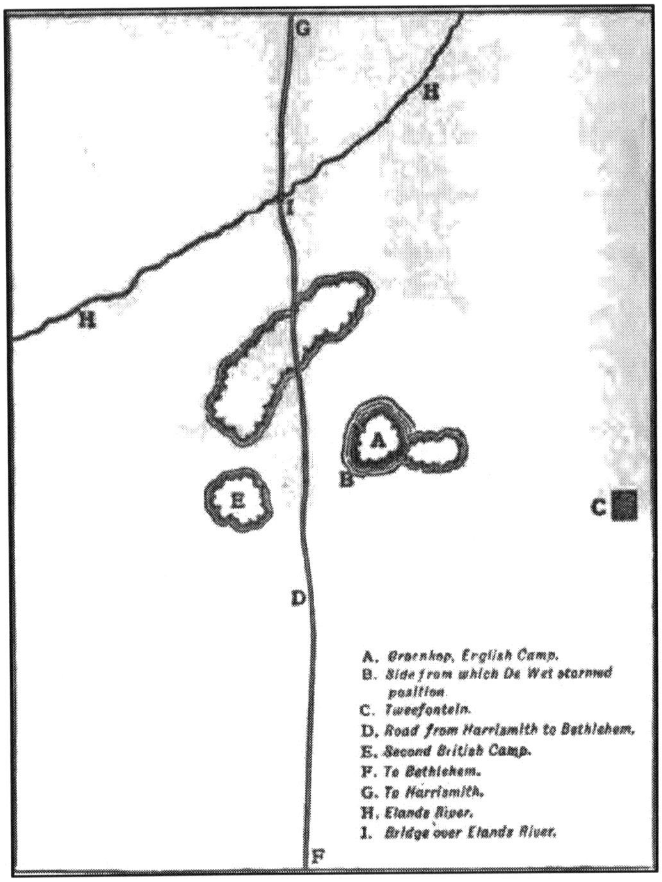

Tweefontein, from a Sketch by the Author.

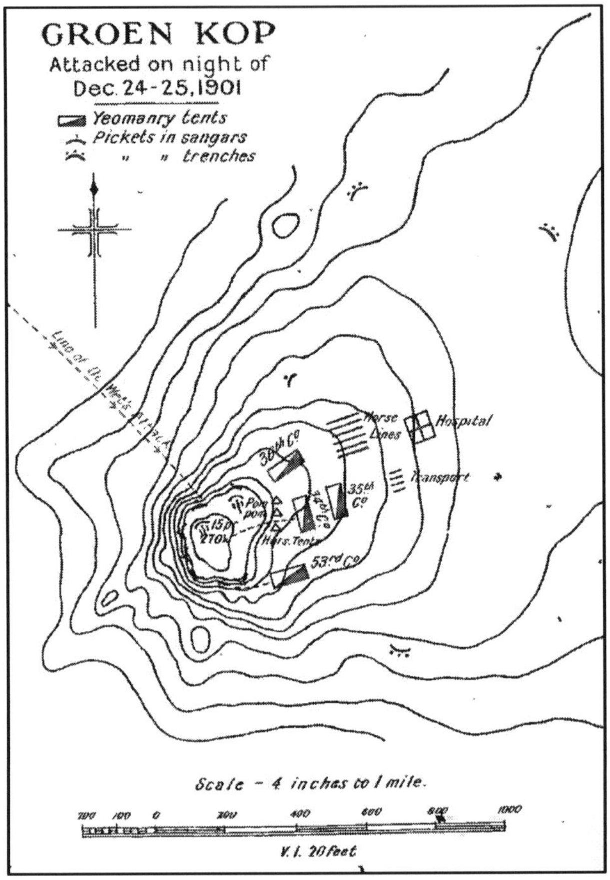

Groen Kop Times History Vol 5.

It was a party of Yeomanry with whom we had been dealing, and I must say they behaved very gallantly under exceptionally trying circumstances; for it is always to be expected that when men are attacked during the night a certain amount of confusion must ensue.

It was heartrending to hear the moaning of the wounded in the dark. The burghers helped the doctors to bring the wounded into the tents, where they could be attended to; I gave the doctors as much water as they liked to take for the wounded.

It was greatly to be deplored that the ambulance had been placed in the centre of the camp, for this was the cause of Dr. Reid being fatally wounded.

When the day began to dawn we brought the waggons and guns down the mountain. I sent them in the direction of Langberg, to the west of Groenkop.

The enemy lost about one hundred and sixteen dead and wounded, and two hundred and forty prisoners of war.

Our loss was also heavy—fourteen dead and thirty wounded; among the dead were Commandant Olivier from Bethlehem and Vice-Veldtcornet Jan Dalebout from Harrismith; among the wounded was one of my own staff, Gert de Wet. Later on two more died, one of them being Veldtcornet Louwrens. I appointed Mr. A.J. Bester as Commandant in the place of Commandant Olivier.

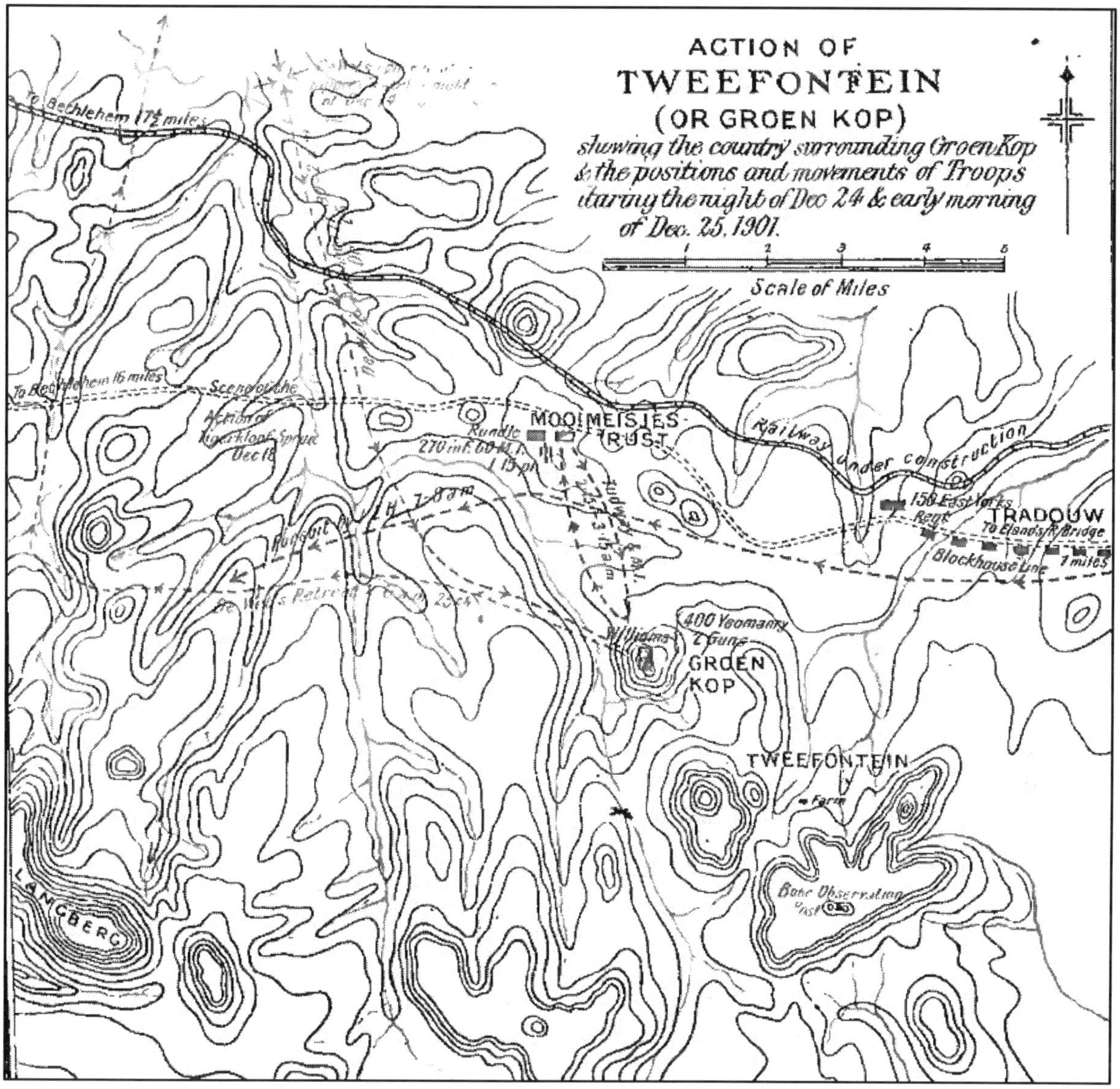

Times History Vol 5.

Besides one Armstrong and one Maxim-Nordenfeldt, our booty consisted of twenty waggons, mostly ox-waggons, a great quantity of rifle and gun ammunition, guns, tents, five hundred horses and mules, and one waggon laden with spirits, so that the burghers, who were not averse to this, could now satisfy their thirst.

The sun had hardly risen when the enemy opened fire from a mountain two miles to the north-east of Groenkop, where there was a little camp with one gun. If I still had had the same numbers as were with me at the storming of Groenkop, then I could also have taken this little camp. But it was not to be thought of, for some of my men had been sent away with the waggons, and the others—well, every one had a horse that he had taken from the English, and as these horses were in the pink of condition for rapid retreat, I thought it wiser not to call upon the burghers to attack. I ordered them, therefore, to go back after the waggons, and in the evening we camped to the north of Bethlehem. From here, on the following day, I sent the prisoners of war through Naauwpoort into Basutoland.

On the same day I gave orders to General Michal Prinsloo to take the commando and to strike a course between Reitz and Heilbron. I myself paid a visit to President Steyn and General Wessel Wessels, after which I put matters straight in our hospital at Bezuidenhoutsdrift, which was under the charge of Dr. H.J. Poutsma.[287]

[287] In the Battle of Groenkop (Battle of Tweefontein) on 25 December 1901, Head Commandant Christiaan de Wet's Boer commando surprised and defeated a force of Imperial Yeomanry under the command of Major Williams.
The next morning, one of the 206 British prisoners of the Boers noted that his foes were so short of clothing that some wore women's attire. The 250 unwounded British prisoners of war were stripped literally naked before they were turned loose the next day. Evans, Martin Marix. The Boer War: South Africa 1899–1902. Oxford: Osprey, 1999. ISBN 1-85532-851-8 p.154.

34. I Cut My Way through Sixty Thousand Troops

The English could not endure the thought that we had their guns in our possession. And, accordingly, when General Michal Prinsloo came near the Liebenbergsvlei, on the road between Reitz and Heilbron, he met a strong force of the enemy which had come from Kroonstad. The English then had a taste of what it was like to be under the fire of our artillery; and so well did the gunners do their work that the enemy were forced to retreat. This occurred shortly before sunset on the afternoon of the 28th of December.

But the forces in front of General Prinsloo were too strong for him, and so when night came he marched past, and the following morning was twelve miles to the south-west of them.

The enemy advanced against the position which General Prinsloo had occupied the previous day, quite unaware that he was now in their rear. In the meantime the General was watching their movements from behind, and quietly enjoying their mistake.

I left the hospital that afternoon, and crossing the Liebenbergsvlei to the rear of the English, I joined the Heilbron commando.

The following day the enemy retreated to the farm of Groenvlei, which lies just to the north of Lindley. They remained there for a few days awaiting large reinforcements.

"I quite understand your plan," I said to myself, as I set to work to split up the great force which the enemy were concentrating. And with this object in view I sent each Commandant to his own district, believing that by dispersing my own men I should again induce the English to divide their troops into smaller parties. Commandant Mears, with his fifty men, I ordered to remain with the guns and the artillery, and to guard them by very careful scouting.

In less than a fortnight seven large columns of the enemy were operating in the district between Heilbron and Bethlehem and Harrismith. These columns burnt all the houses within their reach, and those which had been spared before were now given over to the flames. And not only were the houses destroyed, but every head of cattle was taken.

Towards the end of January, 1902, still more columns arrived and a "drive" began.

I remained in the neighbourhood until the 2nd of February and stationed Commandant Mears with the guns to the east of the Wilge River. The English formed a circle round him, but he succeeded in getting the guns away in safety.[288] When he was out of their clutches, I sent him orders to bring the guns through the blockhouse line between Lindley and Bethlehem, and then to push on towards Winburg.

It was my intention, on arriving there, to collect as rapidly as possible a commando from the men of Bethlehem, Kroonstad, and Winburg, and to attack the first column that gave me a chance of doing so.

Commandant Mears carried out my orders at once. A force of the enemy had been waiting for him for three or four days at the farm of Fanny's Home, on the Liebenbergsvlei. But before the sun had risen, a strong force under Colonel [Julian] Byng had surrounded him and forced him to abandon the guns. And not only were the guns lost, but Captain Muller and thirteen gunners were taken prisoner.

Thus the guns had not been of much benefit to us, for the English had kept us so constantly on the move that it had been impossible to use them.

The forces of the enemy between Harrismith and Vrede had formed a line extending from the Harrismith-Bethlehem blockhouses to the blockhouses between Vrede, Frankfort and Heilbron. And now the troops were advancing in close contact with each other, hoping thus to force us against one or other line of blockhouses.

Nearer and nearer they came, until at noon on February the 5th we saw them to the east of Liebenbergsvlei. As I was watching their movements from the top of Elandskop, I was informed by heliogram[289] from Blaauwkop and Verkijkerskop that there was a cordon of the English from Frankfort to a spot between Bethlehem and Lindley.

Kitchener wrote, "It is very sad and depressing that the boers are able to strike such blows, but I fear ... we shall always be liable to something of the sort from the unchecked rush of desperate men at night." Pakenham, p 577.
https://en.wikipedia.org/wiki/Battle_of_Groenkop

[288] 2/2/1902 De Wet's commando gallantly charged by New Zealanders, Queensland Imperial Bushmen, and South African Light Horse.

[289] We had heliographic communication between Elandskop and Blaauwkop, which formed a connecting link between Bethlehem and Lindley; and from Blaauwkop we had communication with Verkijkerskop. There was also heliographic communication between Bethlehem and Lindley, and Biddulphsberg, across the line of blockhouses.

The intention of the enemy appeared to be to drive us against the Heilbron-Kroonstad blockhouses and the railway line. We had therefore to be prepared to fight our way through the blockhouses. And these, as I found out lately, had been greatly strengthened.

On the 6th of February I was on the march, intending to advance to Slangfontein, to the west of Heilbron. I sent orders to Commandants Mentz, Van der Merwe, and Van Coller, to take a portion of Commandant Bester's burghers, telling them to go to Slangfontein. For I hoped to break through at some point or other that night.

Still nearer the enemy came, marching almost shoulder to shoulder.

The Commandants Van Coller and Van der Merwe did not go to Slangfontein. They broke through the English columns near Jagersrust, and crossed the Heilbron-Frankfort blockhouse line, where they put a few soldiers to flight, not, however, without a loss of two burghers, who were killed.

Neither did the burghers under Veldtcornets Tal jaart and Prinsloo arrive. They preferred to go their own way—and all were captured with the exception of twenty-eight men. But this misfortune was not due to the blockhouses. On the contrary, they were taken prisoners when they were attempting to hide themselves in small bodies. In this way more than a hundred burghers fell into the hands of the English.

There were now with me Commandant Mentz, and portions of the commandos of Commandants Bester, Cilliers, and Mears. That afternoon we marched to a farm which was twelve miles from the Lindley-Kroonstad line of blockhouses. When it was quite dark, we left the farm with the intention of breaking through this line before daybreak. There had been five or six hundred head of cattle with us, but, without my being aware of it, they had gone astray in the darkness.

We intentionally left the path, because we thought that the English would be most vigilant at points where paths crossed the line.

Suddenly we found ourselves at a wire fence. The darkness was so thick, that it was only after we had cut the wire that we discovered that we were close to a blockhouse. Although the house was not more than a hundred paces from us, we could hear and see nothing. When we were some four hundred paces on the other side of the line of the blockhouses, I sent a burgher back to see if all the men and cattle had crossed safely—for we were riding in a long trail, and amongst us were old men and youngsters of only ten years, or even less. These boys would have been taken away from their mothers had they stayed at home; and thus the only way to keep them from captivity was to let them join the commandos.

The burgher soon returned, and told me that the whole commando and all the cattle had crossed the line. Then I marched forward again.

At break of day we were close to the Valsch River. Here I made a short halt, in order to allow the stragglers to come up. It was then that a man came to me who had been riding far behind, and had thus not seen that we had cut the wire. He was probably one of those who quite needlessly feared a blockhouse line.

"General, when shall we come to the blockhouses?" he asked me.

"Oh! we are through long ago!" I answered.

It did not require any deep insight, I can assure you, to see how delighted this burgher was that we were safely out of it!

We discovered now that the cattle had not crossed the line. When I investigated the matter more closely, I found that they had gone astray before we reached the blockhouses. But it was impossible to wait for them, and there was nothing left but to proceed without them.

When we arrived at the Valsch River, there was a sound of shouting behind us, and presently the cattle appeared coming over a rise. I heard from the drivers that they had lost their way, and had only reached the blockhouses at daylight. But they had succeeded in breaking through under a fierce rifle fire. Twenty head of cattle had been killed or wounded, and one of the men's horses had been shot under him.

The burghers who had accomplished this valiant deed were: Jan Potgieter, Gert Potgieter, Jzoon, and Wessel Potgieter—all from the district of Heilbron.

I have, myself, seen a report in an English paper of my breaking through the blockhouse line. This paper declared that I had driven a great herd of cattle in front of me to break down the fencing!... This is the way the English write the reports.

This breaking through of my cattle inspired the English, at least so I thought, to dig trenches everywhere. But they were again wrong; for although a vehicle might have some difficulty before the trench was filled in, no riders, pedestrians, or cattle would have been stopped for a moment.

And now we marched on, till we reached a spot about fourteen miles to the south of the blockhouse line; and there we remained for three days.

Whilst we were waiting here, I sent two burghers back to the blockhouse line, to discover in what direction the English columns had marched, so that I might know where I should go myself. Now, less than ever, was it advisable to make night marches, for our horses were in a very poor condition.

The day following I received a heliographic message from these burghers, who were now on the other side of the line. They signalled that I could come on with my commando, since the English columns had returned to Kroonstad and Heilbron.

When night came I started on my way back. I did not go (as before) to the east of Lindley, but to the farm of Palmietfontein, which lies to the west. When we were close to the line, I sent some burghers in advance to cut the wire. But this time there was a reception ready for us, which we certainly would rather have been without! This was to be ascribed to the fact that instead of only two scouts, as I had ordered, about ten had gone to reconnoitre. So large a number had attracted the attention of the enemy, and the guards had concentrated at the spot where we wished to break through.

Thus before my commando reached the line a fierce fire was opened on it from two sides. Yet notwithstanding this the wires were cut and we reached the other side, but not without loss. One of my burghers was killed, and one wounded. A boy of ten was also killed, and another of seven severely wounded. We could not ascertain the losses of the enemy.

It was terrible that children should be exposed to such dangers; but, as I have already said, if we had not taken them with us they would have been captured. During the very "drive" I have just described, two children who had remained at home with their mothers were taken prisoner by the English. One of these was a boy of nine, the little son of Jacobus Theron. Notwithstanding the prayers and entreaties of the poor mother, he was torn from her and carried away. In the same way another boy, twelve years old, whose name I do not know, was dragged from his mother's arms.

The chronicling of such inexplicable cruelties I leave to other pens. I have drawn attention to them to make it clear that it was not without good cause that children joined the commandos. Some of these little ones became a prey to the bullets of the enemy, and the South African soil is stained by the blood of children slain by England.

With the exception of the sad incidents I have described, we came through in safety.

I afterwards heard that Lord Kitchener had on this occasion gone to Wolvehoek Station in order to see President Steyn and myself carried away in the train to banishment! But his calculations were not altogether correct.[290]

A Higher Power had willed it otherwise.

The burghers had now returned to their own districts. I myself went to a farm in the neighbourhood of Elandskop belonging to Mr. Hendrick Prinsloo—the rooije.[291] After I had been there a few days I heard that a strong column was approaching Lindley from Kroonstad. During the night of the 17th of February this column attacked some burghers who were posted less than four miles from Elandskop, with the object—as I heard later—of catching me. And they would have been quite successful in their attempt had I been sleeping in the house where their information led them to believe they would find me. But as a matter of fact, I seldom, if ever, slept in a house, for to tell the truth, there were scarcely any houses left to sleep in! The women who had escaped capture lived in narrow shelters, which had been made by placing corrugated iron sheets on what was left standing of the walls that remained.

I crossed the Liebenbergsvlei on the 18th of February, and proceeded to the farm of Rondebosch, which stands to the north-east of Reitz. There I met the Government.

[290] 2/7/1902 De Wet, by brilliant manoeuvre, ruptured the British cordon and escaped.

[291] "Rooije" is the Taal for "red."

And now another big "drive" took place. The English columns marched to the south of the Kroonstad-Lindley blockhouse line in the direction of Bethlehem. Other troops came from Heilbron, and advanced to the north of the Heilbron-Frankfort line, driving Commandant Ross across this line to the south.

Nearer and nearer these two great divisions approached each other, until at last they stretched without any break from the Bethlehem-Lindley to the Frankfort-Vrede line of blockhouses. On the 21st of February the whole column moved towards Vrede and Harrismith.

It seemed to me that my best plan would be to go with President Steyn and the Government to the Witkopjes, which lay between Harrismith and Vrede, and then to break through the English columns near Vrede or Harrismith, or, if it proved impossible to do so at these points, at least to force a way through somewhere.

On this occasion we had a great deal more difficulty in escaping from the English than we had had during the previous "drive." Not only had we to deal with these large forces behind, but also with thousands of troops which were now approaching from Villiersdorp, Standerton, Volksrust, and Laingsnek, and which were extended across the country in one continuous line. The whole cordon thus formed consisted, as the English themselves acknowledge, of sixty thousand men.

And again on this occasion they did not attempt to drive us against one or other of the blockhouse lines, but they came, column on column, from all sides, and formed a big circle round us. They thus made it quite apparent that they had lost all faith in their blockhouses.

I only received news of the approach of these reinforcements on the evening of the 22nd of February, after they had passed the blockhouses. The report was brought to me by Commandant Hermanus Botha, a party of whose burghers had been driven across the Vrede-Frankfort line during the previous night. I have already stated that some of the burghers under Commandant Ross had shared the same experience, and now they were retreating before the English. I also heard that Commandant Mentz had gone eastwards, in the belief that the forces behind him would move to the west, but that unfortunately the columns also moved to the east, so that he jumped into the lion's mouth, which was only too ready to close!

We marched that night to Cornelius River, and the day following to Mr. James Howell's farm at Brakfontein. It was my intention to break through somewhere between Vrede and Bothaspas.

But my scouts brought me word in the evening that there was a very poor chance of success in that neighbourhood, for the columns had concentrated there. Other scouts, however, reported that there was a small opening at Kalkkrans, on the Holspruit; and so I decided to march to Kalkkrans.

When the sun had set I left Brakfontein and started on my road to Kalkkrans, with the firm determination to force my way through there, cost what it might. If I failed in the attempt I knew that it would mean an irretrievable loss, for not only should I myself be captured, but also President Steyn and the whole Government.

I had with me a portion of the Harrismith burghers, the commandos from Vrede and Frankfort, and sections of the commandos from Standerton and Wakkerstroom, these latter under Commandant Alberts. This Commandant had come to these districts to obtain horses for his burghers; he was obliged to be content with the wild horses of the veldt, for there were no others to be had.

Beside the above burghers, I had with me old men and children, and others who were non-combatants. These had joined the commando to escape falling into the enemy's hands.

Altogether I had well-nigh two thousand persons with me. Commandant Mentz was, like myself, enclosed in the "drive," but some distance away. General Wessels, Commandant Beukes, and some of the Bethlehem burghers were in the same predicament to the west of us. I did not know for certain where these officers were placed, and therefore I could not inform them of my plan to break through that night, for I had only come to this determination after the sun had set. But I felt sure that they would at all costs make their way through the cordon.[292]

Commandant Jan Meijer had met me at Brakfontein, but one party of his burghers was still six miles to the south. When I decided to break through, I sent him orders to follow me; and this he was quite capable of doing, as he was well acquainted with this part of the country. My orders were that the mounted men were to proceed in advance, taking with them my little waggon drawn by eight mules.

[292] In this I was correct. They contrived to break through where the enemy were more scattered.

This waggon had accompanied me into Cape Colony, and since that time—for fourteen weary months—had never left me. I had even taken it with me when, a fortnight previously, I had broken through the blockhouse lines.

Behind the horsemen came the aged and the sick, who occupied the remaining vehicles, and lastly the cattle, divided into several herds.

In this order we rode on.

When we were approaching the spot at which I expected to find the enemy, I ordered Commandant Ross and one hundred men, with Hermanus Botha and Alberts, and portions of their commandos, to go on ahead of us.

After passing through Holspruit we inclined to the west, as the road to the east would, according to my scouts, have led us right into the English camp. But it was not with one camp only that we had to deal: the English were everywhere: a whole army lay before us—an army so immense that many Englishmen thought that it would be a task beyond the stupid and illiterate Boer to count it, much less to understand its significance. I will pander to the English conception of us and say, "We have seen them: they are a great big lot!"

We had hardly moved three hundred paces from where we had crossed Holspruit, when the English, lined up about three hundred yards in front of us, and opened fire. We saw that they did not intend our flight to be an easy one.

Before we had reached the "spruit,"[293] and while crossing it, the burghers had kept pushing ahead and crowds had even passed us, but the enemy's fire checked them and they wheeled round.

Only the men under Commandants Ross, Botha, and Alberts did not waver. These officers and their veldtcornets with less than one hundred men stormed the nearest position of the enemy, who were occupying a fort on the brow of a steep bank.

I shouted to my command: "Charge."

I exerted all my powers of persuasion to arrest the flight of my burghers; even bringing the sjambok into the argument.

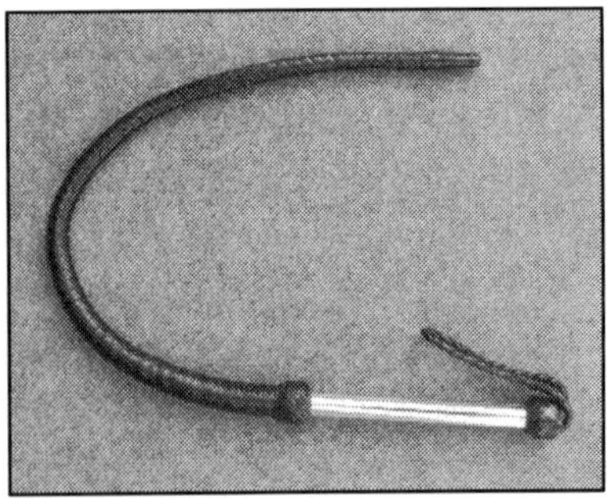

A Sjambok.[294]

Two hundred and fifty were all that I could bring back to the fight, whilst, as I have said, the Commandants had a hundred with them when they charged; the rest, regardless of my attempts to stop them, fled.

I was also without my staff, some of whom had remained under the fire of the enemy awaiting my orders as to what was to be done with my little waggon. Others, amongst whom was my son Kootie, [Jacobus] who was then acting as my secretary, had followed me, but had got lost in the confusion of the moment.

This confusion arose from the fact that the burghers imagined that they had got through at the first attempt, but had found themselves again fired at from the front. Meanwhile, I hurried to and fro, encouraging the burghers in their attempts to break through. When thus engaged I came across two of my staff, Albertus Theunissen and Burt Nissey. To them I gave the order: "Get the waggon through at all costs." I also found my son, Isaac, and kept him with me. The English now were firing not only from in front but also on our right, and there was nothing for it but to clear a road for ourselves, and this we eventually succeeded in doing, and in about forty minutes had at last broken through.

[293] Spruit—rivulet.

[294] The sjambok or litupa, is a heavy leather whip. It is traditionally made from adult hippopotamus or rhinoceros hide, but it is also commonly made out of plastic.

The enemy had dug trenches, thirty to forty paces from each other, which served as schanzes. In each of these trenches were placed ten to thirty men. They had also a Maxim-Nordenfeldt, which, at first, kept up a hot fire; but soon was silenced as the gunners were shot down. The rest of the troops retired with the gun, but had to leave the caisson behind them. It was evident to me from the way in which they fired that the English were retreating, and so I dispatched two men to tell the burghers, who had gone back, to come on; but this they did not do, thinking perhaps that they could discover a safer route on the following evening. This was short-sighted policy on their part, for the circle within which they were caught was daily becoming narrower, and it was plain that on the third day the enemy would be so close that all hope of escape would be gone.

The two burghers did not return, and we went on without them, taking with us our wounded—twelve in number. Two of these, whose injuries were serious, had been placed by some of my staff on my waggon; one was Van der Merwe, a member of President Steyn's bodyguard; the other was a boy of thirteen years old, named Olivier.

We hurried on, and came, shortly after sundown, to the farm called "Bavaria," on the Bothasberg. There Van der Merwe died.

The boy had already been relieved from his sufferings. Thus, once again, the soil drank the blood of a child.

Eleven of my men were left dead on the battlefield. We had to leave them there, for to recover their bodies might have meant the sacrifice of more lives.

When the burghers and I forced our way through the storm of bullets, we had with us President Steyn, the Members of the Government, and the Rev. D. Kestell, minister of the Dutch Reformed Church at Harrismith.

The greater part of the English, indeed all of them, so far as we could observe, remained, during the 24th, on the spot where we had left them. We found out, later on, that we had broken through their lines at the point where Colonel Rimington's force was stationed.

The following day the columns departed. We then went to bury our dead, but found that the enemy had already done so. But as the graves which they had made were very shallow, we dug them deeper.

During that night (the 25th) another force of burghers, to the number of about three hundred and fifty, broke through the English cordon. Our men only lost two killed, and eleven wounded.

Besides those already mentioned, the burghers under General Wessel Wessels and Commandant Mentz were also among those who escaped of the two thousand troops surrounded by the enemy.

With the others it fared but ill.

The English closed in, and the circle became narrower and narrower.

On the 27th of February, 1902—"Majuba Day"—Commandant Van Merwe and four hundred men fell into the hands of the enemy.[295]

On that very day, in the year 1881, the famous battle of Majuba had been fought. Nineteen years afterwards, on the same day of the same month, we suffered a terrible defeat at Paardeberg, where we lost General Piet Cronje and a great force of burghers.

And now the 27th of February had come round again, and this time it was the twenty-first anniversary of Majuba that we were celebrating. The day of our coming of age had thus arrived, if I may be allowed to say so. But instead of the Republics now attaining their majority—as they should have done, according to all precedent—minority would have been a more fitting word to describe the condition in which we now found ourselves—for, through the losses which we had just sustained, we were minus not only a large number of burghers, but also an enormous quantity of cattle, which ought to have served as food to our commandos and families, but which the enemy had captured.[296]

[295] Also my son, Jacobus (Kootie). He has now returned from St. Helena, whither he had been sent as a prisoner, and we have met. He tells me that on the night when I broke through, he wanted to come with me, but was unable to do so, because his horse had been shot under him.

[296] 2/27/1902 Operations for driving Boers against Harrismith-Van Reenan's blockhouse line. Manie Botha killed; 600 Boers killed, wounded, or prisoners. Defence by New Zealanders under Major Bauchop and New South Wales Mounted Infantry under Colonel Cox. 2/28/1902 Capture of Boers near Steynsdorp by Captain Holgate (Steinacker's Horse).

The cattle which had just been taken from us had formed the greater part of our cattle in this district. We had always been able, until now, to get them safely away; the unevenness of the veldt here was greatly in our favour. This time we could not. How am I to explain the inexplicable? We had sinned—but not against England![297]

35. I Go to the Transvaal with President Steyn

On the 26th of February I went with the Government to Duminys Drift, on the Wilge River, and we thus found ourselves again at the farm of Rondebosch.

Damaged bridge over the Wilge River.[281]

The Government remained there for a few days, and then President Steyn decided to go into the western parts of the State, where Generals Badenhorst and Nieuwouwdt were then operating. He thought that if he absented himself from the north-eastern districts the English would cease their devastations in that part of the country, for it was well known that the enemy's concentration of forces was principally aimed at the President and myself.

I, however, did not intend to follow his example, but, on the contrary, got myself ready to join the Heilbron commando. By March 22nd all my preparations were made, and I had, alas! to say farewell to my trusty friend—my little waggon! I saw that it must be relinquished—that I could not carry it about with me any longer. I left it at a farm, first taking out my documents and papers; I ordered these to be concealed for greater safety, in a cave on the farm of General Wessels.

The clothes and ammunition of myself and staff had been hidden in this cave for some time.

The following day I joined President Steyn, who told me that he wished me to accompany him in his march to the west. And although it did not agree with my own ideas—principally, because I did not want the enemy to think that I was running away from them—I consented to this plan, and the more willingly because it was some time since I had visited the western commandos.

It was a long journey that lay before us, and I had only the clothes that I was then wearing. I would have sent for another suit had I not heard that the enemy were encamped close to the cave where our treasures lay hidden.[298][299]

[297] By 5 February 1902, Kitchener's blockhouse lines were completed and he sent 9,000 men on a massive sweep through the countryside. In this first operation, 285 Boers were captured but de Wet and President Marthinus Steyn and their men escaped the trap. The second drive lasted from 16 February to 28. Again, de Wet got away, but this time he had to abandon most of his cattle. On 27 February, Colonel Henry Rawlinson's column encircled and captured a 650-man Boer commando at Lang Reit, a few miles from Tweefontein. This brought the British "bag" in the successful sweep to 778 surrendered Boers. The third drive by Major Elliott's division, from 4 March to 11 March, was a failure, with only about 100 Boers captured. Worse, de Wet escaped to join Fighting General Koos de la Rey in the Western Transvaal. Pakenham, p 579–582. 3/11/1902 Close of big drive in Orange River Colony; 127 Boers taken. Commandant Celliers wounded.

[298] Shortly afterwards I heard that it was Colonel Rimington's column who were encamped there. They discovered the cave, and removed the documents and wearing apparel, leaving me with only a suit of clothes—which I should have liked to preserve as a curiosity!

[299] 3/6/1902 Colonel Ross (Canadian Scouts) made valuable captures in a cave near Tafel Kop.

I had therefore to do the best I could with what I had. There was no clothing to be got in the western districts, so that when my present outfit was worn out, I should be compelled to put on "khaki"—although there was nothing I relished less than to rob a prisoner of war.

We started out that same evening in the direction of the railway line. Our party consisted of about two hundred men, composed as follows: the President, with his bodyguard of thirty men, under Commandant H. Van Niekerk, the Government, Commandant Van de Merwe, of Vredefort, my staff and myself.

Before daybreak we got through the Heilbron-Frankfort line of blockhouses without accident; and on the following night (March 5th) we crossed the railway line, between Wolvehoek and Viljoensdrift. Whilst we were occupied there in cutting the telegraph wires, the enemy fired a few rounds on us, at a distance of five hundred or six hundred paces. We approached nearer, and they then opened fire with a Maxim—but without doing any damage.

We continued on our road, past Parijs and Vredefort, towards Bothaville, and we came upon a blockhouse line which extended from Kroonstad to the Vaal River. We rested for two days, to the north of Bothaville; during this time my scouts captured from the enemy eighteen horses, most of which were in good condition.

On the night of March 12th we broke through the blockhouse line, some five miles to the west of Bothaville. When we were about fifty paces from the line, somebody to our left challenged us:

"Halt! Who goes there?"

He challenged us a second time, and then fired.

At once seven or eight sentries fired upon us. Shots also were directed at us from the right. Nevertheless we cut through the barbed wire and crossed in safety, the firing still continuing, until we were about fifteen hundred paces on the far side of the line. Fortunately no one was hit.

Having thus escaped from the last "White Elephant" that we should have to reckon with, the next obstacle to be encountered was the Vaal River. For the President, since we had crossed the Valsch River, had decided to visit De la Rey, in order to place himself under medical advice. His eyes had become very weak during the last fortnight or so, and he thought that Dr. Van Rennenkampf might be able to do something for them.

Thus we had to cross the Vaal River.

But we heard that there was a military post at Commandodrift, where we wanted to cross, and further, that all the other fords were occupied by the English. We should have been in a great difficulty had not one of our burghers, Pietersen, who knew this district thoroughly, brought us across the river by a footpath ford.

We crossed on March 15th. The current was so strong that in places the horses were almost swimming; in other places the river-bed was strewn with huge boulders, over which our steeds had to climb. However, we all managed to get safely over, and arrived at Witpoort on the evening of the 16th. On the following day we joined General De la Rey.

It was a most interesting occasion. We had a hearty reception, several impromptu "addresses" being presented to the President, who in turn spoke to the burghers with much fire and enthusiasm. They were already in the best of spirits, as they might well be, for their General had but recently won victories over Von Donop and Lord Methuen.[300] [301]

Dr. Van Rennenkampf, having examined the President's eyes, said that he must remain for some time under his care. Accordingly I left President Steyn with De la Rey, and, on the third day after our arrival, set out with my

[300] 2/24/1902 De la Rey pounced on a wagon convoy commanded by Lieutenant Colonel S. B. Von Donop. For the loss of 51 Boers, De la Rey killed, wounded or captured 12 officers and 369 men. In response, Methuen tried to track the Boer leader down. Pakenham, p. 583.

[301] 3/7/1902 Successful attack by Delarey on Methuen's force at Tweebosch. Methuen seriously wounded and taken prisoner. Tweebosch is a farm where the Klein-Harts River flows into the Harts River.

De la Rey ambushed Methuen's column at Tweebosch on the Little Harts River. The British force numbered 1250, including nearly 1000 mounted men and four guns. Methuen's force was largely made up of green troops; these panicked and fled or surrendered. Only the British regulars in the column fought stubbornly in the combat which lasted from dawn until 9:30 am. The British lost 200 killed and wounded, plus 600 men and all four guns captured. After being wounded twice and suffering a broken leg when his horse fell on him, Methuen was captured. He was the only general taken prisoner by the Boers during the war.

De la Rey sent the wounded Methuen to a British hospital in his own carriage under a flag of truce, despite demands from his own troops to execute him. The Boers court-martialed De la Rey for freeing such a valuable prisoner, but after convincing the court that Methuen would withdraw from the war, he was let off. https://en.wikipedia.org/wiki/Battle_of_Tweebosch

staff to join General Badenhorst, who was then in the neighbourhood of Boshof. It was becoming more and more important that I should see Badenhorst and Nieuwouwdt, and discuss with them how best they might collect their forces, for I wished to be able to attack the first English column that should enter the western district of the State.

I had received reports that, with the exception of the garrison at Boshof, the west, for the moment, was free from the enemy; and this information caused me no surprise, for I could well believe that they had just "packed up their trunks" in the north.

On the 25th of March I joined General Badenhorst on the Gannapan,[302] thirty miles to the north-east of Boshof. I at once sent an express to General Nieuwouwdt, ordering him to come to me with all speed, and to bring about four hundred and fifty of his men with him. Meanwhile, General Badenhorst received instructions from me to get all his scattered commandos together.[303]

Before there had been time for these orders to be carried out I received, on March 28th, a letter from President Steyn, giving me the following information:

Mr. S.W. Burger, Vice-States President of the South African Republic, had written to President Steyn, saying that he was at Kroonstad, and that he wished to meet the Government of the Orange Free State. He also said that a copy of the correspondence between the Governments of the Queen of the Netherlands and of the King of England had been sent to him by Lord Kitchener.

From this correspondence it appeared that the Netherlands Government (considering the condition of affairs to be exceptional, in that the Boers who were still fighting were unable to negotiate either with the British Government or with the Deputation in Europe) felt justified in offering to act as an intermediary. In this capacity they were prepared to ask the Deputation if they were willing—supposing that a safe conduct could be obtained from England—to go to South Africa, and discuss matters with the Boers, in order to be able subsequently to return to Europe, empowered to conclude a Treaty of Peace, which would be binding both in South Africa and in Europe.

Lord Lansdowne, in the name of the British Government, replied that his Government highly appreciated the humane intentions of the Government of the Netherlands, but that they had made up their minds to abide by their former decision, and not to accept any foreign intervention. Further, that the Deputation could, if they wished, address a request for a safe conduct to the British Government, but that the matter could not be decided in England, until the precise nature of the request, and the grounds on which it was preferred, were fully understood.

Lord Lansdowne also said that the British Government was not quite clear as to whether the Deputation still retained any influence over the Boer leaders in South Africa; that they thought that the power to negotiate for the Orange Free State lay with President Steyn, and, for the Transvaal, with President Burger; and that they considered that the most satisfactory arrangement would be for the leaders of the Boers to negotiate directly with the Commander-in-Chief of the British forces in South Africa, who had been ordered to transmit at once to the British Government any offers or proposals which might be made to him.

Lord Lansdowne concluded by saying that, if the Boers wished to negotiate, it must be in South Africa, and not in Europe. For, if the Deputation were to go to South Africa, at least three months must elapse before anything could be effected, and, as hostilities must continue during this delay, much suffering would be caused.

Vice-President Burger went on to say that when he received a copy of this correspondence he could only conclude that Lord Kitchener, indirectly at least, if not directly, was asking the Boer leaders to negotiate with him. Accordingly, he wrote to Lord Kitchener for a free pass, and, having obtained it, came with his Government by rail to Kroonstad. He now, accordingly, requested President Steyn to let him know when and where the two Governments could meet. He also intimated that he had written to Lord Kitchener, informing him that he wished—after consulting the Government of the Orange Free State—to make a Peace Proposal.

President Steyn told me that when the Free State Government received this letter from President Burger, they had not been able to see their way to refuse what the latter asked, as the promise of a Peace Proposal had already

[302] A salt lake.

[303] Commandant Jacobsz was somewhere not very far from Kimberley; Commandant Bester, close to Brandfort; Commandant Jacobus Theron, near Smaldeel; Commandant Flemming, near Hoopstad; and Commandant Pieter Erasmus, near the Gannapan.

been sent. They had regretted, however, that the Transvaal Government had made use of a safe conduct, and gone through the English lines—not that they had for one moment distrusted the Government—but simply because the proceeding had seemed to have been ill-advised. Nevertheless the Free State, finding itself not only obliged to discuss the matters in question with the Transvaal, but also, conjointly with the Transvaal, to make a Proposal to Lord Kitchener, had appointed a place of meeting in accordance with the request which had been addressed to it.

This was what I learnt from President Steyn's letter.

On the 5th of April the President received another letter from President [Schalk] Burger, arranging that the meeting should take place at Klerksdorp. A safe conduct for the President and Government of the Orange Free State was sent at the same time.[304] [305] [306]

[304] 3/12/1902 Many prisoners captured by Colonel Ternan and Colonel Pilcher.
3/13/1902 Little garrison of fifty men at Fort Edward surrounded by Beyers' commando.
3/15/1902 Attack on laager near Vryheid by General Bruce-Hamilton. General Cherry Emmett captured.
3/16/1902 Rebels at Sliphock captured by Captain Bowker.
3/17/1902 Some of Bezuidenhout commando captured in Cape Colony by Colonel Baillie.
3/18/1902 Lieutenant Williams, a notorious train-wrecker, captured by National Scouts.
3/21/1902 Colonel Harrison sent out from Pietersburg small force under Colonel Denny to relief of Fort Edward. Advance opposed by Boers.
3/23/1902 Arrival at Pretoria of so-called Acting Transvaal Government to discuss the terms of peace.
3/26/1902 Death of Cecil John Rhodes.
3/28/1902 Colonel Colenbrander from Krugersdorp moved to Pietersburg and from thence accomplished relief of Fort Edward.
3/29/1902 Total defeat of Beyers and dispersal of investing commando.
3/30/1902 Serious railway accident at Barberton.

[305] 3/31/1902 The Battle of Hart's River, also known as the Battle of Boschbult. A British column under Colonel Cookson, dispatched by General Walter Kitchener to pursue a force of 2,500 Boers under the command of generals Kemp and De La Rey, who were believed to be operating in the western Transvaal, came into contact with an advance party of Boers. The main force of the column took off after the Boers, but around 1:30 in the afternoon ran into the full force, 2,500 strong, and became encircled by the larger numbers. The Boers had brought up pieces of artillery and began shelling the British positions, using the artillery as a screen to push forward mounted troops and drive in British defenses.
The 2nd Canadian Mounted Rifles, who had been guarding the baggage train, spurred forward towards the fight and made a series of determined charges on the Boer positions to relieve pressure on the main force. Twenty-one Canadians from 3 and 4 Troops, 'E' Squadron, under the command of Lt Bruce Carruthers became cut off from the main force during a charge, but rather than surrender they determined to fight to the last, eventually running out of ammunition and being overrun. Around 5 pm, the Boers called off the attack and withdrew, having inflicted great damage on the British force and escaping with minor losses.
The Canadian Mounted Rifles suffered the most casualties during the battle, losing 13 men killed and over 40 men wounded, the second largest loss of life in battle for Canada after Paardeberg. A telegram to the Minister of Militia of Canada stated: "The regiment and field hospital have undergone [a severe] test, and have acquitted themselves most creditably. I regret the heavy losses." Lord Roberts telegraphed his congratulations to the Governor-General of Canada as well as to the men in the field
https://en.wikipedia.org/wiki/Battle_of_Hart%27s_River

[306] Battle of Boschbult [Translation from Afrikaans Wikipedia] De la Rey's scouts informed him of the British Forces' movements between Klerksdorp and Lichtenburg. He became aware of a force of 1800 British soldiers with four cannons at a farm called Brakspruit. On March 30, he informed his forces that he was planning an attack for the morning of March 31.
At ten o'clock in the morning, a group of Boers attacked the British Force in an attempt to create disorder. The British, however, do not pursue the British, but dig them in themselves. At half-past three, Generals Du Toit and Kemp's men began to stalk the British forces. At the same time, the Boers were also moving loose with cannons on the British line. The mules are stunned and go on a panicked, wild rush. Generals Du Toit and Kemp decided to exploit the disorder and began their attack on the British camp. Kemp believed that fleeing British soldiers were easier targets than British soldiers who were encircled. However, the British forces repulsed the attack and Kemp fell back by five o'clock. De la Rey also never allowed Liebenberg's force to participate in the attack. Kemp was very angry with De la Rey afterwards; however, he says that it was never his plan to attack the British camp. Kemp doesn't believe him.
The British losses were 27 killed and 70 wounded and 100 British soldiers were taken prisoner. On the Boer side, six soldiers died and 16 were wounded. https://af.wikipedia.org/wiki/Slag_van_Boschbult

Stone commemorating Canadians who died at the Battle of Harts River, 1902.[282] Commandant Potgieter 30 yards from the British line, after the battle of Rooiwal on 11 April 1902.[283] He and 50 of his men died gallantly charging the British line on horseback.

General Christiaan de Wet [Far Right] on his way to peace talks at Vereeniging, on April 13th 1902.[284]

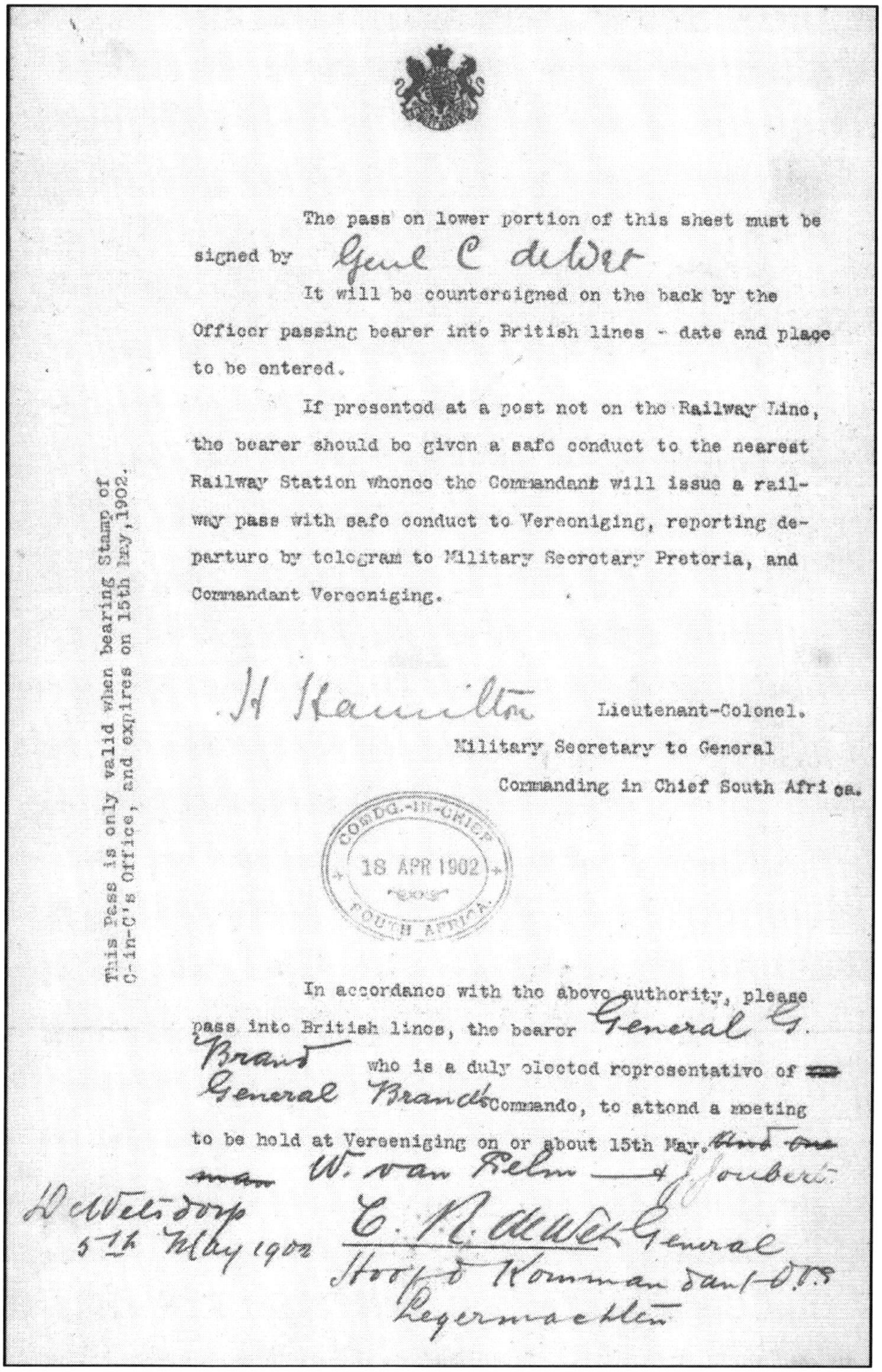

This Pass is only valid when bearing Stamp of C-in-C's Office, and expires on 15th May, 1902.

The pass on lower portion of this sheet must be signed by Genl C de Wet

It will be countersigned on the back by the Officer passing bearer into British lines - date and place to be entered.

If presented at a post not on the Railway Line, the bearer should be given a safe conduct to the nearest Railway Station whence the Commandant will issue a railway pass with safe conduct to Vereeniging, reporting departure by telegram to Military Secretary Pretoria, and Commandant Vereeniging.

H Hamilton Lieutenant-Colonel.
Military Secretary to General
Commanding in Chief South Africa.

COMDG.-IN-CHIEF
18 APR 1902
SOUTH AFRICA

In accordance with the above authority, please pass into British lines, the bearer General G. Brand who is a duly elected representative of ~~the~~ General Brand's Commando, to attend a meeting to be held at Vereeniging on or about 15th May. ~~And one man~~ W. van Eelm — & J Joubert

De Wetsdorp
5th May 1902 C. R. de Wet General
Hoofd Kommandant O.V.S.
Legermachten

Pass countersigned a the bottom by General de Wet's for safe conduct of General Brand during peace negotiations.[285] Dated May 5th, 1902 by De Wet.[307]

Identification from the Conference minutes infra. G.A. Brand, Vice-Commandant-in-Chief; for Bethulie, Rouxville, Caledon River, and Wepener in the eastern part of Bloemfontein.

[307] 3/31/1902 Delarey defeated in engagement with Colonels Keir and Cookson.
4/1/1902 Boschman's Kop, Transvaal. British forces surprised Gens. H.A. Alberts and P.R. Viljoen. They regrouped, attacked the British forces, and drove them out during the day.
4/4/1902 Ookiep invested by Commandant Smuts.
4/8/1902 Successful attack on Beyers' laager near Pietersburg by Colonels Colenbrander and Murray.

36. Peace Negotiations

General De la Rey, who, as a Member of the Transvaal Government, had to be present at the coming deliberations, accompanied the President to Klerksdorp, where they arrived on the 9th of April, and found the Transvaal Government already there awaiting them.[308]

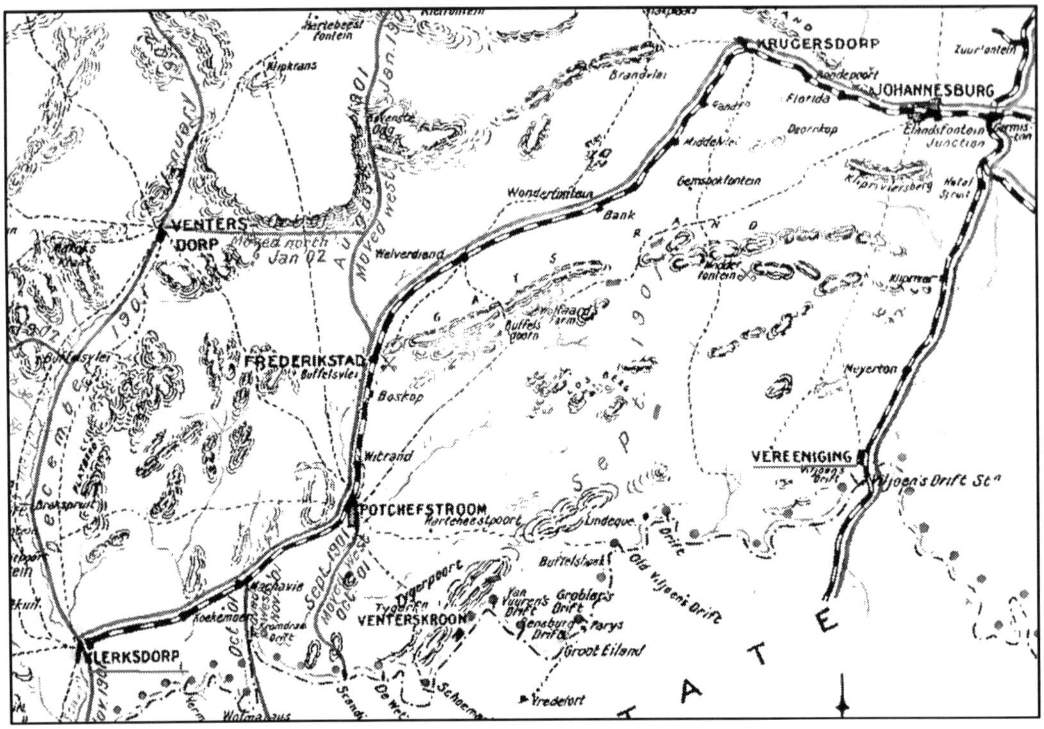

Klerksdorp is just inside the Transvaal Border. Note Vereeniging on the right.

The two Governments held their first meeting in the afternoon of the same day. The South African Republic was represented by:—Vice-States-President S.W. Burger; Commandant-General Louis Botha; Secretary of State F.W. Reitz; General De la Rey; Ex-General L.J. Meijer; and Mr. J.B. Krogh. Although not a member of the Government, the States-Procureur, L. Jacobsz, was also present.

On behalf of the Orange Free State appeared:—States-President M.T. Steyn; Commander-in-Chief C.R. de Wet; Vice-Commander-in-Chief Judge J.B.M. Hertzog; States-Secretary W.J.C. Brebner; and General C.H. Olivier.

It was decided that no minutes should be taken. Accordingly, I am only able to give a summary of the proceedings.

The meeting having been opened with prayer, the Vice-President of the South African Republic said that the fact that Lord Kitchener had sent in a copy of the correspondence between the Governments of the Netherlands and England, was looked upon by himself and his Government as an invitation on the part of England to the two States to discuss the matter dealt with in that correspondence, and to see if peace could not be concluded. Before, however, the meeting could make a proposal, it would be necessary to hear what the state of affairs really was.

Thereupon, firstly, Commandant-General Louis Botha, then I, and lastly, General De la Rey, gave a report of how matters stood.

President Burger now asked whether an interview with Lord Kitchener should be asked for, and (in case Lord Kitchener acceded to this) what we were to demand, and what we should be prepared to sacrifice. He went on to ask President Steyn what he thought of the proposal which the Transvaal had made to the Free State Government in the October of the previous year.

[308] 4/9/1902 Conference between Transvaal and Orange Free State leaders at Klerksdorp in regard to negotiations for peace.

President Steyn answered that he was still of the same opinion as in June, 1901, when the two Governments had agreed to stand by Independence. If the English now refused to grant Independence, then the war must continue. He said that he would rather surrender to the English unconditionally than make terms with them.

The remainder of the day was occupied in listening to speeches from State-Secretary Reitz and President Burger.

On the following day the speakers were:—L.J. Meijer, J.B. Krogh, myself, State-Secretary Reitz, and Judge Hertzog. The last-named made a proposal, which was seconded by General C.H. Olivier. This proposal, after it had been subjected for revision to a Commission, consisting of the two Presidents, Mr. Reitz, and Judge Hertzog, was accepted on the following day. It ran as follows:—

"The Governments of the South African Republic and of the Orange Free State, having met, induced thereto by the receipt, from His Excellency Lord Kitchener, of the correspondence exchanged in Europe between the Government of His Majesty the King of England, and that of Her Majesty the Queen of the Netherlands, referring to the desirability of giving to the Governments of these Republics an opportunity to come into communication with their plenipotentiaries in Europe, who still enjoy the trust of both Governments:

"And taking into consideration the conciliatory spirit which, as it appears from this correspondence, inspires the Government of His Britannic Majesty, and also of the desire therein uttered by Lord Lansdowne, in the name of his Government, to make an end to this strife:

"Are of opinion that it is now a favourable moment to again shew their readiness to do everything possible to bring this war to an end:

"And decide, therefore to make certain proposals to His Excellency Lord Kitchener, as representative of the Government of His Britannic Majesty, which may serve as a basis for further negotiations, having in view the achievement of the desired peace.

"Further, it is the opinion of these two Governments that, in order to expedite the achievement of the desired aim, and to prevent, as far as possible, any misunderstanding, His Excellency Lord Kitchener should be asked to meet personally these Governments at a time and place by him appointed, so that the said Governments may lay before him Peace Proposals (as they will be prepared to do), in order that, by direct conversation and discussion with him, all such questions as shall arise may be solved at once, and also that this meeting may further and bring about the desired result."

A letter was now written to Lord Kitchener (who was at Pretoria) enclosing the above Proposal, and signed by the two Presidents.

In the afternoon the two Governments met again, to consider what proposals they should make to the British Government. After a lengthy discussion, it was decided, on the proposal of General De la Rey, seconded by States-Procureur L. Jacobsz, that the matter in hand should be entrusted to the Commission, which consisted, as I have already said, of the two Presidents, States-Secretary Reitz, and Judge Hertzog: and the next morning this Commission handed in the following report, which was accepted by the meeting:—

"The Commission, after having taken into consideration the wish of the meeting, namely, that proposals should be drafted (in connexion with the letter of yesterday, signed by the two Presidents, to His Excellency Lord Kitchener) for eventual consideration by His Excellency Lord Kitchener, proposes the following points:—

"1. The concluding of a Treaty of Friendship and Peace, including:

"(a) Arrangements re a Customs Union.

"(b) re Post, Telegraph and Railway Union.

"(c) Granting of the Franchise.

"2. Demolition of all States Forts.

"3. Arbitration in any future differences which may arise between the contracting parties; the arbitrators to be nominated in equal numbers from each party from among their own subjects; the said arbitrators to add one to their number, who is to have the casting vote.

"4. Equal rights for the English and Dutch languages in the schools.

"5. Reciprocal amnesty."

The same morning a letter enclosing this proposal was sent to Lord Kitchener, after which Judge Hertzog and Commander Louis Botha addressed the meeting.

After the latter had finished an address of great importance, General Wilson, who had the command at Klerksdorp, entered the room where the meeting was being held and stated that Lord Kitchener was prepared to grant us an interview, and that we could travel to Pretoria that very evening.

Accordingly, on the evening of the 11th of April, we went to Pretoria, where, on the following morning, we met Lord Kitchener and handed in our proposal.[309]

Lord Kitchener wished for a proposal of a very different character from that of the two Governments; but as it would not have been proper for them to make any proposal injurious to Independence, the Presidents declared that they could not do so, and asked him to send to the English Government the proposal which they had already laid before him. Lord Kitchener at last acceded to this request, and the following telegram was accordingly sent to England:

From Lord Kitchener to the Secretary of State.

"Pretoria, April 12th, 1902.

"The Boer Representatives desire to acquaint His Majesty's Government with the fact that they entertain an earnest wish for peace, and that they, therefore, have decided to ask the British Government to bring hostilities to an end, and to proceed to formulate a Treaty of Peace. They are ready to accept an Agreement, by which, in their opinion, all future wars between them and the British Government in South Africa may be avoided. They think that this aim can be attained if provisions are made in relation to the following points:—

"1. Franchise.

"2. Equal rights for the Dutch and English languages in Educational matters.

"3. Customs Union.

"4. Demolition of all the forts in the Transvaal and Free State.

"5. Arbitration in case of future disagreements, and only subjects of the parties to be arbitrators.

"6. Mutual amnesty.

[309] 4/11/1902 Meeting of Boer representatives at Klerksdorp in relation to Peace movement. Colonel Kekewich defeated Boers in Western Transvaal.
The Battle of Rooiwal was an engagement of the Second Boer War. It took place on 11 April 1902 and resulted in a victory by a British force commanded by Colonel Robert Kekewich over a Boer commando led by Generals Ferdinandus Jacobus Potgieter and Jan Kemp.
The action consisted of a Boer attack on horseback on an entrenched British hillside position in the valley of Rooiwal, near Klerksdorp in the Western Transvaal. The Boers were attempting to break out of a British encircling manoeuvre. Their attack was repulsed at some cost to the Boers in killed and injured.
On 6 April, Kitchener put Colonel Ian Hamilton in command of another drive to try to trap De La Rey's fighters. The plan was to 'squeeze' the Boers against the British mobile columns and a line of blockhouses and entrenchments at Klerksdorp. Colonel Robert Kekewich, who was in command of one of Kitchener's columns, dug in at Rooiwal to strengthen his left flank. Having mistakenly gotten tangled up with another British column under Colonel Henry Rawlinson, Kekewich was ordered by Hamilton to proceed to Rooiwal, where he arrived on 10 April. This change of plan proved to be a fortunate one for the British, because the Boers had scouted the Rooiwal position earlier and found it weakly defended. Dug in on the hillside at Rooiwal were about 3,000 mounted infantry, supported by 6 field guns and 2 pom poms.
The Boers, however, did not know of the British deployment and they still believed that Rooiwal was only lightly held. One of their commandos, under Commandant Potgieter and General Kemp, therefore tried to overrun the British position early on the morning of 11 April, in an effort to escape Hamilton's 'drive'. Potgieter had around 1,700 men, all mounted riflemen.
At around 7:15 am on 11 April, they charged the British position on horseback, firing from the saddle. A British picket of 40 mounted infantry was overrun, taking 20 casualties. Kekewich's position was a strong one, but the sight of the charging Boers panicked some of the inexperienced British troops and a number of Yeoman units fled the scene of the battle and were not stopped until they were a mile away from the fighting. A Lieutenant Carlos Hickie managed to stop the stampede with a mixture of pleas and threats. In addition, a number of the regular British officers on the scene were very critical of the 'wild' shooting of their men.
In spite of this, however, the Boer charge was stopped about 30 metres from the British line by artillery and rifle fire. Fifty Boers were killed outright and more were wounded. Among the dead was Potgieter, wearing a distinctive blue shirt. Kekewich later commented that, 'one good company of infantry could have killed 300 Boers'. The surviving Boers made good their retreat. Boer fire, delivered from the saddle, had produced about 50 casualties in the British line.
Ian Hamilton and Rawlinson arrived on the scene just as the fighting was ending. However, Hamilton delayed the pursuit of the beaten Boers as he feared that the retreat was a ruse and that his men would fall into Boer ambushes. At about 9:45, or 90 minutes after the Boer charge had been repulsed, Hamilton sent his mounted troops in pursuit of the enemy. They captured a further 50 Boers and re-captured the artillery lost at Tweebosch.
Kitchener had issued orders that Boers captured wearing British uniforms were to be shot. However, although a number of wounded Boers were indeed wearing captured British khaki, Hamilton ordered that they be spared.
https://en.wikipedia.org/wiki/Battle_of_Rooiwal

"But in case these terms should not be satisfactory, then they wish to know what terms the British Government will give them, so that the result which they all desire may be attained."

On Monday, April 15th, Lord Kitchener sent to the two Governments a copy of the following telegram, which he had received from the Secretary of State:—

From Secretary of State to Lord Kitchener.

"London, April 13th, 1902.

"His Majesty's Government shares with all its heart in the earnest wish of the Boer Representatives, and trusts that the present negotiations will lead thereto. But they have already declared in the clearest manner and have to repeat that they cannot take into consideration any proposals which have as basis the sanction of the Independence of the former Republics, which are now formally annexed to the British Crown. And it would be well if you and Milner were to meet the Boer Representatives, and make this plain to them. You must encourage them to make fresh proposals which we will willingly receive."

In this telegram, as the reader will have observed, the name of Lord Milner is mentioned. Up till now we were dealing with Lord Kitchener alone, but at our next conversation the first-named was also present.

Both Representatives of the British Government insisted that we should negotiate with them, taking the surrender of our Independence for granted. We could not do so. We had repeatedly told Lord Kitchener that, constitutionally, it was beyond the power of our Governments to discuss terms based on the giving up of Independence. Only the nation could do that. Should however, the British Government make a proposal which had, as a basis, the temporary withdrawal only of the Independence, then we would lay this proposal before the nation.

Thereupon the following telegram was drawn up and dispatched:—

From Lord Kitchener to the Secretary of State.

"Pretoria, April 14th, 1902.

"A difficulty has arisen in connexion with the negotiations. The representatives declare that, constitutionally, they are not entitled to discuss terms which are based on the surrender of their independence, as the burghers alone can agree to such a basis. If, however, His Majesty's Government can propose terms by which their independence shall be subsequently given back to them, the representatives, on the matter being fully explained to them, will lay such conditions before the people, without giving expression to their own opinions."

The reply to this was as follows:—

From the Secretary of State to Lord Kitchener.

"London, April 16th.

"With great astonishment we have received the message from the Boer leaders, as contained in your cable. The meeting was arranged in accordance with their desires, and they must have been aware, from our repeated declarations, that we should not be prepared to consider any proposal based on the revival of the independence of the two South African States. We, therefore, were justified in believing that the Boer representatives had abandoned all idea of Independence, and that they would make terms for the surrender of the forces still in the Veldt. They now declare that they are not constitutionally in a position to discuss any terms which do not include the restoration of their Independence, but they ask what conditions would be made if, after consulting their followers, they should abandon the claim for Independence. This does not seem to us a satisfactory way of expediting the end of the hostilities which have caused the loss of so many lives and so much money.

We are, however, as we said before, desirous of preventing any further bloodshed and of accelerating the restoration of peace and prosperity in the countries harassed by the war, and we empower you and Lord Milner to refer the Boer leaders to the offer made by you to General Botha more than twelve months ago, and to inform them that—although the great decrease which has lately taken place in the forces opposed to us, and also the further sacrifices involved by the refusal of that offer, would justify us in dictating harder terms—we are still prepared, in the hope of a lasting peace and reconciliation, to accept a general surrender in the spirit of that offer, with such amendments with regard to details as might be agreed upon mutually."

It was quite self-evident that the Governments could not accept this proposal of the British Government, because by it the independence of the Republics would be sacrificed.

President Steyn pointed out emphatically that it lay beyond our right to decide and conclude anything that would endanger the independence of the two Republics. The nation alone could decide on the question of independence. For this reason, therefore, we asked if we might consult the people, and it was agreed by Lord Kitchener and Lord Milner that we should go back to our commandos and hold meetings in every district, in order to learn thus the will of the nation.

It was further agreed that at the meetings of the nation representatives should be chosen who, on the 15th of May, 1902, at Vereeniging, should inform the Governments what course the nation desired them to take.

On the 18th of April Commandant-General Louis Botha, General De la Rey, and I left Pretoria, provided with a safe conduct for ourselves and for anyone whom we should appoint, and proceeded to our different commandos.

I went first to the burghers of Vrede at Prankop, where I met General Wessel Wessels with his commandos on the 22nd of April. The nation was in a very miserable condition, suffering from the want of all necessaries, and living only on meat and maize, which food was also exceedingly scarce, and would only last for a few months more. Notwithstanding this, the burghers decided, to a man, that they would not be satisfied with anything less than independence, and that if the English would not accede to this they would continue to fight.

Mr. Wessel Wessels, Member of the Volksraad, was elected as chairman, and Mr. Pieter Schravezander as secretary. The representatives chosen were Commandants A. Ross, Hermanus Botha, and Louis Botha (son of Philip Botha).

My second meeting I held at Drupfontein, in the district of Bethlehem, on the 24th of April, with the burghers under the command of Commandants Frans Jacobsz, Mears, and Bruwer. Mr. J.H. Naude was made chairman, and Landdrost J.H.B. Wessels secretary. It was unanimously decided that independence had to be maintained, and Commandants Frans Jacobsz and Bruwer were chosen as representatives.

The next meeting I held on the 26th of April, at Tweepoort Farm, with the commandos under General Michal Prinsloo. Mr. Jan Van Schalkwijk was chosen as chairman, and Mr. B.J. Malan as secretary. Here also the votes were unanimous, and General Michal Prinsloo, Commandant Rautenbach, and Commandant J.J. Van Niekerk were elected as representatives.

After that on Roodekraal Farm. I met the burghers under Commandants Cilliers, Bester, Mentz, and Van Coller. The chairman was B.W. Steyn (Member of the Volksraad), and the secretary Mr. S.J.M. Wessels. Here again it was unanimously decided not to surrender the independence, and Commandants Mentz, Van Coller and Bester were the representatives chosen.

The fifth meeting I held with the commandos under General Johannes Hattingh, on the 1st of May, on the Weltevrede Farm, under the chairmanship of Mr. Jan Lategan, Johannes C. Pietersen being secretary. As representatives we chose General Hattingh and Commandant Philip De Vos. The voting was unanimous that the independence should be maintained.

On the 3rd of May I held my sixth meeting, with the commandos under General C.C. Froneman, at Schaapplaats. Mr. Jan Maree was chairman, and Mr. David Ross secretary.

The result was the same as at the other meetings, and General Froneman, Commandants F. Cronje and J.J. Koen were chosen to represent the commandos.

From there I went to Dewetsdorp, where I met, on the 5th of May, General George Brand's commandos. Mr. C. Smith acted as chairman, and Mr. W.J. Selm as secretary; the representatives chosen were General Brand and Commander J. Rheeder; and the burghers were equally determined to keep their independence.

I went on to Bloemfontein, and thence by rail to Brandfort, and afterwards to the Quaggashoek Farm, where, on the 11th, I held my eighth meeting, with the commandos of C.C.J. Badenhorst. The chairman was Mr. N.B. Gildenhuis, and the secretary Mr. H.M.G. Davis. The elected representatives were General Badenhorst and Commandants A.J. Bester and Jacobsz. This was my last meeting, and it also decided on maintaining the independence.

The commandos under the Commandants Van der Merwe and Van Niekerk (Vredefort and Parijs), Flemming (Hoopstad), Nagel (part of Kroonstad), and General Nieuwouwdt (Fauresmith, Philippolis, and Jacobsdal), were visited by Commander-in-Chief Judge Hertzog, Member of the Executive Council. At meetings held with these commandos the following representatives were chosen:—General Nieuwouwdt, and the Commandants Munnik

Hertzog, J. Van der Merwe, C. Van Niekerk, Flemming, A.J. Bester, F. Jacobsz, H. Pretorius, and Veldtcornet Kritzinger.

At these meetings also the burghers were unanimous in their decision not to give up their independence. I must add that Commandant H. Van Niekerk was chosen as representative of the bodyguard of President Steyn. It had been agreed with Lord Kitchener at Pretoria that if the chief officers of a commando were chosen as representatives, then there would be an armistice between this commando and the English during the time the officers were absent at the meeting at Vereeniging. It was also decided that Lord Kitchener should be informed of the date of the departure of such officers.

This was done. I sent the following telegram on the 25th of April to Pretoria:—

“To His Excellency, Headquarters, Pretoria:

“At meetings held in the districts of Vrede and Harrismith and in that part of Bethlehem east and north-east of the blockhouse lines of Fouriesburg, Bethlehem, and Harrismith, General Wessels and the Commandants were duly chosen as representatives.

“I have decided that all the representatives shall leave their different commandos on the 11th of May, and therefore, in accordance with our mutual agreement, I shall expect an armistice to be granted to the different commandos from that date until the return of their commandants from the meeting at Vereeniging, on or about the 15th of May.

“I should be glad to receive Your Excellency’s sanction to my request that each Representative should have the right to take one man with him.

“Your Excellency will greatly oblige by sending a reply to Kaffirsdorp in the district of Bethlehem, where I am awaiting an answer.

“C.R. De Wet,

General Commander-in-Chief, Orange Free State.

Bethlehem, April 25th, 1902.”

To this I received the following answer from Lord Kitchener:—

“Imperial Residency, Pretoria,

April 25th, 1902.

“To General de Wet, Kaffirsdorp.

“In answer to your message, I agree altogether with your demands that during the absence of the chosen Representatives from their commandos, from the 11th of May until their return, such commandos shall not be troubled by us. I also agree that every Representative, as you propose, shall be accompanied by one man.

“I shall also be glad if you would send an officer, at least two days before the Meeting, in order to let me know about the number, and the necessary arrangements for the treatment of the Representatives at this Meeting.

(Signed) “Kitchener.”

On the 11th of May I sent a telegram to Lord Kitchener, in which I said that, as all my generals and chief officers had been chosen as Representatives, the armistice must begin on the 11th of May. The telegram was as follows:—

From General de Wet to His Excellency Lord Kitchener.

“Pretoria, May 11th, 1902.

“The following chief officers have been chosen as Representatives for the commandos of the districts: Hoopstad, Boshof, and parts of Winburg and Bloemfontein,—districts to the west of the railway line.

“1. General C. Badenhorst.

“2. Commandant J. Jacobsz.

“3. Commandant A. Bester.

“It thus appears that all my generals and chief commanding officers are chosen as Representatives to attend at the Meeting of Vereeniging, on the 15th inst., and according to our mutual agreement at Pretoria, an armistice will be given from to-day (11th May, 1902) in all districts of the Orange Free State up to a date which shall be agreed upon after the close of the Meeting at Vereeniging. Any answer, previous to noon of the 11th inst., will reach me at Brandfort.

"Commander-in-Chief,

Orange Free State Armies."

In answer to this I received the following telegram:—

"Imperial Residency, Pretoria,

May 12th.

"To General de Wet, Brandfort.

"I have given orders, according to our Agreement, that from to-morrow, the 13th inst., all commandos, whose leaders or chief officers have been chosen to attend the Meeting at Vereeniging, shall be exempted from being attacked by my columns during the absence of their leaders, in so far as such commandos withhold from offensive operations. But that does not imply that outposts cannot be taken prisoner in case they should approach our lines.

"Kitchener."

It was rather surprising to me that Lord Kitchener, in this telegram, spoke only of an armistice beginning on the 13th of May, because in his telegram of the 25th he had agreed that there should be an armistice from the 11th of May. I heard also from officers of Heilbron, Vrede, and Bethlehem, whom I met, on the evening of the 14th of May, at Wolvehoek Station, that the English columns had operated in their districts on the 11th, 12th, 13th, and 14th. My order was that my officers should not operate, but should retreat, if the enemy should unexpectedly operate on the 11th. On the above-mentioned dates houses were burnt down, cattle carried away, maize and other grain destroyed, burghers taken prisoner, and (in one instance) shot.

Such a misunderstanding was very regrettable, and all the more so because we were never indemnified for the damage thus done.[310]

[310] 4/12/1902 Laager at Schweizerreneke surprised by Colonel Rochfort. Fifty-five prisoners taken.
5/1/1902 Relief of Ookiep by British troops under Colonels Cooper and Caldwell.
5/2/1902 Lieutenant Murray (District Mounted troops) killed at Tweefontein by Boers in khaki.
5/6/1902 Pieter de Wet sentenced by Treason Court to pay a fine of £1,000 or undergo two years' imprisonment.
5/6/1902 Holkrans, Natal. 56 Boers are killed by Zulus.
5/9/1902 Patrol attacked by Boers near Middelburg, Cape Colony. Captain Hinks killed.

37. The End of the War

On the morning of the 15th of May, I arrived at Vereeniging with some of the Free State delegates.

The others were already there, together with the thirty Transvaal delegates, Commandant-General Louis Botha and General De la Rey. In addition to the above, the following had also arrived: Vice-State President Burger, States-President Steyn, the members of the two Governments, and General J.C. Smuts (from Cape Colony).

I was exceedingly sorry to find that President Steyn was seriously ill. For the last six weeks he had been in the doctor's hands; and, since his arrival at Pretoria, had been under the care of Dr. Van der Merwe, of Krugersdorp. This physician said that serious consequences might ensue if his patient were to attend our meetings, and advised him to go to his home at Krugersdorp, where he could be properly nursed. It was sad for us to receive this news immediately we arrived. We asked ourselves what we should do without the President at our meetings? At this moment he seemed more indispensable to us than ever before.

President Steyn was a statesman in the best sense of the word. He had gained the respect and even the affection of us all. Of him, if of any man, it may be said that he never swerved from his duty to his country. No task was too great for him, no burden too heavy, if thereby he could serve his people. Whatever hardships he had endured, he had never been known to complain—he would endure anything for us. He had fought in our cause until he could fight no longer, until sickness laid him low; and he was worn out, and weak as a child. Weak, did I say? Yes! but only in the body—his mind was still as strong, as brave, as clear as ever.

And thus it was that President Steyn was only able to be present on two occasions at our meetings; for, on the 29th of May—before the National Representatives had come to any decision—he went with Dr. Van der Merwe to Krugersdorp.

As I write these lines—six months after the meetings at Vereeniging—and think that during all the intervening time he has been lying on a bed of sickness—I am cheered by the news which I received in Holland that hopes are now entertained of his ultimate recovery.

The National Representatives began their important deliberations on the morning of the 13th of May, 1902.

For three days we discussed the condition of our country, and then proceeded with Lord Kitchener and Lord Milner to Pretoria. This Commission was composed of Commandant-General L. Botha, Commander-in-Chief C.R. de Wet, Vice-Commandant-General J.H. De la Rey, Vice-Commander-in-Chief Judge J.B.M. Hertzog, and States-Procureur J.C. Smuts.

The negotiations with the representatives of the British Government continued from the 18th to the 29th of May; and upon their conclusion the Commission communicated to the National Representatives the terms on which England was prepared to conclude peace.[311]

On May the 31st we decided to accept the proposals of the English Government.[312] The Independence of the two Republics was at an end!

I will not attempt to describe the struggle it cost us to accept these proposals. Suffice it to say that when it was over, it had left its mark on every face.

There were sixty of us there, and each in turn must answer Yes or No. It was an ultimatum—this proposal of England's.

What were we to do? To continue the struggle meant extermination. Already our women and children were dying by the thousand, and starvation was knocking at the door—and knocking loudly!

In certain districts, such as Boshof and Hoopstad, it was still possible to prolong the war, as was also the case in the districts of Generals Brand and Nieuwouwdt, where the sheep and oxen, which had been captured from the enemy, provided an ample supply of food. But from the last-named districts all the women and children had departed, leaving the burghers free to wander at will in search of food—to Boshof, to Hoopstad, and even into the Colony.

[311] 5/15/1902 Members of the late Governments met together to discuss Peace proposals.
5/17/1902 Surrender of Hinton, the notorious train-wrecker.
5/20/1902 Delegates of late Governments arrived at Pretoria to arrange terms of surrender.
5/27/1902 Malan mortally wounded and captured by Jansenville District Mounted Horse (under Major Collett), and Lovat's Scouts.

[312] A complete report of the various proceedings in connexion with the conclusion of peace will be found in the Appendix of this book.

In other parts of the Free State things were very different. In the north-eastern and northern districts—for instance, in Ladybrand, Winburg, Kroonstad, Heilbron, Bethlehem, Harrismith and Vrede—there were still many families, and these could not be sent to Boshof or to Hoopstad or to the Colony. And when, reduced to dire want, the commandos should be obliged to abandon these districts, their wives and families would have to be left behind—to starve!

The condition of affairs in the Transvaal was no better. We Free-Staters had thought—and I, for one, had supported the view at Vereeniging—that, before sacrificing our independence, we ought to tell the owners of these farms, where there were still women and children, to go and surrender with their families, and thus save them from starvation. But we soon realized that such a course was not practicable—it would involve the loss of too many burghers.

Moreover, even if, by some such scheme as this, we had succeeded in saving the women, we, who remained in the field, would still have been exposed to the dangers of starvation, for many of us, having no horses, could not have left want behind us, by removing to Cape Colony or some other equally prosperous region.

In the large eastern divisions of the Transvaal also, there were many burghers without horses, while the poor jaded creatures that remained were far too feeble and exhausted to carry their masters into Cape Colony, without the certainty of being captured by the enemy.

Our forces were now only twenty thousand in all, of which the Transvaal supplied ten thousand, the Free State six thousand, while the remainder came from Cape Colony. But our numerical weakness would not in itself have caused us to abandon the struggle had we but received encouraging news from the Colony. But alas! reports which we received from there left us no room for hope.

No room for hope! that was the message of Vereeniging—a message which struck a chill in every heart. One after another we painted the destitution, the misery of our districts, and each picture was more gloomy than the last. At length the moment of decision came, and what course remained open to us? This only—to resign ourselves to our fate, intolerable though it appeared, to accept the British proposal, and to lay down our arms.

Most bitter of all was the thought that we must abandon our brethren in Cape Colony and in Natal, who had thrown in their lot with ours. And many a sleepless night has this caused me. But we could not help ourselves. There was nothing else to do.

And as things have turned out, may we not hope that the Cape and Natal Governments, following in the wake of the British Nation, will soon understand that the wiser course is to forgive and forget, and to grant as comprehensive an amnesty as possible? It is surely not unjust to expect this of these Governments, when one remembers that whatever the Colonists may have done, must be ascribed to the tie that binds them to us—the closest of all ties—that of blood.

It is now for the two Governments to strive to realize the situation, and then, by granting a general amnesty, to promote, as far as in them lies, the true progress of South Africa.

On the evening of the 31st of May, 1902, the members of the Government of both Republics met Lord Kitchener and Lord Milner, in the former's house, at Pretoria.

It was there that the Treaty of Peace—the British Proposal which the National Representatives had accepted—was now to be signed.

It was a never-to-be-forgotten evening. In the space of a few short minutes that was done which could never be undone. A decision arrived at in a meeting could always be taken into reconsideration, but a document solemnly signed, as on that night, by two parties, bound them both for ever.

The table in Melrose House where the Treaty of Vereeniging was signed, which ended the Second Boer War in 1902.[313]

Every one of us who put his name to that document knew that he was in honour bound to act in accordance with it. It was a bitter moment, but not so bitter as when, earlier on the same day, the National Representatives had come to the decision that the fatal step must be taken.

On the 2nd June, 1902, the Representatives left Vereeniging, and returned every man to his own commando. It was now their sad duty to tell their brave and patient burghers that the independence which they cherished so dearly was gone, and to prepare them to surrender their arms at the appointed places.

I left Pretoria on the 3rd of June with General Elliott, who had to accompany me to the various centres to receive the burghers' arms.

General L. Botha [With 'Aussie' hat, right side pinned up.] in conversation with General Lyttelton and Staff at Newcastle Station.[286] "The Commissioners travelled to Pretoria, and on the same night, at five minutes past eleven, only an hour before the expiry of the term of grace, the Treaty was signed. "We are good friends now," said Kitchener, as he shook hands with the representatives of the late Republics."[314]

[313] The table in Melrose House where the Treaty of Vereeniging was signed, which ended the Second Boer War in 1902. Author Amanda Calitz. https://commons.wikimedia.org/wiki/File:Table_where_treaty_was_signed.JPG
[314] The Times' History of the War in South Africa 1899–1902 by Amery, L. S. (Leopold Stennett), Volume 5 page 606.

On the 5th of June the first commando laid down their weapons near Vredefort. To every man there, as to myself, this surrender was no more and no less than the sacrifice of our independence. I have often been present at the death-bed and at the burial of those who have been nearest to my heart—father, mother, brother and friend—but the grief which I felt on those occasions was not to be compared with what I now underwent at the burial of my Nation!

It was at Reitz that the commandos of Vrede, Harrismith, Heilbron and Bethlehem laid down their arms. Accordingly I went there on the 7th of June, and again had to be a spectator of what I fain would never have witnessed. Had I then to go on from commando to commando, to undergo everywhere the martyrdom of beholding ceaseless surrenders? No! I had had enough, and could bear no more. I decided, therefore, to visit all the other commandos, in order to acquaint the burghers with what had taken place, and to explain to them why we, however unsatisfactory the Peace Proposal was, had felt bound to accept it, and then to leave each commando before the men handed over their arms to General Elliott. Everywhere I found the men utterly despondent and dissatisfied.

The whole miserable business came to an end on the 16th of June, when the burghers who had fought under Generals Nieuwouwdt and Brand, laid down their arms—the Nation had submitted to its fate!

There was nothing left for us now but to hope that the Power which had conquered us, the Power to which we were compelled to submit, though it cut us to the heart to do so, and which, by the surrender of our arms, we had accepted as our Ruler, would draw us nearer and ever nearer by the strong cords of love.

To my Nation I address one last word.

Be loyal to the new Government! Loyalty pays best in the end. Loyalty alone is worthy of a Nation which has shed its blood for Freedom!

Boer general Christiaan de Wet, photographed at the end of the Second Boer War in Bloemfontein on June 6th 1902.[315]

Christiaan de Wet Cartoon in Vanity Fair July 31st, 1902.[287]

[315] Boer general Christiaan de Wet, photographed at the end of the Second Boer War in Bloemfontein on 6.6.1902, before embarking for Europe. Source: S. King File:Generaal Christiaan de Wet - Bloemfontein 6.6.1902.jpg - Wikimedia Commons

Kitchener's Final Despatch

Despatch by General H. Kitchener, dated 23rd June, 1902, relative to Military Operations in South Africa.[316]

From Lord Kitchener to the Secretary of State for War. Cape Town, 23rd June, 1902.

Sir, I have the honour to submit a final despatch which covers the period between the signature of the terms of peace at Pretoria and my departure from Cape Town. ....

2. In my last despatch dated 1st June, I had the pleasure of reporting the conclusion of peace. Directly the terms were signed I appointed the following three Peace Commissioners to receive the surrenders in the territories of the new Colonies:

General Bruce Hamilton for the Eastern Transvaal.

General W. [Walter] Kitchener for the Western Transvaal.

General Elliot[317] for the Orange River Colony.

I also directed General French to make the necessary arrangements for the same purpose in the Cape Colony.

3. The reports of General French and the three Peace Commissioners are herewith submitted. .... The Boer leaders concerned, played their part with loyalty and in a thoroughly straightforward manner, whilst the burghers and rebels on commando also accepted the situation in a manly spirit which augurs well for the future of South Africa. So strongly did I feel this that I sent the following telegram to Generals Louis Botha, De la Rey, and Christian de Wet:

> "Now that the surrender of the arms in the hands of the burghers in the Transvaal and Orange River Colony has been completed, I desire to place on record my high appreciation of the unflagging energy and unfailing tact with which you have facilitated the work of my Commissioners. I recognize how much it is due to your exertions that the burghers have displayed everywhere such a loyal spirit in accepting the change of Government that has taken place, and I can inform you that the manner in which the burghers have come in has greatly pleased His Majesty, the King, and deeply impressed the British people, who are heartily anxious to welcome them as fellow citizens. I feel confident that a new era of complete reconciliation between all races has now dawned in South Africa."[318]

The total numbers of the surrenders of armed burghers [were]

in the Transvaal were	11,166
in the Orange River Colony	6,455
in the Cape Colony	3,635
Total	21,256

4. I well understand that the accompanying list, ... who have in my opinion rendered exceptionally good service during the campaign, is incomplete, ... there must necessarily be a proportion of Officers omitted, who have rendered equal or even better service than those who have been mentioned. This, however, is the fortune of war, and will, I am sure, be well understood by the Army I have been so proud to command. What follows are 21 pages of lists of individuals and units and their service.]

[316] Proclamations Issued In South Africa: Vol II p.757 https://babel.hathitrust.org/cgi/pt?id=hvd.32044089026009&seq=757&view=1up

[317] Lieutenant General Sir Edward Locke Elliot, (28 January 1850–12 December 1938) was a British Army General during the Second Boer War and in British India. Edward Locke Elliot - Wikipedia

[318] At this time the Dutch and English were thought of and considered different 'races'. It was also British policy and thought that the conclusion of the conflict would materially improve the condition of all people in South Africa.

Boer Declaration & Signing

The Boer Declaration prior to signing the peace treaty in May of 1902. [319]

"We, the national representatives of both the South African Republic and the Orange Free State, at the meeting held at Vereeniging, from the 15th of May till the 31st of May, 1902, have with grief considered the proposal made by His Majesty's Government in connexion with the conclusion of the existing hostilities, and their communication that this proposal had to be accepted, or rejected, unaltered. We are sorry that His Majesty's Government has absolutely declined to negotiate with the Government of the Republics on the basis of their independence, or to allow our Government to enter into communication with our deputation. Our people, however, have always been under the impression that not only on the grounds of justice, but also taking into consideration the great material and personal sacrifices made for their independence, that it had a well-founded claim for that independence.

"We have seriously considered the future of our country, and have specially observed the following facts :

"Firstly, that the military policy pursued by the British military authorities has led to the general devastation of the territory of both Republics by the burning down of farms and towns, by the destruction of all means of existence, and by the exhausting of all resources required for the maintenance of our families, the existence of our armies, and the continuation of the war."

"Secondly, that the placing of our families in the concentration camps has brought on an unheard-of condition of suffering and sickness, so that in a comparatively short time about twenty thousand of our beloved ones have died there, and that the horrid probability has arisen that, by continuing the war, our whole nation may die out in this way."

"Thirdly, that the Kaffir tribe, within and without the frontiers of the territory of the two Republics, are mostly armed and are taking part in the war against us, and through the committing of murders and all sorts of cruelties have caused an unbearable condition of affairs in many districts of both Republics. An instance of this happened not long ago in the district of Vryheid, [320] where fifty-six burghers on one occasion were murdered and mutilated in a fearful manner." [321]

"Fourthly, that by the proclamations of the enemy the burghers still fighting are threatened with the loss of all their movable and landed property—and thus with utter ruin—which proclamations have already been enforced."

"Fifthly, that it has already, through the circumstances of the war, become quite impossible for us to keep the many thousand prisoners of war taken by our forces, and that we have thus been unable to inflict much damage on the British forces (whereas the burghers who are taken prisoners by the British armies are sent out of the country), and that, after war has raged for nearly three years, there only remains an insignificant part of the fighting forces with which we began."

"Sixthly, that this fighting remainder, which is only a small minority of our whole nation, has to fight against an overpowering force of the enemy, and besides is reduced to a condition of starvation, and is destitute of all necessaries, and that notwithstanding our utmost efforts, and the sacrifice of everything that is dear and precious to us, we cannot foresee an eventual victory."

"We are therefore of opinion that there is no justifiable ground for expecting that by continuing the war the nation will retain its independence, and that, under these circumstances, the nation is not justified in continuing the war, because this can only lead to social and material ruin, not for us alone, but also for our posterity. Compelled by the above-named circumstances and motives, we commission both Governments to accept the proposal of His Majesty's Government, and to sign it in the name of the people of both Republics."

"We, the representative delegates, express our confidence that the present circumstances will, by accepting the proposal of His Majesty's Government, be speedily ameliorated in such a way that our nation will be placed in a

[319] The Times' History of the War in South Africa 1899-1902 by Amery, L. S. (Leopold Stennett), Volume 5 pages 604- 606.

[320] Vryheid - Wikipedia And:

[321] The Battle of Holkrans or Holkrantz (6 May 1902) took place at Holkrans, near Vryheid. To stop the raids operated by the Boers on their cattle and crops as a punishment for helping the British, the leader of the abaQulusi, Sikhobobho, responded by assembling his forces and attacked the commando camp at the foot of Holkrans hill. The Zulus attacked at night, and in a mutual bloodbath, the Boers lost 56 killed and 3 wounded, while the Zulus suffered 52 killed and 48 wounded. This battle was the last victory of the abaQulusi to date Battle of Holkrans - Wikipedia

position to enjoy the privileges to which they think they have a just claim, on the ground not only of their past sacrifices, but also of those made in this war."

"We have with great satisfaction taken note of the decision of His Majesty's Government to grant a large measure of amnesty to the British subjects who have taken up arms on our behalf, and to whom we are united by bonds of love and honour; and express our wish that it may please His Majesty to still further extend this amnesty."

The motion was carried by fifty-four votes to six. The irrevocable decision had been made. The time for speech making had gone, but Schalk Burger closed the meeting in a few sentences of simple eloquence.

"We are standing here at the grave of the two Republics. Much yet remains to be done, although we shall not be able to do it in the official capacities which we have formerly occupied."

"Let us not draw our hands back from the work which it is our duty to accomplish. Let us ask God to guide us, and to show us how we shall be enabled to keep our nation together. We must be ready to forgive and forget whenever we meet our brethren."

"That part of our nation which has proved unfaithful we must not reject."

The Commissioners travelled to Pretoria, and on the same night, at five minutes past eleven, only an hour before the expiry of the term of grace, the Treaty was signed.

Treaty of Vereeniging Terms

Quote from The Great Boer War by Arthur Conan Doyle.

On May 31st this resolution was notified to Lord Kitchener, and at half-past ten of the same night the delegates arrived at Pretoria and set their names to the treaty of peace. ....[322] [323]

The following were in brief the terms of surrender:—

1. That the burghers lay down their arms and acknowledge themselves subjects of Edward VII.
2. That all prisoners taking the oath of allegiance be returned.
3. That their liberty and property be inviolate.
4. That an amnesty be granted—save in special cases.
5. That the Dutch language be allowed in schools and law-courts.
6. That rifles be allowed if registered.
7. That self-government be granted as soon as possible.
8. That no franchise be granted for natives until after self-government.
9. That no special land tax be levied.
10. That the people be helped to reoccupy the farms.
11. That 3,000,000 pounds be given to help the farmers.
12. That the rebels be disfranchised and their leaders tried, on condition that no death penalty be inflicted.

These terms were practically the same as those which had been refused by Botha in March 1901. [At the Middleburg Conference] Thirteen months of useless warfare had left the situation as it was."[324]

[322] Doyle, Arthur Conan. The Great Boer War 17th Edition, 1902 with the two volumes as one document: The Great Boer War by Arthur Conan Doyle - Free Ebook (gutenberg.org) Chapter 39. Pages 232-234 of the Word document.

[323] General Article: Treaty of Vereeniging - Wikipedia

[324] See the Century of Wrong volume (9) for additional documentation and the following events. There is simply not sufficient room in this volume for that material.

Christiaan's Later Life Summary and Pictures

At the conclusion of the war he visited Europe with other Boer generals. While in England the generals unsuccessfully sought a modification of the peace terms concluded in Pretoria. In November 1907, he was elected a member of the first parliament of the Orange River Colony and was appointed Minister of Agriculture.

General de Wet, in 1903.[288 289]

Christiaan de Wet at the National Convention in 1908.[290]

De Wet was one of the leaders of the Maritz Rebellion[325] which broke out in 1914. One of his sons was killed in the uprising. De Wet himself was defeated at Mushroom Valley by General Botha on 12 November 1914, taken prisoner by Commandant Jorrie Jordaan (the commanding officer was Colonel Brits) on 1 December on a farm called Waterbury in the Northwest province near Tosca. The general remarked: "Thank God it is not an Englishman who captured me after all." He was sentenced to a term of six years imprisonment, with a fine of £2000. He was released after one year's imprisonment, after giving a written promise to take no further part in politics. The rebellion also earned the nickname The Five Shillings Rebellion after De Wet remarked: "I was charged before [the Magistrate of Reitz] for beating a native boy. I only did it with a small shepherd's whip, and for that I was fined 5 [shillings]."

An illustration of the Battle of Mushroom Valley in 1914 between British forces led by Louis Botha and Boer and German Forces led by Christiaan de Wet.[291] Boer general Christiaan de Wet, after his surrender in the Boer Rebellion of 1914.[292]

[325] The Maritz rebellion, was an armed pro-German insurrection in South Africa in 1914, at the start of World War I. The rebellion failed, with at least 124 out of 12,000 rebels killed in battle, another 300 dying during a retreat into the Kalahari Desert, and at least 229 wounded. The surviving ringleaders received heavy fines and prison terms. One of them, Jopie Fourie, was executed. https://en.wikipedia.org/wiki/Maritz_rebellion

General C.R. de Wet and his wife, Cornelia Margaretha Kruger, sitting in front of their house, 1917.[293]
Parliament building in Bloemfontein, South Africa, with monument to Christiaan de Wet.[294]

Over the following years De Wet aged with dignity and on February 3rd, 1922, he died on his farm.

General Smuts, who had become Prime Minister, cabled his widow: “A prince and a great man has fallen today.”

De Wet was given a state funeral in Bloemfontein and buried next to President Steyn and Emily Hobhouse at the foot of the memorial to the Boers who died in Second Boer War concentration camps.

On the 100th anniversary of his birth, October 7th, 1954, a bronze equestrian statue, by Coert Steynberg, was unveiled at the Raadzaal in Bloemfontein.

De Wet was a personal friend of Helene Kröller-Müller (1869–1939), who commissioned a statue of him in the Hoge Veluwe National Park in the Netherlands.

Statue of Christiaan de Wet in 2012.[295]

Sculpture of Christiaan de Wet Close Up.[296]

The National Scouts

The National Scouts[326] were a military unit in South Africa created by the British authorities in 1900 during the Second Boer War (11 October 1899–31 May 1902). Its membership consisted of former Boer Orange Free State and South African Republic military. They were recruited in significant numbers towards the end of the war from Afrikaner prisoners and defectors, and were commonly known as hensoppers (Afrikaans for Hands-uppers) or joiners among the Boers. Many of their fellow citizens despised them as traitors so that the label of National Scout became a swear word. According to official figures there were 1,480 members in May 1902. After the war, they were largely ostracized by the community and a number of them founded their own church, known as the Kruiskerk (Church of the Cross) in the Transvaal Colony.

A few months after the occupation of Pretoria by the British forces on 5 June 1900, a first public meeting was held in the Rex Bar at Kerkplein in Pretoria to recruit National Scouts from the ranks of the citizens. In most cases, they were initially employed as "cattle guards" (Afrikaans: beeswagters) to collect and protect the livestock of farms against seizure by roaming Boer commandos. However, some took part in military action against them.

The so-called hensoppers were men who had surrendered immediately to the enemy, sometimes even before the actual start of hostilities, or later after only a short time in the field. The joiners on the other hand were Boers who offered their services to the enemy by acting as National Scouts or guides to the British troops in the field.

Prominent members of the National Scouts were former Boer generals Andries Cronjé (1849–1923), brother of general Piet Cronjé, who had surrendered at Paardeberg and was sent to Saint Helena, and Piet de Wet (1861–1929), brother of general Christiaan de Wet, who kept up a guerrilla war against the British.

Andries Petrus Johannes Cronjé.[327]

The hostility ran high among the Boer population against those joiners, which was summed up in a poem sent to the National Scouts' membership during the war:

(Translation:) Afrikaans Original.[328]

Traitors! Monsters! Curse of the Earth:

Despicable creatures of nature:

God's watch that spared you until now:

By Hellfire sometime should eradicate you—Anonymous

After the Boer War, only 10 percent of the total of 1,750 Boers serving on the British side as National Scouts claimed their Queen's South Africa Medals. Ostracized by the Boer Bittereinders and their womenfolk, they had

[326] National Scouts: https://en.wikipedia.org/wiki/National_Scouts

[327] Andries Petrus Johannes Cronjé (1849–1923) File:Andries Petrus Johannes Cronjé (1849-1923) with ammo.jpg - Wikimedia Commons

[328] (Afrikaans original:) Verraders! Monsters! Vloek der Aarde: Verneerde schepsels der natuur: Gods waken die u tot heden spaarde: Verdelg u eens door Helsche Vuur.

to found their own separate Afrikander church organisation, the Kruiskerk (Church of the Cross) in Pretoria. Towards the end of 1902.[329]

General Louis Botha decided to destroy the list of National Scouts compiled by the Boers during the war, to so keep their descendants in ignorance about their actions during the war.[330 331]

Piet de Wet

Pieter Daniel de Wet (18 August 1861–27 February 1929)[332] was a Boer general in the Anglo-Boer War (1899–1902) and a younger brother of Boer general and politician Christiaan de Wet. Piet de Wet participated in the Battle of Poplar Grove (7 March 1900), the Battle of Sanna's Post (31 March 1900) for the waterworks there, and defeated the 13th Battalion Imperial Yeomanry at Lindley (31 May 1900). In July 1900, he surrendered to the British at Kroonstad, Orange Free State. He became a prominent member of the National Scouts helping the British in the last years of the Boer War.

General Peter De Wet.[297]

De Wet was born at the farm Nuwejaarsfontein in Dewetsdorp as one of the fourteen children of Jacobus Ignatius de Wet (13 July 1823, Boonjieskraal, Caledon, Western Cape–1891, Dewetsdorp) and Aletta Susanna Margaretha Strydom (28 February 1828, Caledon–16 May 1870, Dewetsdorp).

With his older brother Christiaan he left the Orange Free State in 1879 for the Heidelberg district in the South African Republic.

Both brothers fought in the First Boer War (1880–1881) at the Battle of Majuba Hill and Piet de Wet participated in the expedition to capture Mampuru II.[333]

[329] Biggins, David (2023). "South African units. National Scouts." angloboerwar.com. Retrieved 13 July 2023. The National Scouts were mainly recruited from Boers who had surrendered to the British or chosen to fight for the imperial forces. QSAs were issued to the National Scouts and they are often appear as renamed medals.

[330] Grundlingh, Albert (1999). Die Hendsoppers en Joiners - Die Rasionaal en Verskynsel van Verraad (in Afrikaans) (2de ed.). Protea Boekhuis. ISBN 978-0620241540.

[331] Wulfsohn, Lionel (December 2001). "Hendsoppers of the Rustenburg Commando. Military History Journal Vol 8 No 6 - December 1991." samilitaryhistory.org. The South African Military History Society/Die Suid-Afrikaanse Krygshistoriese Vereniging. Retrieved 13 July 2023. This is a doleful chapter in the history of certain Rustenburg men, and no pleasure is derived from writing about their failure to perform at a crucial stage of the Anglo-Boer War. What is recorded actually happened and the facts need to be revealed, lest the story be incomplete. As a youth, the writer knew some of these men and wishes to assure their descendants that what follows is not intended as a slur on the good name of their ancestors. In fact it is an attempt to pinpoint responsibility for the state of affairs.

[332] https://en.wikipedia.org/wiki/Piet_de_Wet Permalink: https://en.wikipedia.org/w/index.php?title=Piet_de_Wet&oldid=1271582192 And: Afrikaans Wikipedia, https://af.wikipedia.org/wiki/Piet_de_Wet

[333] Mampuru II (1824–22 November 1883) was a king of the Pedi people in southern Africa. Sekwati died in 1861 and his son, Sekhukhune claimed the throne. Sekhukhune ruled until 1879 when a British-Swazi invasion deposed him and installed Mampuru as king. Mampuru ordered the assassination of Sekhukhune in 1882 for which he was arrested and hanged by the Boer South African Republic the following year. Mampuru is regarded as an early liberation movement icon in South Africa and the prison where he was executed has been renamed in his honour. https://en.wikipedia.org/wiki/Mampuru_II

In 1883, he returned to the Orange Free State, settled at the farm Vinkfontein near Lindley and married Susanna Margaretha de Wet (18 September 1865, Kroonstad, Orange Free State, South Africa–26 August 1924). They had five sons and six daughters. De Wet represented Midden-Liebensbergvlei-wyk in the Volksraad from 1895 to 1897, moved to Pretoria, but came back and became military commander (veldkornet, field cornet) for Lindley.

Military Success.

Piet de Wet led his 200 men of the Bethlehem Commando near Nicholson's Nek outside Ladysmith on 5 October 1899 and during the siege of Ladysmith.

Then he was tasked to march west and attack Vaalkop near Arundel on 16 December 1899.

He impressed president Steyn of the Free State who gave him the command of all troops south of the Orange River with headquarter at Colesberg.

British troops continued their advance and occupied Bloemfontein, the capital of Orange Free State, on 13 March 1900.

On 28 May Orange Free State was annexed by Lord Roberts as the Orange River Colony. But in the mean time the brothers Christiaan and Piet de Wet won skirmishes with the British at Abrahamkraal (10 March),

Sanna's Post (Sannaspost, 31 March) and Dewetsdorp (20 April).

On 31 May 1900 Piet de Wet with General Marthinus Prinsloo captured 13th Battalion Imperial Yeomanry near Lindley. Piet de Wet bagged in total 530 men, including Spragge, Lord Longford, Lords Ennismore, Leitrim and Donoughmore (and the future Lord Craigavon) who were marched off to the eastern Transvaal northwards.

However, after the occupation by Roberts of Kroonstad on 12 May 1900, Free State soldiers were disheartenend and Piet de Wet had started to doubt the use of fighting on. Roberts occupied and Pretoria on 5 June 1900.

On 18 May 1900, British brigade general Robert Broadwood received a message that Piet de Wet was willing to surrender on the condition that he and his men could return to their farms.

However, Commander-in-chief Roberts demanded an unconditional surrender so that the fighting continued. But in July 1900, Piet de Wet and several of his staff surrendered to the British at Kroonstad, becoming "joiners," "hensoppers" (hands-upper), and "wapenneerlêers" (who put down their weapons).

His unavailing letter of 11 January 1901 to his brother Christiaan de Wet to stop his guerrilla war, was published in the Bloemfontein Post and separately as a pamphlet "Broeder tot broeder" (Brother to brother). In March 1902 Piet de Wet established the Orange River Colony Volunteers on the British side. and had joined the National Scouts headed by Andries Cronjé (1849–1923), who served the British troops as Boer auxiliaries.

However, on 9 April 1902, peace negotiations started at Vereeniging leading to the Treaty of Vereeniging on 31 May 1902 ending the war.

References.[334]

[334] Coghlan, Mark (December 2000). "The other De Wet. Piet de Wet and the Boer 'Hendsoppers' in the Anglo-Boer War. Military History Journal Vol. 11 No. 6 - December 2000". *samilitaryhistory.org*. The South African Military History.
Publications: (Brother to Brother) Broeder tot broeder: een prijzenswaardige brief [aan zijn broeder, Kommandant Christiaan De Wet gezonden]; een smeekstem tot De Wet [en] De Wet's verantwoordelijkheid, 1901. with Kroonstad Vredes Comité (translation: Kroonstad Peace Committee), Open brief aan de inwoners der Kaap Kolonie, Die Comité, Kroonstad, 1901.
The Maritz conspiracy : negotiations with German South West during 1913, Wallachs, Pretoria, 1915.

Three Years' Appendices

Summary

There are four sections in the Appendix area of the original work beginning with the Correspondence.

1. Correspondence.

Letter From The States-Secretary of The South African Republic To The British Agent At Pretoria. From Francis Reitz, Ministry of Foreign Affairs, Pretoria, 9th October, 1899. [The Ultimatum]

Mr. Chamberlain's Telegrams:—From Mr. Chamberlain to the High Commissioner, Sir Alfred Milner. (Sent 7.30 p.m. 10th October, 1899)

2. Appendix A: Report Of The Meeting Of The General Representatives Held At Vereeniging, In The South African Republic, On The 15th Of May, 1902, And The Following Days.

3. Appendix B: The Conference At Pretoria Between The Commission of The National Representatives And Lords Kitchener And Milner (May 19th-May 28th, 1902).

4. Appendix C: Minutes of The Meeting of The Special National Representatives At Vereeniging, South African Republic, Thursday, The 29th Of May, 1902, And The Following Days.

To which has been inserted a 5th section, containing the March 1900 Offer & Reply and the Middleburg Proposal.

Ultimatum Correspondence

A Letter from The States-Secretary of The South African Republic to The British Agent at Pretoria

Ministry of Foreign Affairs,

Pretoria, 9th October, 1899.

Sir,—

The Government of the South African Republic feel themselves compelled to again refer the Government of Her Majesty, the Queen of Great Britain, to the London Convention of 1884, concluded between this Republic and the United Kingdom, which in Article XIV. guarantees certain specified rights to the white inhabitants of this Republic, to wit:—

"All those who, although not born in this Country, yet abide by the laws of the South African Republic, (a) shall have full freedom to come with their families into, to travel in, or to reside in any part of the South African Republic; (b) shall be entitled to hold in possession their houses, factories or warehouses, shops, and allotments, either on hire or as their own property; (c) may transact their business, either in person or through agents, to their own satisfaction; (d) shall not be subjected to any other general or local taxation—with regard to their families or properties, or their commerce or trade—than those which shall be laid on the burghers of the said Republics."

Our Government wishes also to draw attention to the fact that the above-mentioned rights are the only ones which Her Majesty's Government, in the above-mentioned Convention, has stipulated for the foreign inhabitants in this Republic, and that only contravention of these rights can give the British Government the right of diplomatic intervention; whereas, further, the adjustment of all other questions concerning the position, or the rights, of the foreign inhabitants under the said Convention is vested in the Government and National Representatives of the South African Republic; among the questions the adjustment of which comes exclusively under the authority of the Government and the Volksraad, are those of the Franchise and representation in this Republic.

Although, therefore, the exclusive right of this Franchise and representation is indisputable, our Government has approved of discussing in a friendly way the Franchise and the representation with Her Majesty's Government; without, however, acknowledging by so doing any right thereto on the side of Her Majesty's Government. Our Government has also, by the wording of the already existing Voting Law, and the decision concerning the representation, always kept this friendly consultation in view.

On the side of Her Majesty's Government, however, the friendly manner of these consultations has made way for a more threatening tone; and the minds of the people of this Republic, and of the whole population of South Africa, have been brought into a state of apprehension; and a state of unusual tension has been created by the action of Her Majesty's Government, in no longer abiding by the laws concerning the voting right, and the decision concerning the representation of this Republic; and lastly, as is expressed in your letter of the 25th of September, 1899, in breaking off all friendly communication, giving us to understand that Her Majesty's Government were about to formulate their own proposals for final arrangement. Our Government can see in the before-mentioned notification nothing less than a new violation of the Convention of 1884, which does not reserve to Her Majesty's Government the right of a one-sided adjustment of a question which belongs exclusively to the inner policy of our Government, and has been already settled by them.

On the grounds of the tension, the considerable loss arising therefrom, and the interruption of business in general, which is caused by the correspondence on the Franchise and the representation of this Republic, Her Majesty's Government has not long ago insisted on a speedy adjustment, and finally, through your intervention, insisted on an answer—within forty-eight hours—(later on somewhat amended)—to your Memorandum of the 12th of September, which was answered by the Memorandum of our Government of the 15th of September, and by the Memorandum of the 25th of September, 1899; on which other friendly negotiations were interrupted, and our Government received notice that the proposal for final arrangement would be made within a short time; but although these promises were repeated, no such proposal has as yet reached our Government. When the friendly correspondence was still going on, a great increase of troops was made by Her Majesty's Government, which troops were drawn up in the neighbourhood of the frontiers of our Republic. Taking into consideration certain events in the history of our Republic, which events need not here be recited, our Government found themselves compelled to look upon the Army in the neighbourhood of the frontier as a threat to the independence of the South African Republic, because they were not aware of any circumstances which could justify the presence of such a force in South Africa and in the neighbourhood of their frontier.

In answer to a question concerning this, addressed to His Excellency the High Commissioner, our Government received, to their great astonishment, the covert accusation that from the State of the Republic an attack on Her Majesty's Colonies was being arranged, and also a mysterious hint of coming possibilities, by which our Government were strengthened in their suspicion, that the independence of the Republic was threatened.

As a measure of defence, they were, therefore, compelled to send a body of burghers to the frontiers in order, if required, to be able to resist such an eventuality. The unlawful interference of Her Majesty's Government in the inner policy of our Republic, in defiance of the London Convention of 1884, which interference consisted in the exceptional strengthening of troops in the neighbourhood of the Republic's borders, has thus created an unbearable state of affairs, of which our Government—not only in the interests of our Republic, but also in the interests of the whole of South Africa,—feel it their duty to bring to an end as speedily as possible, and consider themselves called upon to insist emphatically and energetically on an immediate conclusion of this condition of things, and to ask Her Majesty's Government to give them the assurance (a) that all points of mutual difference shall be adjusted by friendly arbitration, or by any other amicable way that may be agreed upon between our Government and that of Her Majesty; (b) that the troops on the frontiers of the Republic shall be recalled at once, and that all reinforcements which, after the 1st of June, 1899, have arrived in South Africa, shall be removed within a time agreed upon with our Government,—with the counter assurance and guarantee from our Government that no attack on, or hostilities against, any part of the possessions of the British Government shall be undertaken by the Republic during the further negotiations within the time which shall be agreed upon by the Government—our Government shall, in accordance with this, be ready to call back the armed burghers of the Republic from the frontiers; (c) that Her Majesty's troops, which are now on the high sea, shall not be landed in any of the harbours of South Africa.

Our Government has to insist on an immediate and favourable answer on the above four points, and urgently requests Her Majesty's Government to give an answer in this spirit before, or on, Wednesday, October 11th, 1889, before 5 o'clock in the afternoon. They wish to add further, that in case, against their expectations, no satisfactory answer within this time should be received by them, that they, to their great sorrow, would be obliged to look upon the actions of Her Majesty's Government as a formal declaration of war, for the consequences of which they do not consider themselves responsible; and, in case further movements of troops should take place within the above-mentioned time in the direction of our borders, that our Government will be compelled to look upon this also as a formal declaration of war.

I have the honour to be, etc.,

F.W. Reitz,

State-Secretary.

Mr. Chamberlain's Telegrams:—

From Mr. Chamberlain to the High Commissioner, Sir Alfred Milner.

(Sent 7.30 p.m. 10th October, 1899)

"10th October, No. 7. The British Agent has, in answering the demands of the Government of the South African Republic, to say that, as the Government of the South African Republic have declared in their dispatch, that they will look upon a refusal to consent to their demands as a formal declaration of war, he has received orders to demand his passport." [Presumably, his passport was being held by the Transvaal Govt. while he was in residence]

From Mr. Chamberlain to the High Commissioner, Sir Alfred Milner.

(Sent 10.45 p.m. 10th October, 1899)

"10th October, No. 8. The Government of Her Majesty has received with great sorrow the determined demands of the Government of the South African Republic contained in your telegram of the 9th of October, No. 3. You will, as an answer to the Government of the South African Republic, communicate to them that the conditions put forward by the Government of the South African Republic are of such a nature that the Government of Her Majesty cannot possibly think of taking them into consideration."

March 1900 Offer and Reply:

Correspondence Between the Two Presidents and Lord Salisbury

From the States-President of the South African Republic and the Orange Free State to His Excellency Lord Salisbury, London.

"Bloemfontein, 5th March, 1900.

"The blood and tears of the thousands who have suffered through this war, and the prospect of all the moral and material ruin which now threatens South Africa, render it necessary for both parties carrying on the war to ask themselves calmly, and in the faith of the Trinity, for what they are fighting and if the aims of both justify all this horrible misery and devastation.

On this account, and with an eye to the assertion of several English Statesmen that the war was begun and carried on with the determined end to undermine Her Majesty's authority in South Africa, and to establish in the whole of South Africa a Government independent of Her Majesty's Government, we consider it our duty to declare that this War was only commenced as a measure of defence and for the purpose of obtaining a guarantee for the threatened independence of the South African Republic, and was only continued in order to ensure the indisputable independence of both Republics as Sovereign International States, and to obtain the assurance that the subjects of Her Majesty who have taken part with us in the war will not suffer the least hurt either in their lives or their possessions.

On these conditions alone we demand, as in the past, to see peace restored in South Africa, and an end made to the wrong that now exists there. But if Her Majesty's Government has decided upon destroying the independence of the Republic, nothing remains to us and our people but to persist to the bitter end on the road now taken, notwithstanding the overpowering might of the British Empire, trusting that God, who has lit the inextinguishable fire of the love of liberty in our hearts, and in the hearts of our fathers, will not abandon us, but will fulfil His work in us, and in our descendants.

"We hesitated to lay this declaration earlier before Your Excellency, because we were afraid that as long as the advantage was on our side, and our Army had in their occupation positions of defence far into the British Colonies, such a declaration would have hurt the feelings of the English nation; but now that the prestige of the British Empire may be considered to be restored, through the capture of one of our armies, and we are compelled by this to sacrifice other positions which our armies occupied, this difficulty is removed, and we can no longer hesitate to tell you, in the face of the whole civilized world, why we are fighting, and on what conditions we are prepared to make peace."

Lord Salisbury's Reply

From Lord Salisbury to their Excellencies the States-Presidents of the South African Republic and Orange Free State.

"London, 11th March, 1900.

"I have the honour to acknowledge the receipt of your Honour's cable, dated 5th March, from Bloemfontein, of which the purport is principally whether Her Majesty's Government will acknowledge the indisputable independence of the South African Republic and Orange Free State and treat them as Sovereign International States, and will offer to conclude the war on these conditions.

"In the beginning of October of this year, there was peace between the Queen and the two Republics, under the Convention which then held good. There was a discussion carried on during a few months between Her Majesty's Government and the South African Republic, of which the purport was the amendment of very serious grievances under which English inhabitants suffered in the South African Republic. In the course of these negotiations, the South African Republic obtained the knowledge that Her Majesty's Government had made considerable preparations for war, and had taken steps to provide the necessary reinforcements for the English garrisons at Cape Colony and Natal.

No inroad on the rights guaranteed by the Conventions had, until then, taken place on the English side. Suddenly the South African Republic, after having two days previously issued an insulting ultimatum, declared War on Her Majesty; and the Orange Free State, with which there had been no disagreement, took a similar step. Thereupon an inroad was made into Her Majesty's territory by the two Republics; three towns within the British frontier were besieged, a great part of the two Colonies was over-run, with great destruction of property and life, and the Republics claimed the right to treat the inhabitants of Her Majesty's territory as if this territory had been annexed by one of these States. The Transvaal having these actions in view, had for years stored up, on an enormous scale, military provisions, which could only have been destined for use against England.

"Your Excellencies made some remarks of a negative nature concerning the aim for which these preparations were made. I do not consider it necessary to discuss the question which you have thus raised, but the consequences of the preparations, made in great secrecy, have been that the British Empire has found itself forced to repel an inroad which has brought on a costly war, and caused the loss of thousands of valuable lives.

This great misfortune has been the punishment that Great Britain has had to undergo during the last few years for having suffered the two Republics to exist. Keeping in sight the use which the two Republics have made of the position presented to them, and the misfortunes which their unprovoked attacks on Her Majesty's territory have brought, Her Majesty's Government can only reply to Your Honour's telegram by saying that they are not prepared to acknowledge the independence either of the South African Republic, or of the Orange Free State."

The Middelburg Proposal

In early February of 1901, General Kitchener, with the consent of the Government decided to make peace overtures to General Botha well before the February 14th debate in the House of Commons. The actual conference took place from February 28th, until its unsuccessful conclusion on March 16th 1901.

The Middleburg[335] Conference. Group photograph of the participants in the Middelburg peace conference. (February-March 1901)[298 299] Standing (left to right); Colonel G. F. R. Henderson, Dirk Overgauwe ("Dick") van Velden,[336] Major Watson, H. Fraser, Captain Maxwell, H. de Jager. Sitting (left to right); Christiaan de Wet, (sic, likely Wynand Viljoen) General Louis Botha, General Lord Kitchener, Colonel Ian S. M. Hamilton.

Note the black armbands worn by all five British officers to commemorate Queen Victoria's death a month earlier. (January 22nd, 1901).

Annie Botha, the wife of general Louis Botha.[300 337] "Mrs. Botha is descended from Robert Emmet,[338] the famous "United Irishman" who was executed for high treason, September 20, 1803."

Lord Kitchener to Commandant-General Botha. Pretoria, March 7, 1901.[339]

To his Honour Commandant-General Louis Botha.

Your Honour,

With reference to our conversation at Middelburg on the 28th February, I have the honour to inform you that, in the event of a general and complete cessation of hostilities, and the surrender of all rifles, ammunition, cannon, and other munitions of war in the hands of the burghers, or in Government depots, or elsewhere, His Majesty's Government is prepared to adopt the following measures. His Majesty's Government will at once grant an amnesty in the Transvaal and Orange River Colony for all bona fide acts of war committed during the recent hostilities. British subjects belonging to Natal and Cape Colony, while they will not be compelled to return to those Colonies, will, if they do so, be liable to be dealt with by the laws of those Colonies specially passed to meet the circumstances arising out of the present war.

[335] It was initially established as a halfway station between Lydenburg and Pretoria. The British built a large concentration camp in Middelburg during the Second Boer War. A Memorial Museum was built at the site next to the 1,381 graves of women and children who died in the concentration camp. Middelburg, Mpumalanga - Wikipedia

[336] Van Velden Dirk Overgauwe Kaptein Commando: Staf General Botha. Awarded the DTD.

[337] Annie Botha (née Emmett; 3 July 1864–20 May 1937) was a South African civic leader and political hostess. She was the wife of Louis Botha, who served as the first Prime Minister of South Africa. She established an orphanage in South Africa and, with Georgiana Solomon, co-founded and chaired the South African Women's Federation. Annie Botha - Wikipedia

[338] Robert Emmet (4 March 1778–20 September 1803) was an Irish Republican, orator and rebel leader. Emmet's Proclamation of the Provisional Government to the People of Ireland, his Speech from the Dock, and his "sacrificial" end on the gallows inspired later generations of Irish republicans. Robert Emmet - Wikipedia

[339] Part of appendix B from Three Years' War By: Christiaan Rudolf De Wet. Charles Scribner's Sons, (1902).

As you are doubtless aware the special law in the Cape Colony has greatly mitigated the ordinary penalties for High Treason in the present case. All prisoners of war, now in St. Helena, Ceylon, or elsewhere, being burghers or Colonists, will, on the completion of the surrender, be brought back to their country as quickly as arrangements can be made for their transport.

At the earliest practicable date military administration will cease, and will be replaced by civil administration in the form of Crown Colony Government.

There will, therefore, be, in the first instance, in each of the new Colonies, a Governor and an Executive Council, composed of the principal officials, with a Legislative Council consisting of a certain number of official members to whom nominated unofficial element will be added.

But it is the desire of His Majesty's Government, as soon as circumstances permit, to introduce a representative element, and ultimately to concede to the new Colonies the privilege of self-government. Moreover, on the cessation of hostilities a High Court will be established in each of the new Colonies to administer the laws of the land, and this Court will be independent of the Executive.

Church property, public trusts, and orphan funds will be respected. Both the English and Dutch languages will be used and taught in public schools when the parents of the children desire it, and allowed in Courts of Law.

As regards the debts of the late Republican Governments, His Majesty's Government cannot undertake any liability. It is, however, prepared, as an act of grace, to set aside a sum not exceeding one million pounds sterling to repay inhabitants of the Transvaal and Orange River Colony for goods requisitioned from them by the late Republican Governments, or subsequent to annexation, by Commandants in the field being in a position to enforce such requisitions.

But such claims Will have to be established to the satisfaction of a Judge or Judicial Commission, appointed by the Government, to investigate and assess them, and, if exceeding in the aggregate one million pounds, they will be liable to reduction pro rata.

I also beg to inform your Honour that the new Government will take into immediate consideration the possibility of assisting by loan the occupants of farms, who will take the oath of allegiance, to repair any injuries sustained by destruction of buildings or loss of stock during the war, and that no special war tax will be imposed upon farms to defray the expense of the war.

When burghers require the protection of fire-arms, such will be allowed to them by licence, and on due registration, provided they take the oath of allegiance. Licences will also be issued for sporting rifles, guns, &c., but military fire-arms will only be allowed for purposes of protection.

As regards the extension of the franchise to Kaffirs in the Transvaal and Orange River Colony, it is not the intention of His Majesty's Government to give such franchise before representative Government is granted to those Colonies, and if then given it will be so limited as to secure the just predominance of the white race. The legal position of coloured persons will, however, be similar to that which they hold in the Cape Colony.

In conclusion, I must inform your Honour that, if the terms now offered are not accepted after a reasonable delay for consideration they must be regarded as cancelled.

I have, &c.,

Kitchener, General,

Commander-in-Chief, British Forces, South Africa.

But in the meantime the prospects of peace had been declining. French, hampered by rain and scarcity, had for some time been at a complete standstill on the Swazi frontier, and the burghers behind him had had time to recover their spirits. Botha had consulted his Government and found it stubborn; he had corresponded with De la Rey and had met with no encouragement. Before the time came to reply, it was known that de Wet and Steyn had escaped capture, and their opinion of the negotiations scarcely needed asking.[340] Nor would they have held themselves bound by any decision of the Transvaal.

[340] Original footnote: * Botha. wrote to de Wet and Steyn, asking them to come north and confer with him, but the letter arrived too late and the meeting did not take place till April. De Wet's secretary, however, went by train to Pretoria, only to find that negotiations had been broken off. He could have been of no use, for his master's temper, as shown by a circular to his burghers, was as unbending as ever. In this circular, which is dated April 1, he inveighs against various points in the proposed settlement, and asks finally, what is the use of examining it in detail, since nothing short of independence will induce the Boers to make peace .

In a letter dated March 16 Botha, without Botha giving any reasons, declined to go any further with the negotiations.

Botha's Reply. These terms were refused by Botha and reported on March 23rd, in the following message:[341]

> I have the honour to acknowledge the receipt of your Excelleney's letter stating what steps your Excellency's Government is prepared to take in the event of a general and total cessation of hostilities. I have advised my Government of your Excellency's said letter, and, after. mutual exchanges of views at Middleburg on the 28th February, it will certainly not surprise your Excellence to know that I do not feel disposed to recommend that the terms of the said letter shall have the earnest consideration of my Government. I may add also that my Government and my chief officers here entirely agree to my views.

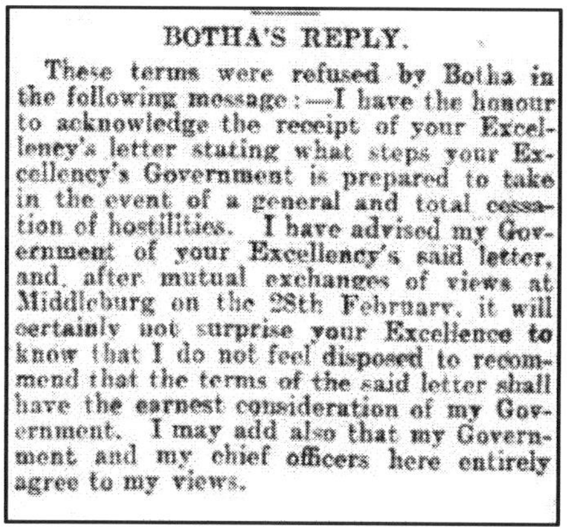

BOTHA'S REPLY.

These terms were refused by Botha in the following message:—I have the honour to acknowledge the receipt of your Excelleney's letter stating what steps your Excellency's Government is prepared to take in the event of a general and total cessation of hostilities. I have advised my Government of your Excellency's said letter, and, after mutual exchanges of views at Middleburg on the 28th February, it will certainly not surprise your Excellence to know that I do not feel disposed to recommend that the terms of the said letter shall have the earnest consideration of my Government. I may add also that my Government and my chief officers here entirely agree to my views.

Kitchener was always inclined to feel that with a little a little stronger support, especially on the question of amnesty, he might have obtained peace. However long the war lasted, it must, he held, eventually end in some such settlement as this. There was much truth in that; but the time had not yet come. His overtures were a bold but premature experiment. The Boer spirit-that new spirit born amidst the disasters of the previous year and matured in an offensive revival still quite recent, was far from being broken.

Although exhaustion and despondency existed in some quarters, when it was put to the Boers in black and white that their cause was lost irretrievably and that it only remained to settle subordinate issues, the national will revolted. The conditions which favoured this revolt are conditions which it can never be easy for a highly civilised industrial nation fully to understand.

The causes which bring such a nation to its knees were absent here. There were no vital industries to dislocate, no teeming population of peaceful workers to plunge into starvation. The people lived on the land, and though its fruits could be destroyed and the farms upon it gutted, the land could not be damaged. A year's crops were a relatively small matter, farm-buildings were of little value, and when a farm was once gutted it signified little whether it was rebuilt a year sooner or later.

Meanwhile, the male population, in other words the army, was inured by habit and inherited instinct to a life of peril and adventure under the stars. These men, moreover, were scattered over a vast expanse of territory under circumstances so varied as to render a common judgment unattainable. A few were suffering acutely; some were fighting with fair success; some were living in perfect safety and comfort. In short, there had never come to the Boers one of those moments of national agony which wring from a people a cry for peace.

As events fell out, Botha had effected a very useful reconnaissance in two directions. He had tested the sentiment of his burghers, made them familiar-and that was something-with the idea of negotiation; and he had found out on what lines an ultimate settlement was to be framed. ....

Commander-in-Chief by right of undoubted military merit, he nevertheless had not risen without effort. A keen ambitious purpose had helped him to gain the high position which he occupied. To crown its tenure by signing away independence as long as there remained the faintest hope of carrying his career to a triumphant military climax, was a hard step to take.

[341] Southport Guardian - Saturday 23 March 1901 page 5. Col 7
https://www.britishnewspaperarchive.co.uk/viewer/bl/0001783/19010323/121/0005

He knew very well that it was impossible to recover the lost territories with the forces at his disposal, but there always remained the chance of a general rising in Cape Colony. The chimera of foreign intervention still survived, as did the dim hope that pro-Boer sentiment in Great Britain would grow and triumph.

Meanwhile he could salve his conscience with the knowledge that the British policy of concentration camps freed him from responsibility for the women and children, whose misery was the most cogent argument for submission. There was little to lose; there was much to gain. For Botha had the shrewd perception that, whatever the ultimate issue, the Boers could fare no worse than they would fare under the proposed settlement.

His mental attitude is revealed clearly in a letter addressed to his burghers on March 15.

"Virtually, "he writes, "Lord Kitchener's letter contains nothing more, but rather less than what the British Government will be obliged to do should our cause go wrong." ....

"The cause is not yet lost," he writes, "and since nothing worse than this can befall us, it is well worth while to fight on." ....

Such were the abortive negotiations of 1901. Their failure caused much disappointment in England.

On public platforms and in Parliament the Government was bitterly attacked by a section of the Opposition for having thrown away the chances of peace by attempting to exact too much, and the differences of opinion, frankly set forth in the despatches, between Mr. Chamberlain, Lord Milner and Lord Kitchener, were used to further the argument. ....

The nation at large instinctively grasped the truth, and resigned itself with unflinching determination to a further struggle of indefinite length and severity.

Meeting on 15th of May, 1902,

Appendix A

Report of the Meeting of the General Representatives Held at Vereeniging, in the South African Republic, on The 15th Of May, 1902, and the Following Days

The first meeting of the representatives of the two Governments took place at 11.30 a.m. on May 15th.

There were present:—

For the South African Republic—His Honour the President, S.W. Burger, F.W. Reitz, Commandant-General L. Botha, Messrs. J.B. Krogh, L.J. Meijer, L.J. Jacobs, and His Honour the Staats-Procureur.

For the Orange Free State—States-President, M.J. Steyn; Judge, J.B.M. Hertzog; Secretary of State, W.J.C. Brebner; Commander-in-Chief, C.R. de Wet; and Mr. C.H. Olivier.

The first matter discussed was the formula for the oath which the delegates were to take, and it was decided that it should run as follows:—

"We, the undersigned, duly swear that we, as special national representatives, will remain true to our people, country, and Government, and that we will serve them to the best of our ability, and fulfil our duties faithfully and with all necessary secrecy, as is the duty of all faithful burghers and representatives of the nation. So help us God."

The question now arose as to whether the representatives had the right to decide, if circumstances rendered it necessary, upon any matter touching the independence of the country, irrespective of the powers given to the various delegates, for at some of the meetings the delegates had only received limited powers, whilst at others full authority had been given them to act according to their own judgment.

After considerable discussion it was decided to lay the matter before the delegates themselves.

The following representatives were called into the tent, and took the oath:—

For the South African Republic.

1. H.A. Alberts, Vechtgeneraal; for Heidelberg.
2. J.J. Alberts, Commandant; for Standerton and Wakkerstroom.
3. J.F. De Beer, Commandant; for Bloemhof.
4. C.F. Beijers, Assistant-Commandant-General; for Waterberg.
5. C. Birkenstock, burgher; for Vrijheid.
6. H.J. Bosman, magistrate; for Wakkerstroom.
7. Christiaan Botha, Assistant-Commandant-General; for Swaziland and the States Artillery.
8. B.H. Breijtenbach, Veldtcornet; for Utrecht.
9. C.J. Brits, Vechtgeneraal; for Standerton.
10. J.B. Cilluos, Vechtgeneraal; for Lichtenburg.
11. J. De Clercq, burgher; for Middelburg.
12. T.A. Dönges, Veldtcornet; for Dorp Middelburg in Regeeringswacht.
13. H.S. Grobler, Commandant; for Bethal.
14. J.L. Grobler, burgher; for Carolina.
15. J.N.H. Grobler, Vechtgeneraal; for Ermelo.
16. B.J. Van Heerden, Veldtcornet; for Rustenburg.
17. J.F. Jordaan, Commandant; for Vrijheid.
18. J. Kemp, Vechtgeneraal; for Krugersdorp.
19. P.J. Liebenberg, Vechtgeneraal; for Potchefstroom.
20. C.H. Muller, Vechtgeneraal; for Boksburg.
21. J.F. Naude, burgher; for Pretoria, late Commandant with General Kemp.
22. D.J.E. Opperman, Veldtcornet; for Pretoria.

23. B.J. Roos, Veldtcornet; for Piet Retief.
24. P.D. Roux, Veldtcornet; for Marico.
25. D.J. Schoeman, Commandant; for Lijdenburg.
26. T.C. Stoffberg, Landdrost; for Zoutpansberg.
27. S.P. Du Toit, Vechtgeneraal; for Wolmaransstad.
28. P.L. Uijs, Commandant; for Pretoria.
29. P.R. Viljoen, burgher; for Heidelberg.
30. W.J. Viljoen, Commandant; for Witwatersrand.

For the Orange Free State.

1. C.C.F. Badenhorst, Vice-Commandant-in-Chief; for Boshof, Hoopstad, West Bloemfontein, Winburg, and Kroonstad.
2. A.J. Bester, Commandant; for Bethlehem.
3. A.J. Bester, Commandant; for Bloemfontein.
4. L.P.H. Botha, Commandant; for Harrismith.
5. G.A. Brand, Vice-Commandant-in-Chief; for Bethulie, Rouxville, Caledon River, and Wepener in the eastern part of Bloemfontein.
6. H.J. Brouwer, Commandant; for Bethlehem.
7. D.H. Van Coller, Commandant; for Heilbron.
8. F.R. Cronje, Commandant; for Winburg.
9. D.F.H. Flemming, Commandant; for Hoopstad.
10. C.C. Froneman, Vice-Commandant-in-Chief; for Winburg and Ladybrand.
11. F.J.W.J. Hattingh, Vice-Commandant-in-Chief; for the eastern part of Kroonstad, in the district of Heilbron.
12. J.B.M. Hertzog, Commandant; for Philippolis.
13. J.N. Jacobs, Commandant; for Boshof.
14. F.P. Jacobsz, Commandant; for Harrismith.
15. A.J. De Kock, Commandant; for Vrede.
16. J.J. Koen, Commandant; for Ladybrand.
17. H.J. Kritzinger, Veldtcornet; for Kroonstad.
18. F.E. Mentz, Commandant; for Heilbron.
19. J.A.P. Van der Merwe, Commandant; for Heilbron.
20. C.A. Van Niekerk, Commandant; for Kroonstad.
21. H. Van Niekerk, Commandant.
22. J.J. Van Niekerk, Commandant; for Ficksburg.
23. I.K. Nieuwouwdt, Vice-Commandant-in-Chief; for Fauresmith, Philippolis, and Jacobsdal.
24. H.P.J. Pretorius, Commandant; for Jacobsdal.
25. A.M. Prinsloo, Vice-Commandant-in-Chief; for Bethlehem in Ficksburg.
26. L.J. Rautenbach, Commandant; for Bethlehem.
27. F.J. Rheeder, Commandant; for Rouxville.
28. A. Ross, Commandant; for Vrede.
29. P.W. De Vos, Commandant; for Kroonstad.
30. W.J. Wessels, Vice-Commandant-in-Chief; for Harrismith and Vrede.

The meeting now proceeded to choose a chairman, and the following were proposed:—J. De Clercq, C.F. Beijers, C.C. Froneman, W.J. Wessels, and G.A. Brand.

The choice of the meeting fell on General C.F. Beijers, who called upon the Rev. Mr. Kestell to offer prayer.

His Honour, S.W. Burger, now declared that the meeting was formally opened, and after the Chairman had spoken a few words, the representatives adjourned until three o'clock.

When they reassembled, the Chairman requested President Burger to explain the objects for which the meeting had been called.

Then the President spoke a few words of welcome to all; he expressed his sorrow for the absence of some who would certainly have been present had they not given their lives for their country. But still there were many left to represent the two Republics.

"The difficulties which confront us," continued the President, "are like a great mountain, at the foot of which we have just arrived. Everything now depends on us who are assembled together here. It is impossible to deny that the state of affairs is very serious, and that the future looms dark before us. Our position requires the most careful consideration, and as there are sure to be differences of opinion, it will be necessary for us to bear with one another, and yet, at the same time, to speak our minds freely."

The President proceeded to refer to the correspondence which had taken place between Holland and England. A copy of this correspondence had been sent, through Lord Kitchener, to the Governments of the two Republics. The opinion of the Transvaal Government (which was the first to receive the correspondence) was that advantage should be taken of this opportunity. It was proposed to ask Lord Kitchener to allow the Transvaal Government to meet that of the Orange Free State, so that they might discuss the desirability of making a peace proposal to England. The two Governments had accordingly met, and had corresponded with Lord Kitchener and Lord Milner. As a result of this, a letter, with the above correspondence annexed, had been sent to the various commandos.

"We felt," continued President Burger, "that we had no power to surrender our independence, and that we were only justified in making such terms of peace as would not endanger our national existence. Whether it is or is not our duty to surrender our independence is a question that must be left to the decision of our people. And it is to represent the people that you are here. It is from your lips, then, that our Governments must learn the opinions of the two nations. It is clear enough that the English Government has no idea of allowing us to remain independent—it expresses surprise that we even dare to speak of such a thing.

"You have now to report upon the condition of the country, and upon the circumstances in which your wives and children are placed. You have also to decide whether you are willing to make any further sacrifices. We have lost so much already that it would be hard, indeed, to lose our independence as well. But, although this matter is so near to our hearts, we must still listen to the voice of reason. The practical question, then, which we have to ask ourselves is, whether we are prepared to watch our people being gradually exterminated before our eyes, or whether we should not rather seek a remedy.

"The Government can do nothing without the support of the nation. You, therefore, must determine our best course. For instance, if you come to the conclusion that we have exhausted every expedient, will you still continue the struggle? Are we not to desist until every man of us is in captivity, in exile, or in his grave? Again let me urge you to speak freely, and yet with consideration for the feelings of others. For myself, I can truly say that my spirit is not yet broken; but I would hear from you what the feeling of the people is."

"At this point, however, a difficulty arises. Some of you, having only received limited powers from your constituencies, appear to think that you would not be justified in exceeding your mandates, while others have been authorized to act as circumstances may seem to require. But I do not think that this difficulty should be insurmountable. At least I beg of you not to allow it to cause any dissension among you. Let us all be of one mind. If we are united, then will the nation be united also; but if we are divided, in what a plight will the nation find itself?"

A letter was then read from the deputation in Europe, which had been written five months previously, and which had been brought through the English lines in safety. It contained little more than an assurance that our cause occupied a better position in Europe than it had ever done before.

The Chairman then asked Commandant L. Botha to address the meeting.

Complying with this request, the Commandant said that he wished to be assured, before anything further was done, that the fact that some of the representatives had been entrusted with limited powers, whereas others had been given a free hand, was not going to prove to be an insurmountable obstacle to united action on their part.

To this Judge Hertzog replied that it was a principle in law that a delegate is not to be regarded as a mere agent or mouthpiece of his constituents, but, on the contrary (when dealing with public affairs), as a plenipotentiary—with the right, whatever his brief might be, of acting to the best of his judgment.

States-Procureur Smuts concurred in this opinion, which appeared to satisfy both the Commandant-General and also all the other representatives, for no further allusion was made to the subject by anybody.

Commandant-General Botha now made his report.

In the districts of Vrijheid and Utrecht, he stated, the store of maize was so small that it could not last for more than a short time; but there was still a great number of slaughter-cattle. In the districts of Wakkerstroom there was hardly sufficient grain for one month's consumption. Two other districts had still a large enough number of slaughter-cattle—enough, in fact, to last for two or three months. In Ermelo, to the west and north-west of the blockhouses, and in Bethal, Standerton, and Middelburg, there was grain for one month. But the Heidelberg and Pretoria commandos had now, for the first time, no corn remaining for food. In the neighbourhood of Boksburg the only grain left was the old maize of the previous year, whilst there were no cattle at all in the district. When he had visited Boksburg he had found that the commandos had had no meat for three days. In the country between Vereeniging and Ermelo there were only thirty-six goats, and no cattle whatsoever. In the Wakkerstroom district, however, there were still a few slaughter-cattle. The horses were everywhere worn out and exhausted. They had been so constantly kept on the move, owing to the enemy's increasing attacks, they could now only cover the shortest distances.

The Kaffir question was becoming from day to day more serious. At Vrijheid, for instance, there was a Kaffir commando which had already made several attacks upon the burghers. This attitude of the Kaffir population was producing a very dispiriting effect upon the burghers.

The women were in a most pitiable state, now that the lines of blockhouses had been extended in all directions over the country. Sometimes the commandos had to break through the lines and leave the women behind alone; and when the burghers later on returned they would perhaps find that the women had been driven from their houses, and, in some instances, treated with atrocious cruelty.

Referring to the numbers in the field, he said that there were, in the whole of the Transvaal, ten thousand eight hundred and sixteen men, and that three thousand two hundred and ninety-six of them had no horses. The enemy during the summer had taken many of the burghers prisoner; and since June, 1901, the commandos had diminished to the extent of six thousand and eighty-four men. The burghers thus lost to them had either been killed, or taken prisoner, or had surrendered their arms.

The number of households was two thousand six hundred and forty.

The Commandant-General concluded by saying that the three greatest difficulties with which they were confronted were their horses, their food supply, and the miserable condition of their women and children.

Commander-in-Chief de Wet then spoke. He said he would leave it to the delegates who were officers to make reports. They had come from far and near, and knew exactly what the condition of things was. He, however, could state that the number of burghers in the Orange Free State was six thousand one hundred and twenty, of whom about four hundred were not available for service. The Basutos, he found, were more favourably inclined to the Boer cause than ever before.

"General De la Rey," continued General de Wet, "like myself, does not quite know what task he has to perform here, but he thinks with me that the duty of making reports belongs to the delegates. However, he feels bound to state that in his divisions there is a great scarcity of everything. But precisely the same state of affairs existed there a year ago. And when his burghers were at that time without food—well, he went and got it for them." (Cheers.)

General Beijers (Waterberg) then addressed the delegates, telling them that he would not detain them long. In Zoutpansberg, he stated, they had still a plentiful supply of food, for they were able to buy from the Kaffirs. At Waterberg the Kaffirs were neutral, but at Zoutpansberg they were getting out of hand. Yet, since no co-operation existed amongst them, they were not to be feared, and any uprising could easily be quelled.

Besides this trouble, they had many difficulties to face, which were produced by horse-sickness and fever.

As to the question of grain, there was food enough for the whole of the Transvaal and the Orange Free State. But now the English were beginning to buy up the maize at £1 a sack.

General Muller (Boksburg) reported that in his division the burghers had never suffered from hunger. He could still hold out for a few months more, as food could be obtained from the Kaffirs. There was, it could not be denied, a tendency to mutiny amongst the Kaffirs, but he did not think that this need cause any anxiety. He believed that he would be able to carry on operations until the end of the winter.

General Froneman (Ladybrand) said that the condition of his divisions, namely Winburg and Ladybrand, gave no cause for uneasiness. There were still eighty families in the districts, but they were able to provide for all their necessities. The Kaffirs were peaceable and well disposed, and were of great service to the burghers, for whom they bought clothing in Basutoland. It was possible for the burghers, he considered, to hold out for more than a year.

General Hattingh (Kroonstad) declared that in one part of the Kroonstad district there were still plenty of sheep and cattle, and that seed had been sown for next year's harvest. But another part of the district was entirely exhausted, and had to obtain its supplies from Bethlehem.

General Badenhorst (Boshof) stated that he could report on the Boshof district and the parts of the Winburg and Bloemfontein districts to the west of the railway. There were enough cattle to last his commandos for years, even if they had no other food at all. Recently he had captured fifteen hundred head of cattle, and he was in a position to give assistance to other districts. Grain, however, was not so plentiful as it had been the previous year, but nevertheless there was still a large enough supply to permit him to send help to others.

General Nieuwouwdt (Fauresmith) reported that his district was entirely devastated, and that for the last seven months there had been a dearth of all provisions; nevertheless, his burghers had contrived to live. There was, moreover, enough corn left to last them for another year. There were now only three women in the whole of his district.

General Prinsloo (Bethlehem) declared that he would be telling a falsehood if he were to say that there was no food in his district. He possessed slaughter-cattle and corn, and could help other districts. One of his commandants had recently found a store of maize (consisting of one hundred and thirty sacks) buried in the ground. The enemy had made many inroads into his district, and especially during the last few months. The blockhouses were a source of constant annoyance to him.

General Brand (Bethulie) reported upon the south-western part of the Orange Free State, where he commanded. There were some parts of his division, he said, which had been entirely laid waste. Everything had been carried off; there was not a sheep left; and the burghers had been without meat for days. But he was able to capture booty, and could still hold out for a year.

General Wessels (Harrismith) drew attention to the constant passage of large Kaffir families through the districts of Harrismith and Vrede. He could tell the delegates that the Kaffirs had been quite astonished that there were still cattle and sheep and supplies of grain in the districts. He had not yet come to the end of his provisions; but, even if everything were taken, he saw a chance of obtaining food from elsewhere.

Commandant C.A. Van Niekerk (Kroonstad) declared that if there was one part of the country which was entirely exhausted it was the part where he was in command, namely Hoopstad and a portion of Kroonstad. But yet, during the last twelve months, they had been able to obtain food, and even to sow for the ensuing year. There were no cattle in his district; but he had taken a thousand sheep and fifty-two cattle from the English.

Commandant Van der Merwe (Heilbron) spoke to the same effect.

General Smuts was the next to address the meeting. He began by saying that his expedition into Cape Colony had been the outcome of the advice which the deputation had given in July, 1901, namely to continue the war. That he had been in command of it had come about in the following way. News had been received in the Transvaal that affairs in Cape Colony were taking a favourable turn, and accordingly General De la Rey had received orders to go thither, and to take over the command there.

But afterwards it was thought wiser to annul these orders, because De la Rey could not well be spared from the western parts of the Transvaal. Owing to this, he (General Smuts) took the task upon his own shoulders, and crossed the Orange River with two hundred men. He had had a difficult task to accomplish. He had marched through Cape Colony to Grahamstad, and from thence he had pushed on towards the coast, through Graaff Reinet. Thence he had proceeded to the neighbourhood where he was now carrying on operations.

He had visited every commando, and as he had seen that there were signs of disorder amongst them he had taken them all under his own command. In this way he had found himself at the head of some fifteen hundred men. During his expeditions Commandant Lotter had been captured with a hundred men; this had reduced his force to only fourteen hundred. But since then the number had nearly doubled, so that they now had two thousand six hundred men (divided into twenty commandos) under arms in Cape Colony. In addition to these men there was a division under General De Villiers operating in Griqualand West, and another under Commandant Van der Merwe in Bechuanaland. The total numbers of these two divisions amounted to about seven hundred men.

Passing on to the question whether help was to be expected from Cape Colony, General Smuts declared that there would be no general rising. The reports which represented such a rising as possible had exaggerated matters. There were great difficulties in the way of a general rising. First, there was the question of horses—and in Cape Colony the want of horses was as great, if not greater, than in the Republics. Secondly, it was exceedingly difficult for Colonials to rise, for they knew that not only would they have to be voetgangers,[342] but also that if they were captured they would be very severely punished by the English. The scarcity of grass was also greatly against any such attempt. The horses had to be fed, and, as the enemy had forbidden any sowing, it was almost impossible to find food for them. A counter proclamation had indeed been issued by the Republics, but it had been of no avail.

He was of opinion that the small commandos which had already been in Cape Colony had done the best they could. The question that now arose was whether the whole of their forces ought to be sent from the Republics into Cape Colony. He himself thought that there was an opening for them, but the difficulty was to find a method of getting them there. The existence of this difficulty, and the facts which he had brought before the delegates, had forced him to the conclusion that a general rising in Cape Colony was an impossibility.

As to the continuation of the war and matters of that nature, they must naturally be settled by the Republics, and not by Cape Colony.

The meeting was then adjourned until eight o'clock in the evening.

Upon its reassembling, Commandant Nijs (Pretoria, North) said that in that part of the district of Pretoria which lay to the north of the Delagoa Bay Railway there were still cattle enough to last for a considerable time, but that the store of grain would be exhausted within a fortnight. The number of horses also was insufficient. The district could muster one hundred and fifty-three mounted men and one hundred and twenty-eight voetgangers. In the division of Onderwijk, Middelburg, there were twenty-six mounted men and thirty-eight voetgangers.

Commandant Grobler (Bethal) stated that in his district they had not been left undisturbed during the summer. Only a short time previously he had lost sixty-three men in an engagement, where he had been besieged in a kraal, out of which he, with one hundred and fifty-three burghers, had managed to escape. Bethal had been laid waste from one end to the other, and he had no provisions for his commandos. He had on his hands three hundred women and children; these were in a serious position, owing to the lack of food; some of the women had also been assaulted by Kaffirs.

General Christiaan Botha (Swaziland) then reported on the condition of the Swaziland commando. They had no provisions in hand, and were simply living by favour of the Kaffirs. They had no women there. His commando of one hundred and thirteen men was still at Piet Retief. As there was no grain to be had, they were compelled to go from kraal to kraal and buy food from the Kaffirs, and this required money. Yet somehow or other they had managed to keep soul and body together. "I have fought for the Transvaal," he concluded, "for two and a half years, and now, since I hear that there is food in the Free State, I shall fight for the Free State for two and a half years more."

General Brits (Standerton) said that he had still provisions for two months, but no cattle. He had sixty-five families with him, and found it very difficult to provide them with the necessaries of life. Altogether, things were in a most critical state.

[342] Infantry.

Mr. Birkenstock (Vrijheid) spoke as follows:

"I shall go deeper into some of the points which the Commandant-General has brought forward in his general report of the matter. At Vrijheid we have been harassed by large forces of the enemy for six or eight months, and the district is now completely devastated. The presence of women and children causes great difficulty, for of late the English have refused to receive the families which, compelled by absolute famine, wished to take refuge with them. There is also continual danger from the Kaffirs, whose attitude towards us is becoming positively hostile. Both horses and grain are scarce; but as far as the latter is concerned there will be sufficient, provided that the enemy does not return. One morning recently a Kaffir commando, shortly before daybreak, attacked a party of our men, who lost fifty-six killed out of a total of seventy.[343]

That peace must be made at all costs is the opinion of all the families in my district, and I feel it my duty to bring this opinion before you."

Commandant Alberts (Pretoria and Middelburg) said that his burghers had had no rest for a year, and that during that period no ploughing or sowing had been done in the district. Consequently a commando would not be able to find the means of subsistence there. On three occasions he had been forced to take refuge in a kraal, but fortunately had always been able to make his escape. They had no cattle which they could use for food, although he had received some, through Commandant Roos, from the Free State. Their horses were in the worst possible condition.

Landdrost Bosman (Wakkerstroom) then gave an account of the condition of affairs in his district. They were dependent for everything, except meat, upon the Kaffirs, giving them meat in exchange. This year there had been a very poor crop of mealies, and, such as it was, it had been much damaged by the enemy. Still the burghers might manage, with what mealies they had, to last out for another two months; but the women and children also needed to be provided for. The cattle were beginning to run short, and the few horses that they had were so weak that they would require a fortnight's rest before they could be used. It might become necessary for the commandos to leave the district, and if so, what was to become of the families?

Mr. De Clercq (Middelburg) regretted that he was unable to give as cheery a report as some of the gentlemen present had done. The part of Middelburg which he represented was in an almost hopeless condition. There were no slaughter-cattle, and only enough grain to last for a very short time. Out of five hundred horses only one hundred now remained, and these could do no work, being too weak even to get away when it became necessary to retreat from the enemy.

The state of the burghers was very discouraging; if they should be compelled to leave the district the question would arise whether, considering the condition of their horses, it would be possible for them to reach their new destination. There were fifty families in Middelburg, and things were going very badly with them. The district would have to be abandoned, and what would then be the fate of the families, which even now could only be scantily provided for? The women had wished to go on foot to the English, but he had advised them to wait until the results of the present negotiations should become known.

Commandant David Schoeman (Lijdenburg) said that although but a short time ago there had been eight hundred head of cattle in his district, they had now all been carried off. Grain there was none. Should fighting be continued, he was at a loss to know how he could provide for the women.

Commandant Opperman (Pretoria, South) reported on that part of the Pretoria district which lies south of the line. What he said agreed substantially with the report of Commandant Alberts. (See page 343.)

Commandant Liebenberg (Potchefstroom) stated that during the last eight or nine months blockhouses had been erected in his district. All that was now left to him was a strip of country about twelve miles long; here he could still exist. A good deal of seed had been sown, but the crops had of late fallen into the hands of the English. The grain was altogether spoilt; some of it had been burnt, the rest trodden down by the horses. There were ninety-three households in his district. Between Lichtenburg and Potchefstroom there were some women from the Orange Free State who were reduced to the most dire straits.

[343] The Battle of Holkrans or Holkrantz (6 May 1902) took place at Holkrans, near Vryheid. To stop the raids operated by the Boers on their cattle and crops as a punishment for helping the British, the leader of the abaQulusi, Sikhobobho, responded by assembling his forces and attacked the commando camp at the foot of Holkrans hill. The Zulus attacked at night, and in a mutual bloodbath, the Boers lost 56 killed and 3 wounded, while the Zulus suffered 52 killed and 48 wounded. This battle was the last victory of the abaQulusi to date Battle of Holkrans - Wikipedia

They had told him that if things did not improve they intended to go on foot to Klerksdorp, and he had replied that they must wait for the result of the negotiations. He had still four hundred mounted men, in addition to one hundred voetgangers. He could hold out for a short time longer, and then would have to look for some way out of his difficulties.

General Du Toit (Wolmaransstad) said that there were five hundred families in his district, but little enough for them to live on. Though his horses were weak, he would be able to save himself by strategy if he should get into a tight corner. His commandos were small—only four hundred and fifty mounted men. The cattle were in good condition, but grain was scarce.

Commandant De Beer (Bloemhof) had still under his command as many as four hundred and forty-four mounted men and one hundred and sixty-five voetgangers. Both grain and cattle were scarce, but then Bloemhof had never possessed many head of cattle. So far the families had not suffered from want. He would be able to hold out for another year.

General Kemp reported that he had under him Krugersdorp, Rustenburg, and parts of Pretoria and Johannesburg. In the district of Krugersdorp no more sowing was possible, and the majority of cattle had been carried away. Yet there was no want. Why should he lack for anything when he was in possession of a great "commissariat" extending as far as the Zoutpansberg, where General Beijers was in command? He took what he wanted from the Kaffirs—it was not their property; he was only taking back what really belonged to the burghers.

Commandant-in-Chief de Wet here asked why the eastern divisions of the Transvaal could not do like General Kemp, and take what they required from the Kaffirs?

General Kemp replied that the fact that in the eastern parts the Kaffirs were united with the English made the difference. The Kaffirs there, he said, gave all they looted to the English, who then sold them the cattle back again. If then cattle were taken in those parts, it would be cattle which was really the property of the Kaffirs. Moreover, the Zulus were Kaffirs of a different sort to those with which he (the General) had to deal. General Botha also had said that among the Kaffirs in the Eastern Transvaal there were not to be found any cattle belonging to the burghers.

Mr. J.L. Grobler (Carolina) had not as yet had to complain of any lack of cattle or grain in his district. The English, however, by their system of blockhouses, had cut the burghers off from the greater part of the crop. If nothing happened, the newly-sown crops ought to produce a good harvest; but he did not like the temper of the Kaffirs. His men could still hold out for another six or seven months. The three hundred horses still remaining to them were in a weak condition; such as they were, there was not one apiece for the burghers.

Mr. J. Naude (Pretoria) said that he represented a part of Pretoria and General Kemp's flying column. In his district sowing and harvesting went on as usual. There were fortunately no women and children. Although the commandos had not a superabundance of cattle, yet no one lacked for any of the necessaries of life.

The meeting was then closed with prayer, and adjourned until the following morning.

Friday, May 16th, 1902.

The meeting opened with prayer a little after nine a.m. The correspondence which the two Governments had addressed to the burghers, in order that it might be communicated to their representatives at one of these meetings, was first read. It was then debated whether the meeting should request Lord Kitchener to put it into communication with the deputation in Europe. After speeches pro and con, it was decided not to do so.

Thereupon General Froneman proposed the following resolution:

"This meeting is of opinion that the Governments should be asked in the first place to thank His Majesty the King of England and Her Majesty the Queen of the Netherlands, through Lord Kitchener, for the efforts which (as appears from the correspondence between the said Governments) they have made to set on foot negotiations for peace; and, in the second place, to express to them the regret of this meeting that His Majesty's Government has not accepted the proposal of Her Majesty's Government that the representatives of the two Republics now in Europe (who still enjoy the full confidence of their fellow-countrymen) should be allowed to return home, and also that Lord Kitchener has declined a similar request addressed to him by the Governments of the two Republics."

This proposal was seconded by Commandant Flemming, and carried.

After another proposal, made by H.J. Bosman, and seconded by J.L. Grobler, had been rejected, the correspondence referred to above came under discussion.

The first speaker was Mr. P.R. Viljoen, who spoke as follows:

"We can apply to our own country those words of Scripture, 'The place whereon thou standest is holy ground.' The soil on which we are now standing, wet as it is with the blood and tears of our forefathers and also of the many who have fallen in this present struggle, may well be regarded as 'holy ground.'

"That we should ever have to surrender this country is a horrible thought. Yet it must be faced. It is certain at least that many districts must be abandoned, for the enemy is doing his utmost to collect us together at a few isolated places, where he will be able to concentrate his forces upon us.

"From the reports which we have received it appears that the state of affairs in the Orange Free State is still hopeful. Not so in the Transvaal. There our prospects are of the gloomiest.

"My opinion is that we must endeavour to bring this war to an end. If there was the least chance of our being able to maintain our independence, we would still fight on, and not even the bitterest sufferings would appear unendurable. But have we any such chance?—that is the question which we have got to answer.

"We know nothing, it will be said, of the present state of affairs in Europe, for the report from our deputation, which has just been read in your presence, is six months old. Nevertheless, if anything favourable to us had occurred since then, we must have heard of it by now.

"It is evident that we must endeavour to obtain peace on terms honourable to ourselves. But how are we to do so? By keeping our independence in view when making terms with the enemy, you will answer. Nevertheless, I think it would be advisable for us to commission our Governments to ask the English Government once more what concessions it is prepared to make to us on condition of our surrendering our independence. Until we know this we can come to no final decision.

"Though it is a bitter thing to have to say, yet I feel it my duty to tell you that I honestly believe it to be impossible for us to carry on the war any longer."

Mr. De Clercq then addressed the meeting in the following words:

"The question before us is, whether or not the war can be continued? To answer it, we must look forward into the future. We must ask ourselves what consequences will ensue from a continuance of hostilities, and what will be the result of their cessation.

"We have only fifteen thousand men against the enemy's quarter of a million. Our food and horses are scarce, and we have other difficulties besides these. It is impossible to go on with the struggle.

"Nevertheless, if I believed that to do so would give us a chance of retaining our independence, I also would be ready for further sacrifices. But as it is impossible to retain our independence, surely we shall only be storing up misery for the future if we continue fighting until every man of us is a prisoner or in his grave. I am of opinion that our most reasonable course is to save what is still left to us—our existence as a nation. It is not too late to save it now, but who can tell what the future holds in store for us? If we are to be still further reduced in number, we shall soon cease to exist as a nation. Can it be right to sacrifice a nation which has fought as the African nation has done?"

Commandant Rheeder (Rouxville) then spoke as follows:

"I know that the times are very dark, but still there are some rays of light. You have been asked whether you will continue fighting until you are exterminated. But there is another alternative. Will you not continue fighting until you are relieved? I maintain that our independence must be a sine qua non of any negotiations that we make—we cannot give it up. So long as we have life we must continue to fight, and we must only lay down our arms when relief arrives."

General Kemp now rose to his feet. "I am fully aware," he said, "of the very serious position in which we are placed. Yet, when the war began, the position was no less grave. We must continue our resistance. When we recall to our minds how much this war has cost us, and what rivers of blood have flowed, we feel that it is impossible to surrender. As far as I am concerned, unless relief comes, I will fight on till I die.

"But one should not look only at the dark side of the picture. It is true enough that in some districts food is scarce, but there are none in which it is absolutely unobtainable. The districts threatened by famine must be abandoned—that is the way to deal with the difficulty.

"It has been pointed out that a large number of our men have been killed or taken prisoners. This fact, however, only fills me with courage. A cause that has cost us so dearly must never be forsaken. To own ourselves beaten would be to dig a grave for the African nation, out of which it would never rise. Why should we lose our trust in God? Up to this moment He has aided us, and He will always be our Helper."

Vice-Commandant Breijtenbach (Utrecht) then spoke as follows:

"The burghers whom I represent have told me to inform them, when these deliberations have come to an end, whether a continuation of the war is possible, and if it be possible, how it is to be accomplished. If I cannot assure them that we are able to continue the struggle, the men of Utrecht will not fight any more. As you know, I can give them no such assurance.

"There are ten districts in the Transvaal which are unable to fight any longer. It surely is not proposed to leave these districts in the lurch! We must not only consult our sentiments, but also our reason. And what does the voice of reason say? This—that the continuation of the war is an impossibility. Should you decide now to continue the war, you would have to start a fresh campaign; and you know that that is beyond our powers.

"A previous speaker has referred to the help of the Lord, but who is able to fathom His counsels? Yet we can understand the answer God has given to our prayer—that prayer which we offered with the Mausers in our hands when the war began. And what was the answer we received ... I leave it to you to reply.

"Yes, we must use our reason. If we continue the struggle we give the death-blow to our existence as a nation. We have been told that there are ten districts that cannot go on fighting. Are we going to say, 'We will continue the struggle and leave these districts to their fate'? No! We must save what we can."

General Liebenberg then spoke. "I am able to give my support," he said, "to all that has fallen from the lips of Messrs. Viljoen and De Clercq. It cannot be doubted that the future is very dark. Yes, we can only trust in God, and use our reason to the best of our ability. I have been commissioned by those whom I represent to retain our independence if possible, and if it be not possible to make peace on the best terms that we can get."

Commandant Uijs was the next speaker. He explained that if the war were to be continued he would have to leave his district and abandon the women and children to the mercy of the Kaffirs. He could see a chance of saving the mounted men if only he could feel certain that they would all follow him, but the case of the women and children would be hopeless.

A serious difficulty confronted the delegates, and it was with them, and no longer with the Government, that its solution rested. Never before had he been called upon to face so gigantic a task. It was not the time now to criticize one another, but to practise mutual forbearance. The Bible had been quoted by one of the speakers, but let them not forget the text in which the king is spoken of who calculated whether he was strong enough with ten thousand to encounter him who marched against him with twenty thousand. Then there was the question as to the disposal of the widows and orphans.

What was to become of them if the burghers, by refusing to come to terms with the enemy, should no longer be able to act as their mutual protectors? Let them make no more widows and orphans, but let them open their eyes and recognize that the hand of God was against them.

The next business was the reading of two letters—one from General Malan and the other from General Kritzinger. Malan reported on his doings in the Cape Colony, while Kritzinger advised that the war should be discontinued.

General Du Toit then spoke, emphasizing the responsibility of the delegates and the importance of the occasion. He went on to say that he represented a part of the nation which had suffered very severely, but which nevertheless had commissioned him to stand up for independence, if by any means it could be retained; if he failed in this, he was to take whatever course seemed best to him. In his district the burghers were not reduced to such a pass as to oblige them to surrender, but the condition of other districts must also be taken into consideration, and if it appeared that the war could not be continued, the delegates must get the best terms they could. In their demands they must be united—this was the principal reason why dissension was so much to be avoided. For himself, he could only say that whether the meeting voted to continue the war or to bring it to a conclusion, he would fall in with the wishes of the majority. Any decision would be better than the failure of this conference, as that would leave everything undecided.

He was followed by Secretary of State Reitz, who said:

"You all know what the Governments have done. The question now is, Is there anything further that we can do? For my part, I think that there is. We might offer to surrender Witwatersrand and Swaziland; we might also relinquish our rights to a foreign policy; we might even accede to an English Protectorate. If France has been able to do without Alsace and Lorraine, surely we can do without the goldfields. What benefit have they ever done us? Did the money they brought ever do us any good? No! rather it did us harm. It was the gold which caused the war. It is then actually to our advantage to cede the goldfields, and moreover by so doing we shall be rid of a very troublesome part of our population."

Mr. Reitz then went on to discuss in detail the position in regard to Swaziland, the question of a British Protectorate, and the surrender of our right to treat with foreign powers.

General Muller (Boksburg) expressed sympathy with the views of the Secretary of State, while Vice-Commandant Roux (Marico) said that he was prepared to sacrifice many things, but that he intended to hold out for independence.

The next speech was made by Landdrost Stoffberg (Zoutpansberg), who said:

"I agree with General Du Toit in what he said about the necessity for unity amongst us. Disunion must not be so much as mentioned. I have a mandate from the burghers of Zoutpansberg not to sacrifice our independence. But if anything short of this will satisfy the English, I am quite prepared to make concessions. Some of the burghers think that it might be well to surrender the goldfields for a certain sum of money, while others point out that the gold was the cause of the war. I also think that we have suffered through the gold, and that we might give up the goldfields without doing ourselves any harm. For what has the gold done for us? It has enriched us, many will say. Yes! but it has also been a stumbling-block to many a man. And is it not better to be a poor but independent nation than to be rich and at the same time subject to another Power. Let the goldfields go. We shall still, with our markets, be rich enough."

Commandant Mentz (Heilbron) then rose.

"I appeal to the forbearance of the delegates," he said, "for making any speech at this meeting. I fear I am unable to give as rose-coloured a report as my brother Free-Staters have done: My district has been continually harassed by the enemy's troops, and great devastation has been wrought. But the greatest trouble I have is the presence of so many families, for there are still two hundred in the district. I have only eighty burghers under my command, and it is clear to me that I shall soon be obliged to leave the district. What will then become of these families?

I received a commission not to sacrifice our independence. But since my burghers met more than half of them have been made prisoners. The remainder have instructed me to do my best to preserve our independence, but if I find that it cannot be maintained to act according to my own judgment. It appears to me that it may be possible to retain our independence by ceding some part of the country; if this be the case it ought most certainly to be done. I can remember the late President Brand saying in connexion with the diamond fields, 'Give them up; you will gain more by giving them up than by keeping them.' This remark may well apply to the present situation."

Commandant Flemming (Cape Town) reported that his district was well-nigh devastated. But they still possessed a fair number of cattle, which they had carried away with them. But even if they had no cattle, that would be no excuse for surrender, for in his district it was possible to live on the game. The view which he and his burghers had taken was that since they had already sacrificed nearly everything they possessed, they would not now sacrifice their independence. For should this also be lost, then there would be nothing left to them. That had been their opinion, but they had not then known how matters stood in the Transvaal. Now that he was aware of the state of affairs, he agreed with State Secretary Reitz that their best course was to cede a part of their territory.

Vice-President Burger now rose from his seat, and said:

"This meeting has to formulate a fresh proposal to the English Government, and to await its answer. If this proposal be rejected, well, you will be no worse off than you are at present. If there be a man who has earnestly considered what the sacrifice of everything means to us, then I am that man. It has been said, we must retain our independence, or else continue to fight; and we are still able to hold out for another six months, or even a year. Now, supposing that we can hold out another year, what should we gain by doing so? Why, we should only grow weaker, whilst the enemy grew stronger! I emphatically state that the war cannot be carried on any longer;

and I ask if there is any man here who can maintain with a clear conscience that the struggle can be successfully continued.

"Some of you may tell me that complications may arise in Europe. But that is a groundless hope. Others may say that it is astonishing enough that we have been able to hold out till now, and that we still have the power of making our voices heard. Yes! that is very surprising; but shall we retain this power long? I heard some delegates say, 'We shall fight till we die!' That is a manly sentiment. But was it not, perhaps, prompted by a desire to make a fine speech, which would go down to posterity? Was not the aim in some cases that future generations might recall these speeches when they were told of the brave fight our men had made?

"Let every one consider this well: Is he prepared to sacrifice the nation on the shrine of his own ambition? Ambition, although it may cost us our lives, can never lead to martyrdom. A martyr is made of finer stuff!

"Have we not arrived at the stage of our history when we must pray, 'Thy will be done'? That prayer, considered rightly, is a prayer of faith. Do not let us imagine that we can compel God to do our will—that is not faith.

"I beg of you to consider what will become of the women and the children and the banished burghers if you still persist until your last shot has been fired. What right shall we have to intercede for these unfortunate ones when we have rejected the proposals of the English Government? We shall have no right whatsoever.

"Perhaps it is God's will that the English nation should oppress us, in order that our pride may be subdued, and that we may come through the fire of our troubles purified.

"My opinion is that we should make a peace proposal to England, yielding as much as we rightly can; and if England rejects our proposal, it will be time enough then to see what other course is open to us.

"There is one fact which we cannot allow ourselves to forget. There are ten districts in the Transvaal which must be abandoned. In the Free State, too, there are districts in a similar plight. It is the opinion of lawyers that so long as the inhabitants remain in a district their property cannot lawfully be confiscated; but if the district be abandoned, then confiscations can take place.

"It is criminal to say, 'Come what may, we will fight till everything is lost and all of us are dead!'"

The following resolution was then proposed by General Kemp, and seconded by Mr. J. Nand:

"This meeting decides, in order to expedite the work in hand, to depart from the original programme; and to constitute a Commission, to be composed of the Hon. Jacob Smits and the Hon. Judge Hertzog, and to give this Commission authority to draw up, conjointly with the two State Presidents, a draft proposal, to be laid before the delegates to-morrow morning."

This resolution was put to the meeting, and accepted by the delegates. The meeting then adjourned.

At half-past seven in the evening the delegates reassembled.

General Cilliers (Lichtenburg and Marico) was the first to make a report. "In my division," he said, "things are in a very favourable condition. Yet we are bound to take the other divisions into consideration. My burghers said to me, 'Stand firm for independence!' But when they gave me the order they did not know about the condition of the other districts. Will those other districts—such of them, I mean, as are in a worse predicament than ourselves—be able to co-operate with us in continuing the war? Some of them have already answered my question in the negative. Must we then not ask ourselves, What will be the best for the nation as a whole? Shall we say continue the war, or shall we approach the enemy and make a proposal?

"But are we really justified in prolonging the struggle, and making still further sacrifices? Some will answer, 'Yes, for we have a God in whom we have trusted from the beginning; shall we not continue to trust in Him who has worked such wonders for us already?' But I have heard a brother say, 'God's hand is against us.' It was bitter to hear these words from him, and for myself I will have none of them. My vote is given here and now for a continuance of the war.

"But we must hear what the rest of the delegates have to say, and if they can point out some other way by which we can retain even a portion of our national independence, we must be ready to follow it."

General Froneman next addressed the meeting.

"I fear," he began, "that too much is being made of the condition of my division: things are not so prosperous with us as some here appear to imagine. But for all that, my burghers are for nothing short of absolute independence. They cannot forget the blood which has already been spilt in our cause. They mean to hold out until they are relieved.

"I sympathize deeply with those districts that are less happily circumstanced than my own, but it pains me to discover that there are some here who doubt that God is for us. For what has supported us up till now save faith in God?—the faith of those who first prayed God to prevent the war, and then, when they saw that this was not His will, fought like men, putting all their trust in Him.

"Up till now the Lord hath been my helper; the enemy has cut us off from everything, and yet we see our two little Republics still full of hope, still holding out."

He concluded his speech by saying that he would like to hear the opinions of Generals Botha, De Wet, and De la Rey. They ought to be able to throw much light upon the matter.

Commandant General Botha then rose, and said:

"I am glad to have an opportunity of giving my views upon the present state of affairs. We know that differences of opinion are to be found everywhere and on every question; when, therefore, a man differs from those who think that this war can and ought to be continued, we must ascribe his opinion to discouragement, weakness, or cowardice. We must acknowledge the truth of the facts from which he draws his conclusions, and which have compelled him to utter it. His object is to make known the true state of the country—which indeed is his plain duty. Were he not to do so on the present occasion he would be accused, later on, of having kept secret what he ought to have revealed. Differences of opinion then need not, and must not, cause a disunion and discord. Whatever our private opinions may be, yet, as delegates of the burghers, we must speak and act as one man.

"The war has now lasted two years. But the question for us to answer is this: Are we going forwards or backwards? My own conviction—a conviction founded upon the views expressed by my commandos and the speeches which I have listened to at this meeting—is that we are not gaining, but losing ground. There is nothing, in my opinion, more evident than that, during the last six months, the tide has been setting steadily against us, and in favour of the enemy.

"A year ago there were no blockhouses. We could cross and recross the country as we wished, and harass the enemy at every turn. But now things wear a very different aspect. We can pass the blockhouses by night indeed, but never by day. They are likely to prove the ruin of our commandos.

"Then, as regards food. We are told that there is food here, and food there; but how are we to get at it? How are we to transport it from one district to another? Outside the frontiers of our Republics there are plenty of provisions, but it becomes daily more difficult to get them into our hands. The cattle, for instance, that used to be at Ladysmith have now been removed to Estcourt. Even the friendly Kaffirs, from whom we are now able to obtain provisions, may quite possibly soon turn against us. The time is coming when we shall be compelled to say, 'Hunger drives us to surrender.'

"The horses have been chased about so incessantly, and have suffered so much from want of forage, that their strength is almost exhausted. They are so weak that it is almost impossible to accomplish any long distance with them.

"As to the Cape Colony, I had always understood that the Colonists were going to rise en bloc, but General Smuts has just told us that there is no chance of such a thing happening. And he speaks from personal knowledge, having just returned from paying them a visit. Moreover, he has seen our horses, and says that it is impossible for them to go into the Colony, so it appears that our successes there are over. This is a severe check indeed; but it could not have been otherwise. We have not enough horses to enable us to give the Colonists effectual help, and they themselves have been cowed by the heavy penalties imposed upon all those who did rise. Many of those who are well disposed towards us dare not join us now.

"Again, there is no chance of European intervention: not one of the Powers will do anything for us. To see this it is only necessary to peruse that correspondence between the Netherlands and England, which was the cause of these negotiations. There we shall find that the Dutch Minister says that our deputation is only accredited to Holland, whereas it had been accredited by the two Republics to all the Governments in Europe. Moreover, the correspondence makes it very plain that England will not tolerate the intervention of any foreign Power whatsoever. But the truth is, that no foreign Power wants to help us. When the women were first made prisoners I thought that European intervention might perhaps be attempted, because to make prisoners of women is a thing quite outside the usual methods of warfare. But nothing was done even then. We were told that we had the sympathy of the nations of Europe—their sympathy, and nothing more!

"I have come to a subject that is very near our hearts—our women-folk. If this meeting decides upon war, it will have to make provision for our wives and children, who will then be exposed to every kind of danger. Throughout this war the presence of the women has caused me anxiety and much distress. At first I managed to get them into the townships, but later on this became impossible, because the English refused to receive them. I then conceived the idea of getting a few of our burghers to surrender, and sending the women in with them. But this plan was not practical, because most of the families were those of prisoners of war, and the men still on commando were not so closely related to these families as to be willing to sacrifice their freedom for them.

"We have heard much talk about fighting 'to the bitter end.' But what is 'the bitter end'? Is it to come when all of us are either banished or in our graves? Or does it mean the time when the nation has fought until it never can fight again? As to myself, personally, I can still continue the struggle. I have horses, my household is well provided for, and as far as my own inclination goes I am all for going on. But am I only to consider myself? Is it not my first duty to look at the interests of my nation?

I have always been, and still am, of the opinion that, before letting the nation go to rack and ruin, it is our duty to parley. We must not let the chance for negotiations slip out of our hands. When our numbers have fallen to only four or five thousand men under arms we shall no longer have that chance, and this will undoubtedly happen if we hold out for another year, or even six months.

"There are some who say, 'We must trust in God and keep on fighting,' and I grant them that miracles are possible at all times. But it is beyond our power to say whether God will work a miracle for us. We do not know what His will may be. If we continue the war, and if it should afterwards appear that everything has been in vain, our responsibility will be only the heavier, the blinder our confidence now is. And over and over again we shall hear, 'He is dead,' 'and he, and he.' Will not this make our remorse all the more bitter? Our commandos are so weak, our country so exhausted, that the loss of one great battle, the surrender of a single strong force, would spell ruin for us.

"'But we have managed to hold out for so long.' Yes, but there is a natural reason, a military reason, why this has been the case. The fact that our commandos have been spread over so large a tract of country has compelled the British, up to the present time, to divide their forces. But things have changed now; we have had to abandon district after district, and must now operate on a far more limited territory. In other words, the British army can at last concentrate its forces upon us.

"I firmly believe that, under like circumstances, no other nation in the world would have fought as our nation has done. Shall such a nation perish? No! we must save it. If we delegates are convinced that we can no longer offer resistance to the enemy, it is our plain duty to tell the people so. We must not let them be exterminated for want of timely advice. More than twenty thousand women and children have died in the camps during this one year.

"There are men of our own kith and kin who are helping to bring us to ruin. If we continue the war, it may be that the Afrikanders against us will outnumber our own men.

"What is there left to hope for? Are we to retain our independence by ceding a part of our territories? Most assuredly yes, if such a compromise is feasible. As regards Swaziland, it is of so little importance to us that we can give it up without a thought. Then there are the goldfields—let them go. They are but a cancerous growth, sapping the very life of our country.

"We must face the fact that things are not at a standstill: we are slipping back every moment. We must all pull together, or everything is lost. If our sacrifices will buy our independence, well and good. But suppose that we are compelled to give it up—well, if it even comes to this, we must never do so unconditionally. An unconditional surrender would be well enough if the leaders only had to be considered. But we must think of the interests of the nation. We must say to our people, 'We have no thought of ourselves: our only desire is to place ourselves in the breach, if so we may save you.'"

General Botha then proceeded to discuss eventualities in the event of independence being lost. Representative government, he said, might perhaps still be retained, and the national language need not necessarily be supplanted. Thus the nation would still retain its old ideals and its old customs. General Roux had been pertinently asked whether it were better to strive for the recuperation of the people now or to wait until they were altogether overpowered and reduced to such straits that it would require some thirty years before they could once more call themselves a nation. He then went into the terms of the proposal by the British Government, and repeated that there must be no idea of unconditional surrender.

The General concluded in the following words:

"Although we do not wish to accept terms, we have no right to refuse them altogether. On the other hand we must not say to the English, 'Do with us as you like.' For then our descendants would eternally reproach us. We should have lost the privilege of looking after our own wives and children. They would be handed over to strangers. No! we must secure by some means or other that we ourselves shall be able to provide for them. The fate of our country is in the hands of the men in this tent. It has been bitter, indeed, for me to have to speak as I have done. But if I have not spoken the truth, convince me of my error, and I will be the first to own it. But do not condemn me, for I have had no other object than to tell you what I believe to be the truth."

General De la Rey spoke.

"I will not detain you long," he began, "but there are a few points to which I wish to draw attention. In regard to the districts under my command, every one will understand that my burghers, after their recent brilliant successes, are firmly resolved not to sacrifice their independence. If I allude to the battles which I have just fought it is with no thought of boasting, but only that you may picture to yourselves the effect which they must have had upon the enemy; and that no one may be angry with myself and my burghers for standing firm when our feet are on such solid ground.

"But since my arrival at Vereeniging I have heard about our districts where matters are in a far less favourable condition than in my own. So far as I myself am concerned, I cannot think of laying down my arms. Yet it appears to me that some parts of the country will be compelled by starvation to give up the struggle. It is well that those who represent these parts have spoken openly, and not left this meeting in ignorance of the state of affairs only to go and lay down their arms.

"I myself have never thought intervention possible. Even before the war broke out I said that nothing would come of it. I saw that South Africa was divided between Germany and England. And that if only the Republics could be extinguished, then England and Germany would be the only Powers left, and Germany would be safe. But if the Republics were victorious, then Germany would be in danger. Why then should Germany interfere in favour of the Republics, when she has everything to lose by such a course of action? No! intervention was entirely out of the question.

"There has been talk about fighting to the bitter end; but has not the bitter end already come? Each man must answer that question for himself.

"You must remember that everything has been sacrificed—cattle, goods, money, wife, and child. Our men are going about naked, and some of our women have nothing but clothes made of skins to wear. Is not this the bitter end?

"I believe that the time has now come to negotiate. England will never again give us the chance of doing so, should we allow this opportunity to slip by. But how shall we negotiate? I must leave it to this meeting to answer that question. If we do not obtain what we ask for, we shall at least stand or fall together. Yet let us act with reason.

"I cannot agree with one of the opinions expressed by Commandant-General Botha and States-Secretary Reitz. They have stated that they are against surrendering the goldfields to England; firstly, because England would never accept such a proposal, for by doing so she would declare to the whole world that she had only been fighting for the goldfields; and, secondly, because if we gave up the goldfields we should lose a source of revenue, without the aid of which we could not repair the damages which the war has wrought."

Commandant-in-Chief de Wet spoke as follows:

"I am of opinion that the circumstances in the Orange Free State are no less critical than those in the Transvaal. Nine districts were entirely ruined; but these, though at one time abandoned by the burghers, have now been reoccupied.

"If I now differ from those who are of opinion that it is useless to prolong the war, it must not be thought that I am lacking in respect for their judgment. By no means. I know that what has been said about the wretched plight of the people is only too true; but they must not take it amiss if I point out that the same condition of affairs was described in the correspondence from the Transvaal which fell into the hands of the English at Reitz. But, granting that the facts have been correctly stated, even then the Orange Free State will refuse to give in. Let me be candid with you, and say frankly that, in my opinion, this is virtually the Transvaal's war. This, however, makes no difference to me. For me the barrier of the Vaal River has never existed. I have always endeavoured to maintain the Nauwere-Vereeniging,[344] and I feel strongly the obligation which the union of the two States casts on each one of us. They are two nations, but their cause is one.

"What, then, is the prevailing feeling in the Orange Free State? Of the six thousand burghers who have been attending meetings, I myself have been in command of five thousand, and I can confidently say that never were five thousand men more unanimous in their opinion than were those I led when they cried, as with one voice, 'Persevere; we have everything to lose, but we have not yet lost it.' What, then, is the answer to be? I am firmly persuaded that we have only one course before us.

If we are unable to obtain what we are asking for, then it only remains for us to alleviate as best we may the lot of those who cannot help themselves. I do not as yet clearly see how this is going to be done, but, at all costs, let us continue fighting. What was our total strength when we began this war? Sixty thousand men all told. Against this the English had a standing army of seven hundred and fifty thousand troops. Of these two hundred and fifty thousand, or one-third, are now in South Africa. We know from experience that they are unable to send more than one-third. And we? Have we not also one-third of our army left?

"I do not wish to imply that I am not prepared to concede something, but nothing will induce me to consent to any part of the country in our territory being given up. It will never do to have an English colony planted in our midst, for England then would have far too firm a hold upon our country.

"It is said, and with some truth, that the goldfields have been a curse to us, but surely there is no reason why they should continue to be so. I fail to see how, without retaining possession of these goldfields, the Republics are to be saved. Swaziland perhaps could be ceded, but never the goldfields. I feel that any intervention is out of the question; but is not the very fact that it has not taken place a sure proof that it was not the will of God? Does it not show that He is minded to form us, by this war, into a nation worthy of the name? Let us then bow to the will of the Almighty.

"My people will perhaps say, 'Our Generals see only the religious side of the question.' They will be right. Without faith we should have been foolish indeed to have embarked on this war and to continue it for so many months. Indeed, it must be a matter of faith, for the future is hidden from us. What has been is within our ken, but what is before is beyond the knowledge of the wisest man.

"Cape Colony is a great disappointment to me. I do not refer so much to what we have learnt about it from the reports as to the fact that no general uprising can be expected in that quarter. So much we have heard from General Smuts. But though there is to be no uprising, we have no reason to think that there has been any falling off in the number of our adherents in the Colony. The little contingent there has been of great help to us: they have kept fifty thousand troops occupied, with which otherwise we should have had to reckon.

"I feel deeply for our women and children; I am giving earnest consideration to their miserable plight. But their sufferings are among what we may call the necessary circumstances of the war. I have nothing to do with the circumstances. For me, this is a war of religion, and thus I can only consider the great principles involved. Circumstances are to me but as obstacles to be cleared out of the road.

"If we own ourselves defeated—if we surrender to the foe—we can expect little mercy from him. We shall at all events have dug the grave of our national independence, and, as things are, what difference is there between this and digging our own graves?"

Mr. Birkenstock said that the question about the goldfields must be carefully considered. This source of income must not be given up.

The meeting was then closed with prayer.

[344] Closer Union.

Saturday, May 17th, 1902.

The Chairman first called upon Chief Commandant de Wet to offer up prayer.

A private report from Mr. J. Schmorderer, who had brought the missive from the deputation in Europe, was then read.

The first delegate to speak was Landdrost Bosman (Wakkerstroom), who said:

"My opinion is that the best way of ascertaining the probable future course of events is to see what has already happened in the past. A year ago there were six hundred burghers in my district, and each man had a horse; now there are not more than half that number, and many of them have to go on foot. Last year we had from three to four thousand bags of maize ready to hand; this year there are not more than as many hundred, and how to get at them is more than I can tell. If such has been the history of the past year, in what sort of condition shall we be at the end of the present one?

"The great difficulty with regard to our families is not how to clothe them, but how to feed them. I know of a woman who has lived for weeks on nothing but fruit. I myself have had to satisfy my hunger with mealies for days together, although I have no wish to complain about it. Even the scanty food we can get has to be obtained from the Kaffirs by persuasion. Moreover, the Kaffirs side with the English, who in their counter-marches are clearing all the food out of the country.

"The men in my district told me that if I came back and reported that the war was to be continued, they would be obliged—for the sake of their wives and children—to go straight to the nearest English camp and lay down their arms. As to the women it is true that they are at present full of hope and courage, but if they knew how matters stood in the veldt, they would think very differently. Even now there are many of them who say that the war ought to be put a stop to, if only for their sakes.

"The Kaffirs are another great source of trouble; in this problem they are a factor which cannot be neglected.

"There is no hope of intervention, nor can we expect anything from the English nation. Facts that have come to my knowledge prove to me that England has become more and more determined to fight to the bitter end.

"I do not see what we can possibly gain by continuing the war. Our own people are helping the English, and every day the enemy are improving their position. What advantage can there then be in persisting in the struggle? We have now a chance of negotiating, and we should seize that chance. For we have the opportunity given us of obtaining some help for our ruined compatriots, who would be entirely unable to make a fresh start without assistance.

"As to the religious side of this matter, I am not ashamed to say that I believe I am serving God in the course which I am taking. We must not attempt to obtain the impossible against all reason. If we make any such attempt, the results will probably be exactly opposite to what we wish. I have the greatest doubt whether it really is in order to give glory to God that the nation wishes to retain its independence. On the contrary I believe that the motive is obstinacy, a vice to which human nature is always prone.

"It has been said that it would be shameful to disregard the blood already spilt; but surely one ought also to consider the blood that might yet be shed in a useless struggle."

The proposal of the Commission was now read, and after some discussion accepted. It ran as follows:

The meeting of national representatives from both Republics—after having considered the correspondence exchanged, and the negotiations conducted, between the Governments of the two Republics and His Excellency Lord Kitchener, on behalf of the British Government; and after having heard the reports of the deputies from the different parts of both Republics; and after having received the latest reports from the representatives of the two Republics in Europe; and having taken into consideration the fact that the British Government has refused to accept the proposal of our Governments made on the same basis; and notwithstanding the above-mentioned refusal of the British Government—still wishes to give expression to the ardent desire of the two Republics to retain their independence, for which already so much material and personal sacrifice has been made, and decides in the name of the people of both Republics to empower both Governments as follows:—To conclude a peace on the following basis, to wit: the retention of a limited independence offering an addition to what has already been offered by the two Governments in their negotiations, dated the 15th of April, 1902.

(a) To give up all foreign relations and embassies.

(b) To accept the Protectorate of Great Britain.

(c) To surrender parts of the territory of the South African Republic.

(d) To conclude a defensive alliance with Great Britain in regard to South Africa.

During the discussion it was clearly explained that the territory which it was suggested should be ceded was the already mentioned goldfields and Swaziland. The question was put whether the South African Republics would have to pay for the damage done during the war. "By all means let us pay," said Mr. De Clercq. "If I could only buy back the independence of the Orange Free State, I would gladly give all I possess."

Several other Transvaal delegates expressed themselves in the same sense, and said that they fully appreciated the sacrifices which the Orange Free State had made. General Froneman thanked them in the name of the Free State.

He felt that the two Republics no longer thought of themselves as having conflicting interests. In the fire of this war they had been firmly welded together.

Commandant Ross (Vrede) thought it wrong even to discuss the possibility of giving up independence. The delegates had received a definite mandate. They had been com missioned to see that the national independence had remained untouched, whatever else might have to be given up. This being the case, they might come to decisions on all other points, so long as they remembered that independence was not an open question.

Commandant J. Van Niekerk (Ficksburg) spoke to the same purpose. He could not even think of sacrificing independence.

After some other delegates had made a few short remarks, General Brand, seconded by Commandant A.J. De Kock, proposed the following resolution, which was accepted by the meeting:

"This meeting of the national representatives of the two Republics hereby charge the Governments to nominate a Commission for the purpose of entering upon negotiations with His Excellency Lord Kitchener, acting on behalf of His Britannic Majesty's Government. The Commission is to endeavour to make peace on satisfactory terms, and is then to lay the result of its negotiations before this meeting, for the sanction of the two Governments."

The meeting was then closed with prayer.

The Conference at Pretoria (May 19th–May 28th, 1902)

Appendix B

The Conference At Pretoria Between The Commission of The National Representatives And Lords Kitchener And Milner (May 19th-May 28th, 1902)

Minutes of the Conference held at Pretoria on May 19th, 1902, between Lord Kitchener and Lord Milner, representatives of the British Government, and Commandant-General L. Botha, Commander-in-Chief C.R. de Wet, General J.H. De la Rey, Judge J.B.M. Hertzog, and General J.C. Smuts, delegates of the national representatives, who had met at Vereeniging on May 15th, 1902.

Mr. N.J. de Wet acted as interpreter; Mr. O. Walrond was secretary for the English Government; and the Rev. J.D. Kestell and D. Van Velden acted in a similar capacity for the Commission.

The Conference met at ten o'clock in the morning at the house of Lord Kitchener. After having greeted each other, the members took their seats at the table in the centre of the room.

Commandant-General L. Botha opened the proceedings in the following words:

"Allow me to state that, although the negotiations have taken a longer time than we expected, I am able to assure your Excellencies that we are acting in good faith, and that everything has been done with the sole aim of concluding the peace which we all desire.

"I must also draw attention to the fact that everything we transact here must be submitted to our national representatives, in order to obtain their sanction."

The suggestion was then made that the proposals which the Commission was prepared to make should be laid before the Conference, whereupon the following letter was read to the meeting:

Pretoria, 19th May, 1902.

To their Excellencies, Lord Kitchener and Lord Milner, Pretoria.

Your Excellencies,—

With a view to finally concluding the existing hostilities, and being fully empowered by the Government of the two Republics, we have the honour to propose the following points—in addition to the conditions already offered in the negotiations of April last—as a basis for negotiations:

(a) We are prepared to cede our independence as regards our foreign relations.

(b) We wish to retain self-government in our country, under British supervision.

(c) We are prepared to cede a part of our territory.

Should your Excellencies be prepared to negotiate on this basis, then the above-mentioned points can be elaborated.

We have the honour to be,

Your Excellencies' most obedient servants,

Louis Botha.

C.R. De Wet.

J.H. De La Rey.

J.B.M. Hertzog.

J.C. Smuts.

When this letter had been read, a discussion followed.

Lord Milner: "Considering the wide difference between this proposal and that made by His Majesty's Government, when we last met, I fear that I can hold out very little hope of any good results following negotiations on the basis you have suggested."

Lord Kitchener: "We can take those proposals into consideration, but I cannot see how it is possible to bring them into harmony with those of His Majesty's Government."

Commandant-General Botha: "If this is the position you take, we should like to receive from you a final answer to our proposals."

Lord Milner: "Do you wish us to refer your proposals to His Majesty's Government?"

Commandant-General Botha: "Yes, unless you have full powers to give us a final reply."

Lord Milner: "I am quite convinced that your proposal will be rejected; and I feel bound to say that to refer it, as it stands, to His Majesty's Government will only do you harm."

Commandant-General Botha: "If you have no power to decide upon this proposal here, we should like you to refer it to His Majesty's Government."

Lord Milner: "I have no objection to taking the responsibility of refusing your proposal on myself. The instructions received by myself and Lord Kitchener are quite clear on this point."

Commandant-General Botha: "I must then understand that when Lord Salisbury said that this war was not carried on with a view to annex territory, he did not mean it."

Lord Kitchener: "It is no longer a question of territory, for annexation is an accomplished fact."

Commandant-General Botha: "I am unable to see how our proposal is inconsistent with annexation."

Lord Milner: "I cannot now recall the exact words used by Lord Salisbury, but it is true that Lord Salisbury declared that his Government did not begin the war with the intention of obtaining territory. But in the course of the war circumstances developed in such a way that the decision to annex the Republics became a necessity, and the British Government have pronounced their firm intention not to withdraw from this decision."

Judge Hertzog: "I should like to be informed as to what the great difference is between the basis now proposed by us and that laid down by His Majesty's Government during the negotiations of last year—I do not mean the difference in details, but in principle."

Lord Kitchener: "Do you mean by your proposal that the Boers will become British citizens?"

General Smuts: "I cannot see that our proposal is necessarily in contradiction to that of last year. Our proposal only makes provision concerning the administration."

Lord Milner then quoted from the terms offered at Middelburg by the British Government the previous year:—

"At the earliest possible date military administration shall cease, and be replaced by civil administration in the form of a Crown Colony Government. At first there will be in each of the new Colonies a Governor, an Executive Council consisting of the highest officials, and a Legislative Council, which latter shall consist of a certain number of official members and also of a nominated non-official element. But it is the wish of His Majesty's Government to introduce a representative element as soon as circumstances permit, and, in course of time, to grant to the new colonies the right of self-government.

"It may be that I do not properly understand your proposal, but it seems to me to differ not only in detail, but also in spirit from the scheme I have just read to you."

Judge Hertzog: "I entirely agree with you that there is a difference in idea between the two proposals; but only such a difference in idea as might well be found between Colonies of the same State. In other words, one constitution is adapted for one colony, whilst another constitution is found fitting for another colony, but yet they all belong to the same Empire."

Lord Milner: "Exactly. There are different constitutions in different Colonies; but it seems to me that the policy laid down in your proposal differs from that laid down by His Majesty's Government."

Judge Hertzog: "I think that I am expressing the opinion of the whole Commission when I say that we wish for peace. I draw attention to this to show the way in which, according to my opinion, we should consider the matter. For if we on both sides are really desirous of coming to a settlement, we should not make too much of theoretical difficulties, so long as the practical aim has been obtained. For instance, the different Colonies which now are joined to form the United States once possessed constitutions differing much from one another. Now the constitution laid down in our proposal does not differ so much from that laid down in yours that a practical difference should arise therefrom; and such a practical difference would arise if you insisted upon carrying on negotiations on your own basis. I imagine that England has a certain object before her in South Africa, and I believe that that object can be as well obtained by our proposal as by that of Middelburg. I therefore ask, Is the difference so great that, in order for England to obtain her object, an entirely new status must be called into existence?"

Lord Milner: "We are comparing two different things. Here in the Middelburg scheme there are a number of definite proposals, which enter upon a great mass of particulars. I do not mean to imply that we have not the power to go into particulars. I perfectly understand that it lies within the power of Lord Kitchener and myself to carry on further deliberations with you about details, so as to throw light on any doubtful points, and, perhaps, to make such changes as would not fundamentally affect the scheme. As you say that your proposals are not in contradiction with those formulated at Middelburg, then there is no reason why you should not lay aside your proposals and discuss the Middelburg proposals, which are definite."

Judge Hertzog: "I quite admit that you, Lord Milner, are entitled to say that there is a fundamental difference between our proposals. But it is another question whether the difficulty that thus arises is of such a nature that we—those of us who on both sides are anxious to conclude peace—should not be able to find a solution to it satisfactory to both parties. I cannot answer that question; nor can I see why the same result would not be reached by negotiating on the basis proposed by us as by carrying on negotiations on the Middelburg proposal."

Lord Milner: "I understand, then, that you acknowledge that there is a fundamental difference between the two bases. Well, I do not think that we are empowered to negotiate on a basis differing from that laid down in the last report of His Majesty's Government, and also differing from the tenor of the Middelburg proposal. I may say that I believe that His Majesty's Government in their latest message went as far as it was possible for them to go with the object of meeting you. The whole spirit of the telegram was to that effect."

Commander-in-Chief de Wet: "I hope you will understand that I do not speak as a lawyer. (Lord Kitchener, laughing: "That's the case with me too!") I fully concur with what General Botha and Judge Hertzog have said in regard to our eagerness to establish peace. In order to be brief, I will only remark that I did not understand His Excellency, Lord Milner, to mean—any more than I myself meant—that we should go to the nation with the Middelburg proposal, with the idea of coming back with it unaltered."

Lord Milner: "No; if I gave that impression, I did not intend to do so. But I believe that when you went to your people with the last message from His Majesty's Government it was with the knowledge—which the message itself made clear—that His Majesty's Government was not prepared to take into consideration any terms which differed widely from the policy laid down in the Middelburg proposal."

Commander-in-Chief de Wet: "That was indeed what I understood; and accordingly we have now come with a proposal which does not differ very much from the Middelburg proposal."

General Smuts: "I thought that the vital principle your Government had in view was the destruction of our independence, and in our proposal the independence of the two Republics with regard to foreign relations is given up. I was therefore of opinion that the two parties might come to an arrangement on this basis. I did not think that for the restoration of peace the Middelburg terms were essential."

Lord Milner: "Not in the details, but in the general ideas. As the British Government has laid down a basis, and you have had weeks in which to consider the matter, it would never do for you now to put it on one side. Lord Kitchener has given your nation considerable time in which to take counsel; and now you come back, and, ignoring the Middelburg terms, you propose entirely different ones of your own, and say, let us negotiate on these. I do not believe that I and Lord Kitchener would be justified in doing this. But in case he is of another opinion, the British Government can be asked if they are prepared to set on one side all the former deliberations and begin again on a new basis."

Commander-in-Chief de Wet: "We cannot, of course, prevent Lord Kitchener from asking his Government any questions he pleases, but, at the same time, we request that you will cable our behests to the English Government."

Commandant-General Botha: "I cannot see that we are beginning again on a new basis, for, in consequence of the negotiations in April last, you were ordered by the British Government to encourage us to make fresh proposals. Our present proposal is the direct result of that order."

Lord Milner: "I did my best to get fresh proposals from you, but you would not make any. You forced the British Government into making proposals."

Commandant-General Botha: "I am of opinion that we must both work together in this matter of formulating proposals."

Lord Kitchener: "You were asked to make proposals, but you did not do so; and now, after the British Government has made a proposal, you yourselves come forward with one of your own."

General De la Rey: "I think that it was the encouragement given us by correspondence between the Netherlands and the British Government that caused us to make our proposals."

Lord Milner: "That correspondence was at the beginning of the negotiations."

Commander-in-Chief de Wet: "If we had been obliged to make a new proposal in April, we would not have been able to make one so fair, and so much to the advantage of the British Government, as our present one, for, not having consulted the nation, we would have been compelled to insist on entire independence."

Lord Milner: "I must remind you of what has taken place; not with the object of putting you in the wrong, but in order to make the position clear, for there are some points about it which are not very clear. You came and made a proposal. The British Government gave you a distinct answer—they refused to accept it. Their answer was perfectly outspoken, and perfectly intelligible. At the same time they said, 'We are anxious for peace; will you make other proposals?' You then said, 'No! we have no power to do so; we must first consult the nation.' We admitted that argument. Then you said, 'Let the British Government make proposals.' The British Government did so, and they are fully entitled to an answer. In what position do you think you are placing Lord Kitchener and myself? You come back with a totally fresh proposal, and do not say anything about ours. This is not fair treatment to the British Government, and we are not bound to take your proposal into consideration."

Judge Hertzog: "I have endeavoured to show that our reply really cannot be taken as ignoring the proposal of the British Government. The great question in the correspondence in April between us and the British Government was the question of independence; and now, after having consulted the nation, we come here and say that we are prepared to sacrifice in some degree our independence, and we indicate how far we will give it up. And, as General Smuts has said, that is the basis which we have laid down in our present proposal."

Lord Milner: "You say that you give up your independence as regards foreign relations."

Judge Hertzog: "Yes. But then you must understand that this is only a general principle, which we treat in detail later on."

General Smuts: "The independence is given up both in regard to our foreign relations and in regard to interior administration, which will be placed under the supervision of the British Government. So that the effect of these two articles is, that the independence is sacrificed, and that the two Republics will not in the future be able to be regarded as Sovereign States."

Lord Milner: "I understand perfectly well that they would not be Sovereign States any longer, but my intellect is not bright enough for me to be able to say what they really would be."

Lord Kitchener: "They would be a new kind of 'international animal.'"

General Smuts: "It has more than once happened in the course of history that difficulties have been solved by compromise. And this draft proposal goes as near as seems possible towards making us a Colony."

Lord Kitchener: "Do you accept the annexation?"

General Smuts: "Not formally; but I do not see in what way this proposal is in opposition to the annexation proclamation."

Lord Kitchener: "I am afraid I am not clever enough to comprehend this. There would be two Governments in one State. And how do you imagine that this arrangement could be carried on?"

General Smuts: "A more ample explanation will have to be given of the word 'supervision'; and I thought that this was just one of the points on which we could carry on further discussions and negotiations."

Lord Milner: "I am certainly not going to give up an explicit basis for a vague proposal."

Lord Kitchener: "I feel convinced that your proposal would never be able to be carried out in the practical governing of a country."

Commander-in-Chief de Wet: "I agree that our proposal has not been fully worked out, but neither have the Middelburg proposals. This was clearly indicated by Lord Kitchener and Lord Milner when these proposals were made, and they were only looked upon as a basis on which we could negotiate, so that the business might be begun. We naturally cannot compel the British Government to accept our proposal; but, at all events, it is a basis."

Lord Milner: "I am very anxious that these discussions should not end in smoke, and I shall not allow any formalities to stand in the way, but to abandon the definite proposals of Middelburg (March 7th) for a thing like this, and to begin a fresh discussion on the basis of something which is so very vague will surely land us in trouble. I believe we are quite entitled to keep you to the Middelburg proposal, which we might modify in regard to details."

Commandant-General Botha: "Perhaps it would be well if you would first give an answer to our proposals."

Commander-in-Chief de Wet: "I think that (unless your Excellencies have power to give a final answer to our terms) it would not be unfair if we were to ask you to lay our proposal before your Government."

Commandant-General Botha: "We are come here with the earnest intention of concluding peace; and I think that if our proposal is carried out Boer and Briton will be able to live side by side in this country. I presume that it is the wish of both parties to be fair and just, and to make a peace by which both can abide, and which will be permanent in South Africa."

Lord Milner: "That is certainly our aim."

Lord Kitchener: "Your proposal would involve important changes in our own—changes which, so far as I understand them, we should be unable to permit."

Commandant-General Botha: "I am of opinion that before a proposal is made from your side you should give a definite answer to ours."

Lord Kitchener and Lord Milner: "Well, then, change your proposal into ours."

Lord Milner: "I do not believe that the British Government is prepared to go any further to meet you than they have done in their last proposal. They think that they have already gone far in their efforts for peace—further, indeed, than the general opinion of the British public would warrant."

Lord Kitchener: "The difference between our proposals seems to be too great."

Commandant-General Botha: "We shall always remain under the supervision of the British Government."

Lord Kitchener: "Will you then consider yourselves British subjects? 'Supervision' is a new word, and 'suzerainty' has already caused us too much trouble."

Judge Hertzog: "The idea is not so very new. There are several kinds of different States, all belonging to the British Empire. For instance, there is Basutoland."

Lord Milner: "There are many different kinds, but this one is a new variety."

Judge Hertzog: "If your Excellencies could only understand us! We have no wish to lose a single minute. We have been to the nation, and we know what the nation wants and what their temper is. If, then, we are to make a proposal here, it must be:—Firstly, a proposal which shall meet the English Government in a fair way; and, secondly, a proposal which we are honestly convinced will be acceptable to our nation. And such a proposal we have laid before you. And now we are placed in a disadvantageous position, for we are here before your Excellencies, who have not full power finally to decide the matter."

Lord Kitchener: "We are in the same position as yourselves."

Judge Hertzog: "We offer you here what we know is in accordance with the mind of the nation; we cannot possibly do anything that is against it."

Lord Milner: "Are we to understand that the Middelburg proposals are not according to the mind of your people?"

General Smuts: "As yet no answer has been given to them. The only decision come to by the national meeting is that which we are now laying before you."

Lord Kitchener: "Are you prepared to set aside your present proposal and to hand in another one bearing a closer resemblance to that of Middelburg? We must try and find some middle course; and as we are here to endeavour to arrive at something definite, let us try to obtain a basis for discussion. Shall we make a new proposal?"

General Smuts: "As soon as there is a final answer to our proposal we shall be able to take a fresh one into consideration."

Lord Milner: "I believe that the fact that you have refused to enter upon the proposal made by the British Government justifies us in not considering your proposal. Let us rather say that your very refusal implies your answer to what we have proposed."

General Smuts: "I understand the position to be as follows—The British Government has declined our proposals, and at the same time holds fast to the old basis, but without prejudice to its power of making a new proposal."

Lord Milner: "The whole difference between you and myself is that I take the letter of 7th March to be the utmost concession that the British Government is able to grant; not that that letter binds us down to every clause of the proposal, but that it is an indication of how far our Government is prepared to go on the general question. Your answer, however, is no answer at all."

Lord Kitchener then read his telegram, dated 14th April. "A difficulty has arisen in getting on with the proceedings; the representatives state that constitutionally they have no power to discuss terms based on the surrender of independence, inasmuch as only the burghers can agree to such a basis. Therefore, if they were to propose terms, it would put them in a false position with regard to the people. If, however, His Majesty's Government could state the terms which, subsequently to a relinquishment of independence, they would be prepared to grant, the representatives, after asking for the necessary explanations, and without any expression of approval or disapproval, would submit such conditions to their people."] "Clearly you have not kept to what you undertook in this telegram."

Commander-in-Chief de Wet: "If it had only been a question of our feelings being hurt by having to give an answer on the basis proposed to us by the British then it would not have been necessary for the people to come together at Vereeniging. But in matter of fact we have come here with a proposal, which, rightly understood, is nearly equivocal to the Middelburg proposal, and which meets the wishes of the English Government as far as possible."

Commandant-General Botha: "I do not see why we should insist so much on our proposal. If it is not to the mind of your Excellencies, if it is an unacceptable proposal, then let us have a definite answer to it."

Lord Milner: "We wish to have an answer to the proposal made by us."

General Smuts: "I do not see that any proposal has been made by the British Government. A certain basis only has been laid down, and therefore no formal answer is required."

Lord Milner: "Our proposal is six times as definite as yours, and I believe that the British Government is justified in wanting to know if your people are inclined to come to terms on the general lines which have been placed before them."

Lord Kitchener: "Here is quite an original suggestion: How would it be if you were to go back to your people and ask them if they would not make a proposal?"

General Smuts: "You must understand that the Middelburg proposal, with all that took place in April, has been read to the people. Their answer was neither 'Yes' nor 'No.' They simply elected the delegates. The delegates as yet have not given any answer. They are still considering the matter, and, in order to gain time, they have commissioned us to see whether we could not come to some arrangement."

Lord Milner: "We are getting away from the subject. Tell us what alterations you want, and then place our proposal before your people."

Lord Kitchener: "Should you agree that your proposal is not in opposition to the annexation, we shall have accomplished something."

General Smuts: "Is it your opinion that our proposal must be set aside?"

Lord Kitchener: "Yes, surely. It is impossible for us to act on it."

Lord Milner: "It is impossible for us to take your proposal into consideration. We can send it to England, but this would certainly tend to hinder the negotiations. This is my personal opinion, which naturally you are not bound to accept. All that we can say is, that this is the only answer that we can give you."

Lord Kitchener: "It would be better to draw up a new document, in which everything of importance would be noted down, and all unimportant matters left out."

General Smuts: "But paragraph 3 of our proposal has not even been mentioned. We are prepared to cede a part of our territory."

Lord Milner: "This would be in contradiction to the annexation of the whole. If the whole becomes annexed by us, how then can a part be ceded by you?"

General Smuts: "The ceded part would then become a Crown Colony, the remaining part being governed as is here proposed."

Lord Milner: "You mean that one part would become a British Colony of the ordinary type, and another part a protected Republic?"

Lord Kitchener: "Two forms of government in the same country would lead to great friction. Our proposals are too divergent. From a military point of view, the two forms of government could not co-exist. Before a year was over we should be at war again."

The meeting was then adjourned till the afternoon.

During the interval the Commission discussed the situation, and sent General J.C. Smuts to deliberate on several points with Lord Kitchener and Lord Milner.

The meeting opened again at four o'clock.

Lord Milner: "In consequence of an informal conversation with General Smuts, Lord Kitchener and I have drawn up a document, which will show the form in which, as we think, the only agreement that can be arrived at must be worded. It is a draft document, and we believe the Governments will be able to sign it. Our idea is that after it has been taken into consideration here it might be laid before the burghers, and you could ask them, 'Are you willing that we should put our signatures to it?'"

This document ran as follows:—"The undersigned, leaders of the Boer forces in the Veldt, accepting, in their own name, and in that of the said burghers, the annexations as mentioned in the proclamations of Lord Roberts, dated respectively the 24th May, in the year of our Lord nineteen hundred, and number 15, dated 1st day of September, in the year of our Lord nineteen hundred, and accepting as a consequence thereof their status of British citizens, agree herewith immediately to lay down their weapons, and to hand over all guns, small arms, ammunition, and stores in their possession, or under their hold, and to cease all further resistance against the Government of His Majesty King Edward Seventh, or his successors. They do this trusting in the assurance of His Majesty's Government that neither their personal freedom nor their property shall be taken away from them, or from the burghers who surrender with them; and that the future action of His Majesty's Government in relation to the consequences of the war shall be in harmony with the declaration mentioned below. It is clearly understood that all burghers who at present are prisoners of war, in order to be able to enjoy the above-mentioned assurance, will have to notify their acceptance of the status of British citizens."

Commandant-General Botha: "Are we to understand that our proposal is now altogether rejected?"

Lord Milner and Lord Kitchener: "Yes."

Commandant-General Botha: "Then I understand that you are going to be guided only by the Middelburg proposals?"

Lord Kitchener: "No; we can alter them."

Lord Milner: "This draft document was originally written out in order to be annexed to the Middelburg proposals. But instead of the Middelburg proposals, this document is now drawn up, in order to place us in the position to formulate the proposals differently."

General Smuts: "If the idea is then that the Middelburg proposals should be amended, would it not be best to do so now, and then to annex them to this document?"

Lord Milner: "That which will take the place of the Middelburg proposals has to be added as a schedule to this document, and we have to work out this schedule together."

General Smuts: "I think it would be far better if you were to alter the proposal yourselves, and then lay it before us for consideration; we could then see what we could do to meet you."

Lord Kitchener: "I think that a sub-committee should be formed by you in order to draw up the schedule."

Lord Milner: "My idea is that the schedule should be drawn up, so that it and the document could be taken into consideration together."

General Smuts: "We should like to consider first whether we will help in drawing it up."

Lord Milner: "I am willing to draw it up in conjunction with you, or to let it be drawn up by you alone, but, from past experience, I must decline to draw it up by myself."

General Smuts: "If we were to sign this document, would not the outcome be that we leaders made ourselves responsible for the laying down of arms by our burghers."

Lord Milner: "Yes. And should your men not lay down their arms it would be a great misfortune."

Lord Kitchener: "I do not think so, for if some of the burghers refused to lay down their arms, the signatories could not help it. There are sure to be some who are dissatisfied."

General Smuts: "The document does not mention this."

Lord Kitchener: "It can be amended."

General De la Rey: "Well, then, there can be no peace, for one part of the burghers will hold back and continue the war."

Lord Milner: "If the national meeting agrees to give you power to sign this document, it will certainly mean that the burghers as a whole are agreeable; and those who after this do not submit will be—well, I do not know what I can call them—outlaws. But we will not consider such an eventuality possible."

General Botha: "We desire a peace that will be honourable to both parties. And, as I understand this document, we are leaving honour behind us, for we are now not only surrendering our independence, but we are allowing every burgher to be fettered hand and foot. Where is the 'honourable peace' for us? If we conclude peace, we have to do it as men who have to live and die here. We must not agree to a peace which leaves behind in the hearts of one party a wound that will never heal. I will do everything in my power to obtain peace. But it seems to me that this document asks too much of us, because, if I interpret it aright, it means that we must surrender our independence, that every one must give up his weapons, and that the leaders, in addition, must sign an undertaking to this effect."

Lord Milner: "All that we wish is that the people should live peacefully together as British citizens. If we do not obtain this, then I do not know what we do obtain."

Lord Kitchener: "I do not think that the Commandant-General realizes what the schedule contains. In it we state what we are ready to grant. Perhaps it would be best that the schedule should be arranged now, and then you will see that an honourable peace is proposed."

General Botha: "Well, then, explain the document."

Lords Kitchener and Milner: "You are to help us: we do not know what the burghers demand."

Commander-in-Chief de Wet: "By signing this document we shall place ourselves in the position which the Commandant-General has so clearly described."

General De la Rey: "We cannot form a judgment on anything that is not properly elaborated. I have no objection to the constitution of a sub-committee with the duty of helping in the work."

Commandant-General Botha: "I also have no objection, since I understand that it binds nobody to anything."

Lord Kitchener: "No, nobody will be bound."

General De la Rey: "We wish to have the matter concluded, so that we may know what is before us."

Commander-in-Chief de Wet: "I should like to have it clearly understood that I do not think there is the least chance of a Government of which Lords Kitchener and Milner are the heads being accepted. An arrangement of this nature would, it seems to me, be an insurmountable difficulty. When I feel so strongly in this matter, it would not be fair to their Excellencies for me to remain silent."

Lord Kitchener: "I think it would be better if General de Wet were to wait until he has seen the whole document before he gives his opinion."

It was then agreed that Judge Hertzog and General Smuts should act as a sub-committee, in order to draw up a complete draft with Lord Kitchener, who was to be assisted by Sir Richard Solomon.

The meeting then adjourned.

On Wednesday, 21st May, 1902, the Conference reassembled.

Lord Milner laid before the meeting the documents which he had drawn up with the help of the sub-committee. It was in the form of a contract, and the names of the members of both Governments were now filled in. The document was the same as that telegraphed, with the exception of Article 11, dealing with the notes and receipts and the sum of three million pounds.

It was read in Dutch and English, and ran as follows:—

"General Lord Kitchener of Khartoum, Commander-in-Chief, and His Excellency Lord Milner, High Commissioner, on behalf of the British Government;

"Messrs. S.D. Burger, F.W. Reitz, Louis Botha, J.H. De la Rey, L.J. Meijer, and J.C. Krogh, on behalf of the Government of the South African Republic and its burghers;

"Messrs. M.T. Steyn, W.J.C. Brebner, C.R. de Wet, J.B.M. Hertzog, and C.H. Olivier, on behalf of the Government of the Orange Free State and its burghers, being anxious to put an end to the existing hostilities, agree on the following points:—

"Firstly, the burgher forces now in the Veldt shall at once lay down their arms, and surrender all the guns, small arms and war stores in their actual possession, or of which they have cognizance; and shall refrain from any further opposition to the authority of His Majesty King Edward VII., whom they acknowledge as their lawful sovereign.

"The manner and details of this surrender shall be arranged by Lord Kitchener, Commandant-General Botha, Assistant-Commandant-General J.H. De la Rey, and Commander-in-Chief de Wet.

"Secondly, burghers in the Veldt beyond the frontiers of the Transvaal and of the Orange River Colony shall, on their surrender, be brought back to their homes.

"Thirdly, all prisoners of war, being at the time burghers out of South Africa, shall, on their declaring that they accept this status of subjects of His Majesty King Edward VII., be brought back to the farms on which they were living before the war.

"Fourthly, the burghers who thus surrender, or who thus return, shall lose neither their personal freedom nor their property.

"Fifthly, no judicial proceedings, civil or criminal, shall be taken against any of the burghers who thus return for any action of theirs in connexion with the carrying on of the war.

"Sixthly, the Dutch language shall be taught in the public schools of the Transvaal and of the Orange River Colony, where the parents of the children demand it; and shall be admitted in the courts of justice, wherever this is required for the better and more effective administration of justice.

"Seventhly, the possession of rifles shall, on taking out a license in accordance with the law, be permitted in the Transvaal and in the Orange River Colony, to persons who require them for their protection.

"Eighthly, military administration in the Transvaal and in the Orange River Colony shall, as soon as possible, be followed by civil government; and, as soon as circumstances permit it, a representative system tending towards autonomy shall be introduced.

"Ninthly, the question of granting the franchise to the natives shall not be decided until a representative constitution has been granted.

"Tenthly, no special tax shall be laid on landed property in the Transvaal and Orange River Colony to meet the expenses of the war.

"Eleventhly, a judicial Commission shall be appointed, to which the government bank notes, issued under Law No. 1 of the South African Republic, may be presented within six months. All such notes, if found to have been duly issued in conformity with the terms of the law, and if the presenting party shall have given consideration in value, shall be honoured, but without interest.

"All receipts issued in the Veldt by the officers of the late Republics, or by their orders, may also be presented to the said Commission within six months; and if they have been given bona fide in exchange for goods used by the burghers in the Veldt, they shall be paid in full to the persons to whom they were originally issued.

"The amount payable on account of the said Government's notes and receipts shall not exceed £3,000,000; and in case the whole amount of such notes and receipts accepted by the Commission should exceed that amount, a pro rata reduction shall be made.

"The prisoners of war shall be given facilities to present their notes and receipts within the above-mentioned six months.

"Twelfthly, as soon as circumstances shall permit, there shall be appointed in each district of the Transvaal and of the Orange River Colony a Commission, in which the inhabitants of that district shall be represented, under the chairmanship of a magistrate or other official, with a view to assist in the bringing back of the people to their farms, and in procuring for those who, on account of losses through the war, are unable to provide for themselves, food, shelter, and such quantities of seed, cattle, implements, etc., as are necessary for the resuming of their previous callings. Funds for this purpose, repayable by instalments extending over a number of years, shall be advanced—free of interest—by the Government."

Lord Milner: "If we come to an agreement, it will be the English document which will be wired to England, on which His Majesty's Government will decide, and which will be signed."

Commandant-General Botha: "Will not a Dutch translation be annexed?"

Lord Milner: "I have no objection to the addition of a Dutch translation. This, then, is the document which we are prepared to lay before the English Government."

Commandant-General Botha: "There are a few points on which I wish to speak. The first is in reference to the receipts given by our officers. It seems to me quite right that they should be mentioned in the paragraph about government notes. These receipts were issued, in accordance with instructions given by our Government, for the purchase of cattle, grain, and other necessaries for the support of our commandos; and the chief officers now present, as well as all other officers, have acted according to these instructions and issued receipts. Therefore I make this request. Some of these receipts were afterwards paid in part, and others in full, in government notes. But many were not paid at all. I do not believe that the amount is great, but it will strengthen our hands to be able to take up this affair honourably, for our honour is concerned in so far as we have signed the receipts. It will be a great point in our favour to be able to go before our delegates and tell them that they are guaranteed on this point, for most of them are officers."

Lord Kitchener: "I understand that General Botha refers not to commandeer or requisition notes, but only to actual receipts issued on the Treasury."

Lord Milner: "I do not see any difference between these receipts and commandeer notes. The willingness of persons to sell goods makes no difference in a legal document."

Lord Kitchener: "I mean that it makes a difference whether it is an order on the Treasury or a requisition note. I should limit this (guarantee) to receipts on the Treasury, issued in consequence of a law that permitted a certain sum to be issued."

Commander-in-Chief de Wet: "No decision was come to in the Free State as to how much was to be issued."

Lord Kitchener: "Am I to understand by this that it is an unlimited amount, or does it come within the amount decided on by the Volksraad?"

General Smuts: "While the Government existed the Volksraad empowered it to issue notes up to a certain amount. And this was done. Moreover the officers in the Veldt had the right to make purchases for the commandos and to give receipts for them."

Lord Milner: "I can see no difference between receipts and requisition notes, and they have been issued for an unlimited amount."

General Smuts: "These receipts were issued under a totally different law. They were not paid out of the credit voted by the Volksraad."

Commander-in-Chief de Wet: "I would have it clearly understood that I quite agree with what has been said by the Commandant-General, namely that the honour of every officer is engaged for these documents, and if your Excellencies agree it will give us a strong weapon with which to return to the delegates."

Lord Milner: "The proposal is de facto that the British Government shall repay all the monies which the Republics borrowed with the object of fighting against England."

Commander-in-Chief de Wet: "Yet we have fought honourably, and if we give up our independence it is no more than fair that you should meet us in this matter."

Commandant-General Botha: "Am I to understand your position to be that we must surrender everything, and that whilst you take away the freedom of our country (which amounts to many millions) you at the same time refuse all responsibility for our debts. We had been recognized by you as belligerent, and so are entirely in our rights in asking that when you seize the riches of the country you shall also take its debts upon your shoulders. So long as the British Government reaches the great goal at which it is aiming, a matter so easily arranged as this should not cause any difficulty: we are not bickering about trifles, but are bringing forward what to us is a real hardship, and you must take it for granted that when we say something here we really mean it. And now we tell you that this matter is an obstacle in our way. Personally, we have not signed many receipts: it was the officers of lower rank who signed the greater number, and it is these very officers who form the majority of the national meeting at Vereeniging. In some instances, I may add, special persons were appointed for the purpose of carrying out this work."

Lord Milner: "We do not take over the assets without taking also the liabilities. We take over all the debts owed by the country before the war, and we have even agreed to take over a debt—a legal debt—in the shape of notes, which notes we are fully aware it only became necessary to issue on account of the war, and thus we are already paying a part of the cost incurred in fighting us. I think this is a very great concession; and when I agreed that it should be put down I said that I believed (and I still am of the same opinion) that the English Government would take exception to it, although I hope that this will not be the case. But to go further than this, and to ask us to pay not only a debt contracted under a law for the furtherance of the war, but also every debt contracted by every officer in the armies of both Republics, for the purpose of fighting us, is to my mind a most extravagant proposal. In answer to what General Botha has said, I may observe that the Commission appears to think that we have no persons behind us whose feelings and prejudices (if you use that word) we are bound to take into consideration. If this matter causes a difficulty among your burghers, I can only say that I am sure that your proposal will cause the British Government the greatest trouble when dealing with the nation, with whose feelings they have to reckon."

Commander-in-Chief de Wet: "I should like to explain the position of the Orange Free State. In the Transvaal a law was passed empowering the Government to issue £1,000,000; but in the Orange Free State nothing was done, as the Government possessed the right to pay with receipts, and we thought that a receipt was as good and as legal as a note; and therefore, from my point of view, the two are of equal importance."

Commandant-General Botha: "I might point out that we should not insist so much on the technical meaning of words—and this is especially true for your side, because we have assembled here with the aim of stopping the hostilities which cause you such great expenses every month; and our meeting may be able to bring these expenses to an end. Therefore, if you accept our proposal and pay these receipts, you might save almost enough to cover the cost you incur. It would be much cheaper to make an end of the war by co-operation than to let matters drift on. Therefore I believe that it is the duty of both parties to be willing to make concessions when obstacles appear."

General de Wet: "I can assure His Excellency, Lord Milner, that the people always believed that should everything be lost they still would be able to obtain this money due on receipts. If this is not granted, I cannot imagine what the results will be. I am afraid of the consequences; and I trust that you will do your best to meet our wishes."

Commandant-General Botha: "It will not be a very large sum, but we cannot give you the exact amount."

Commander-in-Chief de Wet: "You can well understand that our expenses are only a drop in the ocean compared with yours. If I am right, the Orange Free State had three quarters of a million when the war began, and the issue of receipts only started when that sum was exhausted. Your Excellencies must acknowledge that we have the same obligation of creditor through these receipts as we should have in any other case."

Commandant-General Botha: "You have already many of our notes in your possession. In one case alone there were fifty thousand hidden away, and found by you. I have stated privately to Lord Milner that what we are now striving to obtain has already been granted to us de facto by Lord Kitchener. In Lord Kitchener's Middelburg proposal the paying of the Government notes was refused, but there was a proviso that the receipts should be paid to the amount of one million. Should this now be withdrawn, surely such a withdrawal would form a deviation from the Middelburg proposal. The paying of notes is legal, and is on quite another footing, and I cannot understand how it could have been refused in the Middelburg proposal. That it should be granted now is only reasonable. But as regards the payment of receipts, although it was allowed then up to a certain amount, it is now withdrawn. At this present stage of the proceedings I think that a point which had already been practically conceded in the previous negotiations should not be allowed to form a stumbling-block to a final agreement. I believe that the amount is only small; I was for one year in conjunction with De la Rey in command of the forces of the South African Republic. During that period of time an account was kept of all the receipts, and only a short time back the books were still in our possession. These receipts were issued in an orderly manner, and each of them was duly entered in a book, as far as I was able to judge. These receipts amounted to quite a small sum; and although Lord Milner would draw back if the sum was very big, the question how far he will go can be settled when the proposal is accepted. Yet I personally think that there are no grounds for fear, and the amount is really far smaller than you imagine."

Lord Milner: "I do not think it is so much a question of amount. This paying of notes and requisition notes appears to me very unreasonable. I believe that in this matter I am only voicing the opinion of the great majority of the British nation when I say that my countrymen would much prefer to pay a large sum at the conclusion of hostilities with the object of bettering the condition of the people who have been fighting against them than to pay a much smaller sum to meet the costs incurred by the Republics during the war. Whether such a view is right or wrong, it is a view you have to reckon with. We do not wish to pay the accounts of both parties; and my opinion of the clause quoted from the Middelburg proposal is that that clause was one of its faults. But should anything of the kind become necessary, then I think that the paying of the notes is less objectionable than the paying of the requisition notes. I placed this point about the payment of notes in the draft because I thought that if it came to a choice between paying one or the other you would prefer that the notes should be paid. However, if it should be thought better to return on this point to the Middelburg proposal, although I am greatly against the clause, I will waive my objection to it if Lord Kitchener is agreeable."

General Smuts: "I am afraid that we cannot agree to this, for we thought that the notes would be beyond all dispute."

Judge Hertzog: "I do not think that your Excellency is representing the matter fairly when you say that you will not pay the bills of both parties. There is one thing to be taken into consideration as regards the Orange Free State, and which must be considered before everything else, and that is, that we have made no loans nor have we given any government notes. The notes we used were notes of the South African Republic, which had been sent to the Orange Free State. Our law was formed on the idea that in case of war all the costs should be paid by commission notes. The Orange Free State acted on this principle, and receipts were issued. If we take into consideration at the same time that we have been and still are recognized by you as belligerent, then we can only say: On our side we surrender everything that we possess, and we only ask the other party to acknowledge the fact that if we had contracted a loan it would have been to the charge of the British Government, who, in taking everything from us, renders itself responsible for our public loans. Lord Milner should understand that it is of just as much importance to us for the receipts to be paid as it is to the South African Republic for the loan, which it contracted before the war, to be taken over by the British Government. But I can even go further and give Lord Milner the assurance that we have acted more economically when issuing these receipts than we should have done had we contracted the loan previous to the war. Now we have only what is absolutely necessary to meet our present needs. So that Lord Milner must own that we find ourselves in the same position towards those who are in possession of receipts, as we should have occupied towards any other creditor we might have had before the war began. I must give my support to what the Commandant-General has said; and I can only repeat what I have already informally told Lord Milner, namely, that this difficulty is almost insurmountable."

Lord Milner: "We can refer this to our Government. But your proposal is altogether antagonistic to the Middelburg proposal, which absolutely rejected the idea of taking over all the debts of the two States."

Lord Kitchener: "I should like to know the amount."

General De la Rey: "My issue of notes amounts to between twenty and fifty thousand pounds; but I cannot say what the issue in receipt has been."

Lord Milner: "There really is a feasible compromise, namely, to allow the notes and receipts to come in and to establish the suggested limit of £1,000,000."

Lord Kitchener: "Would that meet your difficulty?"

Commandant-General Botha: "No."

Lord Kitchener: "Well, would two or three million be sufficient? We must have a limit before we can do anything."

Commander-in-Chief de Wet: "It is impossible to stipulate the amount."

Lord Kitchener: "If you were in a position to give a limit, it would simplify matters."

Commander-in-Chief de Wet: "I agree with that entirely, and I can quite understand the position in which you are placed. Yet it is absolutely impossible to assign an amount. Will you give us your permission to adjourn for a moment in order to discuss the matter?"

The meeting was then adjourned. It reassembled at 2.30 p.m.

Commander-in-Chief de Wet: "We have agreed to fix on a sum of £3,000,000 for the government notes and receipts; their amount paid pro rata can be lowered should this sum prove insufficient. We have drawn up an article to lay before the meeting."

General Smuts then read a draft which was inserted at the end of Article 11 in the draft agreement.

In answer to a question by Lord Kitchener, Commander- in-Chief de Wet said: "The prisoners of war on the different islands who are in possession of such notes should be given an opportunity of sending them in for payment."

Lord Milner: "What is the next point you wish to raise? We now understand what your position is."

Commandant-General Botha: "Am I to understand that you mean that we are getting away from the point in discussion?"

Lord Milner: "This document contains your view of the matter, so we are now aware of your idea."

Commandant-General Botha: "We must know what to say to the delegates."

Lord Kitchener: "Is this the only point you wish to bring forward, or are there others in addition?"

Commandant-General Botha: "There is another concerning the protection of debtors, which is a vital question for us."

Lord Milner: "We must not have any beating about the bush. Everything must appear in the document."

General Smuts: "Most of the debts contracted before the war will have to be paid after the war; and if the debtors cannot pay we are afraid that it will result in the ruin of a great part of the inhabitants. We should like to see steps taken to prevent this. If Lord Milner intends to take such steps, we should like to be informed what they are."

Lord Milner: "I think it would be best if you were to make a proposal on this point."

General Smuts: "Our proposal is roughly that all interest which became payable during the war should be joined to the principal, and that this should be payable six months after the war."

Lord Kitchener: "Is it necessary to make a proposal about this?"

General Smuts: "If the Government is prepared to meet us in this difficulty it will be unnecessary to place a formal clause in the draft agreement."

Lord Milner: "As I look at the matter, the Government is making certain promises in this document, and I consider that all promises to which a reference may be made later should appear in it. Everything to which the Government is asked to bind itself should appear in this document, and nothing else. I do not object to clauses being added, but I wish to prevent any possible misunderstanding."

General Smuts: "Well, in that case we are quite willing to propose such a paragraph."

Commandant-General Botha: "We waive this question, so that early measures may be taken to arrive at an understanding. In case a great number of the inhabitants become subjects of His Majesty, it is to every one's interest, and principally to that of the Government, that these people should not be ruined. They will be thrown upon the mercy of a Government, whose duty it is to study their interests. If steps are not taken to prevent it, speculators who have been buying up the liabilities will, as soon as peace is concluded, enforce them, and directly the Courts of Justice are opened they will issue summonses. Against this we have to be on our guard."

Lord Milner: "I agree with the Commandant-General. I think that as these people become subjects of His Majesty, then some provision will have to be made for them. But I believe it to be neither necessary nor advisable to point out in every particular case the way in which His Majesty's Government has to provide for these people. I think that an idea exists—perhaps it is a very natural idea—because we have been fighting against the burghers that, therefore, after peace has been concluded we shall still retain a feeling of enmity against them. Just the opposite, however, is the truth. Our endeavour will naturally be, from the moment hostilities cease, to gain the confidence of the people and to do our best to promote their welfare. But if we have to bind ourselves beforehand in regard to the manner in which we shall deal with all sorts of involved legal questions, further misunderstandings are certain to occur. If you have not confidence in us—that we shall try to be a righteous Government, and to maintain the balance between the different classes of His Majesty's subjects—then you must put in writing every point that strikes you, and let them be laid before His Majesty's Government, to see what they think about them."

Commander-in-Chief de Wet: "I trust that you will not think that we are trying to tie the hands of His Majesty's Government. There are many other points which will give the Government opportunity to win the confidence of the people. But about things which concern the financial position of burghers who are entirely ruined we feel it our duty to obtain definite promises. They will be a weapon in our hands when we return to the delegates."

Commandant-General Botha: "I do not quite understand, Lord Milner. I did not interpret Mr. Chamberlain's telegram in the sense that we had to present new proposals in order to bind our hands further. I thought that proposals were to be made with a view to establishing peace."

Lord Kitchener: "I do not think that it is altogether necessary to include this proposal in the document. It concerns the very involved legal questions as to what the rights of creditor and debtor shall be, and as to what the law in the Transvaal may be on the matter. I think that every one can rest assured that the interests of the Boers will be protected by the Government in every way; and this, whether the point is put down now or left in the hands of the Government with the recommendation from this Commission to take the matter into serious consideration.

"I think that I know of a better way to deal with this involved question. Let this matter be brought under the consideration of the Government. I may be mistaken, but, as far as I can see, it will prove a very thorny question for the lawyers, and will take a long time before it can be clearly stated. It is, however, the wish of us all that you should return to the delegates equipped in such a way that you will be able to arrive at a decision. You may rest assured that the matter which you have brought before us has been included in the minutes of this meeting. I do not think that it is necessary for you to go further than this. The matter can now be carefully considered, not only here, but also in England; and you may be quite sure that your interests will receive, in every way, full consideration."

General De la Rey: "I think that the matter has been sufficiently discussed in the presence of your Excellencies, and that it need not be placed in the draft contract, for by so doing one might stumble on legal questions."

Commander-in-Chief de Wet: "This is my point of view: There are two parties, and one of them is about to cease to exist. It is, therefore, natural that this party cannot allow a vital question to pass unnoticed. It is for this reason that I cannot agree that this matter should be omitted from the draft contract. It will not be necessary that the military Government which now exists should continue after the war."

Lord Kitchener: "But the question will have to be settled by the Civil Government. It is a matter for lawyers, and must be laid before them, and will require much consideration."

Commandant-General Botha: "When hostilities are concluded it will be possible to summon a burgher for a debt contracted before the war. I put this request because our law states that no burgher can be summoned till sixty days have elapsed since the conclusion of peace."

Lord Kitchener: "You may entirely rely upon this, that whenever the war is over each burgher will have the absolute right to obtain consideration for his position in every way, and that his interests will be protected under the new as under the old regime."

Commandant-General Botha: "I understand that perfectly. But the possibility exists that syndicates may be formed to buy up all the debts, and the people may be ruined before a single burgher is in the position to earn anything or to have his position restored."

Lord Kitchener: "I quite agree with what the Commandant-General has said, and he is quite right to bring the question up. Yet I do not think that the draft contract is the best place in which to bring it forward. Once peace is a fact, then it will be the duty of every one to draw the attention of the Government to what is required to aid the nation; but to bring up difficulties at the present moment, and to attempt to right them, seems to be an endless task, and one for which this document was not destined."

Commander-in-Chief de Wet: "I am of opinion that this is a matter which should be settled by a proclamation; but I want to have as many weapons as possible in my hands when I return to the national delegates, and one of the first questions that will be asked me is this, 'What guarantee do we possess that we shall not be ruined by our creditors?' It would not be much trouble to you to give us now a draft of the proclamation which would be issued as soon as peace is concluded."

Lord Kitchener: "But this would be something quite apart from the matter under discussion."

Commander-in-Chief de Wet: "Yes."

Lord Milner: "What is the good then?"

Commander-in-Chief de Wet: "It is such a vital question for us that you cannot take it amiss if we insist upon it, for we have to give up everything."

Lord Kitchener: "Of course, no one is blaming you."

Lord Milner: "But without any thought of blame, I must point out that the effect of their proposal would be that another clause would have to be inserted in the draft contract, undertaking that such a proclamation would be issued."

Lord Kitchener: "I think that as long as the delegates receive an assurance that the Government will take this matter into consideration, in the interests of their subjects, whom they are bound to protect, that such an assurance ought to suffice. There should be no written undertaking, but only a promise that the matter shall receive attention. It is not advisable after the subject has been brought before the Government to press the matter further. The feelings of the burghers, moreover, in other ways than this, will be brought before Lord Milner."

Commander-in-Chief de Wet: "If we wished to do so, we could insist upon many other little points, but we only bring up vital questions."

Lord Kitchener: "This is one of the questions which, when once brought under the consideration of the Government cannot be put aside; and you may tell the burghers that their interests will be protected as fully as is possible. I think that, in so complicated a matter, this ought to be sufficient for them. All that is debated here is recorded in the minutes, and these minutes will be considered not only here, but also in England. Are you satisfied with this?"

Commandant-General Botha: "Yes, so far as I am concerned."

Commander-in-Chief de Wet: "I also am satisfied."

Lord Milner: "I hope it is quite understood that if the matter is allowed to remain where it is, my Government will be under no obligation to treat the matter in any particular way."

Lord Kitchener: "But there is a pledge that the matter will be properly considered."

Lord Milner: "Yes, naturally; if we put anything down in writing. I am convinced that it is necessary to make it quite clear that this document must contain everything about which there is anything in the form of a pledge."

Lord Kitchener: "There is, then, a pledge that the point upon which you have touched will be considered in your interests."

General Smuts: "There still remains the question of the payment of receipts."

Lord Kitchener: "That will be placed before the Government. The sum is an essential point; I believe the amount to be considerable. I should now like to know that it is understood that we are agreed about all these draft proposals, including your amendments, and that there are no further questions to be brought forward—it is necessary to know this, as they would have to be telegraphed to England."

Commander-in-Chief de Wet: "We have no further points to raise."

Lord Milner: "The telegram that I shall despatch is as follows:

'The Commission is prepared to lay before their burgher meeting the following document (in the event of it being sanctioned by His Majesty's Government), and to ask of the meeting a "Yes" or "No."'

"Is that satisfactory?"

Commander-in-Chief de Wet: "Yes, naturally. Only I cannot say that this document has my approval. Yet I shall be content to abide by the decision of the delegates."

Judge Hertzog: "I should not like to think that we are bound to use our influence with the delegates."

Lord Milner: "I think that is understood. I understand that the members of the Commission are not bound in respect of the opinions they may express before the burghers. They are only bound, if the British Government approves of the document, to lay it before the people. I propose to send the following telegram:

'The Commission is prepared to lay the following document before the burgher meeting at Vereeniging, for a "Yes" or "No" vote, in the event of His Majesty's Government approving of it.'

"I want also to state that we have completely deviated from the Middelburg proposal. I believe everyone is fully aware that the Middelburg proposal has been annulled altogether. Should an agreement be arranged in conformity with this document, and signed, then no attempt must be made to explain the document, or its terms, by anything in the Middelburg proposal."

The meeting was now adjourned.

Wednesday, May 28th, 1902.

The Commission met Lord Kitchener and Lord Milner at eleven o'clock with the purpose of hearing the British Government's answer to the draft proposal sent by their Lordships.

Lord Milner read the following memorandum:

"In answer to the telegram composed at our last meeting with the consent of the Commission and of which the members have received a copy, the following message has been received from His Majesty's Government:—

'His Majesty's Government sanctions the laying before the meeting for a "Yes" or "No" vote the document drawn up by the Commission and sent by Lord Kitchener on the 21st May to the Secretary of War, with the following amendments:

'The final proposal made by the British Government, on which the national representatives at Vereeniging have to answer "Yes" or "No."

'General Lord Kitchener of Khartoum, Commander-in-Chief, and His Excellency Lord Milner, High Commissioner, on behalf of the British Government;

'Messrs. S.W. Burger, F.W. Reitz, Louis Botha, J.H. De la Rey, L.J. Meijer, and J.C. Krogh on behalf of the Government of the South African Republic and its burghers;

'Messrs. M.T. Steyn, W.J.C. Brebner, C.R. de Wet, J.B.M. Hertzog, and C.H. Olivier on behalf of the Government of the Orange Free State and its burghers, being anxious to put an end to the existing hostilities, agree on the following points:

'Firstly, the burgher forces now in the Veldt shall at once lay down their arms, and surrender all the guns, small arms, and war stores in their actual possession, or of which they have cognizance, and shall abstain from any further opposition to the authority of His Majesty King Edward VII., whom they acknowledge as their lawful sovereign.

'The manner and details of this surrender shall be arranged by Lord Kitchener, Commandant-General Botha, Assistant-Commandant-General J.H. De la Rey, and Commander-in-Chief de Wet.

'Secondly, burghers in the Veldt beyond the frontiers of the Transvaal and of the Orange River Colony, and all prisoners of war who are out of South Africa, who are burghers, shall, on their declaration that they accept the status of subjects of His Majesty King Edward VII., be brought back to their homes, as soon as transport and means of subsistence can be assured.

'Thirdly, the burghers who thus surrender, or who thus return, shall lose neither their personal freedom nor their property.

'Fourthly, no judicial proceedings, civil or criminal, shall be taken against any of the burghers who thus return for any action in connexion with the carrying on of the war. The benefit of this clause shall, however, not extend to certain deeds antagonistic to the usages of warfare, which have been communicated by the Commander-in-Chief to the Boer Generals, and which shall be heard before a court martial immediately after the cessation of hostilities.

'Fifthly, the Dutch language shall be taught in the public schools of the Transvaal and of the Orange River Colony when the parents of children demand it; and shall be admitted in the Courts of Justice, whenever this is required for the better and more effective administration of justice.

'Sixthly, the possession of rifles shall, on taking out a licence in accordance with the law, be permitted in the Transvaal and the Orange River Colony to persons who require them for their protection.

'Seventhly, military administration in the Transvaal and in the Orange River Colony shall, as soon as it is possible, be followed by civil government; and, as soon as circumstances permit it, a representative system tending towards autonomy shall be introduced.

'Eighthly, the question of granting a franchise to the native shall not be decided until a representative constitution has been granted.

'Ninthly, no special tax shall be laid on landed property in the Transvaal and Orange River Colony, to meet the expenses of the war.

'Tenthly, as soon as circumstances permit there shall be appointed in each district in the Transvaal and the Orange River Colony a Commission, in which the inhabitants of that district shall be represented, under the chairmanship of a magistrate or other official, with the view to assist in the bringing back of the people to their farms, and in procuring for those who, on account of losses in the war are unable to provide for themselves, food, shelter, and such quantities of seed, cattle, implements, etc., as are necessary for the resuming of their previous callings.

'His Majesty's Government shall place at the disposal of these Commissions the sum of £3,000,000 for the above-mentioned purposes, and shall allow that all notes issued in conformity with Law No. 1, 1900, of the Government of the South African Republic, and all receipts given by the officers in the Veldt of the late Republics, or by their order, may be presented to a judicial Commission by the Government, and in case such notes and receipts are found by this Commission to have been duly issued for consideration in value, then they shall be accepted by the said Commission as proof of war losses, suffered by the persons to whom they had originally been given. In addition to the above-named free gift of £3,000,000, His Majesty's Government will be prepared to grant advances, in the shape of loans, for the same ends, free of interest for two years, and afterwards repayable over a period of years with three per cent. interest. No foreigner or rebel shall be entitled to benefit by this clause.'

Lord Milner: "In making this communication to the Commission we are instructed to add that if this opportunity of concluding an honourable peace is not taken advantage of within a time to be fixed by us, then this conference shall be regarded as closed, and His Majesty's Government shall not be bound in any way by the present terms. I have, in order that there may be no mistake about these terms, made a copy of the documents and of Lord Kitchener's telegram, also of the amendments and additions determined on by His Majesty's Government, and of the memorandum to which I have just drawn your attention."

A debate now followed on the time that should be allowed for the discussion of the proposals at Vereeniging, and it was agreed that Commandant-General Botha should propose a term that very day before the Commission left Pretoria.

It was subsequently settled that the delegates must arrive at a decision before Saturday evening, May 31st.

General Botha asked if there were any objection to the delegates erasing any paragraph of the proposal sent by the British Government.

Lord Milner: "There must be no alteration. Only 'Yes' or 'No' is to be answered."

Commandant-General Botha: "I think that the burghers have the right to erase any article they may wish, for they have the right to surrender unconditionally."

Lord Milner replied that the burghers certainly had the power to do so, but the document of the British Government could not be changed.

There now followed an informal discussion about the colonists who had been fighting on the side of the Republics.

Lord Milner communicated what the British Government's intentions were with regard to these colonists; and read the following document:—

"His Majesty's Government has to formally place on record that the colonists of Natal and the Cape Colony who have been engaged in fighting and who now surrender shall, on their return, be dealt with by the Colonial Governments in accordance with the laws of the Colonies, and that all British subjects who have joined the enemy shall be liable to be tried under the law of that part of the British Empire to which they belong.

"His Majesty's Government has received from the Government of Cape Colony a statement of their opinion as regards the terms to be offered to British subjects of the Cape Colony who are still in the Veldt or who have surrendered since April 12th, 1901. The terms are as follows:—In regard to the burghers, they all, on their surrender, after having laid down their arms, shall sign a document before a resident magistrate of the district in which their surrender has taken place, in which document they shall declare themselves guilty of high treason; and their punishment, in the event of their not having been guilty of murder, or of other deeds in contradiction to the customs of civilized warfare, shall be that for the rest of their lives they shall not be registered as voters, nor shall they be able to vote in Parliamentary, district, or municipal elections. As regards justices and veldtcornets of the Cape Colony, and all other persons who had occupied official positions under the Government of Cape Colony, and all who held the rank of commandant in the rebel or burgher forces, they shall be brought on the charge of high treason before the ordinary Courts of the country, or before such special Courts as later on may legally be constituted. The punishment for their misdeeds shall be left to the discretion of the Court, with this reservation, that in no case shall capital punishment be inflicted.

"The Government of Natal is of opinion that the rebels should be judged by the laws of the Colony."

The meeting now adjourned.

The secretaries and Messrs. de Wet and J. Ferreira, with the help of lawyers, set themselves the task of making copies of the proposal of the British Government for the use of the national representatives at Vereeniging. This work kept them engaged until the evening.

At seven o'clock the Commission left Pretoria and returned to Vereeniging.

Special Representatives Meeting, 29th of May, 1902

Appendix C

Thursday, May 29th, 1902, And The Following Days

Minutes of the Meeting of the Special National Representatives at Vereeniging, South African Republic, Thursday, the 29th of May, 1902, and the Following Days

May 29th, 1902.

The Rev. J.D. Kestell having offered prayer, the Chairman requested Vice-President Burger to address the meeting.

Vice-President Burger said that the documents laid before the Governments by the Commission would now be read to the meeting. Thereupon Mr. D. Van Velden read the following letter:

Report of the Commission.

Pretoria, 28th May, 1902.

To the Governments of the Orange Free State and the South African Republic:

Honble. Gentlemen,—

In accordance with instructions received from you, we went to Pretoria in order to negotiate with the British authorities on the question of peace. We have the honour to make the following report:

The meetings lasted from Monday, May 19th, to Wednesday, May 28th, its prolongation having been principally caused by the length of time taken up by the cable correspondence with the British Government.

We first handed in a proposal (annexed under A)[345] in which we attempted to negotiate on the basis of a limited independence with surrender of part of our territory. Lords Kitchener and Milner refused emphatically to negotiate on this basis, and expressed the opinion that to cable this proposal to the British Government would be detrimental to the objects of these negotiations. They told us they had already informed the two Governments that the British Government would only negotiate on the basis of an amended form of the Middelburg proposal. In order finally to formulate this proposal, Lord Milner asked the assistance of some members of the Commission; and this was granted, on the understanding that the assistance of these members of the Commission should be given without prejudice to themselves.

As the result of the deliberations of this sub-committee, Lord Milner produced a draft proposal, in which we insisted that a fresh clause (No. 11) should be inserted; and this was done. This draft proposal (annexed under B)[346] was then cabled to the British Government, revised by them, and then communicated to us in its final shape (annexed under B).[347] We were informed by the British Government that no further revision of this proposal would be allowed, but that it must now be either accepted or rejected in its entirety by the delegates of the two Republics; and that this acceptance or rejection must take place within a stipulated time. We then told Lord Kitchener that he should know our final decision by the evening of the next Saturday at latest.

During our formal negotiations certain informal conversations took place in reference to the British subjects (in Cape Colony and Natal) who have been fighting on our side. As a result of these informal conversations a communication from the British Government was imparted to us (annexed under B).[348]

We have the honour to remain, etc.,

Louis Botha.

J.H. De La Rey.

C.R. De Wet.

J.B.M. Hertzog.

J.D. Smuts.

Vice-President Burger said that the delegates must proceed to discuss this document, and that they would then be asked to decide—firstly, whether the struggle should be continued; secondly, whether the proposal of the British Government should be accepted; and, thirdly, whether they were prepared to surrender unconditionally.

It was decided that minutes of the meeting should be kept, and the delegates then proceeded to discuss the different articles of the British Government's proposal. The whole of the morning and a part of the afternoon sitting were devoted to questions dealing with the meaning of the several clauses, the members of the Commission answering to the best of their ability.

After these questions had been disposed of, Mr. De Clercq rose to speak. He said that he had already given his own opinion, but that now it was for the whole meeting to decide whether they would give up the war, and, if they resolved to do so, whether they would accept the proposal unconditionally. As to the proposal, it could not be denied that it did not give all that they themselves desired, but that could not have been expected. Should they now return to their commandos and be asked by their burghers what they had effected, they would have to reply, "Nothing." How would they be able to meet their burghers with such an answer as that? It would therefore be better to get terms from the British Government; and by doing so they would also gratify the British nation. As for himself, he was for accepting the proposal, unless it could be proved to him that unconditional surrender would be a still better course to take.

[345] See Appendix A.

[346] See Appendix B.

[347] See Appendix B.

[348] See Appendix B.

General Nieuwouwdt then proposed that the meeting should, without further delay, proceed to vote whether the war should be terminated, and whether the terms offered to them should be accepted.

General Froneman seconded this proposal.

Mr. Birkenstock (Vrijheid) felt that this was too important a matter to be treated with such haste. A decision about such a document as the one now lying before the meeting could not be come to in a moment. The delegates would hardly agree with the last speaker in his opinion that they should at once proceed to vote whether the war should or should not be continued. Time was required before coming to such a decision. Moreover it had to be proved whether it were possible to continue the war. There were some districts where it certainly could no longer be carried on. Was it possible for one part of the nation to continue fighting without the other? Then there was the question whether their resources and the troops which they still had were sufficient to justify them in prolonging the struggle. If they were insufficient the war must be discontinued and terms must be accepted. It would not be an easy thing to do; one could not, with a light heart, give up the independence of their country; but half a loaf was better than no bread,[349] and even such a sacrifice as this might be necessary if the nation was to be saved.

Commandant Jacobsz (Harrismith) was at one with the last speaker in holding that they must not be in too great a hurry to vote on the proposal.

Mr. P.R. Viljoen (Heidelberg) felt that the proposal of the British Government would so tightly bind them that they would never again be free. They were knee-haltered[350] now, but under certain circumstances they might even be hobbled.[351]

He considered that the meeting should ask the Governments to stop the war.

General Du Toit (Wolmaransstad) said that the times through which they were passing were very critical; every one ought to say exactly what he thought, and no one ought to be condemned for doing so. A delegate who should say that the war could not be continued must not be considered disloyal to his country because he did so. As regarded the three questions before the meeting, according to the opinion of his burghers the war ought to be continued. The views of his burghers when he left the commandos had been clearly expressed. "Let us retain our independence, or go on fighting," they had said. But why were they of this mind? Because they were unaware how matters stood in other districts. The eyes of the delegates, however, while directed towards God, were also able to observe the condition of the eastern parts of their country. If the burghers in those parts could not hold out, it would be impossible for the other commandos to do so. It could not be denied that some of the commandos were no longer able to continue fighting. That being the case, even if there were a majority in favour of prolonging the struggle, that majority would have to yield to the wishes of the minority, and for this reason: if the war were to be continued in conformity with the wishes of the majority, and if the minority were to be compelled to surrender (and nobody would be surprised at this), then the majority would find themselves too weak to go on fighting. Thus there were clear reasons why the war must be ended. Moreover, its continuation would involve not only the national but also the moral death of the Republics. But it was still to be proved that a continuation of the war was even possible; for himself he feared that it was not so, and if fight he must he could only fight without hope and without heart. If he were now to go back to his burghers, and they were to ask him why he persisted in the war, and he was compelled to reply that he was doing so on the strength of opinions expressed in newspapers, and on the encouragement given to the cause of the Republics in their pages, he would be told that he was building on sand. Again, he feared that if the war were to be continued, detached parties would be formed which would try to obtain terms from the English for themselves. And should the commandos in time become so weak as to be forced to surrender unconditionally, what then would be the fate of the officers? Would they not lose everything, and be banished into the bargain? Let no one think, however, that he was trying merely to do what was best for himself. No. There was now a chance for negotiating; should the meeting let slip that chance, unconditional surrender would most certainly result, and that would be disastrous to all. He hoped that he would not be misunderstood; if the meeting decided to go on with the war, he, for one, would not lay down his arms. No, he would actively prosecute the war, and operate in

[349] The Boer form of this proverb is: Half an egg is better than an empty shell.

[350] The head fastened to the knee.

[351] Having two legs fastened together.

conjunction with the other generals. But what would be the use of it: he sided with those who held that the struggle could no longer be carried on.

Commandant Rheeder (Rouxville) wished to reply to those who demanded reasons for the continuation of the war. One reason, he said, was to be found in the fact that England would not allow them to have any communication with the deputation in Europe; that meant that something advantageous to us was being held back. Another was the consideration of what their descendants in time to come would say. "How is it," they would ask, "that we are not now free men? There were a large number of burghers in the veldt to continue the war—what has become of our independence?" And what answer shall we be able to make?—we whose courage failed us before such tremendous odds, and who laid down our arms when victory was still possible? The speaker would only be satisfied if the meeting were unanimous for stopping the war, not otherwise. He thought of the families. How would the delegates face their families on their return, after the sacrifice of independence? He considered that the commandos should leave those districts where resistance was no longer possible and go to others. If to discontinue the war meant to surrender independence, then the war must not be discontinued.

Vice-President Burger said that he had not heard from the last speaker any reasons whatsoever for continuing the war.

Commandant Rheeder then remarked that if they wanted to surrender their country they should have done so earlier, when the burghers were not entirely destitute. But now nothing was left to them. As to the narrowness of the field of operations, there was still room enough to fight.

Commandant P.L. Uijs (Pretoria) referred to the frequent allusion which had been made to their European deputation. That deputation was now in Holland, and must know if anything was going on there to the advantage of the Republics. If there were any hopeful signs there, their comrades would certainly have informed them. They had not done so, and therefore the meeting should dismiss this subject from its thoughts.

The meeting then adjourned until 7.15 p.m.

Upon reassembling, Commandant Cronje (Winburg) said that he would not detain the meeting for long; he only wished to say a very few words. It had been rightly said that they were passing through a momentous period of their history. To his mind the present was the critical epoch in the existence of the African nation, whose destinies they had now to decide. Delegates were asking what hopes they could now entertain. But what grounds for hope were there when the war began? In his opinion there were none. It was only that men believed then that Right was Might, and put their trust in God. And God had helped them. When the enemy had entered their country everything was dark. There had been a day on which more than four thousand men had surrendered. Then, even as now, they had been without hope. Then, even as now, those who wanted to continue the war had been told that they were mad.

That had been some two years ago, and yet the war was still going on. Then, even as now, there had been no food, and yet they had managed to live. The delegates represented a free people; let them not take a step of which they would afterwards repent. As regarded intervention, he had often said that one could not rely on it. But they could rely on God. When he returned to his burghers, and was questioned as to his reason for the course of action which he had advocated, he hoped to be able to answer, "Belief in God." There had always been times when there was no food, and yet they had always managed to live. A deputation had been officially sent to Europe, and was now there to represent their interests. Had the meeting lost its confidence in that deputation? Did it not realize that if the case of the Republic was hopeless in Europe the deputation would send word to that effect? It had been said that by continuing the war they would be exterminating the nation. He did not believe this. The way to exterminate the nation was to accept the British proposal. To go on with the war was their only policy, and it was a very good policy. The deputation had claimed that their advice should be taken before any negotiations were attempted. What right, then, had the delegates to give up the war on the basis of the proposal now before them? To do so was to give the death blow to their national existence; later on they would have cause to rue it. Moreover, the proposal did not safeguard the interests of their brethren in Cape Colony. Again, landed property belonging to burghers had already been sold, and in all probability these burghers would never see any of the proceeds. The sum (£3,000,000) which the proposal offered to compensate for all damages, was not sufficient to cover damage already done. For these and other reasons the proposal could not be accepted. No other course was open to them except to reject the proposal and to continue hostilities.

General Froneman (Ladybrand) agreed with the last speaker. He loved his country, and could not think of surrendering it. The reasons which had induced them to begin the war were still in force. He had been through the whole campaign, and saw stronger reasons now than ever before for the continuing of the war. His districts, like those of others, were exhausted, and yet his burghers remained in the veldt. He had been present at the surrender of the four thousand; he had seen General Cronje give up his sword. Those had been dark days, but the struggle still went on; they could still keep on their legs. It had been God's will that this war should take place. Prayers had been offered that it might be averted, but God had ruled it otherwise. Therefore they must carry the war through, and never think of surrender. They were Republicans. What would it be to have to give up that name for ever? He had consulted his burghers and their women-folk; he had asked them, "What conditions of peace will you accept?" They had answered, "No peace at all, if it means any loss of independence." And so, before he could vote for peace, he would have again to take the opinion of his burghers.

Veldtcornet B.H. Breijtenbach (Utrecht) urged that a definite yes or no must be given to the question, Is the war to continue? The general condition of the country had been laid before the meeting, and it had been clearly shown that its condition made the carrying on of the war impossible. One could not escape from that fact. Why then should they argue any longer? What reason had they for wishing to prolong this struggle? They surely would not do so blindfold. Unless good reasons could be alleged for continuing it, the war would have to be stopped. As those good reasons were not forthcoming, he would vote with those who were for peace. To continue the war would be a crime. Some of the last few speakers had stated that there had been no sufficient reasons for commencing the war. That might be true. They might have been over-confident then. Be that as it might, they certainly had lost so much ground since then that they must now give up the struggle. This was his irrevocable opinion. It had been clearly shown that fourteen commandos were unable to continue in the veldt. This made peace a necessity, for what was to be gained by continuing a struggle without a proper army. The war might last a few months longer, but it must end then—and end in disaster.

Commandant W.J. Viljoen (Witwatersrand) said that some speakers were for and others against the continuation of hostilities. The first were guided by faith alone; the second had brought forward definite grounds for their opinion. A year ago both parties had been inspired by faith, but what had been the result? He would be glad enough to be convinced, but those who wished to continue the war must show grounds for such a line of action.

General De la Rey would only say a few words. He had received definite instructions before he went to his burghers neither to encourage nor discourage them, whatever they might say at their meetings. He had strictly observed these instructions, and had never attempted to influence them. There were present among the delegates nine men (one being from Cape Colony) who represented his burghers, and who would testify as to their state of mind and temper; he need not therefore say anything. The delegates could bear witness how full of courage the men were. Nevertheless, the war could not be continued. Say or do what they would at that meeting, the war must cease. Some had talked about faith. But what was faith? True faith consisted in saying, "Lord, Thy will, not mine, be done." They must bow before the will of God. The delegates, he continued, must choose one of the three courses which were open to them. It would be a great calamity if they were to decide to surrender unconditionally. Had it been necessary to do so it should have been done while they still possessed something. Should they then continue the war? But the question as to what would become of the people under those circumstances must be faced—to continue fighting would be the ruin of the nation. The delegates might go away determined to fight, but the burghers would lay down their arms, and the state of affairs which would thus ensue would not redound to their honour. But the British Government offered guarantees; it would help the nation so that the nation might help itself. If any one were to say now, "Continue fighting," he and his generals might have the heart to do so if they kept their minds fixed on their recent exploits. For himself, however, he would refuse absolutely to accede to that request. And what real advantage had accrued from his successes in the veldt? What had followed on them? All his cattle had been taken away, some three hundred of his men had been killed, wounded, or taken prisoner. Some of the delegates set their hopes on the European deputation, but what did that deputation say a year ago? It said that all depended on their continuing to fight. They had continued to fight. What more, then, was there left for them to do? Some gentlemen present had definite mandates from their burghers, who very likely had no knowledge of the actual state of affairs when they gave those mandates. He himself had not known at that time in what a plight the country was. He challenged each and all of the delegates to show their burghers the proposal of the British Government, and then to see if those burghers were not in favour of unconditional surrender. But if the meeting insisted on the continuation of hostilities, the nation would be driven into hands-upping; thus the war would end in dishonour and disgrace.

Landdrost Bosman (Wakkerstroom) was glad that General De la Rey had spoken out so boldly; it was every one's duty to do so. He himself also was against the continuance of the war.

Although it had been said that the war had been begun in faith, it ought not to be forgotten that it had also been begun with hope of intervention, as was shown by the sending of the deputation to Europe—that deputation which, as they had often heard, had done so much good work. Another proof that there had then been hope of intervention was that the burghers had ordered the delegates to keep them in communication with the deputation. And that they had not relied exclusively on faith at the beginning of the war was shown by the fact that they had founded great hopes on what their brethren in Cape Colony might accomplish. These hopes had now been dissipated by General Smuts, who had just said that there was no chance of a general insurrection.

Again, could the war be continued when their commandos were so much weakened, and when food was so scarce? It was nonsense to say that food had been scarce a year ago; there had been a sufficiency then, and at the present time there was not. One could ride from Vereeniging to Piet Retief without seeing more than two or three herds of cattle. Moreover, the women and children were in a most pitiable condition. One delegate had spoken against any scheme which would be as it were a trampling on the blood which had already been spilt—he shared that delegate's sentiments; but he considered that to shed yet more blood in a cause which was to all appearance hopeless would be still more reprehensible. He should prefer not to enter into the religious aspect of the question. It was difficult to fathom the purposes of God; perhaps it might be the Divine will that they should lose their independence. All that they could do was to follow the course which seemed to be good and right. Were they, then, to surrender unconditionally? He would say no. It would be giving the enemy opportunities for doing things from which they might otherwise desist. Moreover, by voting for such a policy the leaders would incur the displeasure of the nation. In choosing what course they would pursue the delegates should let nothing else sway them save the good of the nation. They must not be carried away by their feelings; they must listen only to the voice of reason.

Commandant H.S. Grobler (Bethal) felt that, under the circumstances, the war could not be continued. It had already reduced them to such straits that they would soon have to fly to the utmost borders of their territories, leaving the enemy unopposed in the very heart of the country. At the beginning of the war they had not relied on faith alone; there had also been guns, war material and provisions. But now none of these things were left to them. It was terrible to him to think that they must sacrifice the independence of their country. He was a true son of his country, and could not consent to the surrender of her independence unless that were the only way of saving the women and children from starvation. But it was not only the women and children who were on the verge of starvation; the burghers still left in the laagers were in the same predicament. What, moreover, was to happen to the prisoners of war, if the struggle were to be continued? And to the families in the camps? The delegates must not forget those families. If the people generally were dying a national, the families were dying a moral, death. It was a sad thought that there were among their women in the camps, many who were thus losing their moral vitality. It was a thought which should make them determined to conclude the war.

Commandant Van Niekerk (Ficksburg) said that his commandos had commissioned him to hold out for independence. The proposal of the British Government could not be accepted. They must take no hasty step. If they persevered in the war, the enemy would grant them better terms. All they had to do was to act like brave men.

General J.G. Celliers (Lichtenburg) had already told the meeting what mandate he had received from his burghers. But he was there to do the best he could for the nation as a whole. The condition of the country was very critical. The fact that his own commandos were faring well was not a sufficient reason for continuing the war. He must take all circumstances into consideration. He had said that he was in favour of an arrangement by which peace should be made without the sacrifice of independence. Such an arrangement they had attempted to bring about. They had elected a Commission, which had done all in its power to give effect to their wishes in this matter. And the result was the proposal of the British Government now lying before them. That was what the Commission had obtained for them. Which of them could say that he could have obtained better terms for the people than those contained in that proposal? Or that, if the war were to be continued, the people would gain any advantage which that proposal did not give them? It had been said that the deputation in Europe had encouraged the burghers in their prolonged struggle. The last message they had received from the deputation had been: "Go on till every remedy has been tried." Could that be called encouragement? It had also been said that the nation must have faith. He admitted the necessity—but it must not be the sort of faith which chose what it would believe, and what it would disbelieve. They must be prepared to believe that it might be the will of God that they should yield to the enemy. As he had more insight into the state of affairs than his burghers, and therefore was better qualified to form a judgment, he did not feel himself bound by their mandate. Had the burghers known what he now knew, they would have given him a very different commission. He felt that it was a serious thing to continue sacrificing the lives of his fellow-countrymen. Moreover, however dear independence might be, it was useless to attempt impossibilities. Their one aim should be to safeguard the interests of the nation. His vote would be with those who were for accepting the proposal of the British Government.

Commander-in-Chief de Wet was the next to address the meeting. His speech was as follows:—

"As I feel it to be my duty to speak out all my mind before this meeting, I shall go back to the very beginning of the war. And recalling my feelings at that period, I can say that I had less hope then for intervention than I have now. I do not mean to say that I am sanguine about it even now; but I know to-day, what I did not know then, that great sympathy is felt for us by other nations. Even in England this sympathy is to be found, as is shown by the largely-attended 'Pro-Boer' meetings which have been held in that country. And that the feeling in our favour is widespread is evident from the reports which we received by word of mouth from the messenger to whom the deputation entrusted its recent letter, for we cannot believe that the deputation would have employed an unreliable person. And what did that messenger say? Among other things, he said that our cause was winning new adherents every day. It may be asked, however, why the deputation did not send a report of its own? I reply that it had its hand upon the pulse of the Governments, and that the information it was thus gaining was of such a character that it could not be entrusted to any messenger whatsoever. Perhaps the deputation was unable in any way to communicate what it knew to us—it would never do to noise abroad the secrets of European policy. The silence of the delegates ought not, then, to discourage us; on the contrary, we should regard it as a hopeful sign.

"If there is any one man who feels deeply for the critical condition of our country, I am that man. And critical our condition certainly is; so that I am not surprised that some of us are asking, 'What hope have we now in continuing the struggle?' But I would ask another question: 'What hope had we at the beginning of the war?' Our faith in God—we had nothing else to rely on! At the very outset of the war I knew that we, with our forty-five thousand troops, were engaged in a contest against a nation that had no less than seven hundred and fifty thousand men under arms, and who could easily send against us a third of that number. And to counterbalance the terrible odds against us, we had nothing, as I knew, but our faith. At that time there were some who expected that effectual help would come from Cape Colony. I was never deluded by this hope. I knew of course that there were men there who would fight with us against England; I knew how much those men sympathized with our cause; but I also knew that the circumstances of that country would make it impossible for the colonists to help us more than they have, as a matter of fact, done. No! God was our one Hope when the war began. And if, when the war is over, victory lies with us, it will not be the first time that faith in God has enabled the weaker nation to overthrow the stronger.

"Those of you who urge that the war should be discontinued, ask us, who are for carrying it on, what tangible reason we have for our hope. But what tangible reason for hope was there at the beginning of the war? Are our affairs darker now? Quite the contrary—miracles have been worked in our favour during the last twenty-two months. General Botha wrote to me some time ago, saying that the scarcity of ammunition was causing him much anxiety. And he had good cause for that anxiety—ammunition was exhausted. When a burgher came to me at that time with an empty bandolier, it absolutely terrified me. But now, to use an expression of General Joubert's, my pleasure is tempered with shame when I think of the plentiful store of ammunition which we possess. I am not angry with those of my compatriots who ask for reasons—I give my reasons—nor have I given a thousandth part of them.

"The enemy has already made us some concessions. There was a time when Lord Salisbury said that the English Government would be satisfied with nothing short of unconditional surrender. He does not say so to-day. England is negotiating with us—that is to say, she shows signs of yielding to our demands. If we continue the war, England will negotiate again; she will offer still more favourable terms; she will not even stick at independence.

"Do you want more of my reasons? Look back once more upon our past history, and you shall find them. Recall the time when the Transvaal was at war with England. At that time we did not know the English so well as we now know them; we had only thirteen cartridges for each man; and there were the so-called 'Loyalists'—a chicken-hearted crew—to hamper us. Faith was our only support then—and you all know how that war resulted.

“I am asked what I mean to do with the women and children. That is a very difficult question to answer. We must have faith. I think also that we might meet the emergency in this way—a part of the men should be told off to lay down their arms for the sake of the women, and then they could take the women with them to the English in the towns. This would be a hard expedient, but it may be the only one possible.

“America has been referred to by some of the speakers, who have compared our circumstances with those of the United States, when they made war upon England. The comparison is, in one respect at least, an apt one, for we also have large territories to which we can always retreat.

“As to Europe—we know little of the condition of things there. Our information about Europe comes only from newspapers, and ‘Jingo’ newspapers at that. If there is not a great deal going on in Europe which England wants to hide from us, why is she so careful not to let us see European journals? If there were anything in them unfavourable to our cause, England would flood our country with them in her own interests. We must also note that England will not permit our deputation to return to us.

“Taking all these facts into consideration, and remembering that the sympathy for us, which is to be found in England itself, may be regarded as being, for all practical purposes, a sort of indirect intervention, I maintain that this terrible struggle must be continued. We must fight on, no matter how long, until our independence is absolutely secure.”

General Beijers (Waterberg) said that he had to give an answer to the question whether he ought to follow his reason or his conscience; he could only reply that conscience had the first claim upon him. If he were to perish whilst following the guidance of reason, he would feel that he had been unfaithful; whereas, were he to die whilst obeying the dictates of conscience, he would not fear death. Martyrs of old had died for their faith; but he feared that the martyr spirit was now only to be met with in books! Those martyrs had died, and with their death it had seemed that all was lost; but the truth, for which they had given up their lives, had lived!

But how is it now with us? We think our cause a righteous one, but are we willing to die for it? Some spoke of our existence as a nation—but whether that were to be preserved or lost, did not lie with us—it was in the hands of God—He would take care of it. Right must conquer in the end. They must take care to be on the side of right, should it even cost them their lives. He agreed with those who said that, even if the present deliberations were to come to nothing, they would have another chance, later on, of negotiating. This had been proved by what had already happened. General de Wet had shown them how Lord Salisbury had gone back upon his first demands; he (General Beijers) could tell them that on one occasion Lord Roberts had declined even to speak to General Botha—and yet the English were negotiating with them now. He was quite open to conviction, but at present he could not see that the war ought to be stopped. Nevertheless he was not blind to the critical state of their affairs. But their case was not yet hopeless; their anxiety about food, their lack of horses—these were not insurmountable difficulties. They might even find some means by which to save their womenfolk.

No. These difficulties were not insuperable; but there was one difficulty which was insuperable—the present spirit of the nation. When a spirit, be it what it might, inspired or ruled a man, then that man would submit to no other sway. The spirit that now ruled the burghers was a spirit that was driving them over to the enemy. Against that spirit it was impossible to contend. General De la Rey had said that, if the proposal now before the meeting were to be shown to the burghers, they would at once accept it—that was the sort of spirit that was in them, and one must take it into consideration, for he was convinced that it presented an insurmountable obstacle to the continuation of the war.

The meeting was then closed with prayer.

Friday, May 30th, 1902.

After the preliminary prayer had been offered, Vice-President Burger said that before beginning the business of the day, it was his sad duty to inform the meeting that the President of the Orange Free State had been obliged to resign, on account of serious illness. President Steyn had been compelled, in order to obtain medical assistance, to put himself in the hands of the enemy. He had further to communicate that Commander-in-Chief de Wet had been appointed Vice-President of the Orange Free State. He wished to express his deep sympathy with the representatives in the severe loss which they had sustained. President Steyn, he said, had been a rock and pillar to their great cause.

Vice-President de Wet having thanked the Vice-President of the South African Republic for his kind and sympathetic words, Mr. J. Naude (the representative of Pretoria, and of General Kemp's flying columns) put some questions with regard to the colonists who had been fighting on the Boer side. These questions were answered by General Smuts. Mr. Naude then asked if the delegates were expected to come to any decision about independence.

General Botha replied that the Governments had informed Lords Kitchener and Milner that they were not in a condition to decide that question—that it was a matter for the nation to settle. The delegates had then gone to their burghers, and now had returned, and were present.

Mr. Naude said that it must therefore have been known at Klerksdorp that the delegates had to decide upon the question of independence. If that were so, he found himself in a difficulty. Either the delegates had been misled, or they were the victims of a mistake, for they had never been told that they had been elected as plenipotentiaries. Notwithstanding all that the lawyers might say, he considered himself as having a certain definite mission. He had obtained the votes of his burghers on the understanding that he would take up a certain position. He had asked them whether independence was to be given up, and they had answered in the negative. He could not therefore vote for the acceptance of the proposal now before the meeting, for that proposal demanded the surrender of independence. His burghers had also insisted on being allowed to keep their arms, and on the use of their language in schools and Courts of Justice, both of which conditions were refused by the British proposal. Since, therefore, he could not agree to the proposal, he was for continuing the war. Some asked what were the chances of success? He remembered the state of feeling among the burghers at Warmebad—that was a dark time indeed. The Commandant-General had paid those burghers a visit, and had told them that they had nothing to lose, but everything to win, by continuing the struggle. That had been enough for them. They had not had much prospect then; they could not see whither their road was leading. But they had found out afterwards. It had been a dark time too when Pretoria was taken, but most of the burghers had remained steadfast. And after the darkness the light had come back. Again a dark cloud was over them—it would pass away, and the light would reappear.

General De la Rey explained that he had not intended to mislead anybody at the gatherings of the burghers. Every document which the Government had handed over to him had been laid before those gatherings. Mr. Naude had asked whether the delegates at that meeting had to decide about independence. Most certainly they had. And to do so was a duty devolving upon Mr. Naude as much as on any other delegate present. They would have to decide, not for their own districts alone, but for the whole country.

Mr. Naude said that he had no wish to free himself from his responsibility, but he could not forget that he had come there with a definite mission.

Judge Hertzog wished again to explain the rights of the question from a legal point of view. One must ask: If the nation were here, what would it wish to be done? And one must act in conformity with what one thinks its answer would be. The Judge then proceeded to speak on the matter in general. What, he asked, were the arguments in favour of continuing the war? In the first place, England was growing weaker just as their own nation was. Any one could see that with their own eyes. It was true as regarded the financial side of the question. No doubt England could still collect millions of pounds, if she wished, but the time would come when she would have trouble with her tax-payers. Already the British Government found it difficult to pay the interest on the sum borrowed for war expenses, as was proved by the fact that a corn tax had been levied in England. That tax would not have been levied unless things had been in a serious condition. In the second place, he would ask how it was they had not been allowed to meet their deputation? It would only have taken the deputation fourteen days to perform the journey; by now it would have been among them. But permission had been refused them. And why? It was said that to grant a permission would have been a military irregularity. But the present meeting was also a military irregularity. There must be something more behind that refusal. But what were the arguments against going on with the war? He would enumerate them—the situation in which they found themselves was critical; the country as a whole was exhausted. Nearly all the horses had died or had been captured. The strongest argument of all, however, was that some of their own people had turned against them, and were fighting in the ranks of the enemy. Then the condition of the women caused great anxiety; a fear had been expressed that a moral decay might set in among the families in the camps. That consideration had great weight with him. No one with any heart could remain indifferent to it. If there was one thing which more than anything else made him respect Commandant-General Botha, it was that the Commandant-General had the heart to feel, and the courage to express, the importance of that consideration. The present war was one of the saddest that had ever been waged. He doubted if there had ever been a war in which a nation had suffered as they had. But all those sufferings, horrible though they were, did not influence his decision. Did he but see the chance of finally securing freedom for the nation, he would put all such considerations on one side, and go on fighting till death. No; it was not the horror of the situation which influenced him; there was something that weighed upon his heart yet more heavily—it was the holding of that meeting at Vereeniging. He reproached no one. Every one had acted with the best intentions. Nevertheless that meeting was a fatal error; it would give them their death blow. For what had it produced—a statement from the lips of the Commandant-General himself that the condition of the country was hopeless. If there were yet any burghers whose courage was not gone, would they not be utterly disheartened when they heard what their leaders had said at that meeting? That was the saddest thought of all. He could understand that those burghers who had already lost heart should be leaving the commandos, but now those who had never yet been disheartened would become so. But notwithstanding all this, it was difficult to feel certain which was the right course to pursue—to give up the war or to continue it. He could only suggest that those who were now in doubt on the matter should support the line of action which, before their doubt began, had appeared to them to be best.

Mr. L.J. Meijer (a member of the Government of the South African Republic) then gave some account of the devastation of that part of the country which lay to the north of the Eastern Railway, and on the further side of the Sabi River. (This report coincided with those already given by the delegates.) He went on to say that as they were all in the dark, and could not see the road they were travelling along, they must take reason and conscience for their guide. They had already lost much: let them not lose everything. And what could they hope to gain by continuing the struggle? To do so might be to throw away their last chance of peace. What would their progeny say of them if they were to persist in the struggle and thus lose everything they had possessed? They would say, "Our forefathers were brave, but they had no brains." Whereas, if they were to stop the war, their progeny would say, "Our forefathers did not fight for their own glory." He pointed out that however little the British proposal contained of what they desired, it nevertheless promised them representative government. In the past he had been against the war; he had wished that the five years' franchise should be granted. Although the people had opposed this measure he had always supported it. And why? Because he had feared that were that measure not conceded African blood would stain the ground. Must they still continue to shed blood? After the capture of Bloemfontein there had been a secret meeting of the council of war at Pretoria. His Government had then been willing to surrender, but the Free State had refused. The two Governments had therefore decided to go on with the war. A year later, in the month of June, there had been another meeting. A letter had been sent to the Free State. The two Governments had met at Waterval, and had once more decided to continue the struggle. Later on, again, the Government of the South African Republic wrote another letter to the Free State; but there had been no opportunity of meeting until the present occasion, which saw them assembled together at Vereeniging. Were they again going to decide to continue their resistance? It was a matter for serious consideration. There was but little seed-corn left. This must, if they had to go on fighting, be preserved from the enemy at all costs; were it to be destroyed, the African nation must cease to exist. But they could not continue the war. It was the Boers now who were teaching the English how to fight against us; Boers now were with the enemy's forces, showing them how to march by night, and pointing out to them all the foot passes.

Commandant Van Niekerk (Kroonstad) pointed out that the Colonists had already rendered them valuable aid, and could still do so. Were they now to abandon these Colonists, and—thinking only about saving themselves—leave them to fight on alone? It would be sad indeed if the burghers were compelled to lay down their arms.

Commandant-General L. Botha said that in regard to the holding of a national meeting, he had already chosen delegates with power to act. He spoke of the state of affairs at the beginning of the war—the two Republics had then at least sixty thousand men under arms. In reference to the Cape Colony, he said that it had never been expected that that country would allow its railways to be used for the transport of troops. The Commandant-General then proceeded as follows:—

"I used to entertain hopes that the European Powers would interfere on our behalf. All that they have done, however, has been to look on while England was introducing all sorts of new methods of warfare, methods, too, which are contrary to all international law.

"When the war began we had plenty of provisions, and a commando could remain for weeks in one spot without the local food supply running out. Our families, too, were then well provided for. But all this is now changed. One is only too thankful nowadays to know that our wives are under English protection. This question of our womenfolk is one of our greatest difficulties. What are we to do with them? One man answers that some of the burghers should surrender themselves to the English, and take the women with them. But most of the women now amongst us are the wives of men already prisoners. And how can we expect those not their own kith and kin to be willing to give up liberty for their sakes?

"As to the deputation, we must remember that it was accredited to all the Powers of Europe. And yet it has only been able to hand in its credentials to the Netherlands Government. Does not this prove that no other Government is willing to receive it? If you need further proof, I refer you to the letter in which the deputation—they were still allowed to write to us then—said: 'There is no chance for us in Europe.' The deputation wanted to be allowed to return home, but our Government advised them to remain in Europe, because their arrival in South Africa would be a death blow to the hopes of many. That is why the deputation is still in Europe. Later on they said that, although they knew that there was no chance of intervention, yet they felt that they ought to persist in their efforts, because of the sacrifices which we had already made. It is possible that a war may arise in Europe from which we shall gain something, but what right have we to expect such a contingency? Moreover, great nations take but little interest in the fate of small ones—indeed, it is to the advantage of the former that the small nations should be wiped out of existence.

"I cannot refrain from alluding to the faithlessness of some of our burghers, who are to be found in the ranks of the enemy. But this is not the only sign of the way in which affairs are trending—I look back on the past. I remember that we have been fighting a full year since we last heard of our deputation. What have we gained since June, 1901? Nothing. On the contrary, we have been going backwards so fast that, if this weakening process goes on much longer, we shall soon find ourselves unable any more to call ourselves a fighting nation. What have we not undergone in the course of this year which is just over! In the concentration camps alone, twenty thousand women and children have died. When I was in Pretoria I received reports from our information office, and otherwise, of our losses. I found that there were thirty-one thousand six hundred prisoners of war, of whom six hundred had died, and that three thousand eight hundred of our burghers had been killed in the war. Is not a loss such as this, in so short a time as two and a half years, a serious matter? Think, too, of the sufferings which those twenty thousand women who died in the camps must have endured!

"I am not deaf to the claims of the colonists who have been fighting for us. I have said that if we surrender our independence, we must provide for them. Should we serve their interests by continuing the war? No, indeed! The best thing for them would be that we should bring it to a close. But if we are absolutely determined to go on fighting, let us at least say to them, 'We advise you to desist.'

"What I am saying now is in substance what I said at Warmbad at a time when there were two thousand men of that district in the Veldt. How many are there now? Four hundred and eighty! On that occasion I also said that we must continue the war until we were driven by sheer starvation to make peace. Well, in some divisions starvation has already come. The delegates themselves have had to confess that our strength up till now has lain in the fact that we have been able to continue the struggle in every district. In this way we have divided the enemy's forces. But if we are compelled to abandon some of our districts, and to concentrate on certain points, then the English also will concentrate, and attack us with an irresistible force.

"It has been suggested that we ought to march into Cape Colony. I know, however, what that would mean—Commander-in-Chief de Wet marched into the colonies. He had a large force, and the season of the year was auspicious for his attempt, and yet he failed. How, then, shall we succeed in winter, and with horses so weak that they can only go op-een-stap.[352]

[352] The step of a tired horse.

"What, then, are we to do? Some will reply, 'Go on with the war,' Yes, but for how long? For ten or twelve years? But would that be possible? If in two years we have been reduced from sixty thousand fighting men to half that number, where will our army be after another ten years of war? It is clear enough to me that if we go on any longer, we shall be compelled to surrender. Would it not be better to come to some agreement with the enemy, while we have the opportunity? We have all received the gift of reason; let us use it on the present occasion.

"As far as I and my own burghers are concerned, to continue the struggle is still possible. But we must not only think of ourselves. We must almost think of others. There are, for instance, the widows and orphans. If we accept the terms now offered to us, they will remain under our care. But if we go on with the war until we are forced to surrender, who will then take care of them? Or if we were all killed, what could we do for them? We should not even be able to send a deputation to Europe, to ask for money to help us to rebuild our farms, and to feed our burghers.

"There are three questions now before us—three alternatives between which we have to choose—the continuing of the war, unconditional surrender, and the acceptance of the British proposal. With regard to the first, I fail to see what satisfactory result can come to us from persisting in this unequal contest, which must result in the end in our extermination. As to the choice between the other two, in many ways unconditional surrender would be the better. But, for the sake of the nation, we may not choose it. Although to reject it may involve us in many hardships, yet we must think of nothing else but the interests of the nation. Our only course, then, is to accept the proposal of the English Government. Its terms may not be very advantageous to us, but nevertheless they rescue us from an almost impossible position."

After a short adjournment the delegates again assembled at about 2 p.m.

General C.H. Muller (Boksburg) said that his burghers had sent him to defend their menaced independence. One part of them had authorized him to act as his judgment should dictate; another part had ordered him to hold out for independence and to try to get into communication with the European deputation. He had long ago told his burghers that they must trust in God if they wished to continue the war, for they could not do so by relying only on their guns and rifles. He did not like to think of what they would say if he were to go back to them and tell them that he had not been in communication with the deputation, and that the proposal of the English Government had been accepted. He could not bring himself to surrender. Nevertheless, having in view what the Commandant-General and others had said, he felt that he must do so, for it was impossible for him to prosecute the war single-handed. But could not the delegates continue to stand by one another, and make a covenant with the Lord? The district which he represented was one of the poorest in the whole country, and the £3,000,000 offered by the enemy did not include any provision for those who, like his burghers, could do nothing to help themselves. He would again suggest that the delegates should make a vow unto the Lord. For himself, he could not vote for the acceptance of the British proposal.

General J.H. Smuts then spoke as follows:—

"Up till now I have taken no part in this discussion, but my opinions are not unknown to my Government; we have arrived at a dark period both in the history of our war, and in the course of our national development. To me it is all the darker because I am one of those who, as members of the Government of the South African Republic, provoked the war with England. A man, however, may not draw back from the consequences of his deeds. We must therefore keep back all private feeling, and decide solely with a view to the lasting interests of our nation. This is an important occasion for us—it is perhaps the last time that we shall meet as a free people with a free government. Let us then rise to the height of this occasion; let us arrive at a decision for which our posterity shall bless, and not curse us.

"The great danger for this meeting is that of deciding the questions before it on purely military grounds. Nearly all the delegates here are officers who in the past have never quailed before the overwhelming forces of the enemy, and who therefore are never likely to do so in the future. They do not know what fear is, and they are ready to shed the last drop of their blood in the defence of their country.

"Now if we look at the matter from their point of view, that is to say, if we look at it merely as a military question, I am bound to admit that we shall come to the conclusion that the war can be continued. We are still an unconquered power; we have still about eighteen thousand men in the field—veterans, with whom one can accomplish almost anything. From a purely military standpoint, our cause is not yet lost. But it is as a nation, and not as an army, that we are met here, and it is therefore for the nation principally that we must consult. No one sits here to represent this or that commando. One and all, we represent the African nation, and not only those members of it which are now in the field, but also those who rest beneath the soil, and those yet unborn, who shall succeed us.

"No! We do not only represent our burghers on commando, the troops over which we are placed in command; we represent also the thousands who have passed away, after making the last sacrifice for their country; the prisoners scattered all the world over; the women and children dying by the thousand in the prison camps of the enemy; we represent the blood and the tears of the whole African nation. From the prisons, the camps, the graves, the veldt, and from the womb of the future, that nation cries out to us to make a wise decision now, to take no step which might lead to the downfall or even to the extermination of their race, and thus make all their sacrifices of no avail. Our struggle, up to the present, has not been an aimless one. We have not been fighting in mere desperation. We began this strife, and we have continued it, because we wanted to maintain our independence and were prepared to sacrifice everything for it. But we must not sacrifice the African nation itself upon the altar of independence. So soon as we are convinced that our chance of maintaining our autonomous position as Republics is, humanly speaking, at an end, it becomes our clear duty to desist from our efforts. We must not run the risk of sacrificing our nation and its future to a mere idea which can no longer be realized.

"And ought we not to be convinced that independence is now irretrievably lost? We have been fighting without cessation for nearly three years. It is no exaggeration to say that during that period we have been employing all the strength and all the means which we possess, in the furtherance of our cause. We have sacrificed thousands of lives; we have lost all our earthly goods; our dear country is become one continuous desert; more than twenty thousand of our women and children have perished in the camps of the enemy. And has this brought us independence? Just the reverse; it is receding further and further from us every day. The longer we fight, the greater will be the distance between us and the aim for which we are fighting.

"The manner in which the enemy has been conducting, and still continues to conduct, this war, has reduced our country to such a state of exhaustion, that it will soon be a physical impossibility for us to fight any longer. Our only hope lies in the chance of help from outside. A year ago I, in the name of my Government, communicated the condition of our nation to His Honour States-President Kruger, in Europe. He wrote in reply that we must rely on the state of affairs in Cape Colony—and the sympathy of European nations—and that we must continue the war until all other means were exhausted."

The speaker here enlarged upon the political developments which had taken place in the United States and in the principal European countries during the preceding two years, and then continued:—

"So far as we are concerned, the sum total of the foreign situation is that we obtain a great deal of sympathy, for which we are naturally most grateful. More than this we do not obtain, nor shall obtain for many a long year. Europe will go on expressing sympathy with us until the last Boer hero has died on the field and the last Boer woman has gone down to her grave—until, in fact, the whole Boer nation has been sacrificed on the altar of history and of humanity.

"I have already, on a former occasion, told you what I think about the situation in Cape Colony. We have made great mistakes there; perhaps even now Cape Colony is not ripe for the sort of policy which we have been pursuing with regard to it. At all events, we cannot entertain any hopes of a general rising of the Colonists. We cannot, however, give too much honour to those three thousand heroes in the Colony who have sacrificed all in our behalf, even though they have not succeeded in securing our independence for us.

"Thus we have given President Kruger's advice a fair trial. For twelve months we have been testing the value of the methods which he urged upon us. And, as a result of it all, we have become convinced that those methods are of no avail—that if we wish to remain independent we must depend upon ourselves alone. But the facts which the various delegates have brought before our notice show that we cannot thus depend upon ourselves; that, unless we obtain outside help, the struggle must come to an end. We have, then, no hope of success. Our country is already devastated and in ruins; let us stop before our people are ruined also.

"And now the enemy approaches with a proposal, which, however unacceptable it may be to us in other respects, includes the promise of amnesty for our Colonial brethren who have been fighting side by side with us. I fear that the day will come when we shall no longer be able to save these so-called rebels, and then it will be a just ground for reproach that we sacrificed their interests in a cause that was already hopeless. Moreover, if we refused the proposal which the British Government now makes to us, I am afraid that we shall considerably weaken our position in the eyes of the world, and thus lose much of the sympathy which to-day it evinces in our favour.

"Brethren, we have vowed to stand fast to the bitter end; but let us be men, and acknowledge that that end has now come, and that it is more bitter than ever we thought it could be. For death itself would be sweet compared with the step which we must now take. But let us bow before the will of God.

"The future is dark indeed, but we will not give up courage, and hope, and trust in God. No one shall ever convince me that this unparalleled sacrifice which the African nation has laid upon the altar of freedom will be in vain. It has been a war for freedom—not only for the freedom of the Boers, but for the freedom of all the nations of South Africa. Its results we leave in God's hands. Perhaps it is His will to lead our nation through defeat, through abasement, yes, and even through the valley of the shadow of death, to the glory of a nobler future, to the light of a brighter day."

Commandant A.J. Bester (Bloemfontein) said that at the meeting at which he had been elected his burghers had told him that they were resolved not to become the subjects of England. The arguments now urged against the continuation of the war were not new—they had been used in former times of depression. History gave many instances in which their nation had been delivered out of the most critical positions. One could not help believing that Right would conquer. How was it to be explained that two hundred and forty thousand troops had failed to exterminate two small Republics? Then there had been miraculous escapes; surely the thoughts of these ought to encourage them. They must all be of one mind. His own decision was to stand or to fall for his freedom.

Mr. Birkenstock (Vrijheid) asked whether the proposal could not be accepted under protest.

General J.C. Smuts answered that the meeting could empower the Governments to accept the proposal, and to add that they did so with such and such provisos.

Commandant A.J. Bester (Bloemfontein) thought that there had been enough said, and recommended that the discussion be closed.

Commandant F.E. Mentz (Heilbron) also thought that it was not necessary to argue any more. He believed that the war could not be continued. In Heilbron, Bloemfontein, and part of Bethlehem there were not five head of cattle left. The helpless condition of the women and children also demanded consideration. The state of the country was becoming so desperate that they were now obliged to break away from the kraals. He himself had been compelled to this not long ago, and had lost forty men in one day. He would have to leave his district, but could not bring it to his heart to leave the women behind. It was quite clear to him that the war must be stopped, for some parts of the Transvaal were absolutely unable to go on fighting. Moreover, were the war to continue, commando after commando would go over to the enemy.

General Kemp (Krugersdorp) took a more encouraging view of affairs. He would stand or fall with the independence. His mandate was to that effect. His conscience also would not justify him in taking any other course. He thought that the proposal of the English Government was vague, that there was not sufficient provision for the Boer losses in it, and that it treated the Dutch language as a foreign tongue. Circumstances had often been dark, and the darkness would pass away this time as it had done before. Remembering the commission which had been given to him by the burghers, he could not do otherwise than vote for a continuation of the war.

Vice-President Burger: "I have already given my opinion. I am sorry that the meeting seems to be divided. It is necessary for the welfare of our nation that we should be of one mind. Are we to continue the war? From what I have seen and heard, it is clear to me that we cannot do so. I repeat that there is no possibility of it, neither does any real hope exist that by doing so we should benefit the nation. It is idle to compare our condition in the struggle in 1877–1881 with that in which we now find ourselves; I speak from experience.

"It is true that the victory was then ours; that it was so is due to the help which we received from outside. The Orange Free State remained neutral, but assistance came from President Brand in South Africa and from Gladstone in England: thus it was not by our own sword that we were enabled to win.

"It will be asked why, if we have kept up the struggle for two years and a half, can we not still continue to do so?

"Because, in the meantime, we have become weaker and weaker, and if we persist the end must be fatal. What grounds have we for expecting that we may yet be victorious? Each man we lose renders us weaker; every hundred men we lose means a similar gain to the enemy. England's numerical strength does not diminish; on the contrary, there are even more troops in the country at this moment than when Lord Roberts had the command. England also has used our own men against us, and has not been ashamed of arming the Kaffirs; the enemy are learning from our own men in what way they should fight—he must be blind indeed who cannot see these facts.

"I do not think we can appropriately call this altogether a 'war of faith.' Undoubtedly we began this war strong in the faith of God, but there were also two or three other things to rely upon. We had considerable confidence in our own weapons; we under-estimated the enemy; the fighting spirit had seized upon our people; and the thought of victory had banished that of the possibility of defeat.

"The question still remains, What are we to do? I have no great opinion of the document which lies before us: to me it holds out no inducement to stop the war. If I feel compelled to treat for peace it is not on account of any advantages that this proposal offers me: it is the weight of my own responsibility which drives me to it.

"If I think that by holding out I should dig the nation's grave, nothing must induce me to continue the struggle.

"Therefore I consider it my duty, as leader of our nation, to do my utmost that not one man more shall be killed, that not one woman more shall die.

"The sacrifice must be made; is not this also a trial of our faith? What shall we gain by going on? Nothing! It is obvious that further surrenders will take place—here of a few, there of many—and our weakness will increase.

"We shall also be obliged to abandon large areas of the country. Will this make us stronger? Rather, will it not enable the enemy to concentrate still more? And the abandoned tracts—to whom will they belong? To the enemy!

"In all probability this is our last meeting. I do not believe that we shall be given another chance to negotiate: we shall be deemed too insignificant. If we reject this proposal, what prospects have we in the future? If we accept it, we can, like a child, increase in size and strength, but with its rejection goes our last opportunity.

"Fell a tree and it will sprout again; uproot it and there is an end of it. What has the nation done to deserve extinction?

"Those who wish to continue the war are influenced chiefly by hope; but on what is this hope founded? On our arms? No. On intervention? By no means. On what then? No one can say.

"I am sorry that the Transvaal and the Orange Free State are at variance on this point, and I regret that it is the Transvaal which has to declare itself unable to proceed further; but the enemy have concentrated all their forces in this State, and we can hold out no longer."

Mr. L. Jacobsz: "I have hitherto not spoken, because I am a non-combatant. I have also suffered much, although less than others. I have listened to what has been said, but my opinion is not changed by the views I have heard expressed.

"I repeat now what I said at Klerksdorp, namely that the struggle cannot continue. I have noted the condition of the country, which is such that the commandos can no longer be supported. I would point out the condition of the women and children, of whom many are dying, and all are exposed to great dangers. If there was a chance of succeeding in the end, then we might hold out, but there is no such chance; there is no possibility of intervention, and the silence of the deputation is ominous.

"I sympathize with the heroes present at this meeting; we must have a foundation for our faith, and we cannot altogether compare our people with the people of Israel. Israel had promises made to them; we have none. I would further point out that, in the interests of the nation, it will not do to surrender unconditionally: the terms before us may be deceptive, but they are the best obtainable.

"With regard to the difficulty of those delegates who consider that they are bound to act as they have been commissioned, I am of the same opinion as Judge Hertzog and General Smuts."

Commandant J.J. Alberts (Standerton) spoke more or less in the same strain. He was of opinion that the war should be finished by ceding territory, but, failing this, that it should be ended on any terms obtainable.

Vice-President de Wet expressed his opinion that, considering the short time at their disposal, they should proceed, if possible, to make some proposal.

General D.A. Brand said that he would have spoken if he had not thought that enough had been said; he considered it desirable to close the discussion, and was willing to make a proposal.

Veldtcornet D.J.E. Opperman (Pretoria South) considered that the difficulties of continuing the war, and of accepting the proposal, were equal. Some of his burghers would fight no longer. What troubled him most was the condition of the women; it went to his heart to see these families perish. He was of opinion that, for the sake of the women and children who were suffering so intensely, the proposal should be accepted under protest.

Veldtcornet J. Van Steedden, seconded by Veldtcornet B.J. Roos, moved that the discussion be now closed.

The meeting was adjourned after prayer.

Saturday, May 31st, 1902.

The meeting was opened with prayer.

General Nieuwouwdt, seconded by General Brand, made the following proposal:—

"This meeting of special deputies from the two Republics, after considering the proposal of His Majesty's Government for the re-establishment of peace, and taking into consideration (a) the demands of the burghers in the veldt and the commissions which they had given to their representatives; (b) that they do not consider themselves justified in concluding peace on the basis laid down by His Majesty's Government before having been placed in communication with the delegates of the Republic now in Europe, decides that it cannot accept the proposal of His Majesty's Government, and orders the Governments of the two Republics to communicate this decision to His Majesty's Government through its representatives."

Mr. P.R. Viljoen, seconded by General H.A. Alberts, made a proposal, amended afterwards by General Smuts and Judge Hertzog, which appears later on under the proposal of H.P.J. Pretorius and C. Botha.

A third proposal by General E. Botha and General J.G. Celliers was laid upon the table, but subsequently withdrawn.

Mr. F.W. Reitz considered it to be his duty not only to the nation but also to himself as a citizen, to say that, in case the proposal of the British Government should be accepted, it would be necessary for the meeting to make provisions as to whose signatures should be attached to the necessary documents. He himself would not sign any document by which the independence would be given up.

Remarks were made by several members on the first proposal, and Mr. P.R. Viljoen asked that no division should arise.

Vice-President de Wet then said that, as the time was limited, and all could not speak, he would propose that a Commission should be nominated in order to draw up a third proposal in which various opinions of the members should be set down; and that, whilst the Commission was occupied in this way, the Orange Free State delegates on their part and those of the South African Republic on their part, should meet in order that an understanding might be come to between them. They must endeavour to come to a decision, for it would be of the greatest possible advantage to them.

Commandant-General Botha thought that this hint should be taken. They had suffered and fought together: let them not part in anger.

The above-mentioned Commission was then decided upon, and Judge Hertzog and General Smuts were elected.

Then the Orange Free State delegates went to the tent of Vice-President de Wet, whilst those of the South African Republic remained in the tent in which the meeting was held.

After a time of heated dispute—for every man was preparing himself for the bitter end—they came to an agreement, and Judge Hertzog read the following proposal:—

"We, the national representatives of both the South African Republic and the Orange Free State, at the meeting held at Vereeniging, from the 15th of May till the 31st of May, 1902, have with grief considered the proposal made by His Majesty's Government in connexion with the conclusion of the existing hostilities, and their communication that this proposal had to be accepted, or rejected, unaltered. We are sorry that His Majesty's Government has absolutely declined to negotiate with the Governments of the Republics on the basis of their independence, or to allow our Governments to enter into communication with our deputations. Our people, however, have always been under the impression that not only on the grounds of justice, but also taking into consideration the great material and personal sacrifices made for their independence, that it had a well-founded claim for that independence.

"We have seriously considered the future of our country, and have specially observed the following facts:—

"Firstly, that the military policy pursued by the British military authorities has led to the general devastation of the territory of both Republics by the burning down of farms and towns, by the destruction of all means of subsistence, and by the exhausting of all resources required for the maintenance of our families, the subsistence of our armies, and the continuation of the war.

"Secondly, that the placing of our families in the concentration camps has brought on an unheard-of condition of suffering and sickness, so that in a comparatively short time about twenty thousand of our beloved ones have died there, and that the horrid probability has arisen that, by continuing the war, our whole nation may die out in this way.

"Thirdly, that the Kaffir tribe, within and without the frontiers of the territory of the two Republics, are mostly armed and are taking part in the war against us, and through the committing of murders and all sorts of cruelties have caused an unbearable condition of affairs in many districts of both Republics. An instance of this happened not long ago in the district of Vrijheid, where fifty-six burghers on one occasion were murdered and mutilated in a fearful manner.

"Fourthly, that by the proclamations of the enemy the burghers still fighting are threatened with the loss of all their movable and landed property—and thus with utter ruin—which proclamations have already been enforced.

"Fifthly, that it has already, through the circumstances of the war, become quite impossible for us to keep the many thousand prisoners of war taken by our forces, and that we have thus been unable to inflict much damage on the British forces (whereas the burghers who are taken prisoners by the British armies are sent out of the country), and that, after war has raged for nearly three years, there only remains an insignificant part of the fighting forces with which we began.

"Sixthly, that this fighting remainder, which is only a small minority of our whole nation, has to fight against an overpowering force of the enemy, and besides is reduced to a condition of starvation, and is destitute of all necessaries, and that notwithstanding our utmost efforts, and the sacrifice of everything that is dear and precious to us, we cannot foresee an eventual victory.

"We are therefore of opinion that there is no justifiable ground for expecting that by continuing the war the nation will retain its independence, and that, under these circumstances, the nation is not justified in continuing the war, because this can only lead to social and material ruin, not for us alone, but also for our posterity. Compelled by the above-named circumstances and motives, we commission both Governments to accept the proposal of His Majesty's Government, and to sign it in the name of the people of both Republics.

"We, the representative delegates, express our confidence that the present circumstances will, by accepting the proposal of His Majesty's Government, be speedily ameliorated in such a way that our nation will be placed in a position to enjoy the privileges to which they think they have a just claim, on the ground not only of their past sacrifices, but also of those made in this war.

"We have with great satisfaction taken note of the decision of His Majesty's Government to grant a large measure of amnesty to the British subjects who have taken up arms on our behalf, and to whom we are united by bonds of love and honour; and express our wish that it may please His Majesty to still further extend this amnesty."

Mr. P.R. Viljoen then withdrew his proposal.

Commandant H.P.J. Pretorius, seconded by General C. Botha, presented the proposal, as read by the Commission.

General Nieuwouwdt also withdrew his proposal, but it was at once taken over by General C.C.J. Badenhorst, seconded by Commandant A.J. Bester, of Bloemfontein.

The meeting then adjourned till the afternoon.

In the afternoon at 2.05 it again met.

Proceeding to the voting, the proposal of H.P.J. Pretorius, seconded by General C. Botha, was accepted, by fifty-four votes against six. Then Vice-President Burger spoke a few words suitable to the occasion as follows:—"We are standing here at the grave of the two Republics. Much yet remains to be done, although we shall not be able to do it in the official capacities which we have formerly occupied. Let us not draw our hands back from the work which it is our duty to accomplish. Let us ask God to guide us, and to show us how we shall be enabled to keep our nation together. We must be ready to forgive and forget, whenever we meet our brethren. That part of our nation which has proved unfaithful we must not reject."

Later, Vice-President Burger spoke a few words of farewell to the Commandant-General, to the Members of the Executive Councils, and to the delegates.

In the afternoon, as it turned out for the last time, Commandant Jacobsz, seconded by General Muller, made the following proposal, which was unanimously accepted by the meeting:—

"This meeting of Delegates, having in view the necessity of collecting means to provide for the wants of the suffering women and children, widows and orphans, and other destitute persons, who have through this war come to a condition of want, and also having in view the desirability of nominating a Committee, whose duty it shall be to arrange the necessary steps in this matter, and to finally decide on the management and distribution of the donations received, decides:—

"To nominate the Hon. Messrs. M.J. Steyn, S.W. Burger, L. Botha, C.R. de Wet, J.H. De la Rey, A.P. Kriel, and J.D. Kestell, as the Committee, to carry out all arrangements for the above-mentioned purposes, that may seem desirable and expedient to them, and also to appoint new Members, Sub-Committees and working Committees; and the said Committee is empowered to draw up regulations, and to amend them from time to time as shall seem to them expedient.

"This meeting further decides to send abroad from the above-mentioned Committee, Messrs. C.R. de Wet, L. Botha, and J.H. De la Rey, in order that they may help in collecting the above-mentioned donations."

Then this—the last meeting of the two Republics—was closed with prayer.

End Three Years' Appendices

Roberts' Proclamations

Proclamation I

To the Burghers of the Orange Free State.[353] [354] February, 1900

The British troops under my command having entered the Orange Free State, I feel it my duty to make known to all Burghers the cause of our coming, as well as to do all in my power to put an end to the devastation caused by this war, so that, should they continue the war, the inhabitants of the Orange Free State may not do so ignorantly, but with full knowledge of their responsibility before God for the lives lost in the campaign. Before the war began the British Government, which had always desired and cultivated peace and friendship with the people of the Orange Free State, gave a solemn assurance to President Steyn that, if the Orange Free State. remained neutral, its territory would not be invaded, and its independence would be at all times fully respected by Her Majesty's Government.

In spite of that declaration, the Government of the Orange Free State was guilty of a wanton and unjustifiable invasion of British territory.

The British Government believes that this act of aggression was not committed with the general approval and free will of a people with whom it has lived in complete amity for so many years It believes that the responsibility rests wholly with the Government of the Orange Free State, acting, not in the interests of the country, but under mischievous influences from without.

The British Government, therefore, wishes the people of the Orange Free State to understand that it bears them no ill will, and, as far as is compatible with the successful conduct of the war, and the reestablishment of peace in South Africa, it is anxious to preserve them from the evils brought upon them by the wrongful action of their Government.

I therefore warn all Burghers to desist from any further hostility towards Her Majesty's Government and the troops under my command, and I undertake that any of them, who may so desist and who are found staying in their homes and quietly pursuing their ordinary occupations, will not be made to suffer in their persons or property on account of their having taken up arms in obedience to the order of their Government. Those, however, who oppose the forces under my command, or furnish the enemy with supplies or information, will be dealt with according to the customs of war.

Requisitions for food, forage, fuel, or shelter, made on the authority of the Officers in command of Her Majesty's troops, must be at once complied with; but everything will be paid for on the spot, prices being regulated by the local market rates. If the inhabitants of any district refuse to comply with the demands made on them, the supplies will be taken by force, a full receipt being given.

Should any inhabitant of the country consider that he or any member of his household has been unjustly treated by any Officer, soldier, or civilian attached to the British Army, he should submit his complaint, either personally or in writing, to my Headquarters or to the Headquarters of the nearest General Officer. Should the complaint on enquiry, be substantiated, redress will be given.

Orders have been issued by me, prohibiting soldiers from entering private houses, or molesting the civil population on any pretext whatever, and every precaution has been taken against injury to property on the part of any person belonging to or connected with, the Army.

Roberts, Field-Marshal, Commanding in Chief, South Africa. February 1900

[353] All citations from: Robert, Frederick Sleigh Roberts, 1st earl. Proclamations Issued in South Africa. London, 1900. Vol I & II 1901 820 pages. https://babel.hathitrust.org/cgi/pt?id=hvd.32044089026009&seq=15

[354] Proclamation 19 on page 356 https://babel.hathitrust.org/cgi/pt?id=hvd.32044089026009&seq=356&view=1up

Proclamation IXX

Army Head-quarters, South Africa, Pretoria, 16th June 1900. [355]

Whereas small parties of raiders have recently been doing wanton damage to public property in the Orange River Colony and South African Republic by destroying railway bridges and culverts and cutting the telegraph wires, and whereas such damage cannot be done without the knowledge and connivance of the neighbouring inhabitants and the principal civil residents in the districts concerned;

Now, therefore, I, Frederick Sleigh, Baron Roberts, of Kandahar and Waterford, K.P., G.C.B., G.C.S.I., G.C.I.E., V.C., Field-Marshal, Commander-in-Chief of Her Majesty's troops in South Africa, warn the said inhabitants and principal civil residents that, whenever public property is destroyed or injured in the manner specified above, they will he held responsible for aiding and abetting the offenders. The houses in the vicinity of the place where the damage is done will be burnt and the principal civil residents will be made prisoners of war.

Roberts, Field-Marshall, Commander-in-Chief, South Africa.

Proclamation XX

Army Head-quarters, South Africa, Pretoria, 19th June 1900.[356]

Referring to my Proclamation, dated Pretoria, 16th June 1900, I, Frederick Sleigh, Baron Roberts, of Kandahar and Waterford, K.P., G.C.B., G.C.S.I., G.C.I.E., V.C., Field-Marshal, Commander-in-Chief of Her Majesty's troops in South Africa, do hereby declare, proclaim, and make known, that, should any damage be done to any of the lines of railway or to any of the railway bridges, culverts, or buildings, or to any telegraph lines or other railway or public property in the Orange River Colony, or in that portion of the South African Republic for the time being within the sphere of my military operations, the following punishment will be inflicted;

1. The principal residents of the town and districts will be held, jointly and severally, responsible for the amount of damage done in their district.

2. In addition to the payment of the damage above mentioned, a penalty depending upon the circumstances of each case, but which will in no event be less than a sum of 2s. 6d. per morgen[357] on the area of each farm, will be levied and recovered from each burgher of the district in which the damage is done in respect of the land owned or occupied by him in such district. Furthermore, all receipts for goods requisitioned in such district on behalf of the military authorities will be cancelled, and no payment whatsoever will be made in respect of the same.

3. As a further precautionary measure, the Director of Military Railways has been authorized to order that one or more of the residents, who will be selected by him from each district, shall from time to time personally accompany the trains while travelling through their district.[358]

4. The houses and farms in the vicinity of the place where the damage is done will be destroyed, and the residents in the neighbourhood dealt with under Martial Law.

5. The military authorities will render every facility to the principal residents to enable them to communicate the purport of this Proclamation to the other residents in their district, so that all persons may become fully cognizant of the responsibility resting upon them.

Roberts, Field-Marshal, Commanding in Chief in South Africa.

[355] Proclamation 19 on page 364: https://babel.hathitrust.org/cgi/pt?id=hvd.32044089026009&seq=364&view=1up

[356] Proclamation 20 on Page 365. https://babel.hathitrust.org/cgi/pt?id=hvd.32044089026009&seq=365&view=1up

[357] The South African Law Society published a conversion factor of 1 morgen = 0.856532 hectares to be used for the conversion. Morgen - Wikipedia

[358] Just over a month later, on July 27th, this portion of the directive was repealed. See Proclamation XXIII. https://babel.hathitrust.org/cgi/pt?id=hvd.32044089026009&seq=366&view=1up&q1=accompany

Proclamation XXVI

Proclamation[359] [360] 14th day of August, 1900.

To the Inhabitants of the South African Republic

Whereas by Proclamation No. 1 of 1900, burghers who had not taken a prominent part in the hostilities were allowed, upon taking an oath, to return to their homes, and were not dealt with as prisoners of war, and

Whereas by Proclamation No. 2 of 1900 , burghers to whom passes and permits had been granted might retain their stock , or take them to the winter veld, and

Whereas many burghers have taken the said oath, but have , notwithstanding this oath, taken up arms against the forces of Her Majesty the Queen , and

Whereas many burghers who have taken the said oath have aided and abetted the enemy in raiding trains and destroying property belonging to the Forces of Her Majesty the Queen, or have acted as spies for the enemy, and

Whereas the Government of the South African Republic considers such oath immoral, and has issued a notice warning all burghers against taking the said oath, and

Whereas it is manifest that the leniency which has been extended to the burghers of the South African Republic is not appreciated by them, but, on the contrary, is being used as a cloak to continue the resistance against the Forces of Her Majesty the Queen, and Whereas there are no means of distinguishing the combatant from the non-combatant portion of the population:

Now, therefore, I, Frederick Sleigh, Baron Roberts, of Kandahar and Waterford, K.P., G.C.B., G.C.S.I., G.C.I.E., V.C., Field-Marshal, Commander-in-Chief of Her Majesty's troops in South Africa, do hereby proclaim and make known as follows:

1. That from and after this date Sections 1 and 2 of Proclamation No. 1 of 1900 are repealed, except in respect of those burghers who have already taken the said oath.

2. That Proclamation No. 2 of 1900 is repealed.

3. That all such persons who have taken the said oath, and who have in any way broken such oath, will be punished with either death, imprisonment, or fine.

4. That all burghers in the districts occupied by Her Majesty's Forces, except such as have already taken the said oath, shall be regarded as prisoners of war, and shall be transported or otherwise dealt with as I may determine.

5. That all buildings and structures on farms on which the scouts or other forces of the enemy are harboured will be liable to be razed to the ground.

6. That the fine mentioned in Proclamation No. 6 of 1900, Section 2, shall be rigorously exacted where any damage is done to the railway, and persons are hereby warned to acquaint Her Majesty's Forces with the presence of the enemy upon their farms, and, failing to do so they will be regarded as aiding and abetting the enemy.

God Save the Queen.

Given under my hand and seal, at Pretoria, this 14th day of August, 1900.

Roberts, Field-Marshal, Commander in Chief in South Africa

With reference to paragraph 6, the fine therein mentioned shall in no event be less than a sum of .2s .6d per morgen on the area of each farm.

[359] Proclamation 26 on page 368: https://babel.hathitrust.org/cgi/pt?id=hvd.32044089026009&seq=368&view=1up&q1=Proclamation
[360] See Also Proclamation 28 on page 369 regarding the status of forces in the field in the Orange River Colony: https://babel.hathitrust.org/cgi/pt?id=hvd.32044089026009&seq=369&view=1up&q1=Proclamation

Proclamation XXVIIII

Proclamation.

To the Inhabitants of the Orange River Colony. [361]

Whereas by Proclamation a warning was issued on the 1st June 1900, that persons in arms against Her Majesty within the Orange River Colony would be liable to be dealt with as rebels. And whereas it was directed that all arms and ammunition should be surrendered. Now, therefore, I, Frederick Sleigh, Baron Roberts, of Kandahar and Waterford, K.P., G.C.B., G.C.S.I., G.C.I.E., V.C., Field- Marshal, Commander in-Chief of Her Majesty's Forces in South Africa, do hereby proclaim and make known as follows:-

1. That all persons residing in the Orange River Colony, heretofore burghers of the Orange Free State, have, in consequence of the annexation of the territories of that State to the dominions of Her Majesty the Queen, become the subjects of Her Majesty, except those burghers who were attached to some commando prior to the annexation, and who have continuously since then been in arms against Her Majesty.

2. If any burghers of the Orange Free State be found in arms against Her Majesty, the burden of proving that they were attached to some commando prior to the annexation, and that they have continuously since then been in arms against Her Majesty, will rest upon such burghers.

3. Those burghers of the Orange Free State who were attached to some commando prior to the annexation, and who have been continuously in arms against Her Majesty, shall be regarded as prisoners of war, and be transported or otherwise dealt with as I may determine.

4. Those burghers of the Orange Free State who have taken the so-called Oath of Neutrality or submission, and who have in any way broken such oath, shall be punished with either death, imprisonment, or fine.

5. All buildings and structures on farms on which the scouts or other forces of the enemy are harboured will be liable to be razed to the ground.

6. Where any damage is done to the railway, a fine of two shillings and sixpence sterling (2s. 6d.) per morgen will be levied on the area of the farm or farms on which the damage was done, and the farms in the immediate neighbourhood of the damaged place may also, if I deem fit be subjected to a similar fine.

7. All persons are hereby warned to acquaint Her Majesty's Forces of the presence of the enemy upon, or in the neighbourhood of their farms, and if they fail so to do they will be regarded as aiding and abetting the enemy, and will be liable to be treated as rebels. God Save the Queen.

Given under my hand and seal, at the Head- quarters of the Army in South Africa at Belfast, this first day of September, nineteen hundred.

Roberts, Field- Marshal, Commander in Chief in South Africa

Proclamation XXXI

September 14, 1900.

Resignation of Mr. S. J. P. Kruger.[362]

The late President, Mr. Kruger, and Mr. Reitz, with the archives of the South African Republic, have crossed the Portuguese frontier, and arrived at Lourenco Marques with a view to sailing for Europe at an early date.

Mr. Kruger has formally resigned the position he held as President of the· South African Republic, thus severing his official connection with the Transvaal.

Mr. Kruger's action shows how hopeless in his opinion is the war which has now been carried on for nearly a year, and his desertion of the Boer cause should make it clear to his fellow burghers that it is useless for them to continue the struggle any longer.

[361] Proclamation 28 on page 369 https://babel.hathitrust.org/cgi/pt?id=hvd.32044089026009&seq=369&view=1up

[362] Proclamation 31 on Page 371 https://babel.hathitrust.org/cgi/pt?id=hvd.32044089026009&seq=371&view=1up&q1=Proclamation

It is probably unknown to the inhabitants of the Transvaal and Orange River Colony that nearly 15,000 of their fellow-subjects are now prisoners of war, not one of whom will be released until those now in arms against us surrender unconditionally.

The burghers: must by this time be cognisant of the fact that no intervention on their behalf can come from any of the Great Powers, and, further, that the British Empire is determined to complete the work which has already cost her so many valuable lives, and to carry to its conclusion the war declared against her by the late Governments of the South African Republic and Orange Free State, a war to which there can be but one ending.

If any further doubts remain in the minds of the burghers as to Her Britannic Majesty's intentions, they should be dispelled by the permanent manner in which the country is gradually being occupied by Her Majesty's Forces, and by the issue of the Proclamations signed by me on the 24th May and 1st September 1900, annexing the Orange Free State and the South African Republic respectively, in the name of Her Majesty.

I take this opportunity of pointing out that, except in the small area occupied-by the Boer army under the personal command of Commandant General Botha, the war is degenerating, and has degenerated, into operations carried on in an irregular and irresponsible manner by small, and in very many cases insignificant bodies of men. I should be failing in my duty to Her Majesty's Government and to Her Majesty's Army in South Africa if I neglected to use every means in my power to bring such irregular warfare to an early conclusion.

The means which I am compelled to adopt are those which the customs of war prescribe as being applicable to such cases. They are ruinous to the country, and entail endless suffering on the burghers and their families, and the longer this guerilla warfare continues the more vigorously must they be enforced.

Roberts, Field-Marshal, Commanding in Chief in South Africa. Pretoria, September 14, 1900.

Proclamation XXXV

Government Notice No. 115 OF 1900.[363] [28th September 1900]

Uniformity of Treatment.

As it seems to me advisable to insure uniformity in the treatment of the inhabitants of the various districts of the Transvaal and Orange River Colony, the following instructions are hereby published to assist General Officers Commanding in carrying out the conditions of recent Proclamations issued by me:

1. By Military Secretary's telegram, No. C, 4825, dated 20th September, General Officers Commanding have been empowered to promise that Burghers who surrender voluntarily will not be sent out of South Africa, provided they have been guilty of no acts, other than fighting against us which should debar them from this privilege. This concession does not apply to: those who have taken the prominent military or political part in the war, nor to those who have broken their oath of neutrality, nor to foreigners. In the event of a military or political leader enquiring as to terms of surrender, the question is to be referred direct to Army Head–quarters.

2. All stock, supplies, &c., of those on commando, or of those who have broken their oath, are to be taken and no receipt given.

3. In cases where some members of a family, who all live on one farm, have broken their oath and gone on commando, those remaining are to be warned that, unless the former surrender within a reasonable period, all stock, supplies, &c., will be taken and no receipt given.

4. In cases of sniping, leaders of bands are to be informed that, unless it ceases, their (the leaders') houses will be burned, A few days should be allowed for this to become known, notices being sent to the resorts of the leaders and to their wives. In the event of its being necessary to burn the farms, further notice should be issued stating why this has been done, and giving a list of other houses that will next be burned.

5. All stock, supplies, &c., of snipers are to be taken and no receipt given.

[363] Proclamations in SA, Page 373–4: https://babel.hathitrust.org/cgi/pt?id=hvd.32044089026009&seq=373&view=1up&q1=customs

6. Protection is to be given to all inhabitants who have kept the oath of neutrality, remaining quietly at home and taking no further part in the war. Any stock or supplies taken from them are to be paid for or receipts given in full for them.

7. Wives and families of men who are prisoners of war are to be protected, and receipts in full given for anything taken When a man surrenders a protection pass is to be given to his wife.

8. Widows of men killed in the war and all lone women are to receive protection passes, and to be paid or given receipts for every thing taken.

9. Burghers are to be informed that as soon as their leaders submit, and when every cannon has been surrendered, peace will be declared, and all prisoners of war will then be sent back to their homes. Exceptions only will be made in the cases of members of the late Governments of the South African Republic and Orange Free State, who are responsible for the war and its present disastrous prolongation, and of those who may be proved to have been guilty of acts contrary to the customs of war.

Roberts, Field-Marshal, Commanding in Chief in South Africa.

Army Head-quarters, South Africa, 28th September 1900

Proclamation XL

Order issued by: Field-Marshal Lord Roberts, dated 18th :November. [1900][364]

As there appears to be some misunderstanding with reference to burning of farms and breaking of dams, Commander-in-Chief wishes following to be lines on which General Officers Commanding are to act:- No farm is to be burnt except for act of treachery, or when troops have been fired on from premises, or as punishment for breaking of telegraph or railway line, or when they have been used as bases of operation for raids, and then only with direct of General Officer Commanding, which is to be given in writing, the mere fact of a burgher being absent on commando is on no account to be used as a reason for burning the house. All cattle, wagons, and foodstuffs are to be removed from all farms; if that is found to be impossible, they are to be destroyed, whether owner be present or not.

[364] Proclamation 40 on Page 377. https://babel.hathitrust.org/cgi/pt?id=hvd.32044089026009&seq=377&view=1up

Bibliography

Many works have been footnoted in the text, enthusiastically if also somewhat idiosyncratically as to form. Restating those references again here would extend this work unnecessarily.

The full and equally idiomatic bibliography for this series is over sixty pages long and will be included in an Appendix volume to be published at the end of the series. For this volume an abbreviated bibliography is as follows. Note that works are cited by topic rather than alphabetically, i.e. works relating to Christiaan are grouped together to be close to their subject matter.

Three Years' War

Three Years' War By: Christiaan Rudolf De Wet. Charles Scribner's Sons, (1902) Trow Directory Printing and Bookbinding Company New York.

Internet Archive Catalogue Record:
https://archive.org/search?query=creator%3A%28Christiaan%20Rudolf%20De%20Wet%29

Harvard Copy: https://archive.org/details/threeyearswar00wetgoog/mode/2up

Princeton Copy: https://archive.org/details/threeyearswar00dewe/mode/2u

Anglo Boer War Copy: https://www.angloboerwar.com/books/74-de-wet-three-years-war

Project Gutenberg Copy: https://gutenberg.org/ebooks/18794 Release Date: July 8, 2006 [EBook #18794] E-text prepared by Marilynda Fraser-Cunliffe, Taavi Kalju, and the Project Gutenberg Online Distributed Proofreading Team. (http://www.pgdp.net/)

Copyright, 1902, By Charles Scribner's Sons. All Rights Reserved. Published, December, 1902

On the Heels of De Wet

On the Heels of De Wet By The Intelligence Officer. Possibly Lionel James, is the author. W. Blackwood and Sons, London, 1902. First published as a series of papers in Blackwood's magazine, 1902. Catalogue Search:

https://archive.org/search?query=title%3A%28on+the+heels%29+AND+date%3A%5B1898-01-01+TO+1910-01-01%5D

U. Cal Copy: https://archive.org/details/onheelsdewet01unkngoog

Americana Collection: https://archive.org/details/onheelsdewet00unkngoog

Project Gutenberg Edition: On the Heels of De Wet by Lionel James - Free Ebook (gutenberg.org)

https://gutenberg.org/ebooks/20400

With Steyn & De Wet.

With Steyn and De Wet by Pienaar, Philip Methuen & Co, London, 1902
https://archive.org/details/withsteynanddew00piengoog

The capture of De Wet

Phillip J. Sampson, The capture of De Wet, the South African rebellion, 1914, London, E. Arnold Catalog Search: https://archive.org/search?query=title%3A%28The%20capture%20of%20De%20Wet%29

U.Cal copy 1: https://archive.org/details/capturedewetsou00sampgoog/mode/2up

U.Cal copy 2: https://archive.org/details/captureofdewetso00sampiala

Wepener Account and Medal Roll by David and Alexander Biggins. Anglo Boer War - Wepener Account and Medal Roll Published by Token Publishing: Coin, Banknote and Medal Collector's Magazines. Token Publishing–Numismatic Interest

The Times' History of the War in South Africa 1899–1902 by Amery, L. S. (Leopold Stennett), 1873–1955; Childers, Erskine, 1870–1922; Tallboy, G. P; Williams, Basil, 1867-1950 Sampson Low. Marston and Company, Ltd, 1900–1902

Catalog Search: Internet Archive: Digital Library of Free & Borrowable Texts, Movies, Music & Wayback Machine

University of Toronto Copies:

Times History Vol 1 (1901) https://archive.org/details/timeshistoryofwa01ameruoft History up until October 19th, 1899

Times History Vol 2 (1902) https://archive.org/details/timeshistoryofwa02amer Up until Colenso, December 15th 1899

Times History Vol 3 (1905) https://archive.org/details/timeshistoryofwa03amer December 1899 to March 1900

Times History Vol 4 (1906) https://archive.org/details/timeshistoryofwa04amer To Relief of Kimberley and Mafeking May 17th 1900

Times History Vol 5 (1907) https://archive.org/details/timeshistoryofwa05amer Concludes with the peace treaty in June 1902

Times History Vol 6 (1909) https://archive.org/details/timeshistoryofwa06ameruoft Up until 1909

Times History Vol 7 (1909) https://archive.org/details/timeshistoryofwa07ameruoft Appendices, Chronology and a thirty page bibliography followed by an Index (226 pages)

Jacoba de la Rey

Jacoba de la Rey Myne (Spouse of Jacobis 'koos')

Title: Omzwervingen en Beproevingen Gedurende den Oorlog (1903), which was translated into English as "A Woman's Wanderings and Trials During the Anglo-Boer War" translated by Lucy Hotz, and published in London by T. Fisher Unwin.

University of Pretoria, Library Services: africanaspecialcollectionsduplessis;universityofpretorialibraryservices; africanlibraries Location: https://archive.org/details/womanswanderings00reyj/page/n9/mode/2up

Our Military Officers. Album of portraits with life sketches.

Original Publication: Onze krijgs-officieren: album van portretten met levens-schetsen der. by Gustav Schoeman Preller, Frans Vredenrijk Engelenburg 1904 Publisher "Volksstem" kantore

Catalogue Record: https://archive.org/search?query=title%3A%28Onze%20Krijgs-Officieren%29

Direct Link: https://archive.org/details/onzekrijgsoffic00engegoog

Reproduction: Our Military Officers. Album of portraits with life sketches of the Transvaal Generals and Commandants), Volksstem, Pretoria 1904. In Dutch with a preface by Louis Botha. Transcribed and published by: Mr. Pieter van der Merwe.

Onze Krijgs-Officieren Album Van Portretten Met Levens-Schetsen Der Transvaalse Generaals En Kommandanten E. A. Uitgegeven Ten "Volksstem" Kantore, Pretoria. 1904.

File:Onze Krijgs-Officieren. Album van portretten met levens-schetsen, Pretoria, Volksstem, 1904.pdf

Download Access: https://en.wikipedia.org/wiki/File:Onze_Krijgs-Officieren._Album_van_portretten_met_levens-schetsen,_Pretoria,_Volksstem,_1904.pdf

Direct Link to File: https://upload.wikimedia.org/wikipedia/commons/5/52/Onze_Krijgs-Officieren._Album_van_portretten_met_levens-schetsen%2C_Pretoria%2C_Volksstem%2C_1904.pdf

Source Page: https://en.wikipedia.org/wiki/Annie_Botha#cite_note-4

License Links

My many thanks to all the authors, libraries, conservators, and scanners/uploaders who have allowed me to populate, illustrate and annotate this compilation. Responsible efforts have been made to ensure that material in this work public domain material and otherwise is governed by the following licenses.

Creative Commons 1.0: CC0 1.0 Deed | CC0 1.0 Universal | Creative Commons
Creative Commons 1.0: Creative Commons—Attribution 1.0 Generic—CC BY 1.0
Creative Commons 2.0: Creative Commons—Attribution 2.0 Generic—CC BY 2.0
Creative Commons 3.0: Creative Commons—Attribution 3.0 Unported—CC BY 3.0
Creative Commons 4.0: Creative Commons—Attribution 4.0 International—CC BY 4.0
And: CC BY-SA 4.0 Deed | Attribution-ShareAlike 4.0 International | Creative Commons
See Generally: Creative Commons license,[365] Please note that all images have been modified in one way or another during the course of publication.

Also included is material authorized under one or more of the following licenses:

UK Open Government License: This work contains public sector information licensed under the Open Government Licence v3.0. https://www.nationalarchives.gov.uk/doc/open-government-licence/version/3/

UK Open Parliament License: This work contains public sector information licensed under the Open Government Licence v3.0. https://www.parliament.uk/site-information/copyright-parliament/open-parliament-licence/

Some documents are reproduced pursuant to the UK Open Parliament License and this applies not only material in which either House owns the copyright but also material published before 1 August 1989 which is the operative Commencement date of the Copyright and Designs act of 1988; and 50 years having elapsed subsequent to such publication. The same terms applying to previously published material that at one time subsisted under Crown Copyright.[366]

Also, This work contains public sector information made available under the South African Promotion of Access to Information Act, 2000 ("PAIA"). https://www.gov.za/documents/promotion-access-information-act

Google Earth License Links

Google Earth License: https://www.google.com/intl/en-GB_ALL/permissions/geoguidelines/
Sales of these annotated, illustrated and supplemented historical narratives are anticipated to be less than the 5,000 copies permitted. If that is not the case the author will contact the Google Cloud Customer Team for additional permission(s). https://cloud.google.com/contact/

Copyright Discussion

This work is a comprehensive republication of the included works which have been compiled, annotated, illustrated and expanded in order to provide historical context and clarity.

Images and text used by permission or by law are clearly identified by citation and should not be reproduced; as to the remainder there is no copyright claimed by myself on any public domain images, text or any common property information contained in this work.

Future creators should continue to follow the permissions, guidelines and requirements from the original source and author/uploader who is cited. Links to Creative Commons, Google Earth and other licenses are noted in the license links section.

So far as can be determined, as of 2025, the text and images used to illustrate the first-person accounts included in this volume reside in the public domain in the United States having all been first published prior to January 1st, 1930 as well as in all other jurisdictions which employ any of the following copyright regimens:

1. A copyright regimen that grants an author's works subsisting copyright for seventy (70) years pma[367] or less.

[365] Creative Commons License: Creative Commons license - Wikipedia And About CC Licenses - Creative Commons
[366] If the work was commercially published within 75 years of the end of the calendar year in which it was made, Crown copyright will last for 50 years from the end of the calendar year in which it was published.
https://copyrightservice.co.uk/copyright/p01_uk_copyright_law Flow Chart: Crown copyright flowchart 2015
[367] post mortem auctoris or "pma" [post mortem author]

2. A copyright term subsisting for 120 years from the date of creation or first publication.

3. And such jurisdictions that apply the Rule of the Shorter term to an author's pma date or publication date.

This work will not be marketed or deliberately sold in any jurisdiction with a longer copyright term. See Also: 2025 in public domain - Wikipedia

Copyright for the assembled compilation of this material, or substantial portions thereof that would comprise separate articles or works unto themselves is retained. This work and the series in sum comprise a derivative work. Literary, dramatic, and adaptation rights of the saga demonstrated and explicated by this series is retained.

Otherwise, with respect to the underlying texts and images recast in a more permanently accessible form I choose to pay forward my creative energy to the people who lived through these experiences and it has been a pleasure to find and assemble their collective words and images in a new synoptic and synergistic arrangement.

Except where noted otherwise, every effort has been made to ensure that no material has been used without an attributed source. However, if any non-attributed source, incorrect source, or copyrighted material used without permission or justification, has been included then I would be grateful for information that would enable any errors or inappropriate inclusions to be corrected or omitted in subsequent impressions.

Starting from creation, to publication, through to purchase, archiving, and digitizing, this annotated, illustrated and expanded edition is the sixth step in the journey of these works. My thanks go to all who have helped preserve and perpetuate these important historical records.

Endnotes

[1] Christiaan de Wet Vanity Fair July 31st, 1902.
https://commons.wikimedia.org/wiki/File:Christiaan_de_Wet_Vanity_Fair_1902-07-31.jpg

[2] Description: South Africa. General Christian Rudolf de Wet (1854-1922) In South Africa the Boer general during the Second War of Independence. Place unknown. 1903.
https://commons.wikimedia.org/wiki/File:Zuid_Afrika,_SFA006012030.jpg

[3] https://en.wikipedia.org/wiki/Christiaan_de_Wet And Permanent Link, 5/26/2025:
https://en.wikipedia.org/w/index.php?title=Christiaan_de_Wet&oldid=1285150617

[4] SA War Vol 6. And File:Christiaan de Wet4.jpg - Wikimedia Commons Author: Reinhold Thiele (1856–1921) The photo was taken in Potchefstroom on the 27th of August 1900, six days after he and his commando of 246 men escaped from near encirclement north of the Magaliesberg File:Christiaan de Wet4.jpg - Wikimedia Commons

[5] General Christiaan de Wet and staff. - Atom site for DRISA P2921_01.jpg (1400×1131) (drisa.co.za)

[6] Christiaan de Wet Vanity Fair July 31st, 1902.
https://commons.wikimedia.org/wiki/File:Christiaan_de_Wet_Vanity_Fair_1902-07-31.jpg

[7] Description: South Africa. General Christian Rudolf de Wet (1854-1922) In South Africa the Boer general during the Second War of Independence. Place unknown. 1903.
https://commons.wikimedia.org/wiki/File:Zuid_Afrika,_SFA006012030.jpg

[8] Boer general Christiaan de Wet, photographed at the end of the Second Boer War in Bloemfontein on 6.6.1902, before embarking for Europe. Source: S. King File:Generaal Christiaan de Wet - Bloemfontein 6.6.1902.jpg - Wikimedia Commons

[9] Christiaan Rudolf de Wet, De strijd tusschen Boer en Brit. című könyvének borítója, Date: 1902
https://commons.wikimedia.org/wiki/File:De_Wet_De_strijd_tusschen_Boer_en_Brit.gif Author: Catharina Johanna Höveker (1842–1923) & Johan Adam Wormser (1845–1916) Amsterdam / Pretoria 1902

[10] Digitized by Google and by the Internet Archive in 2015. https://archive.org/details/threeyearswarOOdewe

[11] General Christiaan de Wet and staff. - Atom site for DRISA P2921_01.jpg (1400×1131) (drisa.co.za)

[12] Doyle, Arthur Conan. The Great Boer War 17th Edition, 1902

[13] Doyle, Arthur Conan. The Great Boer War 17th Edition, 1902.

[14] Creswicke SA War Vol 5 Map page VIII.

[15] Howard Clement Hillegas, The Boers in War. Published by, Appleton and company, New York 1900. Pages 320 to 322 of the US Edition: https://archive.org/details/boersinwarstoryo00howarich Hathi Trust: https://catalog.hathitrust.org/Record/009262432

[16] Davitt, Boer Fight for Freedom, 3rd ed. 1902. (Duquesene Scan) Michael Davitt The Boer Fight For Freedom, Funk & Wagnalls Company 1902 Duquesene Scan https://archive.org/details/boerfightforfree00mich/page/n623/mode/2up

[17] Times History Vol 4.
[18] Times History Vol 4
[19] Times History Vol 5.
[20] Times History Vol 5.
[21] Times History Vol 5.
[22] Times History Vol 5.
[23] Library of Congress Maps and Charts # 2877 Copyright 1899 by Geo. H. Walker & Co. Boston. 12.5 inches by 5.25 Inches. Search Bing for Old Maps of South Africa. Link(s) to George H. Walker Map of SA. Link to Yellow Map. But it is a slightly sharper scan. Take out 70% of colour and up contrast ~70%. https://relicmaps.com/products/south-africa-catoosa-county-georgia-united-states-walker-county-whitfield-county-1899-poster-print-gift-33342?variant=47622134497587&msclkid=6d07220b96831debcf184a3085bf2c4d&utm_source=bing&utm_medium=cpc&utm_campaign=Shopping_Search_051025_142025&utm_term=4586337883546955&utm_content=Ad%20group%20%231
[24] Times History Vol 5.
[25] The Times' History of the War in South Africa 1899–1902 by Amery, L. S. (Leopold Stennett), 1873–1955; Childers, Erskine, 1870–1922; Tallboy, G. P; Williams, Basil, 1867-1950 Sampson Low. Marston and Company, Ltd, 1900–1902 Catalog Search: Internet Archive: Digital Library of Free & Borrowable Texts, Movies, Music & Wayback Machine University of Toronto Copy: Times History Vol 5 (1907) https://archive.org/details/timeshistoryofwa05amer
[26] With the flag to Pretoria Vol 2
[27] Davitt, Boer Fight for Freedom, 3rd ed. 1902. (Duke Scan)
[28] File:GiantsCastlePanoramaSmall.jpg - Wikimedia Commons
[29] Hillegas, Boers in War, 1900.
[30] Davitt, Boer Fight for Freedom, 3rd ed. 1902. (Duke Scan)
[31] Black & White Weekly Vol 1-2 1899-1900
[32] Davitt, Boer Fight for Freedom, 3rd ed. 1902. (Duke Scan)
[33] Black & White Weekly Vol 1-2 1899-1900
[34] BritishBattles.com Map by John Fawkes
[35] War with the Boers, H. Brown, London [1902?] Vol 2
[36] Times History Vol 2
[37] Anglo-Boer War Album, Cape Town, D. Edwards & co. [1902?]
[38] Times History Vol 2
[39] Hillegas, Boers in War, 1900.
[40] Cunliffe, History of the Boer War Vol 1 (NY Library Scan)
[41] Viljoen Ben, My Reminiscences, 1902. Maps from Drawings by P. Van Breda.
[42] BritishBattles.com Map by John Fawkes.
[43] Image source: With the Flag to Pretoria, H. W. Wilson Vol 1 1900 and britishbattles.com
[44] Courtesy of Neville Constantine: Anglo Boer War - Artillery and Ammunition - Page 25 - Boer War Forum
[45] Times History Vol 2.
[46] With The Flag to Pretoria Vol 1 Preface page vi
[47] Image Source: BritishBattles.com
[48] Creswicke SA War Vol 2
[49] File:Positions round Ladysmith - November 1899 - Project Gutenberg eText 16466.jpg - Wikimedia Commons
[50] With the Flag to Pretoria, H. W. Wilson Vol 2 Dec., 1899. PF: The end of the barrel was damaged and it was cut down and returned to action.
[51] File:Photograph; 'Long Tom', RAMC; 801-22-3 Wellcome L0021295.jpg - Wikimedia Commons
[52] Heroes of the Boer War. Review of Reviews, 1903.
[53] Hillegas, With the Boer forces, 1901 (Photograph by the Author.)
[54] Confirmed his identity from these articles. 06 Dec 1899 - THE COMMANDER IN THE O.F. STATE. - Trove (nla.gov.au) And 07 Apr 1900 - Commandant Grobler. - Trove (nla.gov.au))
[55] Davitt, Boer Fight for Freedom, 3rd ed. 1902. (Duke Scan)
[56] File:The Second Boer War, 1899-1902 Q72020.jpg - Wikimedia Commons Author: William Skeoch Cumming
[57] File:The Second Boer War, 1899-1902 Q72019.jpg - Wikimedia Commons
[58] Davitt, Boer Fight for Freedom, 1902 (U. Cal. Scan)

[59] Doyle, Arthur Conan. The Great Boer War 17th Edition, 1902
[60] Anglo-Boer War Album, Cape Town, D. Edwards & co. [1902?]
[61] In the Orange River trenches holding back the Boers. Digital ID: (b&w film copy neg.) cph 3c02311 https://www.loc.gov/pictures/item/91705268/resource/
[62] Cunliffe, History of the Boer War Vol 1 (NY Library Scan)
[63] War pictures. v. 1 (no. 1-16) 1900 #1 - War pictures. v. 1 (no. 1-16) 1900. - Full View | HathiTrust Digital Library And: Catalog Record: War pictures | HathiTrust Digital Library
[64] Anglo-Boer War Album, Cape Town, D. Edwards & co. [1902?]
[65] Anglo-Boer War Album, Cape Town, D. Edwards & co. [1902?]
[66] Anglo-Boer War Album, Cape Town, D. Edwards & co. [1902?]
[67] Doyle, Arthur Conan. The Great Boer War 17th Edition, 1902
[68] A Handbook of the Boer War with General Map of South Africa and 18 Sketch Maps and Plans Gale And Polden Limited London 1910 A Handbook of the Boer War - Free Ebook (gutenberg.org) Original by Stanford's Geography
[69] Times History Vol 3
[70] Cunliffe, History of the Boer War Vol 2 (Composite of Mich. St. & U. Cal. Scans)
[71] A Handbook of the Boer War - Gale And Polden Limited, London 1910 Source: Anonymous ISBN 978-1-37-497455-5. https://babel.hathitrust.org/cgi/pt?id=uc1.l0061277489&seq=7
[72] Cunliffe, History of the Boer War Vol 2 (Composite of Mich. St. & U. Cal. Scans)
[73] Davitt, Boer Fight for Freedom, 1902 (U. Cal. Scan)
[74] Times History Vol 3
[75] Times History Vol 3
[76] Cunliffe, History of the Boer War Vol 2 (Composite of Mich. St. & U. Cal. Scans)
[77] Cunliffe, History of the Boer War Vol 2 (Composite of Mich. St. & U. Cal. Scans)
[78] Drakensberg. Ox wagon with 16 oxen crossing a drift near Mont-Aux-Sources. - Atom site for DRISA N08623.jpg (1400×1054) (drisa.co.za)
[79] Ox wagon stuck in Crocodile River - Atom site for DRISA N00061.jpg (1400×925) (drisa.co.za)
[80] Warwick Trading Company catalogue number 5525. Warwick Trading Company - Wikipedia and British Pathe.
[81] Warwick Trading Company catalogue number 5525. Warwick Trading Company - Wikipedia and British Pathe.
[82]October 21 Folder Anglo Boer War - Small Arms and Ammunition - Page 4 - Boer War Forum
[83] Per Neville Constantine: Anglo Boer War - Artillery and Ammunition - Page 10 - Boer War Forum
October 21 Folder
[84] File:Philip Rudolph Botha (1851-1901), Boer War general.jpg - Wikimedia Commons
[85] Reports on military operations in South Africa and China. U.S. Adjutant-General's Office. 1901
[86] A Handbook of the Boer War - Gale And Polden Limited, London 1910 Source: Anonymous ISBN 978-1-37-497455-5. https://babel.hathitrust.org/cgi/pt?id=uc1.l0061277489&seq=7
[87] Times History War in S.A Vol 3
[88] Hillegas, Boers in War, 1900.
[89] Cunliffe, History of the Boer War Vol 2 (Composite of Mich. St. & U. Cal. Scans)
[90] Daniel (Danie) Johannes Stephanus Theron File:Danie Theron.jpg - Wikipedia
[91] Heroes of the Boer War. Review of Reviews, 1903.
[92] Appendix to Col. Otter's Report Canadian War Museum: https://www.warmuseum.ca/cwm/exhibitions/boer/map-paardeberg3_lrg_e.html
[93] Times History War in S.A Vol 3 Photo by Mr. Perceval Landon.
[94] File:The Second Boer War, 1899-1902 Q71654.jpg - Wikimedia Commons
[95] File:The Second Boer War, 1899-1902 Q72252.jpg - Wikimedia Commons Author: William Skeoch Cumming
[96] File:The Second Boer War, 1899-1902 Q72174.jpg - Wikimedia Commons
[97] Anglo-Boer War Album, Cape Town, D. Edwards & co. [1902?]
[98] Anglo-Boer War Album, Cape Town, D. Edwards & co. [1902?]
[99] Anglo-Boer War Album, Cape Town, D. Edwards & co. [1902?]
[100] File:The Second Boer War, 1899-1902 Q71944.jpg - Wikimedia Commons Author: William Skeoch Cumming
[101] File:Cronje's Chief Commandants after the surrender, Feb. 27th, at Paaderberg, S. Africa, RP-F-F09176.jpg - Wikimedia Commons
[102] With the Flag to Pretoria, H. W. Wilson Vol 2 Feb. 1900

[103] Black & White Weekly Vol 1-2 1899-1900

[104] After Pretoria: The Guerilla War Vol 3 Feb., 1901

[105] Schomburg Center for Research in Black Culture, Photographs and Prints Division, The New York Public Library. "The Boer prisoners (Cronje's men) resting on the road from Paardeberg to Modder River, S. Africa." The New York Public Library Digital Collections. 1900. https://digitalcollections.nypl.org/items/510d47df-fcb9-a3d9-e040-e00a18064a99

[106] Doyle, Arthur Conan. The Great Boer War: Seventh Impression 1900 (September) London, Smith, Elder & Co. Casualties from Outbreak of War until Saturday, September 8th, 1900 Pages 535 & 536

[107] Image Source: Exiles | Saint Helena Island Info: All about St Helena, in the South Atlantic Ocean

[108] Doyle, Arthur Conan. The Great Boer War 17th Edition, 1902

[109] A Handbook of the Boer War - Gale And Polden Limited, London 1910 Source: Anonymous ISBN 978-1-37-497455-5. https://babel.hathitrust.org/cgi/pt?id=uc1.l0061277489&seq=7

[110] Reports on military operations in South Africa and China. U.S. Adjutant-General's Office. 1901

[111] Reports on military operations in South Africa and China. U.S. Adjutant-General's Office. 1901

[112] Times History Vol 3

[113] Reports on military operations in South Africa and China. U.S. Adjutant-General's Office. 1901

[114] Cunliffe, History of the Boer War Vol 2 (Composite of Mich. St. & U. Cal. Scans)

[115] Times History Vol 3.

[116] Elliott & Fry Portrait: File:Robert-Gascoyne-Cecil-3rd-Marquess-of-Salisbury (cropped).jpg - Wikimedia Commons

[117] File:Boer Legation VA0949.jpg - Wikimedia Commons Author Atelier Straugh, formerly Kameke, The Netherlands

[118] Cunliffe, History of the Boer War Vol 2 (Composite of Mich. St. & U. Cal. Scans)

[119] Cunliffe, History of the Boer War Vol 2 (Composite of Mich. St. & U. Cal. Scans)

[120] With the Flag to Pretoria, H. W. Wilson Vol 2 Dec., 1899.

[121] Anglo-Boer War Album, Cape Town, D. Edwards & co. [1902] Bloemfontein, the Post Office is recognizable in the background.

[122] Anglo-Boer War Album, Cape Town, D. Edwards & co. [1902]

[123] Hillegas, With the Boer forces, 1901 (Photograph by the Author.)

[124] Davitt, Boer Fight for Freedom, 3rd ed. 1902. (Duke Scan)

[125] Hillegas, Boers in War, 1900.

[126] A Handbook of the Boer War - Gale and Polden Limited, London 1910 Source: Anonymous ISBN 978-1-37-497455-5. https://babel.hathitrust.org/cgi/pt?id=uc1.l0061277489&seq=7

[127] Reports on military operations in South Africa and China. U.S. Adjutant-General's Office. 1901

[128] Hillegas, With the Boer forces, 1901 (Drawn by the Author under supervision of General)

[129] Times History Vol 4.

[130] Times History Vol 4.

[131] Hillegas, With the Boer forces, 1901 (Photograph by the Author.)

[132] File:State artillery corps, Orange Free State - Korps van die Staatsartillerie, Oranje-Vrystaat, near Colesberg.jpg - Wikimedia Commons

[133] Andries Petrus Johannes Cronjé (1849-1923) File:Andries Petrus Johannes Cronjé (1849-1923) with ammo.jpg - Wikimedia Commons

[134] Hillegas, With the Boer forces, 1901 (Photograph by the Author.)

[135] Doyle, Arthur Conan. The Great Boer War 17th Edition, 1902

[136] Title: A Case of Root Hog or Die. Boer General Christiaan De Wet prods the boar 'The Cape Boer' with a bayonet. De Wet was attacking the Cape Colony of British South Africa in hopes of provoking a general uprising of the Boers resident there. Collection: Hennepin County Library wikidata:Q530512. By: Bartholomew, Charles Lewis 'Bart', 1869-1949. https://commons.wikimedia.org/wiki/File:A_Case_Of_Root_Hog_Or_Die_-_DPLA_-_331d8036ac6d6a8473f065a07090164b.jpg

[137] Picture: https://www.angloboerwar.com/forum/11-research/31054-a-hand-written-oath-of-neutrality#gallery-2 Topic: https://www.angloboerwar.com/forum/11-research/31054-a-hand-written-oath-of-neutrality Credit: Rory Location: de Wet\Three Years' War

[138] Michael Davitt The Boer Fight for Freedom, 3rd ed., Funk & Wagnalls Company 1902. P.580-581.

[139] The Oranje Vrijstaat Artillerie Corps made use of the "export model" of the German horse artillery's 8-cm leichtes Feldgedchütz Material C/73. Officially it was denoted as: 7.5cm Krupp: Feldkanone L/27, but during the Boer War it became known as the "Free State Krupp". Courtesy of Neville Constantine: Anglo Boer War - Artillery and Ammunition - Page 26 - Boer War Forum

[140] Courtesy of Neville Constantine: Anglo Boer War - Artillery and Ammunition - Page 13 - Boer War Forum
[141] Courtesy of Neville Constantine: Anglo Boer War - Artillery and Ammunition - Page 17 - Boer War Forum
[142] Courtesy of Neville Constantine: Anglo Boer War - Artillery and Ammunition - Page 17 - Boer War Forum
[143] Times History Vol 4.
[144] War pictures. v. 1 (no. 1-16) 1900 #1 - War pictures. v. 1 (no. 1-16) 1900. - Full View | HathiTrust Digital Library And: Catalog Record: War pictures | HathiTrust Digital Library
[145] Cunliffe, History of the Boer War Vol 1 (NY Library Scan)
[146] Hillegas, With the Boer forces, 1901 (Photograph by the Author)
[147] Heroes of the Boer War. Review of Reviews, 1903.
[148] Times History Vol 4.
[149] With the Flag to Pretoria. Vol 2.
[150] Sand River (Free State) - Wikipedia
[151] Ian Hamilton's March 1900 The Copp, Clark Co. Toronto. And Frontiers & Wars.
[152] Times History Vol 4
[153] Richard Harding Davis With Both Armies in South Africa. Charles Scribner's Sons, New York 1900.
[154] Anglo Boer War - Small Arms and Ammunition - Page 14 - Boer War Forum Credit: Rob Davidson
[155] File:The Second Boer War, 1899-1902 Q68470.jpg - Wikimedia Commons Author Unknown author
[156] Bridge across the Valsch River at Kroonstadt, blown up by the Boers. File:Gezicht op een opgeblazen brug over de Valsrivier bij Kroonstad, Zuid-Afrika Railway Bridge across the Valsch River at Kroonstadt, blown up by the Boers, S.A. (titel op object), RP-F-F09001.jpg - Wikimedia Commons
[157] Convoy of the 11th Division "outspanned" on the march to Kroonstadt. File:Gezicht op transportkarren onderweg naar Kroonstad, Zuid-Afrika Convoy of the 11th Division outspanned on the march to Kroonstadt, Boer War, South Africa (titel op object), RP-F-F08997.jpg - Wikimedia Commons
[158] File:The Boer War 1899-1902 Q71943.jpg - Wikimedia Commons Author: William Skeoch Cumming
[159] File:Gezicht op een rivieroever met paardenkarren bij Kroonstad, Zuid-Afrika History Recorders and History Makers, Correspondents' Carts waiting Lord Roberts' entry to Kroonstadt, S.A., crop.jpg - Wikimedia Commons
[160] Anglo-Boer War Album, Cape Town, D. Edwards & co. [1902?]
[161] Anglo-Boer War Album, Cape Town, D. Edwards & co. [1902?]
[162] File:The Second Boer War, 1899-1902 Q68477.jpg - Wikimedia Commons
[163] File:The Second Boer War, 1899-1902 Q68484.jpg - Wikimedia Commons
[164] Anglo-Boer War Album, Cape Town, D. Edwards & co. [1902?]
[165] Times History War in S.A Vol 4 (1906)
[166] Davitt, Boer Fight for Freedom, 3rd ed. 1902. (Duke Scan)
[167] Heroes of the Boer War. Review of Reviews, 1903.
[168] Webster Davis, John Bull's crime; or, Assaults on Republics. 1901 New York The Abbey Press
[169] Davitt, Boer Fight for Freedom, 1902 (U. Cal. Scan)
[170] The Biograph in battle: its story in the South African War related with personal experiences. by William Kennedy Laurie Dickson, London: T. Fisher Unwin, 1901. Catalog Search: https://archive.org/search?query=title%3A%28The+Biograph+in+Battle%29
[171] Times History Vol 4.
[172] Times History, Vol 4.
[173] After Pretoria: The Guerilla War Vol 3 April, 1901
[174] Anglo-Boer War Album, Cape Town, D. Edwards & co. [1902?]
[175] File:The Second Boer War, 1899-1902 Q71949.jpg - Wikimedia Commons Author: William Skeoch Cumming
[176] File:The Second Boer War, 1899-1902 Q72266.jpg - Wikimedia Commons Public Domain
[177] The_Second_Boer_War,_1899-1902_Q71963.jpg (800×525) (wikimedia.org) Author: William Skeoch Cumming
[178] File:The Second Boer War, 1899-1902 Q71965.jpg - Wikimedia Commons Author: William Skeoch Cumming
[179] Richard Harding Davis With Both Armies in South Africa. Charles Scribner's Sons, New York 1900
[180] British troops guarding the Rhenoster River bridge whilst repair work is being done after Boers dynamited it ABW British troops guarding the Rhenoster River bridge whilst repair work is being done after Boers dynamited it ....ABW | Troops, War, Armed conflict (pinterest.com)
[181] File:The Second Boer War, 1899-1902 Q72008.jpg - Wikimedia Commons
Author William Skeoch Cumming

[182] Anglo Boer War - Images from 'On the Transvaal border' published by W H Gill, 1900 - Page 4 - Boer War Forum Credit Dr. David Biggins
[183] Anglo Boer War - Images from 'On the Transvaal border' published by W H Gill, 1900 - Page 4 - Boer War Forum Credit Dr. David Biggins
[184] File:Camps-british-armyRFA.jpg - Wikimedia Commons Unknown Author
[185] Davitt, Boer Fight for Freedom, 3rd ed. 1902. (Duke Scan)
[186] The Sphere - Saturday 09 June 1900 p. 16
https://www.britishnewspaperarchive.co.uk/viewer/bl/0001861/19000609/015/0016
[187] Titled: Father and son going to the front by Boer War Colourised Photographs. File:Boer Family with a wagon.jpg - Wikimedia Commons Tinus le Roux (Photos Redux - Colourising History)
[188] File:The Second Boer War, 1899-1902 Q72284.jpg - Wikimedia Commons Author: William Skeoch Cumming
[189] File:The Second Boer War, 1899-1902 Q72160.jpg - Wikimedia Commons Author: William Skeoch Cumming.
[190] A Handbook of the Boer War - Gale And Polden Limited, London 1910 Source: Anonymous ISBN 978-1-37-497455-5. https://babel.hathitrust.org/cgi/pt?id=uc1.l0061277489&seq=7
[191] Author: Utilisateur: Bel Adone https://commons.wikimedia.org/wiki/File:Lundean%27s_Nek04.JPG
[192] Times History Vol 4.
[193] Times History War in S.A. Vol 4.
[194] Times History War in S.A. Vol 4.
[195] Anglo-Boer War Album, Cape Town, D. Edwards & co. [1902?] T.J. Hall Photo.
[196] Times History War in S.A. Vol 4.
[197] Louis Creswicke: South Africa and the Transvaal War, in six volumes. Vol. VI. File:Position of troops round the Brandwater Basin before the surrender of Prinsloo, South Africa, 1900.jpg - Wikimedia Commons Brandwater Basin - Wikipedia
[198] https://commons.wikimedia.org/wiki/File:Sandkol_by_Surrender_Hill,_R711,_Slaapkrans,_oos-Vrystaat.jpg Author: JMK 2015
[199] https://commons.wikimedia.org/wiki/File:Surrender_Hill_01,_monument.jpg Author: JMK 2011
[200] Navy & Army Illustrated Vol 11
[201] File:Resa del bacino del Brandewater.jpg - Wikimedia Commons
[202] Navy & Army Illustrated Vol 11 Oct. 20th, 1900
[203] https://commons.wikimedia.org/wiki/File:Boer_War_General_Jonathan_Crowther_(1853-1926),_cutout_from_group_photograph.jpg
[204] Navy & Army Illustrated Vol 11
[205] Anglo-Boer War Album, Cape Town, D. Edwards & co. [1902?]
[206] Anglo-Boer War Album, Cape Town, D. Edwards & co. [1902?]
[207] https://commons.wikimedia.org/wiki/File:Paul_Hendrik_Roux_onder_Boere-krygsgevangenes_op_Ceylon,_hoofsaaklik_van_Brandwaterkom_onder_genl_Prinsloo,_a_(cropped).jpg
[208] File:Boere-krygsgevangenes op Ceylon, hoofsaaklik van Brandwaterkom onder genl Prinsloo, a.jpg - Wikimedia Commons
[209] A Handbook of the Boer War - Gale And Polden Limited, London 1910 Source: Anonymous ISBN 978-1-37-497455-5. https://babel.hathitrust.org/cgi/pt?id=uc1.l0061277489&seq=7
[210] Navy & Army Illustrated Vol 9
[211] 7/21/1900 A supply train, with 100 Welsh Fusiliers, captured near Honing Spruit.
[212] Times History Vol 4
[213] File:Schoemansville panorama from cable-way hill.jpg - Wikimedia Commons Author: Farawayman
[214] File:Magaliesberg08.jpg - Wikimedia Commons Credit: Androstachys
[215] A Handbook of the Boer War - Gale and Polden Limited, London 1910 Source: Anonymous ISBN 978-1-37-497455-5. https://babel.hathitrust.org/cgi/pt?id=uc1.l0061277489&seq=7
[216] File:Magaliesberg te Skeerpoort, a,jpg.jpg - Wikimedia Commons Author: JMK
[217] File:Magaliesberg wes van Dewetsnek, c.jpg - Wikimedia Commons Author: JMK
[218] https://commons.wikimedia.org/wiki/File:Kruin_van_Magaliesberg_naby_Dewetsnek.jpg Author: JMK
[219] Creswicke, Louis South Africa and the Transvaal War, Vol. 2 1900 Photo by Cribb, Southsea.
[220] SA War Vol 6. And File:Christiaan de Wet4.jpg - Wikimedia Commons Author: Reinhold Thiele (1856–1921)

[221] Christiaan De Wet (1854–1922), Boer general, politician and rebel leader. The photo was taken in Potchefstroom on the 27th of August 1900, six days after he and his commando of 246 men escaped from near encirclement north of the Magaliesberg. He is posing with a restored rifle that was handed in to the English by one of many "hensoppers" [Hands-Uppers]. This was usually done in return for having their property and farms spared by the enemy. De Wet wrote of this photo: Three Years' War by Christiaan Rudolf De Wet - Free Ebook (gutenberg.org) https://gutenberg.org/ebooks/18794
[222] Navy & Army Illustrated Vol 11 Sept. 29th, 1900
[223] Hiley and Hassell, The Mobile Boer, Grafton Press, New York, (c. 1902)
[224] Heroes of the Boer War. Review of Reviews, 1903.
[225] Heroes of the Boer War. Review of Reviews, 1903.
[226] Davitt, Boer Fight for Freedom, 3rd ed. 1902. (Duke Scan)
[227] Rhenoster River, South Africa. 1900. Members of the Royal Engineers construct a temporary railway bridge beside that destroyed by the retreating Boers. (Original stereocard housed in AWM Archive Store) See Rhenoster River, South Africa. 1900. Members of the Royal Engineers construct a temporary railway ... | Australian War Memorial (awm.gov.au)
[228] South Africa; its History, Heroes and Wars. Mackenzie [1899]
[229] Engineers repairing a Railway destroyed by the Boers near Kroonstadt. File:Reparaties aan de spoorlijn naar Kroonstad, Zuid-Afrika Royal Engineers repairing a Railway destroyed by the Boers - Kroonstadt, South africa (titel op object), RP-F-F09002.jpg - Wikimedia Commons
[230] Hillegas, Boers in War, 1900.
[231] (With Facsimile of his Signature.) Hillegas, With the Boer forces, 1901 (Photograph by Leo Weinthal.)
[232] Navy & Army Illustrated Vol 10 May 9th, 1900 Photo by M. Bennett
[233] Image Source: With the Flag to Pretoria, H. W. Wilson Vol 1
[234] Anglo-Boer War Album, Cape Town, D. Edwards & co. [1902?]
[235] War with the Boers, H. Brown, London [1902?] Vol 2
[236] Times History Vol 5
[237] https://commons.wikimedia.org/wiki/File:Plan_of_the_battle_of_Bothaville_(Battles_of_the_nineteenth_century,_1901).jpg Author Archibald Forbes
[238] Royal Munster Fusiliers fighting from behind the redoubt at Honey Nest Kloof, South Africa. Feb. 16, 1900 Digital ID: (digital file from b&w film copy neg.) cph 3a42932 http://hdl.loc.gov/loc.pnp/cph.3a42932 Reproduction Number: LC-USZ62-42613 (b&w film copy neg. of half stereo) Repository: Library of Congress Prints and Photographs Division Washington, D.C. 20540 USA https://www.loc.gov/pictures/item/2007681600/resource/
[239] Times History Vol 5.
[240] Times History, Vol 5
[241] File:The Second Boer War, 1899-1902 Q72292.jpg - Wikimedia Commons Author: William Skeoch Cumming
[242] Times History Vol 5.
[243] SA National Archives: http://www.nationalarchives.gov.za/node/1236
[244] Navy & Army Vol 13 p. 59 Oct. 5th, 1901.
[245] After Pretoria: The Guerilla War Vol 3 Jan-April 1901
[246] SA National Archives: http://www.nationalarchives.gov.za/node/1236
[247] Navy & Army Vol 13 p. 59 Oct. 5th, 1901
[248] Navy & Army Vol 13 p. 59 Oct. 5th, 1901
[249] Photographical Collection Anglo-Boer War Museum, Bloemfontein SA. Also Available Here: File:VerskroeideAarde1.jpg - Wikimedia Commons
[250] Navy & Army Illustrated Vol 11 p. 123 October 27th, 1900.
[251] File:The Second Boer War, 1899-1902 Q72102.jpg - Wikimedia Commons
[252] Navy & Army Vol 13 p. 59 Oct. 5th, 1901
[253] File:The Second Boer War, 1899-1902 Q72217.jpg - Wikimedia Commons Author: William Skeoch Cumming
[254] Heavily loaded ox wagon. - Atom site for DRISA PB1153_001.jpg (1400×922) (drisa.co.za)
[255] From: Our [Transvaal] Military Officers: Dutch: Onze krijgs-officieren: album van portretten met levens-schetsen der. by Gustav Schoeman Preller, Frans Vredenrijk Engelenburg 1904 Catalogue Record: See his bio Infra. https://archive.org/search?query=title%3A%28Onze%20Krijgs-Officieren%29
[256] Watercolour by L. Calkin, 1899. Image and data provided by Wellcome Collection https://www.jstor.org/stable/community.24840283

[257] Pattern 1 Gun Truck and Train . Circa 1901. First pattern gun truck No 1 armoured train. (Publication on armoured trains in the Anglo Boer War) - Atom site for DRISA
PB3032_005.jpg (1240×971) (drisa.co.za)
[258] Pattern 2 train. Circa 1901. New pattern bogie on No 12 armoured train. (Publication on armoured trains in the Anglo Boer War) - Atom site for DRISA
PB3032_007.jpg (1237×969) (drisa.co.za)
[259] Drakensberg. Ox wagon with 16 oxen crossing a drift near Mont-Aux-Sources. - Atom site for DRISA
N08623.jpg (1400×1054) (drisa.co.za)
[260] Article from: The daily morning Alaskan. (Skagway, Alaska) 1899-1904, March 03, 1901, Image 3 « Chronicling America « Library of Congress Direct link to page 3. 0141.pdf
[261] Anglo-Boer War Album, Cape Town, D. Edwards & co. [1902?]
[262] File:The Second Boer War, 1899-1902 Q72436.jpg - Wikimedia Commons Author: William Skeoch Cumming
[263] De Camp Courant Herdenkings-uitgawe vir Slagoffers van die Tweede Vryheidsoorlog 1899-1902 ** De Camp Courant: A Memorial Publication for Victims of the Anglo-Boer War 1899-1902 (arkpark.net) And: https://decampcourant.arkpark.net/
[264] SA National Archives: http://www.nationalarchives.gov.za/node/1236
[265] File:The Second Boer War, 1899-1902 Q72453.jpg - Wikimedia Commons
[266]Concentration camps as discussed at De Camp Courant (arkpark.net)
[267] Photographical Collection Anglo-Boer War Museum, Bloemfontein SA. Image: Concentration camps in the South African War? Here are the real facts by Fransjohan Pretorius, (The Conversation), February 18, 2019 | South African History Online (sahistory.org.za)
[268] File:Boercamp1.jpg - Wikimedia Commons and Women and Children in White Concentration Camps during the Anglo-Boer War, 1900-1902 | South African History Online (sahistory.org.za)
[269] Photographical Collection Anglo-Boer War Museum, Bloemfontein SA. Image: Concentration camps in the South African War? Here are the real facts by Fransjohan Pretorius, (The Conversation), February 18, 2019 | South African History Online (sahistory.org.za)
[270] War with the Boers, H. Brown, London [1902?] Vol 2.
[271] Hillegas, Boers in War, 1900.
[272] South Africa; its History, Heroes and Wars. Mackenzie [1899]
[273] South Africa; its History, Heroes and Wars. Mackenzie [1899]
[274] Author: Montagu Brownlow Parker Blockhouse 1901 - Category:Blockhouses of the Second Boer War - Wikimedia Commons
[275], Author: Andrew Hall File:Blockhouse, Prieska.JPG - Wikimedia Commons
[276]The distance between blockhouses was calculated to be able to extent the reach as far as possibly but at the same time to prevent the Boers from passing unnoticed between them. The average distance apart was about 1,000 yards (910 m). Anglo Boer War - Blockhouses
[277] War with the Boers, H. Brown, London [1902?] Vol 5
[278] https://en.wikipedia.org/wiki/Martini%E2%80%93Henry Retrieved 2021
[279] Times History War in S.A Vol 5 (1907)
[280] Times History Vol 4.
[281] Damaged bridge during Anglo-Boer War over Wilge River. - Atom site for DRISA Pretoria. Damaged bridge during Anglo-Boer War over Wilge River. - Atom site for DRISA P2446_01.jpg (1400×1050) (drisa.co.za)
[282] https://commons.wikimedia.org/wiki/File:Harts_River_Stone.png Author AxelRavera
[283] https://commons.wikimedia.org/wiki/File:Rooiwal_Potgieter.jpg
[284] Description: Boer general Christiaan de Wet on his way to peace talks at Vereeniging. 13 April 1920, Source Scan of a publicly available post card from 1920. Author Unknown. File:De Wet by Ferreirastasie oppad na Vereeniging.jpg - Wikimedia Commons
[285] General de Wet's pass for safe conduct during peace negotiations.
https://commons.wikimedia.org/wiki/File:Generaal_de_Wet_se_vrypas.jpg
[286] Navy & Army Vol 14 p. 309 June 14th, 1902 Photo Copyright by 'Kodak"
[287] Christiaan de Wet Vanity Fair July 31st, 1902.
https://commons.wikimedia.org/wiki/File:Christiaan_de_Wet_Vanity_Fair_1902-07-31.jpg
[288] Description: South Africa. General Christian Rudolf de Wet (1854-1922) In South Africa the Boer general during the Second War of Independence. Place unknown. 1903.
https://commons.wikimedia.org/wiki/File:Zuid_Afrika,_SFA006012030.jpg

[289] https://en.wikipedia.org/wiki/Christiaan_de_Wet And Permanent Link, 5/26/2025: https://en.wikipedia.org/w/index.php?title=Christiaan_de_Wet&oldid=1285150617
[290] Christiaan de Wet at the National Convention, 1908. https://commons.wikimedia.org/wiki/File:The_National_Convention,_1908,_Christiaan_Rudolf_de_Wet.jpg
[291] South African General Christiaan Frederick Beyers troops surrender at the Vaal River and the Maritz Rebellion is finished December 8 1914. In the drawing Beyers (1869 to 1914) has attempted to escape and he and his companion are drowning. From the War Illustrated Album deLuxe, published London, 1916. 11 November 1914, Source:The British Army and The Illustrated London News. https://commons.wikimedia.org/wiki/File:Battle_of_Mushroom_Valley.png
[292] Boer general Christiaan de Wet, after his surrender in the Boer Rebellion of 1914. https://commons.wikimedia.org/wiki/File:Generaal_De_Wet_-_Rebellie_oorgawe.jpg
[293] General C.R. de Wet and his wife, sitting in front of their house, 1917, Collection Dr. Hendrik Muller, Photo Collection no. VA965.
https://commons.wikimedia.org/wiki/File:CRdeWet_VA0965.jpg
[294]Parliament building in Bloemfontein, South Africa, with monument to Christiaan de Wet. Author: Anne97432 https://commons.wikimedia.org/wiki/File:Bloemfontein01.jpg
[295] Statue of Christiaan de Wet in 2012. Author: Richard Broekhuijzen https://commons.wikimedia.org/wiki/File:Christiaan_de_Wet_statue_2012.JPG
[296] File:Sculpture Christiaan de Wet by Joseph Mendes da Costa (detail), Hoge Veluwe, the Netherlands.jpg Author, Bj.schoenmakers
https://commons.wikimedia.org/wiki/File:Sculpture_Christiaan_de_Wet_by_Joseph_Mendes_da_Costa_(detail),_Hoge_Veluwe,_the_Netherlands.jpg
[297] Hillegas, With the Boer forces, 1901 (Photograph by the Author.)
[298] Original Source: Edwin Sharpe Grew: Field-Marshal Lord Kitchener. His Life and Work for the Empire. London 1916. Volume 2. File:Kitchener at the Peace Conference that ended the Second Boer War.png - Wikimedia Commons On 28 February 1901 British commander Lord Kitchener met Boer commander Louis Botha at Middelburg, Transvaal. The talks failed to bring peace and Kitchener commenced the barbed wire and block house campaign to thwart attacks by Boer guerilla fighters. Category:Middelburg Peace Conference - Wikimedia Commons Also: File:The Second Boer War, 1899-1902 Q68497.jpg - Wikimedia Commons
[299] File:Kitchener at the Peace Conference that ended the Second Boer War.png - Wikimedia Commons
[300] Image: Hillegas, With the Boer forces, 1901 (Photograph by Leo Weinthal, Pretoria.) and quote: With the Flag, After Pretoria: The Guerilla War Vol 3 p. 91 1900.

Made in United States
Orlando, FL
05 November 2026

90230227R00201